Lecture Notes in Computer Science   6640

Commenced Publication in 1973
Founding and Former Series Editors:
Gerhard Goos, Juris Hartmanis, and Jan van Leeuwen

Jordi Domingo-Pascual   Pietro Manzoni
Sergio Palazzo   Ana Pont
Caterina Scoglio (Eds.)

NETWORKING 2011

10th International IFIP TC 6 Networking Conference
Valencia, Spain, May 9-13, 2011
Proceedings, Part I

 Springer

Volume Editors

Jordi Domingo-Pascual
Universitat Politècnica de Catalunya (UPC) - Barcelona TECH
Campus Nord, Mòdul D6, Jordi Girona 1-3, 08034 Barcelona, Spain
E-mail: jordi.domingo@ac.upc.edu

Pietro Manzoni
Ana Pont
Universitat Politècnica de València
Camí de Vera, s/n, 46022 Valencia, Spain
E-mail: {pmanzoni, apont}@disca.upv.es

Sergio Palazzo
University of Catania
V.le A. Doria 6, 95125 Catania, Italy
E-mail: sergio.palazzo@diit.unict.it

Caterina Scoglio
Kansas State University
2069 Rathbone Hall, Manhattan, KS 66506, USA
E-mail: caterina@ksu.edu

ISSN 0302-9743   e-ISSN 1611-3349
ISBN 978-3-642-20756-3   ISBN 978-3-642-20757-0 (eBook)
DOI 10.1007/978-3-642-20757-0
Springer Heidelberg Dordrecht London New York

Library of Congress Control Number: 2011925929

CR Subject Classification (1998): C.2, H.4, D.2, K.6.5, D.4.6, H.3

LNCS Sublibrary: SL 5 – Computer Communication Networks
and Telecommunications

Typesetting: Camera-ready by author, data conversion by Scientific Publishing Services, Chennai, India

Printed on acid-free paper

Springer is part of Springer Science+Business Media (www.springer.com)

Welcome Message from the General Chairs

It is our honor and pleasure to welcome you to the proceedings of the 2011 IFIP Networking Conference. This was the 10th edition of what is already considered one of the best international conferences in computer communications and networks.

The objective of this edition of IFIP Networking conferences was to attract innovative research works in the areas of: applications and services, next-generation Internet, wireless and sensor networks, and network science. This goal was reached and we would like to thank our Technical Program Committee Co-chairs, Jordi Domingo-Pascual, Sergio Palazzo, and Caterina Scoglio, who organized the review of around 300 submissions, for the splendid technical program provided. The selected 64 high-quality papers were organized in two parallel tracks. The conference also included three technical talks from very prestigious scientists: Jim Kurose, José Duato, and Antony Rowstron. We would like to express our gratitude to them for accepting; their presence was a privilege for us all.

The present edition took place at the Computer Engineering School of the Universitat Politècnica of Valencia, in Spain.

All this would not have been possible without the hard and enthusiastic work of a number of people who contributed to making Networking 2011 a successful conference. Thus, we would like to thank all of them, from the Technical Committee Chairs and members, to the Local Organizing Committee, to the authors, and also to the staff of CFP-UPV who dealt with all local arrangements. Thanks also to the Steering Committee of Networking and all the members of the IFIP-TC6 for their support.

And finally, we would also like to encourage current and future authors to continue working in this direction and to participate in forums like this conference for the exchange of knowledge and experiences.

May 2011

Pietro Manzoni
Ana Pont

Technical Program Chairs' Message

It is a great pleasure to welcome you to the proceedings of Networking 2011, which was the 10th event of the series of International Conferences on Networking, sponsored by the IFIP Technical Committee on Communication Systems (TC6). The main objectives of these IFIP conferences are to bring together members of the networking community from both academia and industry, to discuss recent advances in the broad and fast-evolving field of computer and communication networks, and to highlight key issues, identify trends, and develop visions. For this year we had four main areas in the call for papers, namely, applications and services, next-generation Internet, wireless and sensor networks, and network science, which received, respectively, 18.8%, 45.2%, 20.6%, and 15.4% of the submitted papers.

This year, the conference received 294 submissions, representing a huge increase over the figures of the most recent years: it confirms this IFIP-supported initiative as a leading reference conference for the researchers who work in networking. Papers came from Europe, the Middle East and Africa (69.1%), Asia Pacific (16.5%), the USA and Canada (11.6%), and Latin America (2.8%).

With so many papers to choose from, the Technical Program Committee (TPC) job to select the final high-quality technical program was challenging and time consuming. The TPC was formed by 106 researchers from 22 different countries.

All papers were evaluated through a three-phase review process by at least three Program Committee members, who provided their own regular reviews; then, one of the three reviewers, being entitled as a meta-reviewer, opened a discussion among the three reviewers and provided a sum-up recommendation; finally, after a careful analysis of all recommendations, 64 papers were selected for the technical program, organized in 16 sessions, covering the main research aspects of next-generation networks. We would like to thank the members of the TPC, for they had to deal with a significant load reviewing papers due to the increase in the number of submissions. Also, we acknowledge the contribution of the additional reviewers who helped the TPC members in their task.

This event would not have been possible without the hard and enthusiastic work of a number of people who contributed to making Networking 2011 a successful conference. We would especially like to thank the General Co-chairs, Ana Pont and Pietro Manzoni, for their support throughout the whole review process, and the Steering Committee Chair, Guy Leduc, for his invaluable advice and encouragement.

All in all, we would like to thank all participants for attending the conference. We truly hope you enjoy the proceedings of Networking 2011!

Jordi Domingo-Pascual
Sergio Palazzo
Caterina Scoglio

Organization

Executive Committee

Honorary Chair	Ramón Puigjaner, Universitat Illes Balears, Spain
General Chairs	Ana Pont, Universitat Politècnica de València, Spain
	Pietro Manzoni, Universitat Politècnica de València, Spain
Technical Program Chairs	Jordi Domingo-Pascual, Universitat Politècnica de Catalunya, BarcelonaTECH, Spain
	Sergio Palazzo, University of Catania, Italy
	Caterina Scoglio, Kansas State University, USA
Tutorial Chairs	Juan Carlos Cano, Universitat Politècnica de València, Spain
	Dongkyun Kim, Kyungpook National University, South Korea
Publication Chair	Josep Domenech, Universitat Politècnica de València, Spain
Technical Organization Chair	José A. Gil, Universitat Politècnica de València, Spain
Publicity Chair	Carlos T. Calafate, Universitat Politècnica de València, Spain
Financial Chair	Enrique Hernández-Orallo, Universitat Politècnica de València, Spain
Workshops Chair	Vicente Casares, Universitat Politècnica de València, Spain

Steering Committee

George Carle	TU Munich, Germany
Marco Conti	IIT-CNR, Pisa, Italy
Pedro Cuenca	Universidad de Castilla-la-Mancha, Spain
Guy Leduc	University of Liège, Belgium
Henning Schulzrinne	Columbia University, USA

Supporting and Sponsoring Organizations (Alphabetically)

Departamento de Informática de Sistemas y Computadores (DISCA)
Escuela Técnica Superior de Ingeniería Informática
IFIP TC 6

Instituto de Automática e Informática Industrial
Ministerio de Ciencia e Innovación
Telefonica Investigación y Desarrollo
Universitat Politècnica de València

Technical Program Committee

Rui Aguiar	University of Aveiro, Portugal
Ozgur Akan	Koc University, Turkey
Khaldoun Al Agha	University of Paris XI, France
Ehab Al-Shaer	University of North Carolina, Charlotte, USA
Kevin Almeroth	University of California, Santa Barbara, USA
Tricha Anjali	Illinois Institute of Technology, USA
Pere Barlet-Ros	Universitat Politècnica de Catalunya, BarcelonaTECH, Spain
Andrea Bianco	Politecnico di Torino, Italy
Chris Blondia	University of Antwerp, Belgium
Fernando Boavida	University of Coimbra, Portugal
Olivier Bonaventure	Université catholique de Louvain, Belgium
Azzedine Boukerche	University of Ottawa, Canada
Raouf Boutaba	University of Waterloo, Canada
Torsten Braun	University of Bern, Switzerland
Wojciech Burakowski	Warsaw University of Technology, Poland
Albert Cabellos-Aparicio	Universitat Politècnica de Catalunya, BarcelonaTECH, Spain
Eusebi Calle	University of Girona, Spain
Antonio Capone	Politecnico di Milano, Italy
Damiano Carra	University of Verona, Italy
Augusto Casaca	Instituto Superior Técnico, Lisbon, Portugal
Claudio Casetti	Politecnico di Torino, Italy
Baek-Young Choi	University of Missouri, Kansas City, USA
Piotr Cholda	AGH University of Science and Technology, Poland
Marco Conti	IIT-CNR, Italy
Pedro Cuenca	University of Castilla la Mancha, Spain
Alan Davy	Waterford Institute of Technology, Ireland
Marcelo Dias de Amorim	UPMC Paris Universitas, France
Christian Doerr	Delft University of Technology, The Netherlands
Jordi Domingo-Pascual	Universitat Politècnica de Catalunya, BarcelonaTECH Spain
Constantine Dovrolis	Georgia Institute of Technology, USA
Wolfgang Effelsberg	University of Mannheim, Germany
Lars Eggert	Nokia Research Center, Finland
Gunes Ercal	University of California, Los Angeles, USA
Laura Feeney	Swedish Institute of Computer Science, Sweden

Additional Reviewers

Saeed Al-Haj
Carlos Anastasiades
Emilio Ancillotti
Carles Anton
Panayotis Antoniadis
Markus Anwander
Shingo Ata
Baris Atakan
Jeroen Avonts
Mohammad Awal
Serkan Ayaz
Sasitharan
  Balasubramaniam
Pradeep Bangera
Youghourta Benfattoum
Mehdi Bezahaf
Nikolaos Bezirgiannidis
Ozan Bicen
Alex Bikfalvi
Alberto P Blanc
Norbert Blenn
Thomas Bocek
Chiara Boldrini
Roksana Boreli
Raffaele Bruno
Shelley Buchinger
Filipe Caldeira
Valentín Carela-Español
David Carrera
Pietro Cassarà
Egemen K Çetinkaya
Supriyo Chatterjea
Ioannis Chatzigiannakis
Lin Chen
Baozhi Chen
Luca Chiaraviglio
Mosharaf Chowdhury
Delia Ciullo
Florin Coras
Paul Coulton
Joana Dantas
Ignacio de Castro

Marcel Cavalcanti
  de Castro
Sotiris Diamantopoulos
Nikos Dimitriou
Lei Ding
Jerzy Domzal
Falko Dressler
Otto Carlos M.B.
  Duarte
Michael Duelli
Zbigniew Dulinski
Roman A. Dunaytsev
Juergen Eckert
David Eckhoff
Philipp Eittenberger
Ozgur Ergul
David Erman
Chockalingam
  Eswaramurthy
Wissam Fawaz
Adriano Fiorese
Hans Ronald Fischer
Bryan Ford
Anna Förster
Dario Gallucci
Wilfried Gansterer
Xin Ge
Andrea Ghittino
Luca Giraudo
Diogo Gomes
Roberto Gonzalez
Jorge Granjal
Vijay Gurbani
Thomas Haenselmann
Matthias Hartmann
Syed Anwar Ul Hasan
Fabio Hecht
Volker Hilt
David Hock
Michael Hoefling
Philipp Hurni
Fida Hussain

Johnathan Ishmael
Jochen Issing
Eva Jaho
Loránd Jakab
Matthew R Jakeman
Parikshit Juluri
Frank Kargl
Dominik Klein
Murat Kocaoglu
Stavros Kolliopoulos
Ioannis Komnios
Robert Kooij
Efthymios Koutsogiannis
Stein Kristiansen
Michal Kryczka
Adlen Ksentini
Mirja Kuehlewind
Berend W.M. Kuipers
Harsha Kumara
Li-Chung Kuo
Andreas Lavén
Yee Wei Law
Fotis Lazarakis
Eun Kyung Lee
Hendrik Lemelson
Sotirios-Angelos Lenas
Nanfang Li
Yunxin (Jeff) Li
Morten Lindeberg
Teck Chaw Ling
Zeyu Liu
Xuan Liu
Dajie Liu
Dimitris Loukatos
Chris Yu Tak Ma
Francesco Malandrino
Jose Marinho
Angelos K Marnerides
Steven Martin
Alfons Martin
Brian Meskill
Jakub Mikians

Table of Contents – Part I

Energy Efficiency

Mobility Modeling

Network Science

Network Topology Configuration

Next Generation Internet

Path Diversity

Table of Contents – Part II

Resource Allocation Radio

Resource Allocation Wireless

Social Networks

TCP

BotTrack: Tracking Botnets Using NetFlow and PageRank

Jérôme François, Shaonan Wang, Radu State, and Thomas Engel

Interdisciplinary Centre for Security, Reliability and Trust (SnT)
University of Luxembourg - Campus Kircherg, L-1359 Luxembourg, Luxembourg
firstname.lastname@uni.lu

Abstract. With large scale botnets emerging as one of the major current threats, the automatic detection of botnet traffic is of high importance for service providers and large campus network monitoring. Faced with high speed network connections, detecting botnets must be efficient and accurate. This paper proposes a novel approach for this task, where NetFlow related data is correlated and a host dependency model is leveraged for advanced data mining purposes. We extend the popular linkage analysis algorithm PageRank [27] with an additional clustering process in order to efficiently detect stealthy botnets using peer-to-peer communication infrastructures and not exhibiting large volumes of traffic. The key conceptual component in our approach is to analyze communication behavioral patterns and to infer potential botnet activities.

Keywords: Botnets, PageRank, Network security.

1 Introduction

The ever-continuing struggle between malware operators and security officers is a perpetual and parallel development of new attack tools and mitigation techniques. Botnets are known to be a major threat [5]. The recent trend in malware design leverages peer-to-peer communication infrastructures and state-of-the-art cryptography that significantly raises the stakes in their detection [14]. While bots participating in denial of service attacks and spam campaigns can be easily detected by traffic analysis, it is much harder to detect stealthy bots deployed for information stealing and espionage purposes. We address this specific type of botnet in our paper. The contributions of our paper are threefold:

1. We propose a NetFlow [7] monitoring framework that leverages a simple host dependency model for tracking communication patterns.
2. Linkage analysis and clustering techniques that identify groups of hosts sharing similar behavioral/communication patterns
3. Experiments using real NetFlow data in order to assess the performance of detecting botnet traffic in NetFlow data.

This work is related to our previous one [37] but the construction of the dependency does not need any prior knowledge about potential risks in this paper.

J. Domingo-Pascual et al. (Eds.): NETWORKING 2011, Part I, LNCS 6640, pp. 1–14, 2011.
© IFIP International Federation for Information Processing 2011

Thus, it leads to additional clustering techniques for being suitable. Unlike our previous work, this paper tries to detect large-scale shared behavior and particularly botnets.

We start the paper with an introduction to botnets in section 2. Section 3 describes the underlying building blocks of our approach by detailing the host dependency model and data mining techniques. Comprehensive experiments are discussed in section 4. We address related work in section 5 and conclude the paper in section 6.

2   Structured Botnets

2.1   Overview

A botnet is a network of compromised hosts usually called bots or zombies. The botnet owner, also known as the botmaster or controller manages the bots. Currently, botnets make themselves known via the threats they pose as for example: highly distributed denial of service attacks, data stealing or large spam campaign. By not involving all available bots, spam [38] can be generated more stealthily in order to avoid detection. Another feature of bots is the permanent multiplication effort: bots attempt to scan and infect other machines and so grow the network. Attacks are strongly also linked to financial interests [13].

2.2   Botnet Architectures

The botmaster manages her bots through a Command and Control (C&C) channel. In the past, this was generally a pseudo-centralized architecture based IRC [26] (Internet Relay Chat) channel [24]. Since distributed architectures are more scalable and robust, attackers also chose this kind of architecture for constructing botnets. Slapper [3] was the first botnet worm in this category, but structured peer-to-peer (P2P) protocols followed due to their underlying scalability and robust communication facilities [10]. In structured P2P architecture, a huge ID space is defined, a typical value being $N = 2^{160}$. Each host (corresponding to an IP address and a UDP port) has a node ID in this space and a routing table corresponding to hosts that can be directly reached. Each node is also responsible for data information (files) having a key (hashed value) close to its own. For other nodes, a routing process is employed and the main advantage of structured P2P is that reaching a node is guaranteed in a maximal number of

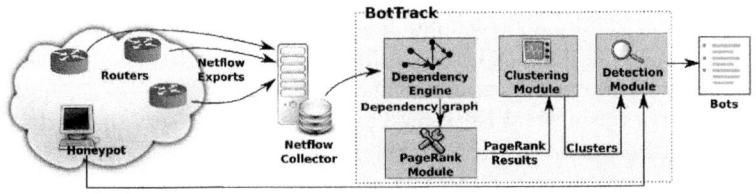

Fig. 1. BotTrack architecture

hops. Chord [33] was one of the pioneering works and guarantees a routing in $log(N)$ hops. Koorde [17] is an optimization of Chord which provides a routing complexity equal to $O(log(N)/log(log(N))$. Storm/Peacomm[28] is a infamous real worm using both cryptographic protection and peer-to-peer communication models [14]. Storm is based on the Kademelia [23] protocol usually used for file sharing. Storm bots look for specific keys within the P2P network corresponding to systems where the botmaster provides necessary information for connecting to a webserver publishing orders to execute. The Kademelia protocol has a searching algorithm that finds these hosts in $O(log(N))$. We argue in this paper that analyzing communication patterns in network traffic can reveal botnets relying on such communication models. By studying Chord which can be seen as the generic protocol for DHT (Distributed Hash Tables) and some extensions in different directions (Kademlia with a high redundancy and Koorde with a very small one), it is enough representative of most of P2P botnets.

3   Bot Detection

3.1   Architecture

Figure 1 highlights the main components of our approach. Routers monitor the traffic and export NetFlow records to a collector. BotTrack then analyzes this data. The first step identifies the interactions between systems by creating a dependency graph between hosts. This graph is then automatically analyzed by a module running the PageRank algorithm [27] to extract the nodes (IP addresses) which are strongly linked to each other. These linkage levels provide authority and hub scores on which a clustering algorithm is applied to find nodes with similar roles within the network. However, nodes playing the same role may not necessarily be P2P nodes or P2P bot nodes. For more precision, data from a honeypot is also leveraged. Hence, some nodes are known a priori to be bots and clusters including such points are considered to be groups of bots to which mitigation techniques can be applied.

3.2   Flow Monitoring

IP flow [29] is a standard tool for today's large-scale network monitoring.

A flow is a series of IP packets sharing the same source, destination address, associated ports, and protocols. For high speed networks, flow monitoring is well used because, firstly, network traffic analysis via flow records is much more efficient than via individual packets (less data to analyze), and secondly, flow records require only a fraction of the storage space needed for full packet capture. The flow collecting architecture consists of probes and collectors. Probes or sensors are devices deployed in different network locations, and are responsible for capturing flow data and forwarding it to the collector. Almost all modern commercial-grade routers support flow record export. In this paper, we use without any distinction the terms IP flow [29] and NetFlow [7] which reflects different standards.

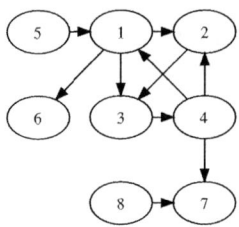 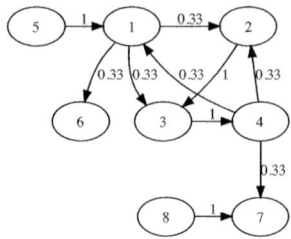

Fig. 2. Sample host communication graph example (trivial example bots on nodes 1,2,3 and 4)

Fig. 3. simplified ranking distribution after 1st iteration

3.3   Host Communication Model

Because of its wide availability and intrinsic host communication information we utilize flow data and link analysis to detect structured botnets. The rationale behind our approach is that P2P structured botnets exhibit a distinguishable communication pattern among bots. Our communication dependency graph engine (see figure 1) generates a directed "who talks to whom" host communication graph. Figure 2 shows a trivial example of a host communication graph of 8 nodes. Each node represents a host, and a directed link from node A to B indicates there is at least one IP flow from host A to host B. It is obvious that nodes 1, 2, 3, and 4, which represent bot nodes, exhibit more intense communications among themselves than with other nodes. The following sections will explain how to leverage the PageRank link analysis algorithm bounded by clustering techniques to detect bots via host communication graphs. The terms host communication graph and dependency graph are used interchangeably in this paper.

3.4   PageRank

The PageRank algorithm [27] is a link analysis algorithm used by the Google web search engine to weight the relative importance of web pages on the Internet. PageRank ranks each web page according to a a score depending on the hyperlink structure of the Internet. A web page is referred to by other pages if other pages contain a hyperlink pointing to it. Intuitively, a frequently-referred page is important, and pages referred to by few important pages are also considered important. For instance, if a web page is linked by only one web site, but this web site is Yahoo, then this page should be considered as important as well. The outstanding scalability and efficiency of PageRank makes it an ideal candidate to analyze the link structures among communicating hosts on large scale networks.

PageRank is computed in an iterative fashion. At each iteration, the current score of each node is distributed through its outgoing links to the nodes it points to. The new rank of each node is the sum of the scores its incoming edges bring. For example, assuming an initial value of each node in Figure 2 of 1, the ranking distribution following the first iteration is shown in Figure 3. Node 1 has three outgoing links, thus each outgoing link takes $\frac{1}{3}$ ranking to its destination node. Node 1 receives two incoming links from node 4 and node 5, thus the new score of node 1 is $\frac{1}{3} + \frac{1}{1} = 1.33$. The iterations continue until the score changes are sufficiently small or the maximum number of iterations is reached.

Based on some knowledge, important nodes scores can be increased by prior defined weight. In the botnet context, it corresponds to nodes that are known to be bots and should have a higher impact on the score calculation and propagation.

Let $P_t(i)$ denote the PageRank score of node i for iteration t, (j, i) denotes a directed edge from node j to node i, O_j represents outgoing link number of node j, and E is the set of directed edges on the whole graph. Based on some knowledge, a subjective preference can be assigned to certain nodes. This is useful in the context of bot detection in the sense that we can promote the rankings of hosts known to be bots. The subjective influence functions via the W component and the impact of this weight and the propagated scores can be adjusted by tuning the d parameter:

$$P_t(i) = (1 - d) \sum_{k=1}^{n} W(k) + d \sum_{(j,i) \in E} \frac{P_{t-1}(j)}{O_j} \qquad (1)$$

For flow monitoring, it is interesting to consider both edge directions. When edges point from source to destination hosts, the ranking scores distribute towards the final destination of traffic, and is named hub rank in this paper; when edges point from destination to source hosts, the ranking scores distribute towards the root cause, and we name this root-cause-selecting ranking authority rank. For the latter, PageRank algorithm is executed again with the graph where the edge directions have been inverted. Obviously, the PageRank algorithm can be seen as a set of matrix computations and theoretical details may be found in [27].

3.5   Clustering

PageRank output is composed of two values: authority and hub. Both are useful for detecting P2P bots since peers in such network have to initiate connections with many others and are also destinations of many connections. A simple approach should consider as a P2P bot a host associated with a high ranks for these two metrics:

$$authority > \theta \text{ and } hub > \gamma \qquad (2)$$

Although defining two thresholds is not easy, section 4 shows that neither can be excluded since good results are obtained only when both are used. Furthermore, infected machines can be clustered in disjoint clusters that are not necessarily adjacent in terms of their traffic flows. This is illustrated in figure 4 by showing the distribution of ingoing and outgoing node degrees. In brief, defining only a simple threshold is not sufficient and investigators should define different ranges of authority and hub values for which the hosts should be considered as bots. Finally, defining such ranges is often dependent of the P2P protocol used (see

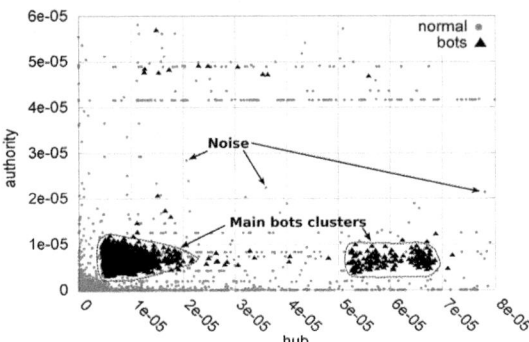

Fig. 4. Kademelia authority and hub values for normal hosts and bot hosts

section 4). For example, figure 4 clearly highlights such ranges for bots, although a minority of normal hosts may have higher values, even though the majority of normal hosts is associated with very low values.

Therefore, applying a clustering technique may be helpful, and limits the number of parameters that must be chosen. Different techniques could be applied but the following criteria have to be considered:

1. few parameters to set
2. no knowledge about the number of clusters, since different kinds of machines could be infected
3. clusters can have irregular shapes because normal communications are also considered
4. since non-infected hosts are also diverse, authority and hub values can be very different.

Several algorithms fulfill these requirements and our method is based on DB-SCAN [4] because only two parameters have to be tuned, and the complexity of this algorithm is acceptable especially when using specific data indexation [4]. In our case, the runtime is slightly above linear. Moreover, there are many extensions for speeding up the clustering process [22] and for using parallel computing [39]. All these works show that DBSCAN is quite fitted for huge databases like usual NetFlow database of real networks.

DBSCAN is a density-based algorithm where the key idea is to consider that each point of a cluster has to be enough close to a minimal number of points, $MinPts$, which represents the neighborhood defined by a maximal distance ϵ. Such a point is qualified as a core point and all points whose the distance from it is lower than ϵ are also included in the cluster and are qualified as a border point if they are not core points. Thus, points which are isolated (neither core nor border) are considered to be noise. More precisely, the algorithm iterates over every point p and does the following for each point that is not yet classified:

1. compute the neighbors: $neighbor = \{p2, dist(p, p2) < \epsilon\}$. In this paper, we use the Euclidian distance.
2. if $|neighbor| \geq MinPts$, create a new cluster containing p and all points of $neighbor$.
3. go back to step 1 by taking previous neighbors as the original node to compute the neighbors and by excluding previously assigned points.

Thus, DBSCAN has only two parameters to tune but we assume $MinPts = 4$ since botnets can sometimes contain only hundreds of hosts and so clusters of bots can be relatively small. However we expect few of them to share similar hub and authority values. Thus, the main parameter to tune in our experiments was ϵ.

4 Experimental Result

4.1 Methodology and Datasets

Our evaluation was made on a real dataset provided by a major Luxembourg Internet operator. We extracted 13.7GB of data described in table 1. Following the approach of [25], we added synthetic botnet flow records without taking into account the duration or the size of flows since our approach does not rely on them. This is the only viable solution since getting a labeled dataset at an operator level is not possible. Hence, the corresponding characteristics are always the same in table 1. To evaluate the robustness of our system with a stealthy botnet, we extracted only 1% of IP addresses for constructing the botnet. As we relies exclusively on IP addresses, other features like timing or the number of bytes are randomly generated for the botnet flows. So, the attack scenario may be simply explained as adding bot to bot communications at the end of the capture by examining routing table entries of each bots.

Moreover, we assume a random distribution of bot IP addresses. By this way, the IP addresses selected as bots can be considered indifferently as the whole botnet or as a part of bigger botnet containing IP addresses also outside form the monitored network. Hence, there is no need to define different scenarios when having a global view of the botnet is unavailable.

Kademlia topology entails many links between hosts since by design it aims to improve performance for file sharing based on multiple paths. Therefore, this topology exhibits the highest number of flows in table 1. Chord is a more compact topology (no multiple paths from a node to a space partition) which implies lower performance. Neighbor node in Koorde ID space share links in order to contact far-distant nodes. Thus each individual node has only to maintain a few links and the number of flows is reduced.

Table 1 also presents general statistics about PageRank results. Just 1% of bots does not impact highly on average values, but the extreme values indicate that there are some changes in the graph topology.

Our evaluation is based mainly on the Receiver Operating Characteristic (ROC) [9] which is well-suited for measuring the efficiency of two classes of classification. The curve measures the True Positive Rate (TPR) against the

Table 1. Flow record statistics (original data + botnet traces)

	chord	Kademlia	Koorde
Flow#	2133887	2399032	1997049
Host#	323610	323610	323610
Bytes#	13.7G	13.7G	13.7G
Duration	18min23sec	18min23sec	18min23sec
Max HR	0.0472	0.0436	0.0442
Min HR	1.264e-08	1.185e-08	1.185e-08
Avg HR	3.0901e-6	3.0902e-06	3.0902e-06
Max AR	0.022897	0.022534	0.023161
Min AR	1.3229e-08	1.1429e-08	1.19e-08
Avg AR	3.090145e-06	3.090138e-06	3.0901e-06

False Positive Rate (FPR). The TPR is the number of bots correctly identified as bots (also called True Positive or TP) divided by the total number of bots. The FPR is the number normal hosts identified as bots (also called False Positive or FP) divided by the total number of benign hosts.

4.2   Pure Link Analysis

Figure 5 illustrates the simple threshold-based method to detect bot links in equation (2). However, in order to measure the efficiency of both metrics (authority and hub), these are considered separately. Hosts with a value higher than a threshold are considered to be bots. Figure 5 displays the ROC curve when the threshold varies. The results are good, particularly for hub detection. This suggests the server role of P2P hosts is more discriminative than the client role. Due to the different linkage level of topologies and so on in the volume of flow records highlighted in table 1, the Kademlia based botnet is the easiest to discover, and Koorde is the best topology for avoiding detection.

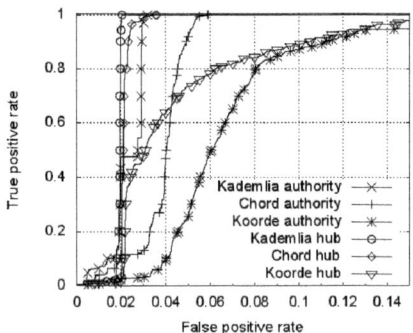

Fig. 5. ROC curves without clustering

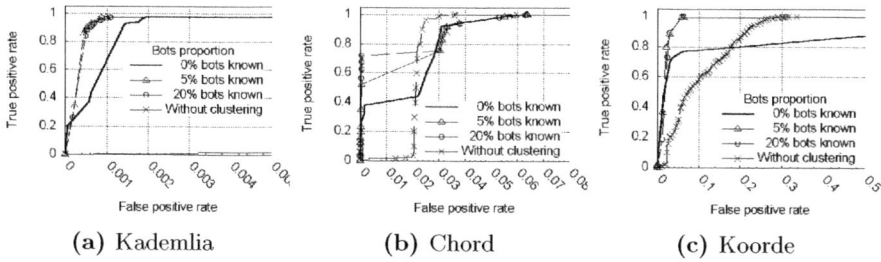

Fig. 6. Clustering impact (without clustering is the results of the simple method based on a threshold and the hub values)

Even if the number of false positives is quite low in most of cases, 0.02 was equivalent to more than 6400 FPs. Thus, taking steps to discard them is essential to avoid blocking too many benign hosts.

4.3 Clustering

In this experiment, clustering was applied to the link analysis results to improve the results by creating clusters of IP addresses. However, once the clusters have been created, it is not clear how to decide which one is a botnet. Hence, we consider that if a cluster contains at least one bot, all points included within it are also bots. This means that we need to know at least one bot for each cluster of bots. This knowledge can be obtained by using intrusion detection systems or honeypots. Meanwhile, it may lead to many false positives but the results described below show that botnet clusters are quite consistent. The difficulty of this task clearly depends on the number and the structure of clusters. Figure 6 highlights good results when applying clustering to the different topologies (curves entitled "0% bots known"). For Kademelia in figure 6a, the results are greatly improved (the original curve is very close to 0 in figure). This means that clustering allows also allows a high TPR to be reached with fewer FP. The same observations can be made for other topologies (figures 6b and 6c) but the only difference is that after a certain threshold of false positive rate, the true positive rate is better without clustering. This is due to bot points which are less distinguishable with such topologies and which are not clustered, because these are considered as noise.

Finally, we consider the inflection point as a good configuration for reducing false positives. It corresponds to the values in table 2, which highlights a large reduction of false positives compared to the threshold-based method, in particular for Kademelia and Chord with about 90% of false positives discarded. Table 2 also highlights that the setting of the parameter ϵ is clearly dependent on the topology. Hence, including a new topology is not straightforward and a priori parameter estimation is needed.

The various performance measures are quite acceptable, except in the case of Chord. From a general point of view, clustering is able to eliminate many false positives, especially in the left part of the curves. However, the results are

Table 2. Clustering impact in best cases (inflexion point)

Topology	ϵ	TPR	FPR	FP reduction	#clusters
Kademlia	2.5e-05	0.97	0.0019	5945 (90.1%)	22
Chord	4e-07	0.38	0.002	6087 (90.3%)	13
Koorde	1e-04	0.76	0.06	32037 (59.1%)	227

still acceptable in the right hand part, even if lower than without clustering. However, sometimes it could be better to detect only a subset of a botnet in order to avoid blacklisting too many benign hosts at the same time.

4.4   Impact of the Number of Known Bots

In this experiment, we assumed that some individual bots had already been detected by other programs such as a honeypots. We influenced PageRank algorithm by such knowledge by personalizing the initial weights on nodes. Figure 6 shows the results when 5 or 20 percent of bots are known. The results are improved significantly. Hence, Chord botnet-based detection is more acceptable when applying clustering with such knowledge. Considering Koorde and Kademlia, clustering always provides the best results as it was also the case without bot knowledge. For instance, detecting around 99% of bots is possible with a FPR of only 0.001. Compared to 0.002 (the FPR without bots known), this means that FP, were halved, representing 320 FPs discarded.

4.5   Cluster Analysis

In our previous experiments, we considered that we were able to identify one bot per created cluster by deploying a probe, such as a honeypot. It can be very easy to obtain good results just by creating small clusters with a minimal number of points. The extreme case is one point per cluster. Such points are considered as noise by DBSCAN and will not be considered further. Figure 7 shows that a high TPR is possible with few clusters. Having too many clusters does not produce the best results, and the TPR increases considerably when the number of clusters is reduced. Obviously, the reader has to refer to figure 6 for the FPR since it is possible to detect all bots with only one big cluster, but this implies a high FPR. For instance, we can consider the configurations in table 2, showing that for Kademlia and Koorde, the number of bots clusters is very limited. Hence, in Kademlia, 22 bots must be known. The number of clusters in Chord is quite high for this particular configuration. Assuming the second inflection point in figure 6 (TPR = 0.96, FPR = 0.04), the number of clusters is reduced to 21.

Unfortunately knowing around 20 bots from different clusters is not always possible. This occurs when some clusters are very small and only contain a limited number of bots. However, in this case, discarding them has a limited impact on TPR. The final experiments assume that clusters with only a minimal number of bots can be detected. As highlighted in figure 8, the TPR is not impacted positively since it slightly decreases. For instance, considering that we

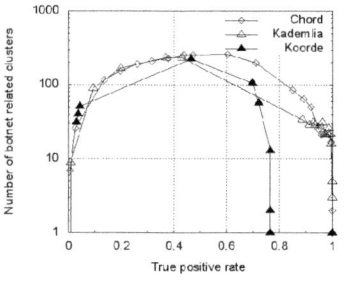

Fig. 7. Number of bot clusters

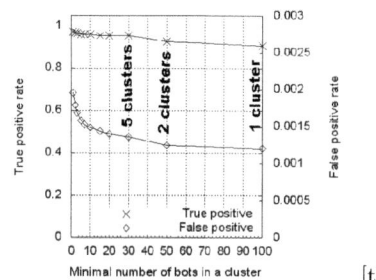

Fig. 8. Kademlia – Small cluster exclusion

have clusters with at least 50 bots yields only 2 clusters while keeping a TPR of 92%. This case correspond to only discovering the main clusters. (See figure 4). Besides, the FPR is more greatly affected and decreases, particularly in the early phases.

5   Related Work

Monitoring of IP flow records was first addressed in [32] by providing statistics on traffic volume and underlying IP addresses in order to apply time-series based anomaly detection. Follow-up works [31] employs Hidden Markov Models (HMM) for traffic modeling. The obvious way to observe a botnet is to deploy a honeypot in order to be infected. This is very efficient with IRC botnets for tracking the servers based on the IP addresses and the DNS (Domain Name System) names [1]. Many detection methods rely on first detecting malicious activities such as spam or scanning activities [20]. Some approaches aim to correlate these malicious activities with C&C detected communications [12,11]. Crawling a P2P botnet is an alternative way to detect it [13]. When the P2P botnet employs a normal P2P network, abnormal ID or IP addresses variation can be used as discriminating features [34]. Connecting components of a graph representing host interactions is presented in [8] and analysis methods can be found in [15]. Building service-level or host-level dependencies has been addressed in the past [6,18,30,2,16] for supporting network management tasks. BLINC [19] also leverages inter-host communication relationships for traffic classification, which is also well-studied in [21]. The closest related work is [25], which is based on random walk and clustering techniques to isolate interaction subgraphs related to P2P communications. Our previous work has leveraged link analysis algorithms on flow dependency graphs [35,36] and host dependency graphs [37] for tracing the root cause flow records corresponding to the malicious traffic. They help to select PageRank as the much suitable algorithm for this paper. Flow dependency graphs [35,36] do not consider hosts as nodes but flows as nodes and try to build causal connections between them based on timing analysis. Besides, our current work differs from the previous works in that we enhanced the link analysis result with clustering techniques to reveal the bots on large-scale networks.

6   Conclusion and Future Work

In this paper we have addressed the automated detection of P2P based botnets. Our approach is based on processing NetFlow-related information to build a host dependency model that captures information about which host is talking to which other host. Linkage analysis on this host dependency model is coupled with a clustering algorithm in order to build clusters of similarly behaving nodes. We show in this paper that bot-infected systems tend to belong to the same cluster. Additional information sources, such as honeypots, can be also used for tagging specific clusters. This will help to distinguish benign P2P applications from botnets. We have shown the viability of our method on large datasets from a major Internet service provider in Luxembourg. Because no labeled datasets exist in the research community, we had to follow a similar approach to [25] and inject synthetic data into the ISP-sourced datasets. We plan to do a complete complexity studies about our solutions for defining optimizations since the current implementation on a single machine handles 18min of Netflow records in around 160 seconds.

Acknowledgment. The authors would like to thank S. Nagaraja for discussions and advice on our work, especially on how to generate realistic datasets [25].

References

1. Abu Rajab, M., Zarfoss, J., Monrose, F., Terzis, A.: A multifaceted approach to understanding the botnet phenomenon. In: ACM SIGCOMM Conference on Internet Measurement (IMC), pp. 41–52 (2006)
2. Aguilera, M., Mogul, J., Wiener, J., Reynolds, P., Muthitacharoen, A.: Performance debugging for distributed systems of black boxes. In: Proceedings of the Nineteenth ACM Symposium on Operating Systems Principles, pp. 74–89 (2003)
3. Arce, I., Levy, E.: An analysis of the slapper worm. IEEE Security and Privacy 1(1), 82–87 (2003)
4. Berkhin, P.: A survey of clustering data mining techniques. In: Grouping Multidimensional Data, pp. 25–71. Springer, Heidelberg (2006)
5. Buxbaum, P.: The fog of cyberwar – to defend... and attack?, http://www.isn.ethz.ch/isn/Current-Affairs/Special-Reports/The-Fog-of-Cyberwar/Botnets/ (accessed on 08/30/10)
6. Chen, X., Zhang, M., Mao, Z.M., Bahl, P.: Automating network application dependency discovery: Experiences, limitations, and new solutions. In: Proceedings of OSDI (2008)
7. Claise, B.: Cisco Systems NetFlow Services Export Version 9. RFC 3954 (Informational) (October 2004)
8. Collins, M.P., Reiter, M.K.: Hit-list worm detection and bot identification in large networks using protocol graphs. In: Kruegel, C., Lippmann, R., Clark, A. (eds.) RAID 2007. LNCS, vol. 4637, pp. 276–295. Springer, Heidelberg (2007)
9. Fawcett, T.: An introduction to roc analysis. Pattern Recogn. Lett. 27(8), 861–874 (2006)

10. François, J., State, R., Festor, O.: Towards malware inspired management frameworks. In: IEEE/IFIP Network Operations and Management Symposium (NOMS), pp. 105–112 (2008)
11. Gu, G., Perdisci, R., Zhang, J., Lee, W.: Botminer: clustering analysis of network traffic for protocol- and structure-independent botnet detection. In: USENIX Security Symposium (SS), July 2008, pp. 139–154. San Jose, CA (2008)
12. Gu, G., Porras, P., Yegneswaran, V., Fong, M., Lee, W.: Bothunter: detecting malware infection through ids-driven dialog correlation. In: USENIX Security Symposium (SS) (August 2007)
13. Holz, T., Steiner, M., Dahl, F., Biersack, E., Freiling, F.: Measurements and mitigation of peer-to-peer-based botnets: a case study on storm worm. In: Workshop on Large-Scale Exploits and Emergent Threats (LEET). USENIX (2008)
14. Hund, R., Hamann, M., Holz, T.: Towards next-generation botnets. In: EC2ND: European Conference on Computer Network Defense, pp. 33–40. IEEE Computer Society, Los Alamitos (2008)
15. Iliofotou, M., Faloutsos, M., Mitzenmacher, M.: Exploiting dynamicity in graph-based traffic analysis: techniques and applications. In: ACM International Conference on Emerging Networking Experiments and Technologies, CoNEXT (2009)
16. Jian-Guang, L., Qiang, F., Wang, J.Y.: Mining dependency in distributed systems through unstructured logs analysis. research.microsoft.com http://research.microsoft.com/pubs/101994/Dependency%252520Camera%Ready.pdf
17. Kaashoek, M.F., Karger, D.R.: Koorde: A simple degree-optimal distributed hash table. In: Kaashoek, M.F., Stoica, I. (eds.) IPTPS 2003. LNCS, vol. 2735. Springer, Heidelberg (2003)
18. Kandula, S., Chandra, R., Katabi, D.: What's going on?: learning communication rules in edge networks. In: Proceedings of the ACM SIGCOMM 2008 Conference on Data Communication, pp. 87–98 (2008)
19. Karagiannis, T., Papagiannaki, K., Faloutsos, M.: BLINC: multilevel traffic classification in the dark. In: ACM Conference on Applications, Technologies, Architectures, and Protocols for Computer Communications, SIGCOMM (2005)
20. Karasaridis, A., Rexroad, B., Hoeflin, D.: Wide-scale botnet detection and characterization. In: First Workshop on Hot Topics in Understanding Botnets (HotBots). USENIX (2007)
21. Kim, H., Claffy, K., Fomenkov, M., Barman, D., Faloutsos, M., Lee, K.: Internet traffic classification demystified: myths, caveats, and the best practices. In: ACM CoNEXT (2008)
22. Kryszkiewicz, M., Skonieczny, Ł.: Faster clustering with dbscan. Intelligent Information Processing and Web Mining, pp. 605–614 (2005)
23. Maymounkov, P., Mazières, D.: Kademlia: A peer-to-peer information system based on the XOR metric. In: IPTPS 2001: International Workshop on Peer-to-Peer Systems, pp. 53–65. Springer, Heidelberg (2002)
24. McLaughlin, L.: Bot software spreads, causes new worries. IEEE Distributed Systems Online 5(6) (2004)
25. Nagaraja, S., Mittal, P., Hong, C., Caesar, M., Borisov, N.: BotGrep: Finding p2p bots with structured graph analysis. In: Security Symposium. USENIX (2010)
26. Oikarinen, J., Reed, D.: rfc 1459: Internet relay chat protocol (1993)
27. Page, L., Brin, S., Motwani, R., Winograd, T.: The pagerank citation ranking: Bringing order to the web (1998)
28. Porras, P., Sadi, H., Yegneswaran, V.: A Multi-perspective Analysis of the Storm (Peacomm) Worm, http://www.cyber-ta.org/pubs/StormWorm/SRITechnical-Report-10-01-Storm-Analysis.pdf

29. Quittek, J., Zseby, T., Claise, B., Zander, S.: Requirements for IP Flow Information Export (IPFIX) (2004), http://www.ietf.org/rfc/rfc3917.txt
30. Reynolds, P., Wiener, J.L., Mogul, J.C., Aguilera, M.K., Vahdat, A.: Wap5: black-box performance debugging for wide-area systems. In: Proceedings of the 15th International Conference on World Wide Web, pp. 347–356 (2006)
31. Sperotto, A., Sadre, R., de Boer, P., Pras, A.: Hidden markov model modeling of ssh brute-force attacks. In: Integrated Management of Systems, Services, Processes and People in IT, pp. 164–176.
32. Sperotto, A., Sadre, R., Pras, A.: Anomaly characterization in flow-based traffic time series. IP Operations and Management, 15–27
33. Stoica, I., Morris, R., Karger, D., Kaashoek, F., Balakrishnan, H.: Chord: A scalable Peer-To-Peer lookup service for internet applications. In: Proceedings of the 2001 ACM SIGCOMM Conference, pp. 149–160 (2001)
34. Wang, B., Li, Z., Tu, H., Hu, Z., Hu, J.: Actively measuring bots in peer-to-peer networks. In: Networks Security, Wireless Communications and Trusted Computing (NSWCTC). IEEE, Wuhan (2009)
35. Wang, S., State, R., Ourdane, M., Engel, T.: FlowRank: Ranking netflow records. In: Proceedings of the 6th International Wireless Communications and Mobile Computing Conference (2010)
36. Wang, S., State, R., Ourdane, M., Engel, T.: Mining netFlow records for critical network activities. In: Stiller, B., De Turck, F. (eds.) AIMS 2010. LNCS, vol. 6155, pp. 135–146. Springer, Heidelberg (2010)
37. Wang, S., State, R., Ourdane, M., Engel, T.: Riskrank: Security risk ranking for ip flow records. In: Proceedings of the 6th International Conference on Network and Services Management, CNSM 2010 (2010) (to appear)
38. Xie, Y., Yu, F., Achan, K., Panigrahy, R., Hulten, G., Osipkov, I.: Spamming botnets: signatures and characteristics. SIGCOMM Comput. Commun. Rev. 38(4), 171–182 (2008)
39. Xu, X., Jäger, J., Kriegel, H.P.: A fast parallel clustering algorithm for large spatial databases. Data Min. Knowl. Discov. 3(3), 263–290 (1999)

Learning Entropy

Lele Zhang and Darryl Veitch

Department of Electrical and Electronic Engineering
The University of Melbourne, Parkville Victoria 3010, Australia
{lz,dveitch}@unimelb.edu.au

Abstract. Entropy has been widely used for anomaly detection in various disciplines. One such is in network attack detection, where its role is to detect significant changes in underlying distribution shape due to anomalous behaviour such as attacks. In this paper, we point out that entropy has significant blind spots, which can be made use by adversaries to evade detection. To illustrate the potential pitfalls, we give an in-principle analysis of network attack detection, in which we design a camouflage technique and show analytically that it can perfectly mask attacks from entropy based detector with low costs in terms of the volume of traffic brought in for camouflage. Finally, we illustrate and apply our technique to both synthetic distributions and ones taken from real traffic traces, and show how attacks undermine the detector.

Keywords: Entropy, Anomaly Detection, Camouflage.

1 Introduction

Entropy is widely used as a summary statistic in diverse application areas, including *anomaly detection*. In recent years anomaly detection has received considerable attention in computer networking both within industry and academia, in particular in relation to security issues such as network based attacks, and entropy based detection is a popular approach.

Generally speaking, in entropy based anomaly detection, the (empirical) entropy, summarising a histogram of some quantity of interest from the underlying network traffic within a time bin, is used as a detector of anomalous events seen across bins. For example, the quantity could be the counts of packets (becoming probabilities after normalisation) with different source port numbers passing a measurement point. Implicitly, it is assumed/believed that entropy will change noticeably when 'significant' changes in the traffic pattern occur due to anomalous behaviours, but change little or not at all when small fluctuations about typical behaviour are encountered. In this vein, [3,5] used entropy of source IP address distributions to capture DDoS attacks, and [10] focused on worm detection using distributions from packet headers. The paper [7] considered entropy of distributions based on the number IP addresses that each host communicates with in addition to those from packet headers. Other work exploiting entropy for anomaly detection can be found in [1,4].

J. Domingo-Pascual et al. (Eds.): NETWORKING 2011, Part I, LNCS 6640, pp. 15–27, 2011.
© IFIP International Federation for Information Processing 2011

These prior works have demonstrated a measure of success of entropy-based anomaly detection, especially in network attack detection. However, this success is founded on two basic assumptions that we challenge here: i) Entropy detects 'significant' changes in distribution well, ii) Attackers are unable and/or unwilling to adopt strategies to evade detection. In fact, since entropy, like all summary statistics, is a compact summary of a complex reality, it must necessarily be blind to many changes in the underlying distribution. For anomaly detection purposes, it is clearly essential to understand the nature of this blindness, however this task has until now not attracted much attention.

In this paper, we shall see that the relationship between changes in distribution details, or 'shape', and entropy can be complex and counter-intuitive. An important consequence for entropy based detectors is that they can be surprisingly easily defeated when the attacker starts to learn and manipulate data. Sophisticated adversaries have been discussed before in the context of detectors (or classifiers) in other domains [2,6,8]. Since entropy based detection is becoming popular, we believe that exploring its limitations and potential countermeasures is both important and timely.

Our main contributions: i) The first quantitative understanding of the behaviour of entropy as a function of the underlying distribution shape. ii) To demonstrate the potential fruits of (i), we apply it to the attack detection problem. We provide the definition and first results on optimal camouflage from entropy based detectors.

The above results bring insights and capabilities at a number of (closely related) levels, including the nature of entropy blindness, how attackers can evade detection at minimal cost, and a meaningful calibration capability for detectors and understanding of their limitations. Although it has been noted before that entropy is not all things to all applications, for example see [3,9,11], this is the first study we are aware of which provides a rigorous quantitative analysis and a systematic investigation. We believe that the insights are very valuable for more general settings and also that the techniques can be extended to analyse more realistic attack scenarios.

In Section 2 examples are used to illustrate some of the key traps one can fall into from assuming that entropy captures changes in distribution shape. Section 3 develops the technical results that enable 'optimal camouflage', which are then used in Section 4 to explore attack detection using an entropy based detector. We conclude and discuss future work in Section 5.

2   Entropy

After introducing the definition of entropy, we explore, using a number of concrete examples, the dangers of simplistic impressions as to the relationship between distribution shape and entropy. In particular, these examples give insight into what detectors might hope to detect and what they might miss, and hence help us better understand why attackers can evade the entropy detector with surprising ease.

2.1   Preliminaries

We view Internet traffic as an infinite stream of packets passing by a passive monitoring point, processed in batches according to contiguous constant width measurement intervals. Within a single such interval consisting of V packets, and given a metric of interest (a function of packet header information), each packet is mapped to one of N classes, resulting in a sequence $\{a_1, a_2, \ldots, a_V\}$ over the bin where $a_j \in \{1, 2, \ldots, N\}$. The associated empirical distribution $\mathcal{D}$ is given by $\mathcal{D} = \{p_1, p_2, \ldots, p_N\}$, where $p_i = f_i / V$, $f_i = |\{j : a_j = i\}|$ and $\sum_{i=1}^{N} p_i = 1$.

The Shannon (empirical) entropy for $\mathcal{D}$ is defined as

$$\mathcal{H}(\mathcal{D}) = -\sum_{i=1}^{N} p_i \log p_i,$$

where $0 \log 0 = 0$, and logarithms are base 2. Provided the alphabet size N is finite, entropy is maximal when the distribution is uniform $\{1/N, 1/N, \ldots, 1/N\}$. In contrast, the minimum possible entropy of zero occurs when probability is maximally concentrated: $p_j = 1$ for some j and $p_i = 0$ otherwise.

2.2   Connecting Entropy and Distribution Shape

As a statistic summarising a distribution, entropy might be expected to be alike for distributions of similar shape, that is, those with only small differences in their probabilities. On the other hand it would be substantially different for ones with radically different shapes. In fact, implicitly or explicitly, this is one of the principles that entropy based detection relies on. Unfortunately, neither of these hold true as we now show.

Order: Although there is no commonly agreed non-parametric definition of distribution 'shape', few would claim that the distributions, $\mathcal{D}_{G1}$ and $\mathcal{D}_{G2}$, shown in Fig 1(a),(b) have similar shapes. However, they share exactly the same entropy due to the fact that $\mathcal{D}_{G1}$ (a truncated, renormalised, geometric distribution with $p = 0.607$), is an ordered version of $\mathcal{D}_{G2}$ and entropy is *blind to order*. Although this is well understood, nonetheless order invariance gives a wealth of examples where very large probability differences are not reflected at all in entropy.

Shape: A natural question is whether all 'dramatic' examples involve ordering. The answer is no. Unless $N = 2$, the set of distributions with a given $\mathcal{H} > 0$ is uncountably infinite, even if they are first ordered. More concretely, consider the examples in Fig 1(c),(d). One can ask "Which distribution is closer to $\mathcal{D}_{G1}$, $\mathcal{D}_{G3}$ or $\mathcal{D}_v$?", to which "$\mathcal{D}_v$" is probably the answer of 99.9% of readers, including statisticians. Not only do the shapes of $\mathcal{D}_{G1}$ and $\mathcal{D}_v$ seem qualitatively very similar, but the probability ranges $[0.0044, 0.3961]$ and $[0.0092, 0.3358]$ are also quite close, compared to $[0.0493, 0.5560]$ from $\mathcal{D}_{G3}$. However, entropy concludes differently. In fact $\mathcal{D}_{G3}$ has the same entropy as $\mathcal{D}_{G1}$, whereas $\mathcal{D}_v$ has an entropy over 10% greater.

Heavy hitters: We would reasonably expect that entropy would be relatively sensitive to large changes in the most significant probabilities, or 'heavy hitters',

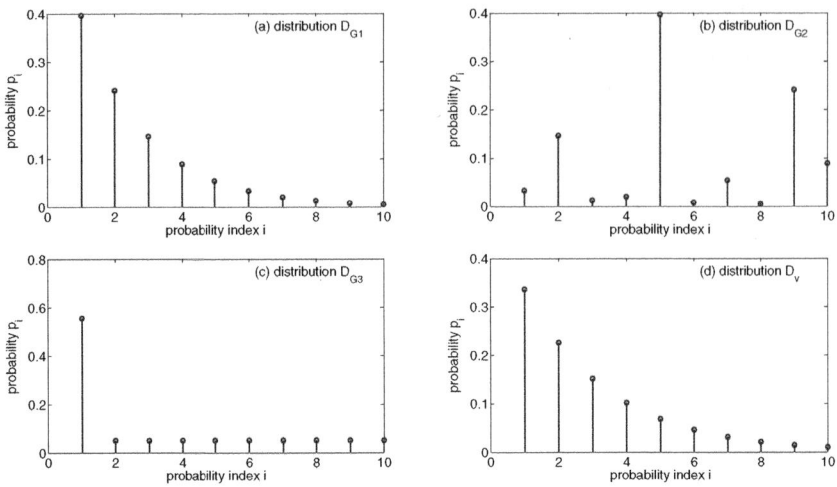

Fig. 1. Four distributions $\mathcal{D}_{G1}, \mathcal{D}_{G2}, \mathcal{D}_{G3}$ and $\mathcal{D}_v$ with $N = 10$ where: $\mathcal{H}(\mathcal{D}_{G1}) = \mathcal{H}(\mathcal{D}_{G2}) = \mathcal{H}(\mathcal{D}_{G3}) = 2.40 < \mathcal{H}(\mathcal{D}_v) = 2.64$

as low probabilities contribute little individually to $\mathcal{H}$ since $p_i \log p_i \to 0$ when $p_i \to 0$. In fact this can be far from the case.

Suppose that outcome $i = 1$ of $\mathcal{D}_{G1}$ disappears for some reason (such as all traffic from a very popular server being rerouted around the monitoring point). Thus its probability p_1, the largest carrying almost 40% of the total, falls to zero, yet despite this the new renormalised distribution has an entropy of 2.37 compared to 2.40 originally, a mere 1.3% decline. The reason lies in the shape of the remainder of the (reordered) distribution combined with the compensating effect of renormalisation.

More generally, the following expression describes the impact of the removal of a single outcome:

$$\mathcal{H}_0 = -p_l \log p_l - (1 - p_l) \log(1 - p_l) + (1 - p_l)\mathcal{H}(\mathcal{D}_R), \qquad (1)$$

where p_l is the missing probability, and $\mathcal{H}(\mathcal{D}_R)$ the entropy of the renormalised distribution $\mathcal{D}_R$ consisting of all the other probabilities except p_l. Clearly there are cases when $\mathcal{H}_0 = \mathcal{H}(\mathcal{D}_R)$, that is when the missing probability has no impact on the entropy at all! In fact, simple algebra shows that the removal of the r largest probabilities of a truncated geometric distribution is the same as that resulting from the removal of the r smallest ones. Thus, failure to track the largest probabilities does not affect entropy much since their absence has the same effect as omitting the same number of small probabilities, which is small (provided r is not too large).

Although the details of the above analysis were based on the special properties of the geometric distribution, it nonetheless highlights a far more general point, that assumptions on how entropy is influenced by distribution features cannot be taken for granted.

It is of course well known that, like other statistics, entropy provides only a coarse summary of data, and so must necessarily be blind to many features. The key message of this section is that the details of the connection between distribution 'shape' and entropy may be counterintuitive, which in the context of anomaly detection strongly motivates a precise quantitative understanding of the connection between the two.

3   Entropy in Attack Detection

In this section we examine the use of entropy in the context of Internet attacks and their measurement based detection. Here an *attacker* sends attack traffic into the network, which he may attempt to disguise, and the *detector* seeks to detect his activities using an entropy based anomaly detector. In this paper our explicit focus is on optimal camouflage strategies for the attacker, but these two sides of the same battle are very closely related as we make clearer below.

Here we flesh out entropy-specific issues using an idealised proof of concept model, rather than attempting to provide a realistic description of network attacks or detection. Our aim is to provide rigorous results and insights, and an approach, which can be used as tools in the study of entropy-based anomaly detection. In Section 4 we give an example showing how more complex scenarios can be built up using the building blocks we provide.

3.1   Model

Detector: We assume that the detector knows a benchmark distribution $\mathcal{D}_0$ with an entropy of $\mathcal{H}_0$ corresponding to normal (and attack free) traffic, and also knows the histogram $\mathcal{D}$ from the current time interval, summarised by its empirical entropy $\mathcal{H}$. The detection mechanism is simple: an attack is declared if $\mathcal{H}$ differs from $\mathcal{H}_0$ by more than a threshold $\theta_{\mathcal{H}} \geq 0$, or the detector is silent if $\mathcal{H}$ falls in the interval $[(1 - \theta_{\mathcal{H}})\mathcal{H}_0, (1 + \theta_{\mathcal{H}})\mathcal{H}_0]$.

In practice there are many issues making the determination of $\mathcal{D}_0$ and $\mathcal{H}_0$ difficult, and furthermore, $\mathcal{H}$ cannot in general be precisely measured. We subsume each of these effects into the need to describe the *sensitivity* of the detector, which we do through $\theta_{\mathcal{H}}$.

Attacker: For our purpose an *attack* is the presence of packets sent by the attacker with malicious intent which pass by the monitoring point during the measurement interval. We model it by the number V_A of attack packets, the set $\mathcal{T}$ of class indices in which they appear for the chosen metric, and their distribution across $\mathcal{T}$. During an attack the measurement interval contains $V_0 + V_A$ packets with an *attack intensity* of V_A, resulting in the modified distribution $\mathcal{D}$.

Attacks require resources to mount. We measure the cost of an attack (per measurement interval) to the attacker by the number of packets he sends. For a given V_A, depending on the nature of the attack may have some flexibility through the choice of $\mathcal{T}$ to reduce his impact on $\mathcal{H}$ and hence the chance of detection. We call this *passive camouflage*. Here we assume that the attacker knows $\mathcal{D}_0$. Although this is a quite conservative assumption in practice, it is plausible that the attacker could learn it over time. When passive camouflage is impossible or insufficient, then the attacker may opt to augment it by using *active camouflage* through sending a number V_C of additional *camouflage packets* in a tailored spread over indices design to further reduce the impact on $\mathcal{H}$. So the traffic volume becomes $V_0 + V_A + V_C$ with a camouflage cost of V_C. The resulting camouflaged distribution is given by $\mathcal{D}_C$. How to camouflage actively and efficiently is one of the main points we focus on.

3.2   Optimal Camouflage

The problem of designing camouflage strategies is formulated as follows. The attacked traffic histogram $\mathcal{D} = \{p_1, \ldots, p_N\}$ (without loss of generality, indexed in non-increasing order) has entropy $\mathcal{H}$ which by definition is outside of the perceived *normal range* $[(1-\theta_\mathcal{H})\mathcal{H}_0, (1+\theta_\mathcal{H})\mathcal{H}_0]$, and so will trigger an alarm. To hide from the detector, the attacker must disguise itself by 'dragging' the entropy back to some *target entropy* $\mathcal{H}_T$ lying within the normal range. According to the difference between $\mathcal{H}$ and $\mathcal{H}_T$ the strategies are different. In general, if $\mathcal{H}_T > \mathcal{H}$ then the attacker needs to equalise the probabilities; otherwise, he should concentrate them.

Changes in the probabilities can be achieved primarily in two ways: through sending extra packets (targeted increments), or by somehow removing normal traffic (decrements). We consider the 'increment only' scenario in this paper, as this is clearly directly feasible for the attacker.

We consider *optimal camouflage*, that is how to achieve a given $\mathcal{H}_T$ at minimal cost, that is, with the smallest possible number V_C of camouflage packets sent. This also reduces the chances that the attack would be captured by other techniques, such as by volume detection.

Formulating the original problem: Let δ_i denote the increment of probability p_i due to camouflage for constant V_0 and V_A. Then the camouflage cost is calculated as $V_C = (V_0 + V_A)\Delta$ with $\Delta = \sum_{i=1}^{N} \delta_i$. Hence the optimisation problem can be formulated as:

$$\mathbf{Min}\Delta: \quad \min_{\{\delta_i\}} \Delta, \quad \text{s.t. } \delta_i \geq 0\, \forall i \text{ and } \mathcal{H}(\mathcal{D}_C) = \mathcal{H}_T,$$

where $\mathcal{D}_C = \{(p_i + \delta_i)/(1 + \Delta), i = 1, 2, \ldots, N\}$ and $1 + \Delta$ renormalises the distribution since all increments are positive.

In other words, we find the smallest increment 'budget' which can achieve the target entropy. Since the objective function Δ is a component of the renomalisation factor, it turns out that this problem is best solved through first solving the inverse problem, where we find the extremal $\mathcal{H}$ using (all of) a fixed budget.

Formulating the inverse problem: If $\mathcal{H}_T > \mathcal{H}$ then the inverse problem is

$$\mathbf{MaxH}: \quad \max_{\{\delta_i\}} \mathcal{H}(\mathcal{D}_C), \quad \text{s.t. } \delta_i \geq 0 \text{ and } \sum_{i=1}^{N} \delta_i = c,$$

for a constant c. Otherwise, if $\mathcal{H}_T < \mathcal{H}$ it becomes

$$\textbf{MinH}: \quad \min_{\{\delta_i\}} \mathcal{H}(\mathcal{D}_C), \quad \text{s.t. } \delta_i \geq 0 \text{ and } \sum_{i=1}^{N} \delta_i = c.$$

As the renormalisation factor, $1+c$, is determined, the above problems are equivalent to the ones below under the same constraints (resp. **MaxH** and **MinH**):

$$\textbf{MinH}^- : \min_{\{\delta_i\}} \sum_{i=1}^{N}(p_i + \delta_i)\log(p_i + \delta_i) \text{ and } \textbf{MaxH}^- : \max_{\{\delta_i\}} \sum_{i=1}^{N}(p_i + \delta_i)\log(p_i + \delta_i).$$

Solving the inverse problems: Consider **MaxH**, that is to solve **MinH**$^-$, whose objective function is convex and constraints belong to a convex set. The global minimum can be solved using Lagrange multipliers and the Karush-Kuhn-Tucker (KKT) conditions. We define the Lagrange function as:

$$\Lambda = \sum_{i=1}^{N}(p_i + \delta_i)\log(p_i + \delta_i) + \mu\left(\sum_{i=1}^{N}\delta_i - c\right) - \sum_{i=1}^{N}\lambda_i\delta_i.$$

For simplicity, we consider the natural logarithm here. The KKT conditions are given by $\sum_{i=1}^{N}\delta_i - c = 0$, $\log(p_i + \delta_i) + 1 + \mu - \lambda_i = 0$, $\lambda_i\delta_i = 0$, $\delta_i \geq 0$ and $\lambda_i \geq 0$, for all i.

Consider the following two cases: i) If $\lambda_i > 0$, then $\delta_i = 0$. Because $\delta_i = e^{\lambda_i - \mu - 1} - p_i > e^{-\mu - 1} - p_i$, we can write $\delta_i = (e^{-\mu - 1} - p_i)^+$. ii) If $\lambda_i = 0$, then $\delta_i = e^{-\mu - 1} - p_i \geq 0$. Overall, δ_i can be written as $\delta_i = (e^{-\mu - 1} - p_i)^+$. Then μ is determined by $\sum_{i=1}^{N}(e^{-\mu - 1} - p_i)^+ - c = 0$, followed by the solution for δ_i.

We see that in the optimal solution $\{\delta_i\}$ decomposes naturally into two subsets. One contains zero δ_i's, which are applied to large probabilities that stay invariant. The other consists of positive δ_i's, which are given to small probabilities in order to raise them to a common level, namely $p_i + \delta_i = e^{-\mu - 1}$.

To solve **MinH** (i.e. **MaxH**$^-$), we make use of the following inequality.

$$(p_x + \delta_x)\log(p_x + \delta_x) + (p_y + \delta_y)\log(p_y + \delta_y) \geq$$
$$(p_x + \delta_x + \delta_y)\log(p_x + \delta_x + \delta_y) + p_y \log p_y \quad (2)$$

if $p_x \geq p_y \geq 0$ and $\delta_x, \delta_y \geq 0$, which follows the fact that the function $F_\epsilon(z) = (z + \epsilon)\log(z + \epsilon) - z\log z$ is strictly monotonically increasing for any $\epsilon > 0$ and $z > 0$. In fact, Equation (2) states that the difference between two probabilities grows (with all others constant), then the entropy drops, whereas the entropy rises when the contrast between them reduces. Following (2), clearly, the optimal solution for **MinH** is $\delta_1 = c$ and $\delta_i = 0 \; \forall i \neq 1$ because moving all increments to the largest probability p_1 reduces the overall entropy.

Solving the original problem: From the solutions to the inverse problems we observe that the maximal entropy increases monotonically as the 'quota' c rises (proof omitted due to space constraints), and it reaches the maximum,

$\log N$, when $c = c_m = \sum_{i=1}^{N}(p_1 - p_i)$. Afterwards, the optimal entropy stays at the maximum, since once the distribution has been made uniform further 'top-ups' can be made evenly to maintain uniformity. Similarly, the minimal entropy decreases as c rises monotonically and it approaches 0 when c goes to infinity. Typically $\mathcal{H}_T$ will be set to either $[(1-\theta_\mathcal{H})\mathcal{H}_0, (1+\theta_\mathcal{H})\mathcal{H}_0]$, the minimum needed to fall under the detector's radar. Note that this inverse problem solution can be used to calibrate the detector, since it provides the largest possible entropy 'response' corresponding to a distribution changing 'signal' of a given size.

Considering now the original problem, for $\mathcal{H}_T > \mathcal{H}$, the minimal value of the total increment required is unique because the inverse solution taking $[0, c_m] \mapsto [\mathcal{H}, \log N]$ is 1-1 onto. The increment should be spread over smallest probabilities to raise them to an uniform value. As for $\mathcal{H}_T < \mathcal{H}_0$, the minimal total increment is also unique because the inverse solution taking $[0, \infty) \mapsto (0, \mathcal{H}]$ is likewise 1-1 onto. The increment is entirely allocated to the largest probabilities. To actually solve for the minimal Δ, the entropy curve can be plotted as a function of Δ based on the solutions of **MaxH** and **MinH**. Then the optimal Δ can be quickly obtained by a numerical search.

Through solving the optimisation problems above, we obtain the technical results for camouflaging attacks from entropy based detection at minimal cost. These results also provide the insight into entropy's behaviour as a function of distribution shape.

4 Empirical Results

In this section we show how the results of the previous section can be used to answer core questions of interest to both the attacker and detector, such as whether an attack can be detected, and whether it can be disguised and at what cost. We begin with distributions from traffic traces, where we explore attacks on a single distribution and their camouflage, and then show how the camouflage technique can be extended to multiple distributions based on multiple traffic metrics. We then use idealised models to cleanly investigate a number of phenomena as a function of parameters. We focus on the case when the attack is *concentrated* on a single class index $i = t$, that is, $\mathcal{T} = \{t\}$. Nevertheless, the results below are generally valid for the concentrated attacks targeting a small number of indices.

4.1 Traffic Traces

We use 24 hours, from 00:00 to 23:59 March 30, 2009, of a 96-hour long trace captured from an OC-3 link, from the "Measurement and Analysis on the WIDE Internet" group (MAWI). The time series of interest were extracted using Wire-Shark and our own C programs, with entropy calculations in MatLab. We focus mainly on a representative 5 minute time interval from 15:30 March 30, 2009. (5 minute intervals are commonly used, e.g. [7,10]).

Concentrated attack detection
We take the packet count per destination IP address histogram (reordered), shown in Fig 2, as the benchmark distribution with $\mathcal{H}_0 = 8.92$ and $N = 40889$,

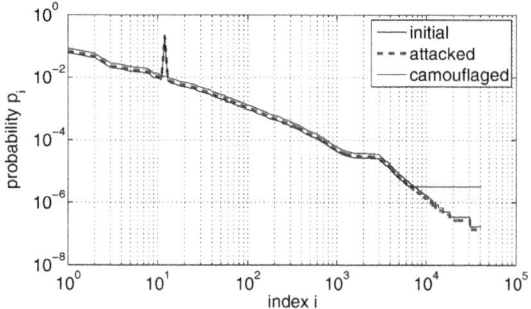

Fig. 2. Distributions of normal traffic, under the attack and after camouflage

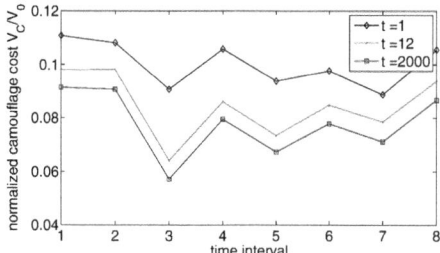

Fig. 3. Camouflage costs while attacking various targets with the intensity $V_A = 0.25V_0$ over 8 time bins with detection sensitivity $\theta_{\mathcal{H}} = 0.05$

and apply a concentrated attack with intensity $V_A = 0.25V_0$ at the target index $t = 12$ (with $p_{12} = 0.0105$). This could arise for example as the result of a DoS attack when the packet metric is the destination IP, since then all V_A packets appear at a single index corresponding to the server under attack.

The attacker is aware that his attack has lowered the original entropy appreciably (by 12.48%). Seeking complete anonymity, he wishes to know the minimal number V_C of active camouflage packets needed to be invisible even to a perfect detector $\theta_{\mathcal{H}} = 0$. Since the attack has lowered entropy from $\mathcal{H}_0$ to $\mathcal{H}$, the camouflage packets must be placed so as to increase it back up to $\mathcal{H}_T = \mathcal{H}_0$. According to the camouflage scheme built in Section 3, the strategy reduces to the following: given the reordered version of $\mathcal{D}$, say $\mathcal{D}' = \{p_i'\}$, the opponent should increase the smallest $\bar{v}$ probabilities to the same value. In the case of Fig 2, which also shows the camouflaged solution $\mathcal{D}_C$ with $\mathcal{H}(\mathcal{D}_C) = 8.92$, $\bar{v} = 34527$ and $V_C = 0.135V$.

We also examined 8 intervals over the 24 hours with the same attack intensity, $V_A = 0.25V$, but various attacking targets, $t = [1, 12, 2000]$ with the sensitivity $\theta_{\mathcal{H}} = 0.05$. The results are similar to those for the representative interval above, and the costs over the 8 bins are shown in Fig 3. We observe that attacking any of these indexes results in an entropy drop. The larger the probability the greater the decrease, and so the higher the camouflage cost.

The above discussion is only an example. The same analysis is applicable to other concentrated attacks and detection based on other metrics. For example, a worm attack may use fixed source port numbers, resulting in significant changes in a few indices of the source port distribution.

Complex detection scenarios

Some studies [7,10] have considered the use of entropy of multiple distributions (e.g. destination IP addresses plus destination ports) in order to improve detection sensitivity. Specifically, an attack is declared if the entropy of any distribution under monitoring is out of its normal range. We now show that this does not increase the difficulty of camouflage compared to the single-metric case. We continue to use the 5 minute interval from 15:30 as our example.

Suppose that there is a DDoS attack with intensity $V_A = 0.25V_0$ targeting index $t = 12$ of the (reordered) destination IP distribution and $t = 3$ of the (reordered) destination port distribution. Individually, the camouflage costs are $0.073V_0$ and $0.052V_0$ at sensitivity $\theta_{\mathcal{H}} = 0.05$. To evade detection based on the distribution pair, the camouflage cost is simply $0.073V_0 = \max\{0.073V_0, 0.052V_0\}$ because each packet can be used to camouflage either metric independently. The camouflage strategy for the address distribution is the same as that for the individual detection, whereas for the port distribution the attacker could use $0.052V_0$ camouflage packets to change the entropy to the desired value as before, and then use the remaining $0.021V_0$ packets to improve the port-camouflage further. In a similar way, the camouflage technique can be applied to other scenarios with more complex detection mechanisms such as in [5].

4.2   Synthetic Distributions

We now provide a more systematic investigation of the attacker-detector battle using a simplified model distribution, specifically a truncated Zipf with $s = 1.5$, $N = 10^4$ and $\mathcal{H}_0 = 4.47$.

Imperfect detector: Clearly, with less sensitive detectors the camouflage possibilities grow, as seen in Fig 4(a), which gives an example of how, for a fixed target $t = 12$ and for each of several different sensitivity levels, the camouflage cost V_C varies as a function of the attack intensity. Not surprisingly, the range of intensities where the cost is zero increases monotonically with $\theta_{\mathcal{H}}$. When $\theta_{\mathcal{H}} = 0$ this is only possible at a single value of intensity ($V_A = 0.09V_0$), but the range expands to $(0, 0.17]$, $(0, 0.25]$ as $\theta_{\mathcal{H}}$ rises through $0.02, 0.05$ respectively. For a fixed attack intensity, whenever camouflage is needed, the volume of camouflage required is monotonically decreasing in $\theta_{\mathcal{H}}$.

Intuitively, we expect that concentrated attacks lower entropy since they concentrate probabilities, but this is not always true. The attack may cause an entropy rise when it is moderate. The active camouflage cost for $V_A \in (0, 0.09V_0)$ and $\theta_{\mathcal{H}} = 0$ in Fig 4(a) is an example. In addition, camouflage volume is monotonic in attack intensity when the resulting entropy reduces. But this monotonicity does not hold when the entropy increases.

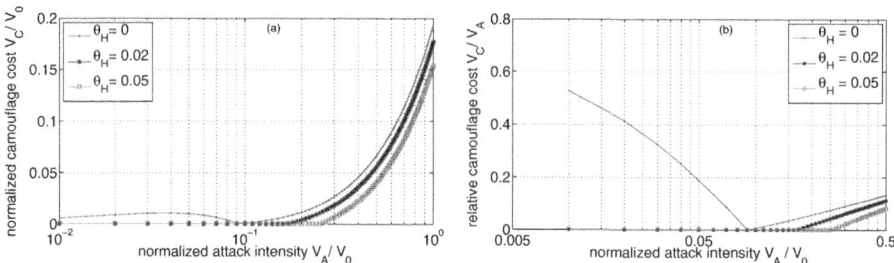

Fig. 4. Camouflage costs when attacking $t = 12$ for different detector sensitivities. (a): absolute cost. (b): relative cost.

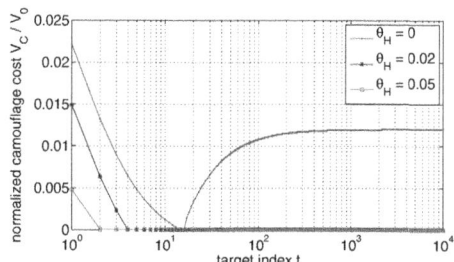

Fig. 5. Camouflage costs while targeting different indices with $V_A = 0.1V_0$

Relative costs: Sometimes the attacker may be more interested in the marginal cost of camouflage rather than the absolute. Fig 4(b) gives the relative cost corresponding to Fig 4(a). We see that for highly intensive attacks like $V_A = V_0$ the relative and absolute costs tell a similar story. However, when the attack size is small, trying to protect oneself against a perfect detector becomes very expensive relative to the attack volume.

Various targets: Finally, we study the camouflage cost as a function of the target t at a constant attack intensity. Fig 5 shows the costs for $V_A = 0.1V_0$ targeting different indices. Entropy decreases as the result of an attack for small t's, and then the behaviour of V_C is simple and monotonic in both t and $\theta_{\mathcal{H}}$. Once t is large enough, the entropy rises rather than drops, and then the camouflage strategy is no longer to raise probabilities in the tail, but to increase the largest: p_1. However, a given absolute change in p_1 influences entropy less than the same change at smaller probabilities, resulting in a larger value of V_C being required to rebalance the entropy. Consequently, the camouflage cost goes up significantly, in particular for a sensitive detector.

In our examples we assumed the attacker was capable of mounting an attack with relatively large V_A, which could constitute a very large amount of traffic on high capacity links. Even then we saw that an $\mathcal{H}$ based detector often failed to detect these attacks. If V_A is much smaller, which is more realistic in many contexts, the attack will be much harder to detect. In any event, we provide the framework and technical results needed to explore these and other related issues.

5   Conclusions and Future Work

We have examined the behaviour of Shannon entropy as a summary statistic, and pointed out that it suffers from a number of significant weaknesses in the context of network attack detection. We formulated and solved optimisation problems yielding the first rigorous results on 'optimal camouflage'. These are of relevance both to detectors and attackers to understand how entropy signatures can be either passively or actively reduced, and to evaluate the cost required to make them invisible to detectors.

Attack and detection strategies are subject to an arms race. We have provided the underlying tools essential to analyse both sides of the battle in simple scenarios, and have shown how more complex cases can be built up using them. We hope our generic approach will be useful as a foundation for the development of new detectors against ever more sophisticated attackers.

There are many directions for future work. These include allowing both decrement and increment based camouflage and discussing distributed attacks. Other questions of interest include investigating optimal camouflage strategies when the attacker has only limited information about the benchmark distribution underlying the detection, and how to overcome the limitations of entropy.

References

1. Celenk, M., Conley, T., Willis, J., Graham, J.: Anomaly Detection and Visualization using Fisher Discriminant Clustering of Network Entropy. In: Pichappan, P., Abraham, A. (eds.) Third IEEE ICDIM, pp. 216–220 (2008)
2. Dalvi, N., Domingos, P., Mausam, Sanghai, S., Verma, D.: Adversarial Classification. In: KDD 2004: Proceedings of the 10th ACM SIGKDD, pp. 99–108 (2004)
3. Feinstein, L., Schnackenberg, D., Balupari, R., Kindred, D.: Statistical Approaches to DDoS Attack Detection and Response. In: DARPA Information Survivability Conference and Exposition, vol. 1, pp. 303–314 (2003)
4. Lee, W., Xiang, D.: Information-Theoretic Measures for Anomaly Detection. In: Proc. of IEEE Symposium on Security and Privacy, pp. 130–143 (2001)
5. Li, L., Zhou, J., Xiao, N.: DDos Attack Detection Algorithms based on Entropy Computing. In: Qing, S., Imai, H., Wang, G. (eds.) ICICS 2007. LNCS, vol. 4861, pp. 452–466. Springer, Heidelberg (2007)
6. Lowd, D., Meek, C.: Adversarial learning. In: KDD 2005: Proceedings of the 11th ACM SIGKDD, pp. 641–647 (2005)
7. Nychis, G., Sekar, V., Andersen, D.G., Kim, H., Zhang, H.: An Empirical Evaluation of Entropy-Based Traffic Anomaly Detection. In: IMC 2008: Proceedings of the 8th ACM SIGCOMM Conference on Internet Measurement, pp. 151–156 (2008)
8. Rubinstein, B.I., Nelson, B., Huang, L., Joseph, A.D., Lau, S.h., Rao, S., Taft, N., Tygar, J.D.: ANTIDOTE: Understanding and Defending Against Poisoning of Anomaly Detectors. In: IMC 2009: Proceedings of the 9th ACM SIGCOMM Conference on Internet Measurement Conference, pp. 1–14 (2009)
9. Tellenbach, B., Burkhart, M., Sornette, D., Maillart, T.: Beyond Shannon: Characterizing Internet Traffic with Generalized Entropy Metrics, pp. 239–248. Springer, Berlin (2009)

10. Wagner, A., Plattner, B.: Entropy based Worm and Anomaly Detection in Fast IP Networks. In: WETICE 2005: Proceedings of the 14th IEEE International Workshops on Enabling Technologies: Infrastructure for Collaborative Enterprise, pp. 172–177 (2005)

11. Ziviani, A., Gomes, A.T.A., Monsores, M.L., Rodrigues, P.S.S.: Network Anomaly Detection using Nonextensive Entropy. IEEE Communications Letters 11(12), 1034–1036 (2007)

Machine Learning Approach for IP-Flow Record Anomaly Detection

Cynthia Wagner, Jérôme François, Radu State, and Thomas Engel

University of Luxembourg - SnT,
Campus Kircherg, L-1359 Luxembourg, Luxembourg
{cynthia.wagner,jerome.francois,
radu.state,thomas.engel}@uni.lu
http://www.securityandtrust.lu

Abstract. Faced to continuous arising new threats, the detection of anomalies in current operational networks has become essential. Network operators have to deal with huge data volumes for analysis purpose. To counter this main issue, dealing with IP flow (also known as Netflow) records is common in network management. However, still in modern networks, Netflow records represent high volume of data. In this paper, we present an approach for evaluating Netflow records by referring to a method of temporal aggregation applied to Machine Learning techniques. We present an approach that leverages support vector machines in order to analyze large volumes of Netflow records. Our approach is using a special kernel function, that takes into account both the contextual and the quantitative information of Netflow records. We assess the viability of our method by practical experimentation on data volumes provided by a major internet service provider in Luxembourg.

Keywords: Netflow Monitoring, Kernel Functions, Machine Learning Techniques, Intrusion Detection.

1 Introduction

In today's business areas, it is nearly impossible to imagine network architectures without monitoring tools, and this for several reasons. Network monitoring has become one of the most important tools in the detection of malicious or unusual events of different genres, and gathering network traffic information is the main requirement for taking appropriate countermeasures. A realistic assumption is to say that nearly all commercially available routers can export monitored data, called Netflow records. These IP flow records, chronologically ordered ranges of packet sequences, are the most important source for the analysis of unidentifiable events, even if a lot of storage capacity and processing time is needed for an accurate analysis.

In this paper, we present a new method for processing Netflow records by referring to Machine Learning techniques. In a first step, we have developed a new kernel function that operates over Netflow records by analyzing its contextual

J. Domingo-Pascual et al. (Eds.): NETWORKING 2011, Part I, LNCS 6640, pp. 28–39, 2011.

and quantitative information. For the detection of IP flow record anomalies, kernel functions have proven their usefulness, but to achieve good classification results, we apply in a second step, our kernel function to Support Vector Machines.

The structure of the paper is as follows: In section 2, we present the architecture of the model. We briefly present fundamental background information and present the anomaly detection part of the tool where we describe its main components. The more we describe the kernel function that has been used for the evaluation of Netflow records. In section 3 we describe the data set and the different attacks, used for the experiments and in section 4 we present our evaluation methods and discusses experimental results. Section 5 discusses related work and conclusions are given in section 6.

2   The Architecture of the Anomaly Detector

In the following section, we present the architecture of our model. The Netflow records, which are exported by the routers from the network to the Netflow collector module are used as input for our Anomaly Detector tool, represented in Figure 1. A Netflow record is a series of packets sent between two entities (hosts) in a chronological order. A Netflow record is composed of a source and destination IP-address, source and destination port numbers, traffic volume, packets and protocol for a session between the previous two endpoints. An argument for using Netflow records is, that they include all relevant network traffic information in a compressed version. It is composed of different modules, which are responsible for the pre-processing, evaluation and interpretation of the results, which means to answer the question, is there an attack or not.

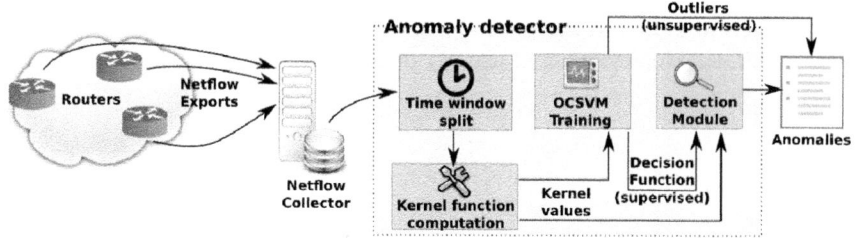

Fig. 1. Anomaly detector architecture

Detecting anomalies is related to classifying data in two different classes: benign and anomalies. Regarding the current context, the range of anomalies is very huge including Spam, Denial of Service (DoS), Scanning, Botnets, etc. Therefore, the anomaly class should regroup many different kinds of data points and so usual classifiers are not well fitted. For this purpose, multi-class classifier are potential solutions, but their main drawback is the necessity to have labeled samples for each class of anomaly you expect to detect. It also means that a system based on such classifier is unable to detect new anomalies which is usually the kind in the current networks and Internet. Therefore, we have decided to use a very specific kind of algorithm which is called *one-class* classification. The aim is to build a classifier that is able to detect new anomalies, i.e. data points which do not follow a general traffic profile which is used for training the classifier.

More precisely, we consider Support Vector Machine (SVM) which provide good accuracy with a low complexity in many domains [22]. Besides, there is a specific one-class method which was initially proposed in [15]. This specific method has also proven its accuracy for intrusion detection in different contexts such as the system calls on a UNIX system. We propose to use OCSVM (One-class SVM) [23] for analyzing Netflow records based on a new kernel function.

2.1   OCSVM

In OCSVM, the main requirement is to have a dataset of samples assumed to represent the single class of data points. This set is denoted X. The general way to define the problem is the following one: assuming that the points from an original space S follow an underlying probability P, the goal is to define a subset space of this original space S such that the probability that a point from P lies outside S is maximized by a given value between 0 and 1 [15]. This means that during the learning phase, the goal is to determine a function which is positive when applied to a point from S and negative otherwise. Therefore, during the testing phase, the sign of this function indicates if the point is classified to the single class or not.

Assume, a labeled sample $X = \{x_1, \ldots, x_n\}$, the objective is to capture a small region enclosing these points by projecting them into a higher dimensional space, such that a better separation between a defined proportion of X and the origin, is obtained. The goal is to find a hyperplane with maximum margin, *i.e.*, by maximizing its distance from the origin. The projection is performed thanks to the function $\phi(x)$ and the proportion of points to separate is defined by $1 - \nu$ with $\nu \in [0, 1]$. This problem is usually defined as follows:

$$min_{w,\rho,\xi_i \ldots n} \frac{1}{2}||w||^2 + \frac{1}{\nu n} \sum_{i=1}^{n} (\xi_i - \rho) \qquad (1)$$

subject to:

$$< w.\phi(x_i)) > \geq \rho - \xi_i, \xi_i \geq 0 \qquad (2)$$

The optimization problem has to be solved to identify two variables: w and ρ. ξ_i variables represent slack variables for allowing some points of S to not be located on the right side of the hyperplane. This avoids to bias the problem with very particular and maybe erroneous points. Since defining a projection function is not obvious, support vector methods traditionally rely on kernel functions. From a general point of view, they can be considered as similarity measure which have to be finitely positive semi-definite functions. The kernel function $K(x_i, x_j)$ is equal to $< \phi(x_i), \phi(x_j) >$. Based on this function and by transforming the problem into its dual form, we obtain:

$$min_{\alpha_i \ldots n} \frac{1}{2} \sum_{i,j=1}^{n} \alpha_i \alpha_j K(x_i, x_j) \tag{3}$$

subject to:

$$0 \leq \alpha_i \leq \frac{1}{\nu n}, \sum_{i=1}^{n} \alpha_i = 1 \tag{4}$$

Once this optimization problem is resolved, the decision function is defined by:

$$f(x) = sgn(\sum_{i=1}^{n} \alpha_i K(x_i, x) - \rho) \tag{5}$$

Therefore, this function is applied to each data point to test and if the sign is positive, it means that the point belongs to the class otherwise it is an anomaly. Obviously, many details were omitted due to space limits: the interested readers should refer to [15,22]. In fact, the main parameter ν represents the maximal proportion of points of X lying on the wrong size of the hyperplane.

2.2   Unsupervised Classification

OCSVM is usually used in a supervised manner as described in the previous section. However, the training can identify outliers thanks to the slack variables ξ_i. Therefore, only by applying this step to an entire dataset, detecting outliers and anomalies is possible. In this case, the parameter ν defines the maximum number of potential anomalies. Thus, it can be regarded as a threshold in a standard method for detecting a deviation from a profile. The main difference is that OCSVM benefit from a better detection, based on a better separation of data points in the high dimensional space.

Thus, we investigate both supervised and unsupervised method in this paper. The main advantage of the supervised one is that it should be more accurate due to the learning stage. Its main drawback is the need of sample data which are free of anomalies while the normal traffic is well represented regarding the different context that may happen. This completeness may be hard to obtain and the unsupervised technique can counter this requirement.

2.3   The Kernel Function for Netflow Processing

Kernel functions have proved their potential in the classification of high dimensional data. A kernel function calculus is homologous to a similarity measure for small data portions, where a mapping of an input space onto a higher dimensional space is performed as already explained in the previous sections. This makes it possible to separate dissimilar data and to calculate the distances, which are derived from a dot product, in this new space. We refer to Vapnik [18], who defines a kernel function K as a mapping of $K : X \times X \in [0, \infty[$ from an original input space X to a similarity score $K(x,y) = \sum_i \phi_i(x)\phi_i(y) = \phi(x) \cdot \phi(y)$, where $\phi_i(x)$ describes a feature function over a data snippet x. A general property of a kernel function is *symmetry*, such that $[K(x,y) = K(y,x)]$ and *positive-definitness*.

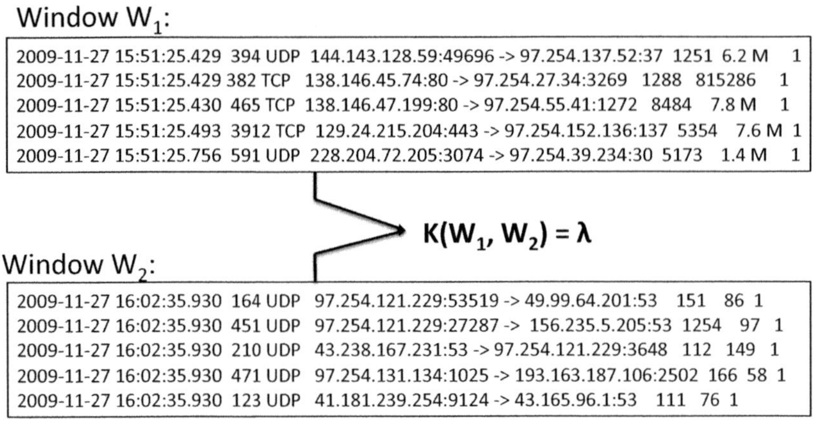

Window W₁:

2009-11-27 15:51:25.429 394 UDP 144.143.128.59:49696 -> 97.254.137.52:37 1251 6.2 M 1	
2009-11-27 15:51:25.429 382 TCP 138.146.45.74:80 -> 97.254.27.34:3269 1288 815286 1	
2009-11-27 15:51:25.430 465 TCP 138.146.47.199:80 -> 97.254.55.41:1272 8484 7.8 M 1	
2009-11-27 15:51:25.493 3912 TCP 129.24.215.204:443 -> 97.254.152.136:137 5354 7.6 M 1	
2009-11-27 15:51:25.756 591 UDP 228.204.72.205:3074 -> 97.254.39.234:30 5173 1.4 M 1	

$$K(W_1, W_2) = \lambda$$

Window W₂:

2009-11-27 16:02:35.930 164 UDP 97.254.121.229:53519 -> 49.99.64.201:53 151 86 1	
2009-11-27 16:02:35.930 451 UDP 97.254.121.229:27287 -> 156.235.5.205:53 1254 97 1	
2009-11-27 16:02:35.930 210 UDP 43.238.167.231:53 -> 97.254.121.229:3648 112 149 1	
2009-11-27 16:02:35.930 471 UDP 97.254.131.134:1025 -> 193.163.187.106:2502 166 58 1	
2009-11-27 16:02:35.930 123 UDP 41.181.239.254:9124 -> 43.165.96.1:53 111 76 1	

Fig. 2. Principle of kernel function and example of Netflow windows

For analyzing monitored windows of Netflow records, we have introduced a new kernel function $K(W_1, W_2)$, which enables to determine the similarity between two windows of IP flow records W_1 and W_2 of n seconds, in order to detect anomalies. An illustrating example of input Netflow windows are shown in Figure 2. As input space we use the Netflow records log files and define two metrics for our kernel function, provided by these flow log files. The first parameter in our kernel function are the IP-addresses from the source (src) and destination (dst) in CIDR[1] format, such that $IP = (prefix, suffixlength)$. While comparing IP-addresses, we consider the $prefix$ as the longest common sequence of two IP-addresses and the remaining bits as the $suffixlength$. As second parameter we have the quantitative factor of the captured Netflow records, which uses the traffic volume, called vol in Bytes. By this, we can model a Netflow record window for our kernel function as a set of IP flows $W = \{f_1, \ldots, f_n\}$ and a IP flow f defined as a $f_i = (prefix(src)_i, suffixlength(src)_i, prefix(dst)_i, suffixlength(dst)_i, vol_i)$.

Our kernel function for Netflow windows $K(W_1, W_2)$ returns as output a similarity score given by the sum over four functions over all flows in a window, considering source and destination information as well as traffic volume and is composed of, a similarity function $s(a_i, b_j) \in [0, \infty[$ for the source and destination information and a matching function $v(a_i, b_j) \in [0, 1]$ for the traffic information. A higher similarity score means the more similar two windows are. Our kernel function K for two windows W_1 and W_2 is defined as

$$K(W_1, W_2) = \sum_{i \in N_{W_1}, j \in N_{W_2}} s_{src}(a_i, b_j) \times v_{src}(a_i, b_j) \times s_{dst}(a_i, b_j) \times v_{dst}(a_i, b_j)$$

(6)

---

[1] CIDR — Classless InterDomain Routing: A standard system for the IP-address allocation and IP packet routing, where IP addresses are described by a network address part and a host identifier part within that network.

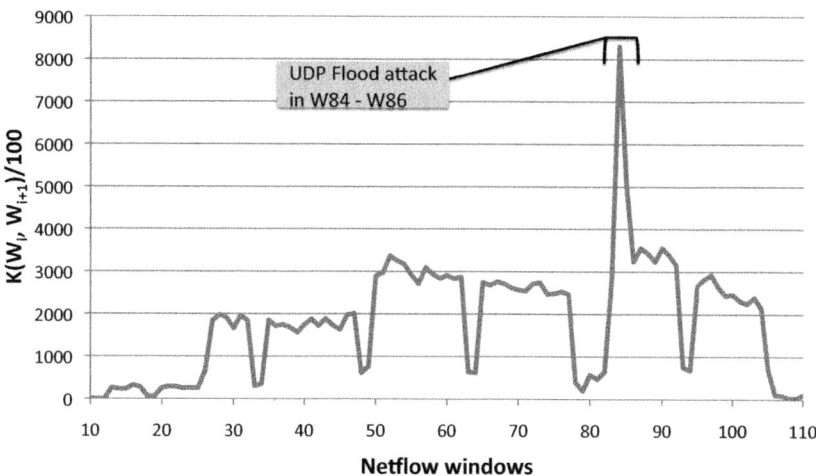

Fig. 3. Visualizing an attack by using the kernel function

where the windows of flows denoted by W_1 and W_2 are represented by N_{W_1} and N_{W_2}. We define the similarity measure parts for source and destination, $s_{src}(a_i, b_j)$, $s_{dst}(a_i, b_j)$ between flows a_i and b_j by

$$s_{src,dst}(a_i, b_j) = \begin{cases} \frac{2^{suffixlength_j}}{2^{suffixlength_i}} & \text{if } prefix_i \text{ prefix of } prefix_j \\ \frac{2^{suffixlength_i}}{2^{suffixlength_j}} & \text{if } prefix_j \text{ prefix of } prefix_i \\ 0 & \text{otherwise} \end{cases} \tag{7}$$

and the matching function for source and destination $v_{src,dst}(a_i, b_j)$ by

$$v_{src,dst}(a_i, b_j) = exp\left(-\frac{|vol_i - vol_j|^2}{\sigma^2}\right) \tag{8}$$

σ is the width scaling factor for the Gaussian kernel [3] and can be estimated on different ways, as for example by a grid search, where a range of values are used to find the optimal values for the kernel. To identify anomalous events, we use the different successive windows $K(W_i, W_{i+1})$ which we compare to each other.

To illustrate the effectiveness of our kernel function without referring to classification, we show on a real example in Figure 3 that the used kernel function detects in this case a UDP flooding attack.

3   Dataset

Anomaly detection is a quite challenging topic and evaluating new innovative solutions is a hard task due to the lack of labeled datasets. There were some works aiming at providing such datasets. A very well-known dataset is the Lincoln data set[2] but it is quite out-dated now. A recent dataset is provided in [17] based on

---

[2] http://www.ll.mit.edu/mission/communications/ist/index.html

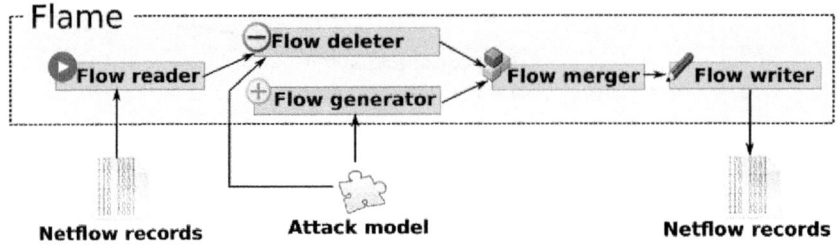

Fig. 4. Dataset generation

Table 1. Data set statistics

Flow #	1,371,194
IP addresses #	128,781 (source), 125,723 (destination)
Duration	13min 31sec
Bytes #	7.5GB
Avg. bytes/flow	5492
Packets #	11.5M
Avg. packets/flow	8.36
UDP Flows #	983,511
TCP Flows #	375,132
ICMP Flows #	11,347
Other protocols Flows #	1204

Netflow records. However, this dataset is only limited in attacks since they use a honeypot. Furthermore, this kind of architecture collects attack traces from end-user point of view and not from the network point of view. The dataset is generated by injecting synthetic attack traces into real traces using the tool Flame [2]. This tool is based on Netflows and is well suited in our case. It takes as input (flow reader) Netflow record files containing traces without malicious activities and creates attacks by generating (flow generator) or deleting (flow deleter) flow records based on attack models. The main idea is that attacks will generate traffic but also affect normal traffic like DDoS. That is why there is a need to merge these activities using the flow merger. Figure 4 highlights this process. A key part of Flame are the attack models which have to be created for the simulation of attacks.

We use as starting point a real dataset provided by a network operator from Luxembourg and assume that it is free of malicious Netflow records. This assumption is made based on a secondary semi-automated screening of the traffic, where an ISP specific solution was used. The general characteristics of this data set are provided in Table 1. Even if the duration of the data set is small, it is sufficient for evaluating our scheme, since we have limited all attacks to last only 30 seconds. The Flame website also provides some attack models which were derived from real observations. At first we used this described attack model and later on, we extended it to create more stealthy attacks. Our data set contains the following attacks:

- **Nachi scan:** the Nachi/Welchia worm was released in 2003. First, it tests the reachability of hosts using ICMP scan which is considered in the attack model. One host of our original dataset sends single ICMP packets of 92 bytes to destination IP addresses following several properties: a shift between 40 to 45 between two consecutive scans, a positive or negative shift of 400 every 200 scans and a shift between 45 to 110 every 800 flows. The inter-arrival time of flows is generally around 2 milliseconds but there is a break of around 61 milliseconds after 5 scans.

- **Netbios scan:** it is a traditional scan for finding vulnerabilities. The corresponding UDP flows contain a single packet to port 137 where the inter-arrival time is generally between 60 and 70 milliseconds. The destination hosts are scanned sequentially while keeping one IP address sometimes. Besides, between 100 and 200 scans there is a shift in the destination IP address between 60 and 70.

- **DDoS UDP flood:** one host of our dataset receives UDP packets on various ports and also sends from multiple IP addresses with various ports. The attack operates by a burst of 40 flows. Then, there is a break between 60 and 120 milliseconds.

- **DDoS TCP flood:** this denial of service attack is against web servers running on TCP port 80 with 3 packets and 128 bytes. There are bursts of 10 packets before a break between 60 and 120 milliseconds.

- **stealthy DDoS UDP flood:** different to a normal UDP flood, we apply here more randomness on flows characteristics (number of packets, size, duration). The inter-arrival time can reach 1 second which represents a very stealthy characteristic for a flooding attack. This attack is more generic in order to improve the completeness of our tests.

- **DDoS UDP flood + traffic deletion:** this is equivalent to the DDoS UDP flood, but each additional flow originated by the victim has a probability of 0.2 to be deleted due to the victim overload.

- **Popup spam:** this kind of spam is similar of sending undesired Windows Messenger popups by using UDP port 1026 and 1027. Only one packet of 925 bytes is needed. The victims IP addresses do not highlight a regular pattern because two consecutive IP addresses have a gap of 200 addresses. The inter-arrivaltime is generally lower than 1 millisecond except every 200 flows where it is around 64 milliseconds and every 550 where it is 250 milliseconds.

- **SSH scan + TCP flood:** the goal of TCP scan is to probe an SSH server by trying to log in. This is by far the most popular attack occuring in the wild.Each flow contains between 1 and 4 packets. The inter-arrival time oscillates between 1 to 50 milliseconds. The destination IP addresses are scanned in a sequential way until 400 scans are executed approximately. After, there is a shift between 200 to 400 IP addresses. In order to test our approach in a real scenario, 5% of the attacks are considered successful and the corresponding victims trigger a TCP flood attack.

4   Evaluation

In the following section we describe the experimental part of this work. We have generated the data sets following the data set generation procedure, as shown in Figure 4. Generated Netflow windows have a duration of 5 seconds each. Then, we have applied our kernel function for the contextual and quantitative evaluation of Netflow record windows. On the similarity values obtained by the kernel function, the OCSVM has been applied in order to see, if our method can detect the different attacks or not. For each data set, we have used for the training phase about 20% of our Netflow record windows and for the testing phase the remaining 80%. The outcomes for our different simulations can be seen in Table 2.

Table 2. Classification results by using OCSVM

Type of Attack	Results		
	Accuracy	False Positive rate	True Negative rate
Nachi scan	0.896	0.004	0.996
Netbios scan	0.938	0.000	1.000
Popup Spam	0.915	0.023	0.97
SSh scan + TCP flood	0.917	0.011	0.989
DDoS UDP flood	0.915	0.022	0.978
DDoS TCP flood	0.907	0.033	0.967
stealthy DDoS UDP flood	0.938	0.000	1.000
DDoS UDP flood + traffic deletion	0.934	0.000	1.000

The results by using the OCSVM are very promising, we can see that we have only a very low false positive rate and that we have an average accuracy over all attack classes of around 92% for the classification results. To compare our work with others, refer to the work of Yuan et al. [24] which used SVM for classification of network traffic of different types having attacks like worms, scans, etc. Yuan et al. achieved a similar classification accuracy (92,8%) as we did, which validates our approach, but there are no indications about the complexity of their algorithms.

5   Related Work

Network monitoring techniques as, for example the work by Bahl et al. [1], are based on Netflow records or related data. One of the main interests of Netflow records is not only their compactness, but they can be used without respecting a complete TCP state machine and are available in most commercial available routers. For example, Karpilovsky et al. [8] analyzed the IPv6 deployment by referring to a Netflow analysis. A major drawback of using Netflow records is storage capacity. Large amounts of Netflow records (30 000 flows/second) are normal for ISP networks or networks with high link loads. In [16], the authors have described the effect of Netflow capture on the accuracy analysis. To reduce the aspect of storage or to perform near real time analysis, it is often referred to Netflow sampling as described for example by C. Estan [5] or Paredes-Oliva [13], but the challenge remains to identify good sampling rates. In the domain of information security and intrusion detection, a lot of relevant work has already been performed for evaluating and processing Netflow related data [10] [7], as for example statistics on packet information.

A new observed trend is to refer to Machine Learning Techniques for evaluating Netflows or IP flow related data, like Flow Mining [11], where main issue relies on the selection of good parameters for achieving high quality. Another aspect of using Machine Learning are the kernel methods, strong mathematical tools, which have found their utility in the evaluation of large and complex data sets. Kernel methods have already been introduced as a mathematical tool in early 1900's by Hilbert and were first introduced as part of Machine Learning Techniques in the late 1990's by Vapnik [18] and further developed in the early 2000's by Schölkopf et al. [14]. In recent past, kernel functions found their applicability in most different various domains, i.e. Genetics [12], Bioinformatics [19], Natural Language Processing [4], and can be applied to most different input formats, as structured format, i.e. graphs, trees, or unstructured format, i.e. texts.

A work that is similar to our work is the evaluation of temporal and spatial IP-flow records by kernel methods presented by Wagner et al. in [20] and [21]. The main difference is that the authors in [20] [21] refer to spatial and temporal aggregated IP-flows, using the Aguri [6] tool and apply a kernel function to detect anomalies. By referring to the aggregation tool, it is only possible to get a view from the source or destination side, but no global view of both sides. Using Netflow record windows without aggregation, we can keep a global view of the whole traffic information. Then, we go further and apply Support Vector Machines in order to detect and classify benign traffic from attacks. Supervised learning has become a common tool for evaluating large data sets on common patterns, as for example by using Support Vector Machines (SVM) [14] [18]. SVMs have proven their utility in most different domains as in natural language processing [4] or bioinformatics [19]. Since a few years, SVMs are also used in computer security, where they are used for intrusion detection [9] [25].

6   Conclusion

In this paper, we have presented a new approach for the evaluation of Netflow records on most various attacks on a real data set obtained from a large ISP in Luxembourg. For validating our approach, we have generated different types of attacks by referring to the trace modification tool called Flame. Our contribution first consists in the evaluation of Netflow records in their quantitative and contextual context, where we have developed a new kernel function for calculating the similarities between Netflow windows. Second, we have applied a new SVM algorithm to our developed kernel function in order to detect different attacks in data sets. We have implemented our tool in a distributed way by using Hadoop[3]. The classification results are very promising, such that most attacks can be identified.

---

[3] http://hadoop.apache.org/

Acknowledgments. This project has been supported by the SnT - Interdisciplinary Centre for Security, Reliability and Trust. Furthermore, we want to address our special thanks to RESTENA Luxembourg for their support.

References

1. Bahl, V., Chandra, R., Greenberg, A., Kandula, S., Maltz, D., Zhang, M.: Towards highly reliable enterprise network services via inference of multi-level dependencies. In: SIGCOMM, pp. 13–24 (2007)
2. Brauckhoff, D., Wagner, A., May, M.: Flame: a flow-level anomaly modeling engine. In: Proceedings of the Conference on Cyber Security Experimentation and Test. USENIX Association (2008)
3. Burges, C.: A tutorial on support vector machines for pattern recognition. Data Mining and Knowledge Discovery 2(2), 121–167 (1998)
4. Collins, M., Duffy, N.: Convolution kernels for natural language. In: Advances in Neural Information Processing Systems, vol. 14, pp. 625–632. MIT Press, Cambridge (2001)
5. Estan, C.: Building better netflow. In: Proceedings of the 2004 conference on Applications, Technologies, Architectures, and Protocols for Computer Communications (2004)
6. Kaizaki, R., Nakamura, O., Murai, J.: Characteristics of denial of service attacks on internet using aguri. In: Kahng, H.-K. (ed.) ICOIN 2003. LNCS, vol. 2662, pp. 849–857. Springer, Heidelberg (2003)
7. Karagiannis, T., Papagiannaki, K., Faloutsos, M.: BLINC: multilevel traffic classification in the dark. In: ACM Conference on Applications, Technologies, Architectures, and Protocols for Ccomputer Ccommunications, SIGCOMM (2005)
8. Karpilovsky, E., Gerber, A., Pei, D., Rexford, J., Shaikh, A.: Quantifying the extent of iPv6 deployment. In: Moon, S.B., Teixeira, R., Uhlig, S. (eds.) PAM 2009. LNCS, vol. 5448, pp. 13–22. Springer, Heidelberg (2009)
9. Khan, L., Awad, M., Thuraisingham, B.: A new intrusion detection system using support vector machines and hierarchical clustering. The VLDB Journal 16(4), 507–521 (2007)
10. Lakhina, A., Crovella, M., Diot, C.: Mining anomalies using traffic feature distributions. In: ACM SIGCOMM 2005 (2005)
11. Lee, W., Stolfo, S., Mok, K.: Mining in a data-flow environment: experience in network intrusion detection. In: 5th International Conference on Knowledge Discovery and Data Mining (1999)
12. Nguyen, H., Ohn, S., Chae, S., Song, D., Lee, I.: Optimizing weighted kernel function for support vector machine by genetic algorithm. In: Gelbukh, A., Reyes-Garcia, C.A. (eds.) MICAI 2006. LNCS (LNAI), vol. 4293, pp. 583–592. Springer, Heidelberg (2006)
13. Paredes-Oliva, I.: Portscan detection with sampled netflow. In: Papadopouli, M., Owezarski, P., Pras, A. (eds.) TMA 2009. LNCS, vol. 5537, Springer, Heidelberg (2009)
14. Schölkopf, B., Smola, A.: Learning with Kernels: Support Vector Machines, Regularization, Optimization, and Beyond. MIT Press, Cambridge (2001)
15. Schölkopf, B., Platt, J.C., Shawe-Taylor, J.C., Smola, A.J., Williamson, R.C.: Estimating the support of a high-dimensional distribution. Neural Comput. 13, 1443–1471 (2001)

16. Sommer, R.: Netflow: Information loss or win? In: Proceedings of the 2nd ACM SIGCOMM Workshop on Internet measurement (2002)
17. Sperotto, A., Sadre, R., van Vliet, D.F., Pras, A.: A labeled data set for flow-based intrusion detection. In: Nunzi, G., Scoglio, C., Li, X. (eds.) IPOM 2009. LNCS, vol. 5843. Springer, Heidelberg (2009)
18. Vapnik, V.: Statistical Learning Theory. Wiley, Chichester (1998)
19. Vert, J.: A tree kernel to analyze phylogenetic profiles (2002)
20. Wagner, C., Wagener, G., State, R., Dulaunoy, A., Engel, T.: Game theory driven monitoring of spatial-aggregated ip-flow records. In: 6th International Conference on Network and services Management (2010)
21. Wagner, C., Wagener, G., State, R., Dulaunoy, A., Engel, T.: Peekkernelflows: Peeking into ip flows. In: 7th International Workshop on Visualization for Cyber Security, pp. 52–57 (2010)
22. Wang, L. (ed.): Support Vector Machines: Theory and Applications, Studies in Fuzziness and Soft Computing, vol. 177. Springer, Heidelberg (2005)
23. Wang, Y., Wong, J., Miner, A.: Anomaly intrusion detection using one class svm. In: Proceedings from the Fifth Annual IEEE SMC, Information Assurance Workshop, pp. 358–364 (June 2004)
24. Yuan, R., Li, Z., Guan, X., Xu, L.: An svm-based machine learning method for accurate internet traffic classification. Information Systems Frontiers 12, 149–156 (2010)
25. Zhang, B.Y., Yin, J.P., Hao, J.B., Zhang, D.X., Wang, S.: Using support vector machine to detect unknown computer viruses. International Journal of Computational Intelligence Research 2(1) (2006)

UNADA: Unsupervised Network Anomaly Detection Using Sub-space Outliers Ranking

Pedro Casas[1,2], Johan Mazel[1,2], and Philippe Owezarski[1,2]

[1]CNRS; LAAS; 7 avenue du colonel Roche, F-31077 Toulouse Cedex 4, France
[2]Université de Toulouse; UPS, INSA, INP, ISAE; UT1, UTM, LAAS; F-31077
Toulouse Cedex 4, France
{pcasashe,jmazel,owe}@laas.fr

Abstract. Current network monitoring systems rely strongly on signature-based and supervised-learning-based detection methods to hunt out network attacks and anomalies. Despite being opposite in nature, both approaches share a common downside: they require the knowledge provided by an expert system, either in terms of anomaly signatures, or as normal-operation profiles. In a diametrically opposite perspective we introduce UNADA, an Unsupervised Network Anomaly Detection Algorithm for knowledge-independent detection of anomalous traffic. UNADA uses a novel clustering technique based on Sub-Space-Density clustering to identify clusters and outliers in multiple low-dimensional spaces. The evidence of traffic structure provided by these multiple clusterings is then combined to produce an abnormality ranking of traffic flows, using a correlation-distance-based approach. We evaluate the ability of UNADA to discover network attacks in real traffic without relying on signatures, learning, or labeled traffic. Additionally, we compare its performance against previous unsupervised detection methods using traffic from two different networks.

Keywords: Unsupervised Anomaly Detection, Sub-Space Clustering, Evidence Accumulation, Outliers Detection, Abnormality Ranking.

1 Introduction

Network anomaly detection has become a vital building-block for any ISP in today's Internet. Ranging from non-malicious unexpected events such as flash-crowds and failures, to network attacks such as Denials-of-Service (DoS/DDoS), network scans, and spreading worms, network traffic anomalies can have serious detrimental effects on the performance and integrity of the network. The principal challenge in automatically detecting and analyzing traffic anomalies is that these are a moving target: new attacks as well a new variants of already known attacks are continuously emerging.

Two different approaches are by far dominant in current research literature and commercial detection systems: signature-based detection and supervised-learning-based detection. Signature-based detection is the de-facto approach

J. Domingo-Pascual et al. (Eds.): NETWORKING 2011, Part I, LNCS 6640, pp. 40–51, 2011.
© IFIP International Federation for Information Processing 2011

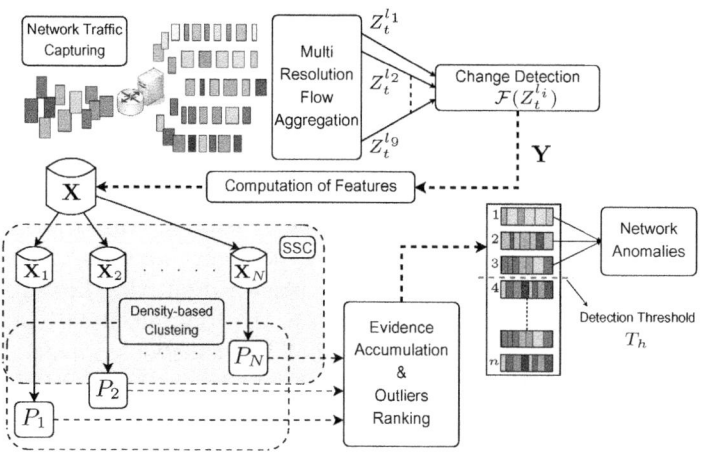

Fig. 1. High-level description of UNADA

used in standard security devices such as IDSs, IPSs, and firewalls. When an attack is discovered, generally after its occurrence during a diagnosis phase, the associated anomalous traffic pattern is coded as a signature by human experts, which is then used to detect a new occurrence of the same attack. Signature-based detection methods are highly effective to detect those attacks which they are programmed to alert on. However, they cannot defend the network against new attacks, because they cannot recognize what they do not know. In addition, building new signatures is a resources-consuming task, as it involves manual traffic inspection by human experts.

On the other hand, supervised-learning-based detection uses labeled data to build normal-operation-traffic profiles, detecting anomalies as activities that deviate from this baseline. Such methods can detect new kinds of network attacks not seen before. Nevertheless, they require training for profiling, which is time-consuming and depends on the availability of anomaly-free traffic data-sets. Labeling traffic is not only time consuming and expensive, but also prone to errors in the practice. In addition, it is not easy to keep an accurate and up-to-date normal-operation profile.

Our thesis is that these two knowledge-based approaches are not sufficient to tackle the anomaly detection problem, and that a holistic solution should also include knowledge-independent analysis techniques. To this aim we propose UNADA, an Unsupervised Network Anomaly Detection Algorithm that detects network traffic anomalies without relying on signatures, training, or labeled traffic of any kind. Based on the observation that network traffic anomalies are, by definition, sparse events that deviate markedly from the majority of the traffic, UNADA relies on robust clustering algorithms to detect *outlying* traffic flows.

UNADA runs in three consecutive steps, analyzing packets captured in contiguous time slots of fixed length. Figure 1 depicts a modular, high-level description of UNADA. The first step consists in detecting an anomalous time slot in which the clustering analysis will be performed. For doing so, captured packets are first aggregated into multi-resolution traffic flows. Different time-series are then built on top of these flows, and any generic change-detection algorithm based on time-series analysis is finally used to flag an anomalous change. The second step takes as input all the flows in the time slot flagged as anomalous. At this step, outlying flows are identified using a robust multi-clustering algorithm, based on a combination of Sub-Space Clustering (SSC) [8], Density-based Clustering [13], and Evidence Accumulation Clustering (EAC) [12] techniques. The evidence of traffic structure provided by this clustering algorithm is used to rank the degree of *abnormality* of all the identified outlying flows, building an *outliers ranking*. In the third and final step, the top-ranked outlying flows are flagged as anomalies, using a simple thresholding detection approach. As we will show through out the paper, the main contribution provided by UNADA relies on its ability to work in a completely unsupervised fashion.

The remainder of the paper is organized as follows. Section 2 presents a short state of the art in the unsupervised anomaly detection field, additionally describing our main contributions. Section 3 describes the multi-resolution traffic aggregation and change-detection procedures used in the first step of UNADA. In section 4 we introduce the core of the proposal, presenting an in depth description of the different clustering techniques used by UNADA to construct the outliers ranking. Section 5 evaluates the ability of UNADA to discover single-source, single-destination, and distributed network anomalies in real network traffic from two different datasets: the public MAWI traffic repository of the WIDE project [16], and the METROSEC project dataset [17]. In this section we also compare the performance of UNADA against previous proposals for unsupervised anomaly detection. Finally, section 6 concludes this paper.

2   Related Work and Contributions

The problem of network anomaly detection has been extensively studied during the last decade. Most approaches analyze statistical variations of traffic volume descriptors (e.g., no. of packets, bytes, or new flows) and/or particular traffic features (e.g., distribution of IP addresses and ports), using either single-link measurements or network-wide data. A non-exhaustive list of standard methods includes the use of signal processing techniques (e.g., ARIMA modeling, wavelets-based filtering) on single-link traffic measurements [1], Kalman filters [4] for network-wide anomaly detection, and Sketches applied to IP-flows [5,6].

Our approach falls within the unsupervised anomaly detection domain. The vast majority of the unsupervised detection schemes proposed in the literature are based on clustering and outliers detection, being [9, 10, 11] some relevant examples. In [9], authors use a single-linkage hierarchical clustering method to cluster data from the KDD'99 data-set, based on the standard Euclidean distance for inter-patterns similarity. Reference [10] reports improved results in the same data-set, using three different clustering algorithms: Fixed-Width clustering, an optimized version of k-NN, and one class SVM. Reference [11] presents a combined density-grid-based clustering algorithm to improve computational complexity, obtaining similar detection results. PCA and the sub-space approach is another well-known unsupervised anomaly detection technique, used in [2,3] to detect network-wide traffic anomalies in highly aggregated traffic flows.

UNADA presents several advantages with respect to current state of the art. First and most important, it works in a completely unsupervised fashion, which means that it can be directly plugged-in to any monitoring system and start to work from scratch, without any kind of calibration and/or training step. Secondly, it uses a robust density-based clustering technique to avoid general clustering problems such as sensitivity to initialization, specification of number of clusters, detection of particular cluster shapes, or structure-masking by irrelevant features. Thirdly, it performs clustering in very-low-dimensional spaces, avoiding sparsity problems when working with high-dimensional data [7]. Finally, we show that UNADA clearly outperforms previously proposed methods for unsupervised anomaly detection in real network traffic.

3   Multi-resolution Flow Aggregation and Change-Detection

UNADA performs unsupervised anomaly detection on single-link packet-level traffic, captured in consecutive time slots of fixed length ΔT and aggregated in IP flows (standard *5-tuples*). IP flows are additionally aggregated at different *flow-resolution* levels, using 9 different *aggregation keys* l_i. These include (from coarser to finer-grained resolution): *traffic per Time Slot* (l_1:tpTS), *source Network Prefixes* ($l_{2,3,4}$: IPsrc/8, /16, /24), *destination Network Prefixes* ($l_{5,6,7}$: IPdst/8, /16, /24), *source IPs* (l_8: IPsrc), and *destination IPs* (l_9: IPdst). The 7 coarsest-grained resolutions are used for change-detection, while the remaining 2 are exclusively used in the clustering step.

To detect an anomalous time slot, time-series $Z_t^{l_i}$ are constructed for simple traffic metrics such as number of bytes, packets, and IP flows per time slot, using aggregation keys $i = 1, \ldots, 7$. Any generic change-detection algorithm $\mathcal{F}(.)$ based on time-series analysis is then applied to $Z_t^{l_i}$. At each new time slot, $\mathcal{F}(.)$ analyses the different time-series associated with each aggregation key, going from coarser (l_1) to finer resolution (l_7). Time slot t_0 is flagged as anomalous if $\mathcal{F}(Z_{t_0}^{l_i})$ triggers an alarm for any of the traffic metrics at any of the 7 aggregation levels. Tracking anomalies at multiple aggregation levels provides additional reliability to the change-detection algorithm, and permits to detect both single source-destination and distributed anomalies of very different intensities.

4   Unsupervised Anomaly Detection through Clustering

The unsupervised anomaly detection step takes as input **all** the IP flows in the flagged time slot. At this step UNADA ranks the degree of abnormality of each flow, using clustering and outliers analysis techniques. For doing so, IP flows are analyzed at two different resolutions, using either IPsrc or IPdst aggregation key. Traffic anomalies can be roughly grouped in two different classes, depending on their spatial structure and number of impacted IP flows: *1-to-N* anomalies and

N-to-1 anomalies. *1-to-N* anomalies involve many IP flows from the same source towards different destinations; examples include network scans and spreading worms/virus. On the other hand, *N-to-1* anomalies involve IP flows from different sources towards a single destination; examples include DDoS attacks and flash-crowds. *1-to-1* anomalies are a particular case of these classes, while *N-to-N* anomalies can be treated as multiple *N-to-1* or *1-to-N* instances. Using IPsrc key permits to highlight *1-to-N* anomalies, while *N-to-1* anomalies are more easily detected with IPdst key. The choice of both keys for clustering analysis ensures that even highly distributed anomalies, which may possibly involve a large number of IP flows, can be represented as outliers. Without loss of generality, let $\mathbf{Y} = \{\mathbf{y}_1, \ldots, \mathbf{y}_n\}$ be the set of n aggregated-flows (at IPsrc or IPdst) in the flagged slot. Each flow $\mathbf{y}_i \in \mathbf{Y}$ is described by a set of m traffic attributes or *features*, like num. of sources, destination ports, or packet rate. Let $\mathbf{x}_i \in \mathbb{R}^m$ be the vector of traffic features describing flow $\mathbf{y}_i$, and $\mathbf{X} = \{\mathbf{x}_1, \ldots, \mathbf{x}_n\} \in \mathbb{R}^{n \times m}$ the complete matrix of features, referred to as the *feature space*.

UNADA is based on clustering techniques applied to $\mathbf{X}$. The objective of clustering is to partition a set of unlabeled samples into homogeneous groups of similar characteristics or *clusters*, based on some measure of similarity. Samples that do not belong to any of these clusters are classified as *outliers*. Our particular goal is to identify those outliers that are remarkably different from the rest of the samples, additionally ranking how much different these are. The most appropriate approach to find outliers is, ironically, to properly identify clusters. After all, an outlier is a sample that does not belong to any cluster. Unfortunately, even if hundreds of clustering algorithms exist [7], it is very difficult to find a single one that can handle all types of cluster shapes and sizes. Different clustering algorithms produce different partitions of data, and even the same clustering algorithm provides different results when using different initializations and/or different algorithm parameters. This is in fact one of the major drawbacks in current cluster analysis techniques: the lack of robustness.

To avoid such a limitation, we have developed a divide & conquer clustering approach, using the notions of *clustering ensemble* and *multiple clusterings combination*. The idea is novel and appealing: why not taking advantage of the information provided by multiple partitions of $\mathbf{X}$ to improve clustering robustness and identification of outliers? A clustering ensemble $\mathbf{P}$ consists of a set of multiple partitions P_i produced for the same data. Each partition provides an independent evidence of data structure, which can be combined to construct a new measure of similarity that better reflects natural groupings and outliers. There are different ways to produce a clustering ensemble. We use Sub-Space Clustering (SSC) [8] to produce multiple data partitions, doing Density-based clustering in N different sub-spaces $\mathbf{X}_i$ of the original space (see figure 1).

4.1   Clustering Ensemble and Sub-space Clustering

Each of the N sub-spaces $\mathbf{X}_i \subset \mathbf{X}$ is obtained by selecting k features from the complete set of m attributes. To deeply explore the complete feature space, the number of sub-spaces N that are analyzed corresponds to the number of

k-combinations-obtained-from-m. Each partition P_i is obtained by applying DBSCAN [13] to sub-space $\mathbf{X}_i$. DBSCAN is a powerful clustering algorithm that discovers clusters of arbitrary shapes and sizes [7], relying on a density-based notion of clusters: clusters are high-density regions of the space, separated by low-density areas. This algorithm perfectly fits our unsupervised traffic analysis, because it is not necessary to specify a-priori difficult to set parameters such as the number of clusters to identify. Results provided by applying DBSCAN to sub-space $\mathbf{X}_i$ are twofold: a set of $p(i)$ clusters $\{C_1^i, C_2^i, .., C_{p(i)}^i\}$ and a set of $q(i)$ outliers $\{o_1^i, o_2^i, .., o_{q(i)}^i\}$. To set the number of dimensions k of each sub-space, we take a very useful property of monotonicity in clustering sets, known as the downward closure property: if a collection of elements is a cluster in a k-dimensional space, then it is also part of a cluster in any $(k-1)$ projections of this space. This directly implies that, if there exists any interesting evidence of density in $\mathbf{X}$, it will certainly be present in its lowest-dimensional sub-spaces. Using small values for k provides several advantages: firstly, doing clustering in low-dimensional spaces is more efficient and faster than clustering in bigger dimensions. Secondly, density-based clustering algorithms such as DBSCAN provide better results in low-dimensional spaces [7], because high-dimensional spaces are usually sparse, making it difficult to distinguish between high and low density regions. Finally, clustering multiple low-dimensional sub-spaces provides a finer-grained analysis, which improves the ability of UNADA to detect anomalies of very different characteristics. We shall therefore use $k = 2$ for SSC, which gives $N = m(m-1)/2$ partitions.

4.2   Ranking Outliers Using Evidence Accumulation

Having produced the N partitions, the question now is how to use the information provided by the multiple clusters and outliers identified by density-based clustering to detect traffic anomalies. A possible answer is provided in [12], where authors introduced the idea of Evidence Accumulation Clustering (EAC). EAC uses the clustering results of multiple partitions P_i to produce a new inter-samples similarity measure that better reflects their natural groupings.

UNADA implements a particular algorithm for Evidence Accumulation, called Evidence Accumulation for Ranking Outliers (EA4RO): instead of producing a similarity measure between the n different aggregated flows described in $\mathbf{X}$, EA4RO constructs a dissimilarity vector $D \in \mathbb{R}^n$ in which it accumulates the distance between the different outliers o_j^i found in each sub-space $i = 1, .., N$ and the centroid of the corresponding sub-space-biggest-cluster $C_{\max}^i$. The idea is to clearly highlight those flows that are far from the normal-operation traffic at each of the different sub-spaces, statistically represented by $C_{\max}^i$.

Algorithm 1 presents a pseudo-code for EA4RO. The different parameters used by EA4RO are automatically set by the algorithm itself. The first two parameters are used by the density-based clustering algorithm: $n_{\min}$ specifies the minimum number of flows that can be classified as a cluster, while δ_i indicates the maximum neighborhood distance of a sample to identify dense regions. $n_{\min}$ is set at the initialization of the algorithm, simply as a fraction α of the total

---

Algorithm 1. Evidence Accumulation for Ranking Outliers (EA4RO)

1: **Initialization:**
2:    Set dissimilarity vector D to a null $n \times 1$ vector
3:    Set smallest cluster-size $n_{\min} = \alpha \cdot n$
4: **for** $i = 1 : N$ **do**
5:    Set density neighborhood δ_i for DBSCAN
6:    $P_i = \text{DBSCAN}(\mathbf{X}_i, \delta_i, n_{\min})$
7:    Update $D(j)$, $\forall$ outlier $o_j^i \in P_i$:
8:    $\quad w_i \leftarrow \dfrac{n}{(n - n_{\max_i}) + \epsilon}$
9:    $\quad D(j) \leftarrow D(j) + \text{d}_{\text{M}}(o_j^i, C_{\max}^i)\, w_i$
10: **end for**
11: Rank flows: $D_{rank} = \text{sort}(D)$
12: Set anomaly detection threshold: $T_h = \text{find-slope-break}(D_{rank})$

---

number of flows n to analyze (we take $\alpha = 5\%$ of n). δ_i is set as a fraction of the average distance between flows in sub-space $\mathbf{X}_i$ (we take a fraction $1/10$), which is estimated from 10% of the flows, randomly selected. This permits to fast-up computations. The weighting factor w_i is used as an outlier-boosting parameter, as it gives more relevance to those outliers that are "less probable": w_i takes bigger values when the size $n_{\max_i}$ of cluster $C_{\max}^i$ is closer to the total number of flows n. Finally, instead of using a simple Euclidean distance as a measure of dissimilarity, we compute the Mahalanobis distance d_{M} between outliers and the centroid of the biggest cluster. The Mahalanobis distance takes into account the correlation between samples, dividing the standard Euclidean distance by the variance of the samples. This permits to boost the degree of abnormality of an outlier when the variance of the samples is smaller.

In the last part of EA4RO, flows are ranked according to the dissimilarity obtained in D, and the anomaly detection threshold T_h is set. The computation of T_h is simply achieved by finding the value for which the slope of the sorted dissimilarity values in D_{rank} presents a major change. In the evaluation section we explain how to perform this computation with an example of real traffic analysis. Anomaly detection is finally done as a binary thresholding operation on D: if $D(i) > T_h$, UNADA flags an anomaly in flow $\mathbf{y}_i$.

5   Experimental Evaluation of UNADA

We evaluate the ability of UNADA to detect different attacks in real traffic traces from the public MAWI repository of the WIDE project [16]. The WIDE operational network provides interconnection between different research institutions in Japan, as well as connection to different commercial ISPs and universities in the U.S.. The traffic repository consists of 15 minutes-long raw packet traces daily collected for the last ten years. The traces we shall work with consist of traffic from one of the trans-pacific links between Japan and the U.S.. MAWI traces are not labeled, but some previous work on anomaly detection has been done on them [6, 15]. In particular, [15] detects network attacks using a signature-based approach, while [6] detects both attacks and anomalous flows using non-Gaussian modeling. We shall therefore refer to the combination of results obtained in both works as our *ground truth* for MAWI traffic.

We shall also test the true positive and false positive rates obtained with UNADA in the detection of flooding attacks in traffic traces from the MET-ROSEC project [17]. These traces consist of real traffic collected on the French RENATER network, containing simulated attacks performed with well-known DDoS attack tools. Traces were collected between 2004 and 2006, and contain DDoS attacks that range from very low intensity (i.e., less than 4% of the over-all traffic volume) to massive attacks (i.e., more than 80% of the overall traffic volume). In addition, we compare the performance of UNADA against some previous methods for unsupervised anomaly detection presented in section 2.

5.1   Features Selection for Detection of Attacks

The selection of the m features used in $\mathbf{X}$ to describe the aggregated flows in $\mathbf{Y}$ is a key issue to any anomaly detection algorithm, but it becomes critical and challenging in the case of unsupervised detection, because there is no additional information to select the most relevant set. In general terms, using different traffic features permits to detect different types of anomalies. In this paper we shall limit our study to detect well-known attacks, using a set of standard traffic features widely used in the literature. However, the reader should note that UNADA can be extended to detect other types of anomalies, considering different sets of traffic features. In fact, more features can be added to any standard list to improve detection results. For example, we could use the set of traffic features generally used in the traffic classification domain [14] for our problem of anomaly detection, as this set is generally broader; if these features are good enough to classify different traffic applications, they should be useful to perform anomaly detection. The main advantage of UNADA is that we have devised an algorithm to highlight outliers respect to any set of features, and this is why we claim that our algorithm is highly applicable.

In this paper we shall use the following list of $m = 9$ traffic features: number of source/destination IP addresses and ports (nSrcs, nDsts, nSrcPorts, nDstPorts), ratio of number of sources to number of destinations, packet rate (nPkts/sec), fraction of ICMP and SYN packets (nICMP/nPkts, nSYN/nPkts), and average packet size (avgPktsSize). According to previous work on signature-based anomaly characterization [15], such simple traffic descriptors permit to describe standard network attacks such as DoS, DDoS, scans, and spreading worms/virus.

Table 1 describes the impacts of different types of attacks on the selected traffic features. All the thresholds used in the description are introduced to better explain the evidence of an attack in some of these features. DoS/DDoS attacks are characterized by many small packets sent from one or more source IPs towards a single destination IP. These attacks generally use particular packets such as TCP SYN or ICMP echo-reply. echo-request, or host-unreachable packets. Port and network scans involve small packets from one source IP to several ports in one or more destination IPs, and are usually performed with SYN

Table 1. Features used by UNADA in the detection of DoS, DDoS, network/port scans, and spreading worms. For each type of attack, we describe its impact on the selected traffic features.

Type of Attack	Class	Agg-Key	Impact on Traffic Features
DoS (ICMP/SYN)	1-to-1	IPdst	nSrcs = nDsts = 1, nPkts/sec > λ_1, avgPktsSize < λ_2, nICMP/nPkts > λ_3, nSYN/nPkts > λ_4.
DDoS (ICMP/SYN)	N-to-1	IPdst	nDsts = 1, nSrcs > α_1, nPkts/sec > α_2, avgPktsSize < α_3, nICMP/nPkts > α_4, nSYN/nPkts > α_5.
Port scan	1-to-1	IPsrc	nSrcs = nDsts = 1, nDstPorts > β_1, avgPktsSize < β_2, nSYN/nPkts > β_3.
Network scan	1-to-N	IPsrc	nSrcs = 1, nDsts > δ_1, nDstPorts > δ_2, avgPktsSize < δ_3, nSYN/nPkts > δ_4.
Spreading worms	1-to-N	IPsrc	nSrcs = 1, nDsts > η_1, nDstPorts < η_2, avgPktsSize < η_3, nSYN/nPkts > η_4.

packets. Spreading worms differ from network scans in that they are directed towards a small specific group of ports for which there is a known vulnerability to exploit (e.g. Blaster on TCP port 135, Slammer on UDP port 1434, Sasser on TCP port 455), and they generally use slightly bigger packets. Some of these attacks can use other types of traffic, such as FIN, PUSH, URG TCP packets or small UDP datagrams.

5.2   Detecting Attacks in MAWI Traffic

We begin by analyzing the performance of UNADA to detect network attacks and other types of anomalies in one of the traces previously analyzed in [6]. IP flows are aggregated with IPsrc key. Figure 2.(a) shows the ordered dissimilarity values in D obtained by the EA4RO method, along with their corresponding manual classification. The first two most dissimilar flows correspond to a highly distributed SYN network scan (more than 500 destination hosts) and an ICMP spoofed flooding attack directed to a small number of victims (ICMP redirect traffic towards port 0). The following two flows correspond to unusual large rates of DNS traffic and HTTP requests; from there on, flows correspond to normal-operation traffic. The ICMP flooding attack and the two unusual flows are also detected in [6]; the SYN scan was missed by their method, but it was correctly detected with accurate signatures [15]. Setting the detection threshold according to the previously discussed approach results in T_{h_1}. Indeed, if we focus on the shape of the ranked dissimilarity in figure 2.(a), we can clearly appreciate a major change in the slope after the $5th$ ranked flow. Note however that both attacks can be easily detected and isolated from the anomalous but yet legitimate traffic without false alarms, using for example the threshold T_{h_2} on D.

Figures 2.(b,c) depict the corresponding four anomalies in two of the N partitions produced by the EA4RO method. Besides showing typical characteristics of the attacks, such as a large value of nPkts/sec or a value 1 for attributes nICMP/nPkts and nSYN/nPkts respectively, both figures permit to appreciate that the detected attacks do not necessarily represent the largest elephant flows in the time slot. This emphasizes the ability of UNADA to detect attacks of low intensity, even of lower intensity than normal traffic.

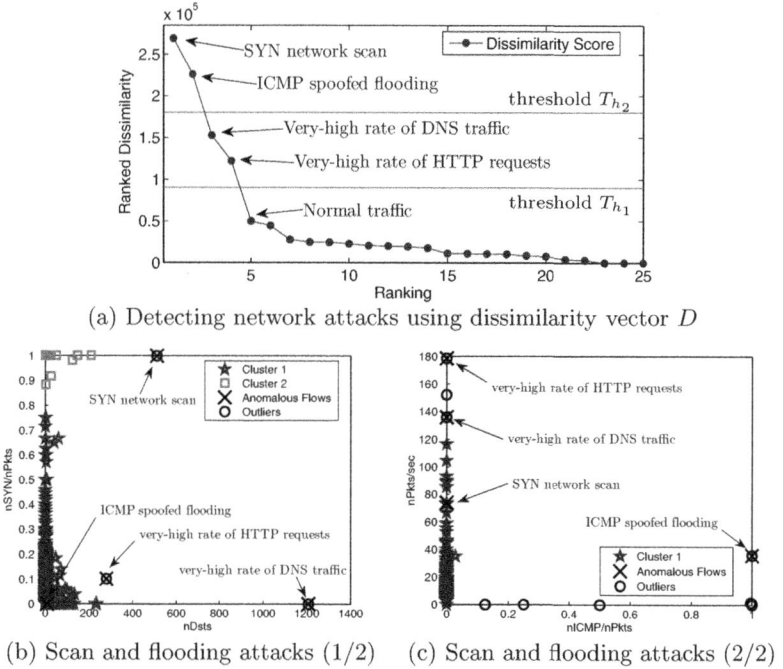

(a) Detecting network attacks using dissimilarity vector D

(b) Scan and flooding attacks (1/2)   (c) Scan and flooding attacks (2/2)

Fig. 2. Detection and analysis of network attacks in MAWI

5.3   Detecting Attacks with Ground Truth

Figure 3 depicts the True Positives Rate (TPR) as a function of the False Positives Rates (FTR) in the detection of different attacks in MAWI and MET-ROSEC. Figure 3.(a) corresponds to the detection of 36 anomalies in MAWI traffic, using IPsrc as key. These anomalies include network and port scans, worm scanning activities (Sasser and Dabber variants), and some anomalous flows consisting on very high volumes of NNTP traffic. Figure 3.(b) also corresponds to anomalies in MAWI traffic, but using IPdst as key. In this case, there are 9 anomalies, including different kinds of flooding DoS/DDoS attacks. Finally, figure 3.(c) corresponds to the detection of 9 DDoS attacks in the MET-ROSEC data-set. From these, 5 correspond to massive attacks (more than 70% of traffic), 1 to a high intensity attack (about 40%), 2 are low intensity attacks (about 10%), and 1 is a very-low intensity attack (about 4%). The detection is performed using traffic aggregated with IPdst key. In the three evaluation scenarios, the ROC plot is obtained by comparing the sorted dissimilarities in D_{rank} to a variable detection threshold.

We compare the performance of UNADA against three previous approaches for unsupervised anomaly detection: DBSCAN-based, k-means-based, and PCA-based outliers detection. The first two consist in applying either DBSCAN or k-means to the complete feature space $\mathbf{X}$, identify the largest cluster $C_{\max}$, and compute the Mahalanobis distance of all the flows lying outside $C_{\max}$ to

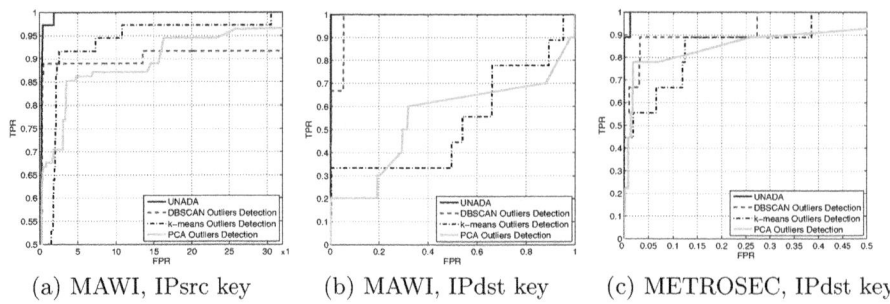

(a) MAWI, IPsrc key   (b) MAWI, IPdst key   (c) METROSEC, IPdst key

Fig. 3. True Positives Rate vs False Alarms in MAWI and METROSEC

its centroid. The ROC is finally obtained by comparing the sorted distances to a variable detection threshold. These approaches are similar to those used in previous work [9, 10, 11]. In the PCA-based approach, PCA and the sub-space methods [2,3] are applied to the complete matrix $\mathbf{X}$, and the attacks are detected by comparing the residuals to a variable threshold. Both the k-means and the PCA-based approaches require fine tuning: in k-means, we repeat the clustering for different values of clusters k, and take the average results. In the case of PCA we present the best performance obtained for each evaluation scenario.

Obtained results permit to evidence the great advantage of using the SSC-Density-based algorithm in the clustering step with respect to previous approaches. In particular, all the approaches used in the comparison generally fail to detect all the attacks with a reasonable false alarm rate. Both the DBSCAN-based and the k-means-based algorithms get confused by masking features when analyzing the complete feature space $\mathbf{X}$. The PCA approach shows to be not sensitive enough to discriminate different kinds of attacks of very different intensities, using the same representation for normal-operation traffic.

6   Conclusions

The Unsupervised Network Anomaly Detection Algorithm that we have proposed presents many interesting advantages with respect to previous proposals in the field of unsupervised anomaly detection. It uses exclusively unlabeled data to detect traffic anomalies, without assuming any particular model or any canonical data distribution, and without using signatures of anomalies or training. Despite using ordinary clustering techniques to identify traffic anomalies, UNADA avoids the lack of robustness of general clustering approaches, by combining the notions of Sub-Space Clustering, Density-based Clustering, and multiple Evidence Accumulation. We have verified the effectiveness of UNADA to detect real single source-destination and distributed network attacks in real traffic traces from different networks, all in a completely blind fashion, without assuming any particular traffic model, clustering parameters, or even clusters structure beyond a basic definition of what an anomaly is. Additionally, we have shown detection results that outperform traditional approaches for outliers detection, providing a stronger evidence of the accuracy of UNADA to detect network anomalies.

Acknowledgments

This work has been done in the framework of the ECODE project, funded by the European commission under grant FP7-ICT-2007-2/223936.

References

1. Barford, P., Kline, J., Plonka, D., Ron, A.: A Signal Analysis of Network Traffic Anomalies. In: Proc. ACM IMW (2002)
2. Lakhina, A., Crovella, M., Diot, C.: Diagnosing Network-Wide Traffic Anomalies. In: Proc. ACM SIGCOMM (2004)
3. Lakhina, A., Crovella, M., Diot, C.: Mining Anomalies Using Traffic Feature Distributions. In: Proc. ACM SIGCOMM (2005)
4. Soule, A., et al.: Combining Filtering and Statistical Methods for Anomaly Detection. In: Proc. ACM IMC (2005)
5. Krishnamurthy, B., et al.: Sketch-based Change Detection: Methods, Evaluation, and Applications. In: Proc. ACM IMC (2003)
6. Dewaele, G., et al.: Extracting Hidden Anomalies using Sketch and non Gaussian Multi-resolution Statistical Detection Procedures. In: Proc. LSAD (2007)
7. Jain, A.K.: Data Clustering: 50 Years Beyond K-Means. Pattern Recognition Letters 31(8), 651–666 (2010)
8. Parsons, L., et al.: Subspace Clustering for High Dimensional Data: a Review. ACM SIGKDD Expl. Newsletter 6(1), 90–105 (2004)
9. Portnoy, L., Eskin, E., Stolfo, S.: Intrusion Detection with Unlabeled Data Using Clustering. In: Proc. ACM DMSA Workshop (2001)
10. Eskin, E., et al.: A Geometric Framework for Unsupervised Anomaly Detection: Detecting Intrusions in Unlabeled Data. In: Applications of Data Mining in Computer Security. Kluwer Publishers, Dordrecht (2002)
11. Leung, K., Leckie, C.: Unsupervised Anomaly Detection in Network Intrusion Detection Using Clustering. In: Proc. ACSC 2005 (2005)
12. Fred, A., Jain, A.K.: Combining Multiple Clusterings Using Evidence Accumulation. IEEE Trans. Pattern Anal. and Machine Int. 27(6) (2005)
13. Ester, M., et al.: A Density-based Algorithm for Discovering Clusters in Large Spatial Databases with Noise. In: Proc. ACM SIGKDD (1996)
14. Williams, N., Zander, S., Armitage, G.: A Preliminary Performance Comparison of Five Machine Learning Algorithms for Practical IP Traffic Flow Classification. ACM SIGCOMM Computer Communication Review 36(5) (2006)
15. Fernandes, G., Owezarski, P.: Automated Classification of Network Traffic Anomalies. In: Chen, Y., Dimitriou, T.D., Zhou, J. (eds.) SecureComm 2009. Lecture Notes of the Institute for Computer Sciences, Social Informatics and Telecommunications Engineering, vol. 19, pp. 91–100. Springer, Heidelberg (2009)
16. Cho, K., Mitsuya, K., Kato, A.: Traffic Data Repository at the WIDE Project. In: USENIX Annual Technical Conference (2000)
17. METROlogy for SECurity and QoS, http://laas.fr/METROSEC

Efficient Processing of Multi-connection Compressed Web Traffic

Yehuda Afek[1], Anat Bremler-Barr[2,*], and Yaron Koral[1]

[1] Blavatnik School of Computer Sciences Tel-Aviv University, Israel
[2] Computer Science Dept. Interdisciplinary Center, Herzliya, Israel
afek@math.tau.ac.il, bremler@idc.ac.il, yaronkor@post.tau.ac.il

Abstract. Compressing web traffic using standard GZIP is becoming both popular and challenging due to the huge increase in wireless web devices, where bandwidth is limited. Security and other content based networking devices are required to decompress the traffic of tens of thousands concurrent connections in order to inspect the content for different signatures. The major limiting factor in this process is the high memory requirements of 32KB per connection that leads to hundreds of megabytes to gigabytes of main memory consumption. This requirement inhibits most devices from handling compressed traffic, which in turn either limits traffic compression or introduces security holes and other dysfunctionalities. In this paper we introduce new algorithms and techniques that drastically reduce this space requirement by over 80%, with only a slight increase in the time overhead, thus making real-time compressed traffic inspection a viable option for network devices.

Keywords: pattern matching, compressed http, network security, deep packet inspection.

1 Introduction

Compressing HTTP text when transferring pages over the web is in sharp increase motivated mostly by the increase in web surfing over mobile cellular devices. Sites such as Yahoo!, Google, MSN, YouTube, Facebook and others use HTTP compression to enhance the speed of their content download. In Section 6.2 we provide statistics on the percentage of top sites using HTTP Compression. Among the top 1000 most popular sites 66% use HTTP compression (see Figure 3). The standard compression method used by HTTP 1.1 is GZIP.

This sharp increase in HTTP compression presents new challenges to networking devices that inspect the content for security hazards and balancing decisions. Those devices reside between the server and the client and perform *Deep Packet Inspection* (DPI). When receiving compressed traffic the networking device needs first to decompress the message in order to inspect its payload. The two major performance penalties associated with this process are *time* and

---

* Supported by European Research Council (ERC) Starting Grant no. 259085.

J. Domingo-Pascual et al. (Eds.): NETWORKING 2011, Part I, LNCS 6640, pp. 52–65, 2011.

space. The time it takes to decompress a packet is a small fraction of the time it then takes to inspect the packet in most DPI applications [1]. However the space complexity cost of decompression is a major obstacle specially when the device is dealing with hundreds and thousands of concurrent connections. Notice that the techniques presented in [1], while reducing the DPI time requirement, it still uses information within the compression, i.e., decompression is still required.

This high memory requirement leaves the vendors and network operators with three bad options: either ignore compressed traffic, forbid compression, or divert the compressed traffic for offline processing. Obviously neither is acceptable as they present security hole or serious performance degradation.

The basic structure of our approach to dealing with the memory problem is to keep the buffers of all the connections compressed, except for the data of the connection whose packet(s) is now being processed. Upon packet arrival, unpack its session buffer and process it. One may naïvely suggest to just keep the appropriate amount of original compressed data as it was received. However this approach fails since the buffer would contain pointers to data more than 32KB backwards. Our technique, called *SOP*, packs the buffer of a connection by combining information from both compressed and uncompressed 32KB buffer to create the new compressed buffer that contains pointers that refer only to locations within itself. We show that by using our technique on real life data we reduce the space requirement by a factor of 5 with a time penalty of 26%. Notice that while our method modifies the compressed data locally, it is transparent to both the client and the server.

We then design an algorithm that combines our *SOP* technique that reduces space with the ACCH algorithm that reduces time complexity. By using the designed algorithm we achieve improvement of 42% of the time and 79% of the space requirements. The time-space tradeoff presented by our technique provides the first solution that enables DPI on compressed traffic in wire speed.

2   Background

Compressed HTTP: HTTP 1.1 [2] supports the usage of content-codings to allow a document to be compressed. The RFC suggests three content-codings: GZIP, COMPRESS and DEFLATE. In fact, GZIP uses DEFLATE with an additional thin shell of meta-data. For the purpose of this paper they are considered the same. Currently the GZIP and DEFLATE compressions are the common codings supported by current browsers and web servers[1].

The GZIP algorithm uses combination of the following compression techniques: first the text is compressed with the LZ77 algorithm and then the output is compressed with the Huffman coding. Let us elaborate on the two algorithms:

LZ77 Compression [3]- The purpose of LZ77 is to reduce the *string presentation size*, by spotting repeated strings within the last 32KB of the uncompressed data. The algorithm replaces the repeated strings by (*distance,length*)

---

[1] Analyzing captured packets from last versions of both Internet Explorer, FireFox and Chrome browsers shows that accept only the GZIP and DEFLATE codings.

pair, where *distance* is a number in [1,32768] (32K) indicating the distance in bytes of the repeated string and *length* is a number in [3,258] indicating the length. For example, the text: 'abcdeabc' can be compressed to: 'abcde(5,3)'; namely, "go back 5 bytes and copy 3 bytes from that point". LZ77 refers to the above pair as "pointer" and to uncompressed bytes as "literals".

Note that the LZ77 compression is a time consuming task, while the decompression is considerably light process (we use this observation later on as a motivation in the design of our algorithm). Experiments in Section 6 show that compression takes around 20 times more than decompression. Roughly speaking, the basic idea of the compression process goes as follows: at each point within the traffic, LZ77 tries to find the longest string that has already appeared in the text within the previous 32KB (the most recent if there are multiples). If such a string is found, LZ77 replaces the current string with a pointer to that occurrence. If no repetition longer than 2 bytes is found, than these bytes are not compressed. To decompress the traffic, one needs to reveal the referred bytes by the pointers which translates to a simple operation of consecutive memory copying directly from the 32KB buffer. Reading consecutive bytes has low per-byte read cost due to the good spatial locality in the cache, i.e., reading consecutive 32 bytes within a cache line costs one main memory access.

Huffman Coding [4]- Recall that the second stage of GZIP is the Huffman coding, that receives the LZ77 symbols as input. The purpose of Huffman coding is to reduce the *symbol coding size* by encoding frequent symbols with fewer bits. The Huffman coding method assigns to symbols from a given alphabet a variable-size *codeword* (coded symbol). *Dictionaries* are provided to facilitate the translation of binary codewords to bytes.

The Huffman decoding process is relatively fast. Common implementation (cf. zlib [5]) extracts the dictionary, with average size of 200B, into a temporary lookup-table that resides in cache. Frequent symbols require only one lookup-table reference, while less frequent symbols require two lookup-table references.

Deep packet inspection (DPI): DPI is the main action taken to inspect traffic, by identifying signatures (patterns or regular expressions) in the packet payload. Today, the performance of security tools is dominated by the speed of the underlying DPI algorithms [6]. The two fundamental paradigms to perform string matching derive from Aho-Corasick (AC) [7] and Boyer-Moore (BM) [8] algorithms. The BM algorithm does not have deterministic time and is prone to denial-of-service attacks using tailored input. Therefore the AC algorithm is the standard.The implementations need to deal with thousands of signatures. For example, ClamAV [9] virus-signature database consists of 27K patterns, and the popular Snort IDS [10] has 6.6K patterns; note that typically the number of patterns considered by IDS systems grows dramatically over time. Implementation of the traditional algorithm translates to dozens of megabytes and may even reach gigabytes of memory. The size of the signatures databases dictates not only the memory requirement but also the speed, since it forces the usage of a larger and slower memory on an order-of-magnitude such as DRAM, instead of using a faster one such as SRAM. That leads to an active research of reducing the memory requirement by compressing the corresponding DFA [11,12,13]; however, all proposed techniques suggest pure-hardware solutions, which usually incur prohibitive deployment and development cost. Still the common case, definitely in a software solution, requires using a very large database of signatures for DPI. Moreover we note that the DPI solutions do not enjoy the spatial locality time boost. Each input byte requires one or two memory reads to different parts of the memory and thus the DPI solutions do not enjoy the benefits of caching.

3   Challenges in Performing DPI on Compressed HTTP

This section provides an overview of the obstacles in performing deep packet inspection (DPI) in compressed HTTP traffic on a multi-connection environment.

As noted in [1], there is no apparent "easy" way to perform DPI over compressed traffic without decompressing the data in some way. This is mainly because $LZ77$ is an *adaptive* compression algorithm.

One of the main problems with the decompression is its memory requirement; the straightforward approach requires a 32KB sliding window for each HTTP connection. Note that this requirement is difficult to avoid, since the back-reference pointer can refer to any point within the sliding window and the pointers may be recursive (i.e., a pointer may point to an area with a pointer). On the other hand, DPI of non-compressed traffic requires storing only a two (or four) bytes variable that holds the DFA state. Hence, dealing with compressed traffic poses a higher memory requirement by a factor of 8 000 to 16 000. Thus, mid-range firewall that handles 100K-200K concurrent sessions needs 3GB-6GB memory while a high-end firewall that supports 500K-10M concurrent sessions needs 15GB-300GB memory only for the task of session decompression. This memory requirement has implication on not only the price and feasibility of the architecture but also on the capability to perform caching or using fast memory chips such as SRAM. Thus reducing the space has also straight implication on the speed. This work deals with the challenges imposed by that space aspect.

Apart from the space penalty described above, the decompression stage also increases the overall *time* penalty. However, we note that DPI requires significantly more time than decompression, since decompression is based on consecutive memory reading and therefore enjoy the cache block architecture and has low per-byte read cost, where DPI employs a very large DFA that is accessed by reads to non-consecutive memory areas therefore requires main memory accesses. Our experimental results in section 6 show that the decompression is 10 times faster than the DPI process in a multi-connection environment.

4   Related Work

There is an extensive research on preforming pattern matching on compressed-files, but very limited is on compressed traffic. Requirements posed in dealing with compressed traffic are: (1) on-line scanning (1-pass), (2) handling of thousands of sessions concurrently and (3) working with LZ77 compression algorithm (as oppose to most papers which deal with LZW/LZ78 compressions).

[1] is the first paper to analyze the obstacles of dealing with compressed traffic but it only accelerated the pattern matching task on compressed traffic and did not handle the space problem, and it still requires the decompression. We show in Section 5.3 that our paper can be combined with the techniques of [1] to achieve a fast pattern matching algorithm for compressed traffic, with moderate space requirement.

There are techniques developed for "in-place decompression", the main one is LZO [14]. While LZO claims to support decompression without memory overhead it works with files and assumes that the uncompressed data is available. We assume decompression of thousands of concurrent sessions on-the-fly, thus what is for free in LZO is considered overhead in our case. Furthermore, while GZIP is considered the standard for web traffic compression, LZO is not supported.

5   Packing Technique

In this section we describe our packing technique to reduce the 32KB buffer space requirement per session. The basic idea is to keep the session buffer in its packed form until the time a new incoming packet arrives for that session. To achieve that we use packing technique to keep a correct updated buffer after each packet processing. It has two parts:

- Swap Out of boundary Pointers (*SOP*) algorithm for packing the buffer.
- Our corresponding algorithm for unpacking the buffer.

Whenever a packet is received, the buffer that belongs to the incoming packet session is unpacked. After the incoming packet processing is finished an updated buffer is packed using the *SOP* algorithm. The next subsections elaborates on those parts of the algorithm.

5.1   Buffer Packing: Swap Out of Boundary Pointers (SOP)

In this subsection we describe our buffer packing technique. The first obvious attempt is to store the buffer in its compressed form using the original received traffic. However this attempt fails since the compressed form of the buffer contains pointers that point to positions prior to the 32KB boundary. Figure 1(a) shows an example of the original compressed traffic. Note that it contains pointer to a part that is no longer within the buffer boundaries. The conclusion from this attempt is that the solution must have the following property: A buffer must contain all information for its pointers extraction.

The second obvious attempt is to compress (each time from scratch) the 32KB buffer using some compression algorithm such as GZIP. That solution follows the above property since the compression is based only on information within the buffer. However, this solution performs compression which is an expensive task, while the memory saving is a negligible 1.5% as compared to *SOP*.

Our suggested solution, called Swap Out-of boundary Pointers (*SOP*), solves the problems of the above two attempts. The technique uses information within the original compressed and uncompressed form of the buffer for a quick packing

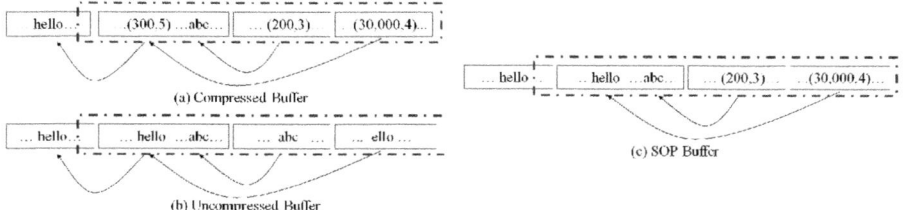

Fig. 1. Sketch of the memory buffer in different scenarios. Each solid box represents a packet. Dashed frame represents the 32KB buffer. Each arrow connects between a pointer and its referred area.

process. *SOP* changes the original GZIP compressed form related to the buffer, so that it contains pointers that refer only to information inside the buffer. To achieve this, *SOP* swaps all the pointers that point outside of the new boundary of the buffer with its referred literals[2]. Figure 1(c) shows an output of this algorithm. The pointer (300,5) that points prior to the buffer new boundary is replaced by the string 'hello' where the others remain untouched.

Since in every stage we maintain the invariant that pointers refer only to information within the buffer - the buffer can be unpacked. *SOP* still has a good compression ratio (as shown in Section 6) since most of the pointers are left untouched because they originally pointed to a location within the 32KB buffer boundary. *SOP* is also fast since it performs only one pass on the uncompressed information and the compressed buffer information, taking advantage of the GZIP compression that the source (server) did on the traffic.

The pseudocode is given in Algorithm 1. The algorithm consists of 4 parts. The first part performs Huffman decoding of the incoming packet in order to determine its uncompressed size which in turn determines the new boundary of the 32KB buffer. Note that pointers cannot be extracted at this point since *oldPacked* is still compressed. Therefore the decoding is kept within a linked-list data structure that contains either sequence of literals or pointer elements.

The other three parts consist of decoding either *oldPacked* or *packet* into *newPacked* using *unPacked*. Part 2 decodes data that would not be packed again since it is located outside of the new boundary of the 32KB buffer (after receiving the new packet). Parts 3 and 4 decode the rest of *oldPacked* and *packet* respectively, and prepare the *newPacked* buffer along the decoding process.

Note that the output is not optimal in terms of space. Figure 2 shows a case where LZ77 algorithm (2(c)) outperforms *SOP* (2(d)). In that case, the original compression did not indicate of any direct connection between the second occurrence of the 'Hello world!' string to the string 'sello world'. The connection can be figured out if one follows the pointers of both strings and finds that both strings share common referred bytes. Since *SOP* performs a straightforward

---

[2] Note that pointers can be recursive, i.e., pointer A points to pointer B. In that case we replace pointer A by the literals that pointer B points to.

swapping without following the pointers recursively it misses this case. However, the loss of space is limited as shown in the experimental results section.

We also tried some other, more sophisticated variants of *SOP* to solve that problem by some kind of a recursive method. However, the complexity of those algorithms is much higher and the gained space reduction is limited.

5.2   Unpacking the Buffer: GZIP Decompression

SOP buffers are packed in a valid GZIP format, hence a regular decompression can be used for unpacking. The main difference is that most of the data is decompressed more than once since it is maintained compressed. Buffer decompression is performed upon each incoming packet. Each byte is decompressed on average 4.2 times (see Section 6) using *SOP* buffers as compared to only once in the original GZIP method.

One may wonder if partial decompression of buffer areas is more affordable than the suggested method that decompresses the entire buffer upon each incoming packet. Since the pointers are recursive, the retrieval of the literals referenced by a pointer is a recursive process touching several locations in the buffer, and requiring to decode more symbols than its own length. For example: a pointer that points to another pointer which in turn points to a third pointer, requires decoding three different areas where the referred pointers and the literals reside. This property explains the poor results.

We designed a method for partially decompression, called *SOP-Indexed*. It is defined as follows:

- Split the buffer to fixed size chunks of bytes and keep indices that hold starting positions of each chunk.
- Recursively extract each pointer and decode only the chunks that contain required information for extraction.

For example: if a 256B chunks are used, referring to the 500th byte of the 32KB uncompressed buffer requires to decode the second chunk that corresponds to the [256-511] byte positions within the uncompressed buffer. If the pointer exceeds the chunk boundary of 511, the next chunk has to be decoded too.

Zero padding needs to be applied at the end of each chunk so the offset would be in terms of bytes and not in terms of bits. Each index is coded with 15 bits in order to represent offsets for up to 32KB. The chunk size poses a time-space tradeoff. Smaller chunks support more precise references and result in less decoding but require more indices that have to be stored along with the buffer, and more padding for each of the smaller chunks, hence cause larger space penalty. The results in Section 6 shows that only a limited time improvement of ratio 36% as compared to *SOP*, is gained by *SOP-Indexed*.

5.3   Combining SOP with ACCH Algorithm

The algorithm, ACCH presented in [1], reduces the time it takes to do pattern matching to less than 26% compared to doing it with Aho-Corasick (AC).

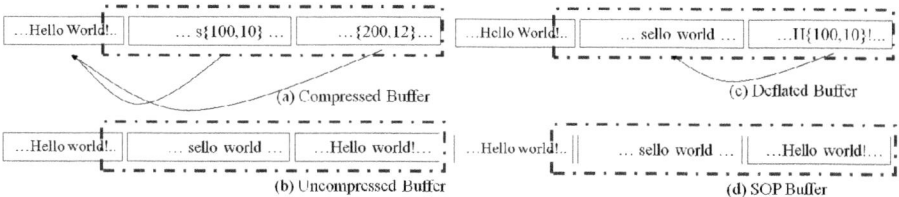

Fig. 2. Sketch of the memory buffer in different scenarios

The general idea is to take advantage of pointer information from LZ77 in order to skip and avoid scanning some bytes. Since most of the compressed bytes are represented as pointers, most of the byte scans can be saved. The data about previous byte scans is stored in a *Status Vector* [1]. Each vector entry is 2 bit long and stores three possible state values: Match, in case that a pattern was located, and two other states; If a prefix longer than a certain threshold was located the status is *Check*, otherwise it is *Uncheck*. An important observation is that the status of pointer bytes can be determined based on the status of the referred bytes only, without further scanning.

The ACCH algorithm is somewhat orthogonal to the *SOP* algorithm, still there are two points that must be addressed. The first one is related to the fact that the *Status Vector* itself has 8KB space requirement which needs to be taken care off in order to continue and enjoy the space benefits of *SOP*. The second point is related to the fact that the *SOP* algorithm changes the structure of some of the pointers, therefore when a pointer is replaced by SOP, the vector needs to be efficiently adapted to the new structure.

We use two techniques to address the above points. The first is to mark only the status changes and the second is to use the pointers to figure out the status. We give here a sketch of the suggested method that handles both points due to space limitations. Let us start with the first point. The general idea is to store only status changes instead of the statuses themselves, among the packed buffer symbols. There is no need to store status changes within pointers since ACCH can restore most of the previous referred statuses from the pointer referred area. The rest of the statuses are maintained using extra bits.

Handling the second point means that whenever a pointer is replaced with literals, the symbols that resemble the statuses within it should remain valid. Achieving this is straightforward. The status symbols are maintained from the referred bytes when they are copied, hence no additional memory references are required for status update after pointer replacing.

Applying the combination of the methods described above enabled us to combine ACCH with *SOP* algorithms and gain space and time improvements. The combination achieves time performance of more than two times faster than performing *SOP* with regular AC (as demonstrated in the next section).

Algorithm 1. Out of Boundary Pointer Swapping

---

packet - input packet.
oldPacked - the old packed buffer received as input. Every cell is either a literal or a pointer.
newPacked - the new packed buffer.
unPacked - temporary array of 32K uncompressed literals.
packetAfterHuffmanDecode - contains the Huffman decoded symbols of the input packet.

1: **procedure** *handleNewPacket(packet, oldPacked)*
   Part 1: calculate *packet* **uncompressed size** n
2: **for all** symbols in *packet* **do**
3:     $S \leftarrow$ next symbol in *packet* after Huffman decode
4:     *packetAfterHuffmanDecode* $\leftarrow S$
5:     **if** S is literal **then**
6:         $n \leftarrow n + 1$
7:     **else**   ▷ S is Pointer
8:         $n \leftarrow n+$ pointer length
9:     **end if**
10: **end for**
    Part 2: decode *oldPacked* **part which is out of boundary**
11: **while** less than n literals were unpacked **do**
12:     $S \leftarrow$ next symbol in *oldPacked* after Huffman decode
13:     **if** S is literal **then**
14:         store literal in *unPacked* buffer
15:     **else**   ▷ S is Pointer
16:         store pointer's referred literals in *unPacked* buffer
17:     **end if**
18: **end while**
    Part 3: decode *oldPacked* **part within boundary**
19: **if** boundary falls within a pointer in *oldPacked* **then**
20:     copy to *newPacked* only the suffix of the referred literals
21: **end if**
22: **for all** symbols in *oldPacked* **do**
23:     $S \leftarrow$ next symbol in *oldPacked* after Huffman decode
24:     **if** S is literal **then**
25:         add symbol to *unPacked* and *newPacked* buffers
26:     **else**   ▷ S is Pointer
27:         store the referred literals in *unPacked*
28:         **if** pointer is out of the boundary **then**
29:             store the coded referred literals in *newPacked*
30:         **else**
31:             store the coded pointer in *newPacked*
32:         **end if**
33:     **end if**
34: **end for**
    Part 4: decode *packet*
35: **for all** symbols in *packetAfterHuffmanDecode* **do**
36:     $S \leftarrow$ next symbol in *packetAfterHuffmanDecode*
37:     This part is the same as Lines [24-33]
38: **end for**

6   Experimental Results

6.1   Experimental Environment

Our time performance results are given relative to the performance of a base algorithm *Plain*, which is the decompression algorithm without packing, therefore the processor type is not an important factor for the experiments. The experiments were performed on a PC platform using Intel® Core™2 Duo CPU running at 1.8GHz using 32-bit Operating System. The system uses 1GB of Main Memory (RAM), a 2MB L2 Cache and 2×32KB write-back L1 data cache.

We base our implementation on the *zlib* [5] software library for the compression/decompression parts of the algorithm.

Our experimental environment does not simulate packets from multiple-connections but only from a single connection at a time. Therefore we flush the entire system's cache upon each packet arrival to create the context switch effect between multiple connections. Note that by flushing the cache we are very conservative in our experimental environment and we suspect that in a real life scenario the time of *SOP* is even better. In the real life environment, some of the writing to the buffers could be performed in the background while handling a different section therefore most of the memory writing penalty could be avoided. Furthermore, sometimes we may receive several consecutive packets from the same flow, hence may be processed without flushing the cache.

6.2   Data Set

The data set consists of 2308 HTML pages taken from the 500 most popular sites that use GZIP compression. The web site list was constructed from the Alexa site [15] that maintains web traffic metrics and top-site lists. Total size of the uncompressed data is 359MB and in its compressed form is 61.3MB.

While gathering the data-set from Alexa, we created an interesting statistics about the percentage of the compressed pages among the sites as shown on Figure 3. The statistics shows high percentage of compression, in particular among the top 1000 sites. As popularity drops the percentage slightly drops. Still, almost 1 out of every 2 sites uses compression.

6.3   Space and Time Results

This subsection reports results concerning the average space requirement for our algorithm as shown in Table 1. We define *Plain* as the basic algorithm that performs decompression and maintains a buffer of plain uncompressed data for each connection. Note that at the beginning of a connection the buffer stores less than 32KB since it stores only as much as was sent. Therefore the average buffer size of *Plain* is 29.9KB which is slightly lower than the maximum value of 32KB. We use *Plain* as a reference for performance comparison to the other proposed methods and set its time and space ratios to 1.

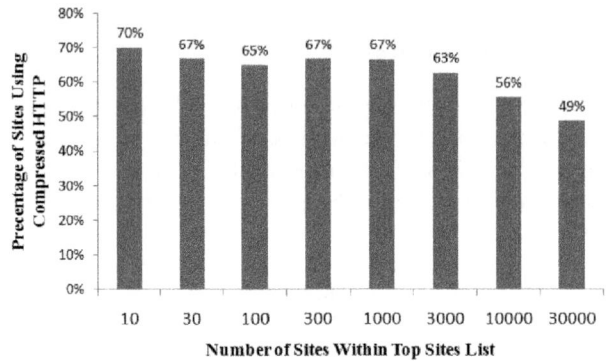

Fig. 3. Statistics of HTTP Compression usage among the Alexa [15] top-site lists

Table 1. Comparison of Time and Space parameters of different algorithms

Packing Method	Average Buffer Size	Compression Ratio per Buffer	Average Time Ratio
Plain	29.9KB	1	1
OrigComp	4.54KB	0.1518	-
Deflate	5.04KB	0.1576	20.77
SOP	7.33KB	0.245	3.91
SOP	6.28KB	0.211	3.89
SOP	5.17KB	0.172	3.85
SOP-Indexed	5.47KB	0.183	3.49

We measure the size of the incoming compressed data representing the buffer and call it *OrigComp*. This data can not be used as a buffer for a *multi-connection environment*, since it contains pointers that point to positions prior to the 32KB buffer range. The average buffer size required by *OrigComp* is 4.54KB, which is considered as a space lower bound.

We define *Deflate* as the method that compresses each buffer from scratch using GZIP. This method represents the best *practical* space result but has the worse time requirements of more than 20 times higher than *Plain*.

SOP takes 3.85 more time than *Plain*. This is a moderate time penalty as compared to *Deflate*, the space requirement of SOP is 5.17KB which is pretty close to the 5.04KB of *Deflate*. The small space advantage gained by *Deflate* in addition to the poor time requirement makes it irrelevant as a solution.

The last method we examined is the *SOP-Indexed* which maintains indices to chunk offsets within the compressed buffer to support a partial decompression of the required chunks only. We used a 256B chunks and got an average of 69% chunk accessed. The time ratio is improved to 3.49 as compared to *Plain*. The space penalty for maintaining the index vector and chunk padding is of 0.3KB.

6.4   Time Results Analysis

As explained in Section 5.2, *SOP* decompresses each byte on average 4.2 times. Hence one would expect *SOP* to take 4.2 times more than *Plain*. Still *SOP* takes only 3.85 times. This can be explained by inspecting the data structures that require main memory accesses by each of the algorithms. *SOP* maintains in main memory the old and new packed buffers (i.e., *oldPacked* and *newPacked* as in Algorithm 1) which are heavily accessed during packet processing. *Plain* on the other hand, uses parts of the 32KB buffer, taken from main memory. When processing a packet, *Plain* touches only parts of the 32KB buffer. We measured the relative part of the buffer which is accessed by *Plain* using a method similar to the one in 5.2 with chunks of 32B. An average 40.3% of the buffer is accessed. *SOP* also uses a 32KB buffer for the uncompressed data, but it keeps it only in the cache and most of it is never written back to the main memory. The key point is that a write-back cache is used, hence this buffer remains in cache. The unpacked 32KB buffer is maintained in a temporary variable and when SOP finishes processing the packet that variable is disposed and *not* written back to main memory. However, working in a write-through cache architecture may harm *SOP* performance. The main memory space used for *SOP* data structures is thus around 10KB upon each incoming packet where as the main memory space accessed by *Plain* is around 12KB (= .4 of the 32KB) which is 20% higher and explains why the time performance of *SOP* is better than the expected 4.2.

6.5   DPI of Compressed Traffic

In this subsection we analyze the performance of DPI. We focus on pattern matching, which is a lighter DPI task than the regular expression matching. We show that the processing time taken by *SOP* is minor compared to the processing time taken by the pattern matching task. Since the regular expression matching task is even more expensive (than pattern matching), the processing time of *SOP* is even less important.

Table 2 summarizes the overall time and space requirements for the different methods that implement pattern-matching of compressed traffic in multi-connection environment. The time parameter is compared to performing Aho-Corasick (AC) on uncompressed traffic, as implemented by snort [10]. Recall that AC uses a DFA and basically performs one or two memory references per scanned byte. The other pattern matching algorithm we use for comparison is ACCH which is based on AC and uses techniques that are adjusted to GZIP compressed input making the pattern matching process work faster. However, the techniques used by ACCH are independent from the actual AC implementation. The space per buffer parameter measures the memory required for every session upon context switch that happens after packet processing is finished, i.e., in *SOP* after packing.

The offline-processing method (*Offline*) represents the only option supported by current network tools, that deals with compressed traffic. The space requirement is calculated by rough estimate of average compressed and uncompressed

Table 2. Overview of Pattern Matching + GZIP processing

Algorithms in Use		Average Time Ratio	Space per Buffer
Packing	Pattern Matching		
Offline	AC	-	170KB
Plain	AC	1.1	29.9KB
Plain	ACCH	0.36	37.4KB
SOP	AC	1.39	5.17KB
SOP	ACCH	0.64	6.19KB

size of 27KB and 156KB respectively. The compressed form is stored during session arrival and the uncompressed data is stored during pattern matching phase. When used in combination with AC as used today, it presents an enormous space requirement in terms of mid-range security tool.

We measured the time ratio of performing *Plain* as compared to pattern matching using AC, both on single and multi-connection environment. On the single-connection environment the ratio is as low as 0.035 where on the multi-connection environment the ratio is 0.101. The difference is due to the context switch upon each packet on the multi-connection environment that harms the decompression spatial locality property. Note that *SOP* is compared to performing *Plain* on a multi-connection environment.

As explained in 5.3, ACCH improves the time requirement of the pattern matching process. Recall that in order to apply the ACCH algorithm we need to store on memory an additional data structure called *Status Vector*. Applying the suggested compression algorithm for the *Status Vector* compressed it to 1.03KB. Therefore the total space requirement of *SOP* combined with ACCH is 6.19KB.

The best time is achieved using *Plain* with ACCH but the space requirement is very high as compared to all other methods apart from *Offline*. It occurs that combining *SOP* and ACCH achieves almost 80% space improvement and above 40% time improvement comparing *Plain* with AC.

As we look at the greater picture that involved also DPI, we need to refer to the space requirement applied by its data structures. As noted before, the DPI process too has large memory requirements. As opposed to the decompression process which its space requirement is proportional to the number of concurrent sessions, the DPI space requirements depend mainly on the number of patterns that it supports. Therefore, the DPI large space requirement is applied from the first session. Hence, assuming that DPI space requirements are at the same order as those of decompression for mid-range network tools, the 80% space improvement for the decompression buffers is translated to a 40% space improvement for the overall process.

7   Conclusion

With the sharp increase in cellular web surfing, HTTP compression becomes common in today web traffic. Yet due to its high memory requirements, most security devices tend to ignore or bypass the compressed traffic and thus introduce either a security hole or a potential for a denial of service attack. This paper presents SOP, a technique that drastically reduces this space requirement (by over 80%) with only a slight increase in the time overhead. It makes real-time compressed traffic inspection a viable option for network devices. We also present an algorithm that combines SOP with ACCH, a technique that reduces the time required in performing DPI on compressed HTTP [1]. The combined technique achieves improvements of almost 80% in space and above 40% in the time requirement for the overall DPI processing of compressed web traffic. Note that ACCH algorithm (thus the combined algorithm) is not intrusive to the Aho-Corasick (AC) algorithm, and it may be replaced and thus enjoy the benefit of, any DFA based algorithm including recent improvements of AC [11, 12, 13].

Acknowledgment

The research leading to these results has received funding from the European Research Council under the European Union's Seventh Framework Programme (FP7/2007-2013)/ERC Grant agreement n°259085.

References

1. Bremler-Barr, A., Koral, Y.: Accelerating multi-patterns matching on compressed HTTP. In: INFOCOM 2009 (April 2009)
2. Hypertext transfer protocol – http/1.1, RFC 2616 (June 1999), http://www.ietf.org/rfc/rfc2616.txt.
3. Ziv, J., Lempel, A.: A universal algorithm for sequential data compression. IEEE Transactions on Information Theory, 337–343 (May 1977)
4. Huffman, D.: A method for the construction of minimum-redundancy codes. In: Proceedings of IRE, pp. 1098–1101 (1952)
5. zlib 1.2.5 (April 2010), http://www.zlib.net
6. Fisk, M., Varghese, G.: An analysis of fast string matching applied to content-based forwarding and intrusion detection. Techical Report CS2001-0670 (updated version) (2002)
7. Aho, A., Corasick, M.: Efficient string matching: an aid to bibliographic search. Communications of the ACM, 333–340 (1975)
8. Boyer, R., Moore, J.: A fast string searching algorithm. Communications of the ACM, 762–772 (October 1977)
9. Clam antivirus. http://www.clamav.net (version 0.82)
10. Snort, http://www.snort.org (accessed on May 2010).
11. Lin, W., Liu, B.: Pipelined parallel ac-based approach for multi-string matching. In: ICPADS (2008)
12. van Lunteren, J.: High-performance pattern-matching for intrusion detection. In: INFOCOM, pp. 1–13 (April 2006)
13. Tan, L., Sherwood, T.: Architectures for bit-split string scanning in intrusion detection. In: Micro, pp. 110–117. IEEE, Los Alamitos (2006)
14. Oberhumer, M.F.: LZO, http://www.oberhumer.com/opensource/lzo
15. Top sites (July 2010), http://www.alexa.com/topsites

The Resource Efficient Forwarding in the Content Centric Network

Yifan Yu and Daqing Gu

Orange Labs, Beijing. 10/f, South Tower Raycom Info Park C, 2 Science Institute
South Rd, Beijing, P.R.C.
{yifan.yu,daqing.gu}@orange-ftgroup.com

Abstract. Today's networking technologies in Internet that can only speak of connections between hosts no longer adapt to the network increasingly dominated by content distribution and retrieval. Some efforts therefore have been made to generalize the Internet architecture by taking content as a primitive- decoupling location in the network instead of focusing on endpoint addresses of hosts. However, one of the representative architectures known as *Content-Centric Networking* (CCN) consumes excessive system resource in terms of router's processing capability, the buffer space in the router and the network bandwidth when deployed as the "universal overlay" in the existing network. We present the solution composed of two methods to improve the resource efficiency based on the CCN approach. The results from the ns2-based simulation disclose that our solution can reduce the resource consumption in routers and network pipeline significantly in the overlay architecture.

Keywords: Content-Centric Networking, Internet architecture, Forwarding, Resource efficient, Data.

1 Introduction

Today's Internet was already designed to be an end-to-end connectivity infrastructure based on packet switching at the beginning of 1960s. The communication solution offered by it targets at solving the telephony's problem: a point-to-point conversation between two hosts using the IP protocol [1]. Thus IP packets exchanged between the two hosts in communication contain two identifiers (addresses), one for the source and one for the destination host. By allowing lower and upper layer technologies to innovate without unnecessary constraints, the Internet achieves the explosive growth to show its power and elegancy.

However, there have been dramatic changes in Internet since its creation [2].

- With the emergency of World Wide Web (WWW) [3] application, the increasing number of Information-intensive business such as travel, banks and financial services is done in the Internet. Since almost anything is available online today, the Internet is becoming the mirror of the real world where more and more people's activities are digitalized.

J. Domingo-Pascual et al. (Eds.): NETWORKING 2011, Part I, LNCS 6640, pp. 66–77, 2011.

- An ever increasing range of content such as voice, images and video etc. can be distributed digitally due to the digital coding advances.

- The cheaper and more ubiquitous digital equipments driven by the Moore's-Law facilitate connecting everything to the Internet: not just computers but also factories, municipal infrastructure, phones, cars, appliances, even light switches. There is considerable consensus that the amount of these equipments could reach the number of billions or even trillions [4] [5].

The above changes make people tend to value the Internet for what content it contains rather than where they can do communication. Then IP protocol, which was designed to act as a key enabler supporting the conversations between communications endpoints, is overwhelmingly used for content distribution. However, the 'conversational' nature of IP makes the existing Internet architecture a poor match to its primary use today.

In order to overcome the poor match between the IP's nature and the practical requirements of people using the Internet, several efforts [6,7,8,9,10,11,12,13,14, 15,16] have been made to pursue the direct and unified way to solve the problems. From the view point of these works, the named data is a better abstraction for today's communication problems than named hosts. Then, the principle of the current innovation aim to interconnect the hosts by using the address that names content instead of the IP addresses traditionally identifying the physical network nodes [18, 19, 20]. In other words, the currently proposed solutions attempt to solve the problem of poor match in the way of replacing *where* with *what* in the core layer that the Internet centers on, which is typically termed as content-centric networking or named data networking.

The architectural principles that guide the design of the new architecture are stemmed from the existing Internet architecture. Substantially speaking, it departs from IP in creating the new universal component, known as content-based "thin waist", in the protocol stack [2] [6].

As illustrated in Fig. 1, the current Internet architecture is built as the hourglass architecture where network layer (IP) acts as a universal component, namely "thin waist", to implement the minimal functionality necessary

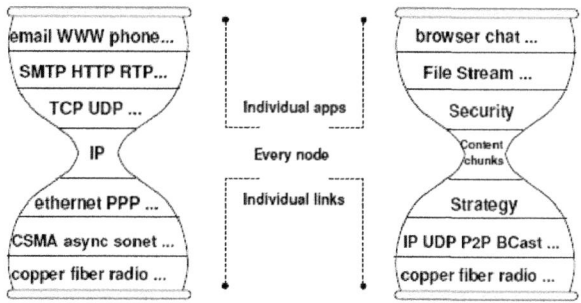

Fig. 1. Today's Internet Architecture Vs. Content Centric Architecture

for global interconnectivity. The content centric architecture keeps the same hourglass-shaped architecture except that the "thin waist" is oriented to the content chunks instead of the physical hosts with the identifier of IP addresses.

Unlike today's Internet architecture that regards the security as an afterthought, the content centric one provides a basic security embedded in the thin waist by signing all named data. In such architecture, security is provided to content itself, rather than the connections over which it travels, thereby avoiding many of the host-based vulnerabilities that plague IP networking. The end-to-end principle that guarantees the robust applications in the face of network failures is retained and expanded by the content centric architecture. Particularly, the multiple simultaneous connections (e.g., Ethernet, 3G, LTE, Zigbee and 802.11) can be utilized to achieve the fine-grained, dynamic optimization for the data under changing conditions with the consideration of traffic self-regulation.

Routing and forwarding plane separation is still stuck to in the content centric architecture. But the new routing system should be carried out to allow the addressing oriented to the content name.

The rest of the paper is organized as following. The motivation of our work in the paper is presented in section 2. In Section 3, our solution is described in detail. We evaluate the performance of our proposal with the extensive ns2-based simulation in section 4. Finally, in Section 5, we give the conclusion of the paper.

2   Related Work and Motivation

The works concerning the content centric architecture can be classified as two categories: hierarchical name-based and flat name-based. The former identifies the content with the hierarchical names much like Uniform/Universal Resource Locator (URL) [21] used in the current web application [6, 7], while the latter would like to define the name of the content using the flat label [8,9,10,11,12,13, 14,15,16] that is similar to the identifiers used in the existing distributed hash table (DHT)-based P2P system [22] [23]. Since our work in the paper focuses on the hierarchical name-based solution, we just describe the existing approaches in this category.

2.1   Content-Centric Networking

The classical approach realizing the content centric networking with the hierarchical name is presented in [6]. It is called as Content-Centric Networking (CCN). The communication in CCN is driven by the data consumer. As shown in Fig. 2, two CCN packet types, Interest and Data, are defined to enable the communication oriented to the specific content chunk.

To receive data, a receiving end, namely the consumer, sends out an Interest packet carrying a name that identifies the desired data. For example, a consumer may send an Interest packet requesting the data named after /parc/videos/ WidgetA.mpg. Fig. 3 gives the example that how the name is defined.

A router maintains a data structure termed as Pending Interest Table (PIT) to remember the face where the request arrives, and then forwards the Interest

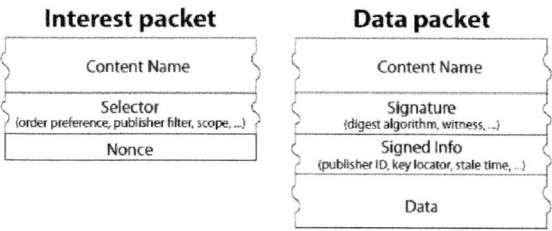

Fig. 2. CCN packet types

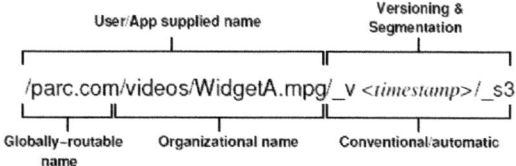

Fig. 3. Example of Data Name

packet by looking up the name in its Forwarding Information Base (FIB) where the list of faces that can serve the request is recorded. FIB is almost identical to an IP FIB except it allows for a list of outgoing interfaces rather than a single one. The name-based routing policy is used for the population of this table. Since the packets introduced in [6] are not only forwarded over hardware network interfaces but also exchanged directly with application processes within a machine, the term face rather than interface is used to denote the place where the packets come from and are forwarded.

Once the Interest reaches a node that has the requested data in its local storage termed as Content Store, a Data containing the requested name and the content of the data packet, together with a signature by the producer's key, is sent back. This Data packet traces in reverse the path created by the Interest packet back to the consumer.

When multiple Interests for the same data are received, only the first Interest is sent upstream towards the data source. When the Data packet arrives, the router finds the matching PIT entry by using longest-match look-up based on the content name and forwards the data to all the interfaces listed in the PIT entry. The router then removes the corresponding PIT entry, and optionally caches the Data in its Content Store to satisfy potential future requests.

2.2 Motivation

In CCN approach, the arrival of each data packet triggers the routers to look up and update its PIT and FIB with the name of received data that is expressed as a structured string similar to URL. However, such method may result in the excessive resource consumption in terms of router processing capability, router's buffer capacity and network bandwidth.

Specifically, CCN's compatibility with today's Internet centering on IP protocol makes it to readily use existing IP infrastructure services. Thus, CCN can run as a "universal overlay" over the current Internet. The routing policy of CCN router then may be regarded as IP+CCN, which means that the CCN router should execute the IP routing prior to the CCN processing of received packet. In other words, there is double load in the processor of router because the arrival of any CCN packet requires the handling both in the IP layer and CCN layer. Especially, if the router is not to cache the received CCN data, which is common because the router is unable to cache each received data due to the limited buffer space, the CCN router will be confronted with extra processing load because the data can be simply forwarded as the IP packet without any CCN operations.

On the other hand, the URL-like name of CCN data implies that the string identifying the contained content is much longer than the physical identifier such as IP address. Thus, it is expected that lots of router buffer space and network bandwidth will be spent in accommodating the name of CCN data. Suppose the CCN name is 200 bytes in length and the router is provisioned with packet buffers 2KB in size with 2 GB of total storage capacity; this means Content Store should allocate the buffer space of 200 MB to store 1 million distinct packet names. In further consideration of the maximum transmission unit (MTU) in the underlying pipeline (i.e. Ethernet), only 85% of the bandwidth is used to transport the content payload given that MTU is 1.5KB.

Our work in this paper is motivated by the above two challenges for the current CCN approach. Briefly, our contributions are summarized as following:

- We propose a method allowing the intermediate router to avoid the unnecessary processing in the CCN layer. Specifically, we design the way of configuring source network identifier (i.e. IP address) in CCN Interest to allow for the direct forwarding of the data not to be cached in the intermediate CCN router without any CCN operations.

- We design the method to reduce the size of content name in the CCN data packet so that more network bandwidth and buffer space in the router can be used by the content payload. The principle of our method is to replace the URL-like content name in CCN data with an integer that can be recognized between the upstream node and downstream node. The control messages enabling such process are presented as well.

3   Resource Efficient Packet Forwarding

The solution proposed in the paper focuses on reducing the system resource consumption in terms of router's processing capability, buffer space in router and the network bandwidth. In details, our proposal includes two innovative methods that involve the enhanced processing of CCN interest data and the header reduction in CCN data, respectively. The former alleviates the processing load in the router, while the latter reduces the buffer space and network bandwidth consumed by the CCN data.

Note that the CCN architecture is assumed to be built as the overlay of the IP-based network in the following sections of our paper. But it does not mean that our solution can merely run in the IP network. In essence, our proposed solution can work in other underlying networks (such as LTE or WiFi etc.).

3.1 Enhanced Processing of CCN Interest and CCN Data

Unlike the CCN approach that creates the PIT entry upon receiving each new CCN Interest, our solution has the router deal with the new CCN Interest based on its decision whether to cache the data requested in the Interest. The router then further determines how to modify the destination IP address and source IP address of the CCN Interest that needs to be forwarded.

Upon the reception of Interest, the router should firstly check whether it can serve the Interest by looking up the Content Store. If the router caches the content requested by the Interest, it will send out the matched data piece whose destination address is set as the source IP address of Interest.

If there is no matched entry in Content Store, the router checks whether the Interest has been pended in PIT. If so, the Interest is discarded and the router adds the source IP address of the Interest to the corresponding PIT entry.

Otherwise, the router is to check whether there is a matched entry in FIB. If not so, the Interest has to be discarded. Otherwise, the router should firstly remove the face where the Interest arrives from the face list of the matched entry. After that, the following steps should be taken:

a) If the router wants to cache the content required by Interest, a new entry for pending the received Interest is firstly created in PIT. Then, the router has to replace the source IP address of the received Interest with its own IP. The destination IP address of the received Interest is then set as the IP address of the node where the Interest is to be forwarded.
b) If the router has no intension to cache the content, it keeps the source IP address of the received Interest unchanged but the destination IP address is set as the IP address of the node where the Interest is to be forwarded.

Finally, the Interest is forwarded over all the faces of the matched FIB entry.

The processing of CCN Interest mentioned above allows the router to only take the CCN-defined reaction to the arriving data that it is ready to cache. It is because that the router is transparent to both its upstream node and downstream node in the sense of CCN routing supposing that the source address of the forwarded Interest is not set as its own IP address.

Then, as the requested CCN data arrives, there will be two consequences: the router deals with the received data in the CCN-defined way and the source IP address of data is set as the router's IP address given that the data was cached. Otherwise, the source IP address of the data remains unchanged and it is treated as the normal IP packets without any CCN operations in the router.

User/App supplied name	Code
/Orange .com/videos/01 .mp4/	33
/Orange .com/news/01.mp3/	75
......	......

(a) Local Code Table

User/App supplied name	Code-Face Pair
/Orange .com/videos/01 .mp4	<A, 33>,<B, 91>
/Orange .com/news /01 .mp3	<C,75>
......	......

(b) Remote Code Table

Fig. 4. Table for Header Reduction

3.2   Header Reduction in CCN Data

The header reduction method in our solution compresses the name of the CCN data by replacing the URL-like identifier with an integer understandable to upstream node and downstream node. It is used in the CCN data delivery between the CCN routers.

In our method, as the downstream CCN router receives a CCN data with the URL-like name, it may generate an integer ranged from 0 to $2^{16} - 1$ to represent the User/App supplied name in the Content Name (in Fig. 3). The generated integer is locally unique in the downstream node. A table, termed as **Local Code Table (LCT)**, as in Fig. 4(a), that records the mapping between the integer and User/App supplied name is maintained in the downstream router.

Since the mapping entries in LCT needs to be known to the upstream router as well, the dedicated message called as **Code Notification Message (CNM)** is sent from the downstream node to the upstream node for LCT duplication that is known as **Remote Code Table (RCT)**. As shown in Fig. 4(b), considering the upstream router may have multiple downstream routers for the dissemination of same data piece, the codes oriented to various faces, namely downstream nodes, should be differentiated in RCT.

The arrival of CNM triggers the upstream router to update its RCT and replace the User/App supplied name in the follow-up data packets composing the same content with the integer announced in the message. Thus, the CCN data with the reduced header appears as Fig. 5 where the component of Versioning & Segmentation is identical to that defined in CCN.

After receiving the CCN data with the reduced header, the downstream router looks up its LCT with the code in the data header. If there is matched entry, the data can be processed in the CCN way. Otherwise, the downstream router has to abandon the data and send a message, called as **Code Error Message (CEM)** where the invalid code is notified, to the upstream router, which then guides the upstream node to remove the invalid entry in its RCT.

The received data can be cached in the router's buffer with the reduced header as well. Upon further forwarding it, the router should lookup its RCT with the User/App supplied name indicated by the code in the compressed data header. The User/App supplied name is easily available to the router due to the mapping recorded in its LCT. If there is matched entry, the code originally contained in the data header is set as that indicated in the matched RCT entry. Otherwise,

Data packet

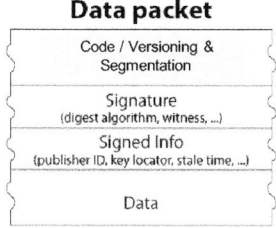

Fig. 5. CCN Data with Reduced Header

the content name in the data header is expressed by the User/App supplied name instead of the code.

The router can optionally decide whether to delete some entry in its LCT given the finite code length (i.e. 16bits). The removal of the LCT entry makes the router to send the **Code Deletion Message (CDM)** to its neighboring CCN routers to inform the removed pair of *<code, User/App supplied name>*.

Although there are several messages introduced in our method, the router can avoid the frequent message exchanges by previously setting up the mappings between the code and User/App supplied name for the popular content in its LCT and notifying them to its upstream node.

While it is feasible to reduce the CCN data header by delivering it using the UDP protocol [6] where the content name can be replaced as the UDP port number, it can only save the network bandwidth but at the expense of introducing the complicated adaption of CCN data into the UDP segment. Especially, there is a potential competence between the CCN data and other existing UDP applications for the limited port numbers resource. However, our method can make the efficient use of the system resource by relying on the single UDP port dedicated for CCN communication in the architecture of CCN over UDP.

It is known that in-packet Bloom Filters [9] can also be used to reduce the data header. But the functions of Bloom Filters should be globally defined in such solution, which sets the barrier for its increasing deployment in the large-scale network. In comparison, our method just requires the negotiation between the routers neighboring to each other rather than the global coordination.

3.3   Integration of Two Proposed Methods

The two proposed methods can be used either together or individually in the CCN architecture. As the enhanced CCN Interest and data processing operates together with the header reduction method, the data delivery based on the compressed header is only implemented between the CCN routers where the data is to be cached. The intermediate routers without the need of caching can still take the traditional IP-policy reaction to the received CCN packets and there are no any operations relevant to the header reduction in these routers as well.

4   Simulation and Discussion

To have some basic indications of the efficiency in our solution, in terms of router's processing load, buffer space consumption and network bandwidth usage, we have run some early ns-2 simulations.

The simulation scenario is illustrated in Fig. 6. The content source disseminates the content item (i.e. a video clip) composed of the 1000 data pieces each with 1KB in length to the client along the path concatenated by 5 content routers. Each router caches the CCN data passing by it with the probability d.

In our simulation, we firstly focus on the router's processing load at the CCN layer that is evaluated as the amount of operations on the received CCN data packets. Then, the router's buffer space and the network bandwidth consumed by the CCN data header are measured.

Fig. 7 illustrates the router's processing load occurring in the CCN approach and our proposal. It is observed that the router running the CCN approach deals with all the data pieces even it does not cache some of them. However, our solution just has the router process the data pieces cached in its buffer. Therefore, there is increasing reduction on processing load in our scheme with the falling cache probability d.

The router buffer space occupied by the content name of the cached data pieces is evaluated in Fig. 8. With the assumption that the content name is defined as the URL-like string with 50 bytes where 45 bytes constitute the User/App supplied name and 5 bytes represent the Versioning & Segmentation, our solution yields the reduction of the buffer space consumption by 86% in comparison to the CCN approach. The reduction comes from the fact that we

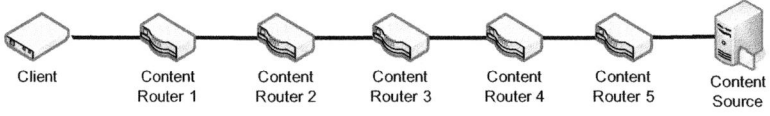

Fig. 6. Simulation Scenario

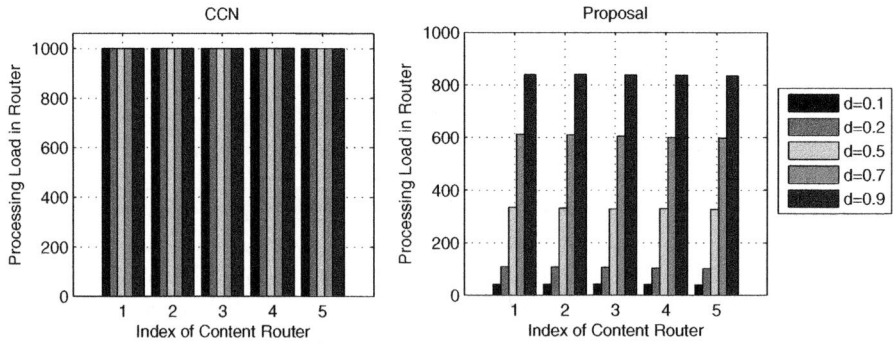

Fig. 7. Processing Load in Content Router

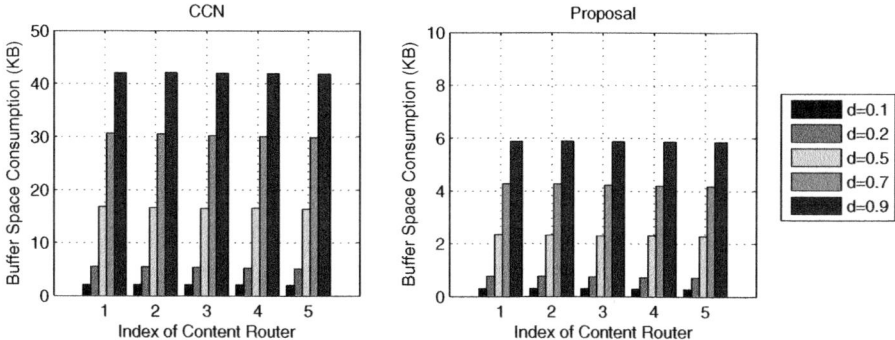

Fig. 8. Buffer Space Consumption of Content Name in Content Router

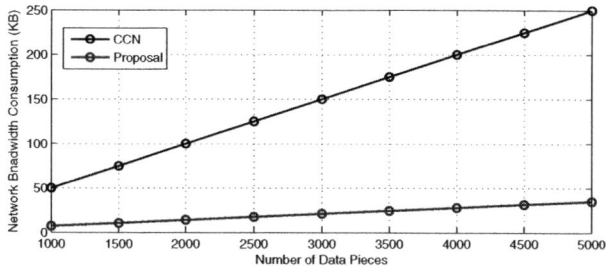

Fig. 9. Network Bandwidth consumed by Content Name

use the 2-byte code to represent the User/App supplied name originally defined in CCN. The improvement is still obvious even taking into account the memory used by the LCT and RCT introduced in our solution because the space occupied by these tables is much less than those accommodating User/App supplied name component in tens or hundreds of data pieces.

The network bandwidth consumed by the content name of data pieces is measured as the total size of the User/App supplied name components sent from the content source in Fig. 9. In order to take the insight into the impact of the content name on the network resource consumption, we illustrate the performance curves with different numbers of data pieces constituting the content item that are denoted by the X-axis in the figure where we still hold the assumption made in Fig. 8. It is noticed that our solution saves more network bandwidth as more data pieces are sent from the content source.

5   Conclusion

The CCN architecture presented in [6] is known as one of the promising candidates of building the future content centric network. However, it may require the potentially respectable resource consumption when operating as the overlay of the existing network architecture such as IP network. In this paper we have presented the solution constituting of two methods that can handle the CCN packets in the resource efficient way. The first method lessens the processing load of the router by making the router to avoid handling the data not to be cached in its buffer. Its principle is to configure the source and destination address in the underlying transmit unit (i.e. IP packet) encapsulating CCN packet based on the router's decision of caching CCN data. The second one replaces the URL-like content name in the CCN header with an integer to reduce the buffer size and network resource consumption. Two tables recording the mapping between the integer and the content name are defined to enable the proposed method can be performed locally without any global cooperation in the network.

In order to evaluate the solution, we created the extensive simulation based on ns-2. The simulation results disclose that our solution can alleviate the processing load in the router especially when the router caches less data in the buffer. The simulation also confirms that there is obvious reduction in the buffer space and network bandwidth consumption in the network with the massive content dissemination.

References

1. Comer, D.E.: Internetworking with TCP/IP. In: Principles, Protocols, and Architecture, 4th edn., vol. 1. Prentice Hall, Englewood Cliffs (2000)
2. Named Data Networking (NDN) Project, http://www.named-data.net
3. Tanenbaum, A.S.: Computer Networks. Prentice-Hall, Englewood Cliffs (2002)
4. AKARI, New Generation Network Architecture AKARI Conceptual Design. Project Description v1.1 (2008)
5. Gluhak, A., Bauer, M., Montagut, F., Stirbu, V., Johansson, M., Vercher, J., Presser, M.: Towards an Architecture for a Real World Internet. In: Towards the Future Internet. IOS Press, Amsterdam (2009)
6. Jacobson, V., Smetters, D.K., Thornton, J.D., Plass, M.F., Briggs, N.H., Braynard, R.L.: Networking Named Content. In: The 5th International Conference on Emerging Networking Experiments and Technologies, pp. 1–12. ACM, New York (2009)
7. Zhu, Y., Chen, M., Nakao, A.: CONIC: Content-Oriented Network with Indexed Caching. In: INFOCOM IEEE Conference on Computer Communications Workshops, pp. 1–6. IEEE Press, New York (2010)
8. Koponen, T., Chawla, M., Chun, B.-G., Ermolinskiy, A., Kim, K.H., Shenker, S., Stoica, I.: A data-oriented (and beyond) network architecture. In: The 2007 Conference on Applications, Technologies, Architectures, and Protocols for Computer Communications, pp. 181–192. ACM, New York (2007)
9. Jokela, P., Zahemszky, A., Rothenberg, C.E., Arianfar, S., Nikander, P.: LIPSIN: Line speeds publish/subscribe inter-networking. In: The ACM SIGCOMM 2009 Conference on Data Communication, pp. 195–206. ACM, New York (2009)
10. Ambrosio, M.D., Marchisio, M., Vercellone, V., et al.: Second NetInf Architecture Description, 4WARD Deliverable D6.2 (January 2010)
11. Freedman, M.: Building a service-centric network with SCAFFOLD Princeton University (2010)
12. Ahmed, R., Boutaba, R.: Distributed pattern matching: a key to flexible and efficient P2P search. IEEE Journal on Selected Areas in Communications 25, 73–83 (2007)

13. Ahmed, R., Boutaba, R.: Plexus: a scalable peer-to-peer protocol enabling efficient subset search. IEEE/ACM Transactions on Networking 17, 130–143 (2009)

14. Moskowitz, R., Nikander, P.: Host Identity Protocol Architecture. RFC 4423, IETF (2006)

15. Trossen, D., Sarela, M., Sollins, K.: Arguments for an information-centric internetworking architecture. SIGCOMM Comput. Commun. Rev. 40, 26–33 (2010)

16. Caesar, M., Condie, T., Kannan, J., Lakshminarayanan, K., Stoica, I.: ROFL: routing on flat labels. In: The 2006 Conference on Applications, Technologies, Architectures, and Protocols for Computer Communications, pp. 363–374. ACM, New York (2006)

17. Eugster, P., Felber, P., Guerraoui, R., Kermarrec, A.-M.: The many faces of publish/subscribe. ACM Computing Surveys 35, 114–131 (2003)

18. Ford, B., Iyengar, J.: Breaking Up the Transport Logjam. In: 7th ACM Workshop on Hot Topics in Networks, HotNets-VII, ACM, New York (2008)

19. Rosenberg, J.: UDP and TCP as the New Waist of the Internet Hourglass, internet draft (2008),
    http://tools.ietf.org/id/draft-rosenberg-internet-waist-hourglass-00.txt

20. Popa, L., Ghodsi, A., Stoica, I.: HTTP as the Narrow Waist of the Future Internet. UCB Technical Report (2010)

21. Berners-Lee, T., Masinter, L., McCahill, M.: RFC1738: Uniform Resource Locators, URL (1994)

22. Mischke, J., Stiller, B.: A methodology for the design of distributed search in P2P middleware. IEEE Network 18, 30–37 (2004)

23. Oram (ed.): Peer-To-Peer: Harnessing the Power of Disruptive Technologies. O'Reilly & Associates, Sebastopol (2001)

Modelling and Evaluation of CCN-Caching Trees*

Ioannis Psaras, Richard G. Clegg, Raul Landa, Wei Koong Chai, and George Pavlou

Department of Electronic & Electrical Engineering
University College London
WC1E 7JE, Torrington Place, London, UK
{i.psaras,r.clegg,r.landa,w.chai,g.pavlou}@ee.ucl.ac.uk

Abstract. *Networking Named Content* (NNC) was recently proposed as a new networking paradigm to realise *Content Centric Networks* (CCNs). The new paradigm changes much about the current Internet, from security and content naming and resolution, to caching at routers, and new flow models. In this paper, we study the caching part of the proposed networking paradigm in isolation from the rest of the suggested features. In CCNs, every router caches packets of content and reuses those that are still in the cache, when subsequently requested. It is this caching feature of CCNs that we model and evaluate in this paper.

Our modelling proceeds both analytically and by simulation. Initially, we develop a mathematical model for a single router, based on continuous time Markov-chains, which assesses the proportion of time a given piece of content is cached. This model is extended to multiple routers with some simple approximations. The mathematical model is complemented by simulations which look at the caching dynamics, at *the packet-level*, in isolation from the rest of the flow.

Keywords: Content-Centric Networks, Packet Caching, Markov Chains.

1   Introduction

The Internet today operates in a *machine-resolution* fashion, that is to say, content is accessed via requests to given hosts. However, the user is not usually interested in a host, but in specific content. Users are essentially interested in *content-resolution*, or *service-resolution*. Recently, the influential paper [8] introduced the *Networking Named Content* (NNC) overlay paradigm as a means to realise *Content-Centric Networks* (CCN). The paper suggests numerous architectural changes to networks based upon the CCN paradigm. For example, according to the NNC approach, names are hierarchical, can be aggregated, are structured, and also human-readable, at least to a certain extent. The approach

---

* The research leading to these results has received funding from the EU FP7 COMET project - Content Mediator Architecture for Content-Aware Networks, under Grant Agreement 248784.

J. Domingo-Pascual et al. (Eds.): NETWORKING 2011, Part I, LNCS 6640, pp. 78–91, 2011.

also brings certain security benefits with named content being signed and secure. The present paper models just one of those proposed changes: the caching strategy used in CCN. This is modelled in isolation from the other proposals within [8], i.e., security, content-resolution scalability and routing are subject to different modelling and evaluation approaches and therefore, of different studies.

The NNC scheme proposed in [8] changes the whole Internet flow model from a sender-driven, TCP-like buffering/congestion controlled environment to a receiver-driven, caching (instead of buffering) scheme. In particular, by exploiting the fact that each data packet can now be identified by name, CCN can take advantage of forwarding the packet to many interested recipients. That is, instead of buffering a packet until it is forwarded to the interested user and then discarding it, as happens today, CCN first forwards the packet to the interested user and then *"remembers"* the packet, until this packet *"expires"* [8].

The *"remembering"* and *"expiration"* of packets is accomplished using caching techniques at the *packet* level. The model requires that every CCN-compatible router is equipped with a cache, which holds all packets for some amount of time; in [8] and in the present study, Least Recently Used (LRU) policies [10] are used, but alternative designs are also possible. If two or more subsequent requests *for the same content* arrive before the content expires *at any one of the routers/caches along the way*, the packet is forwarded again to the new interested user, instead of having to travel back to the content server to retrieve it. Clearly, this will benefit *popular content* (e.g., reduced delivery delay) and will reduce network resource requirements (e.g., bandwidth and server load).

In this study, we focus on the new flow model introduced in [8] and attempt to quantify the potential gains that the paradigm shift will bring with it. To achieve this goal and in order not to violate the semantics in [8], we study the caching part of the CCN paradigm isolated from the rest of the proposed architecture. We carry out a *packet-level* analysis, instead of a flow-level one, since it is our opinion that given the totally unexplored research field, investigations on packet-level dynamics should precede flow-wide conclusions. This study attempts to determine how long a given packet, which is referred to as the Packet of Interest, *PoI*, remains in *any of the caches* along the path from the user back to the server (i.e., not in a single cache), given a system topology and rates of requests. The contributions of the present study to this new field of research are as follows: an analytical model is developed for a single and then multiple caches; the validity of various modelling assumptions is discussed and a Monte-Carlo simulation is used to check some assumptions; a Java model is built based upon the analytical model; this is compared against model-simulator results from *ns-2*.

Our findings indicate that, as expected, there is a clear network-wide performance gain for popular content, but this gain: i) goes down to almost zero for unpopular content and this will be the more common case, ii) depends heavily on the size of the router-cache and iii) is different between routers distant from the original server of the data and those close to the server.

1.1 Macro- vs. Micro-caching

Strategic content replication techniques to reduce response time and the amount of network resources (e.g., bandwidth and server load) needed to send the content back to the user have been investigated in the form of i) Web Proxy Caching (e.g., [15] [3], [7]) and ii) Content Delivery Networks (CDNs) (e.g., [13]).

In both cases, the issue of utmost importance is the choice of the best possible network location to (either statically or dynamically) replicate content. In the case of Web-Proxy caches, the ISP is responsible for deciding where to replicate, while in the case of CDNs, this role is taken by the corresponding company that owns the network of surrogate servers. In both cases, however, *the location of the proxy/surrogate server is always known and fixed in advance.*

Furthermore, IP-multicast has been proposed and investigated (e.g., [14], [6]) to serve multiple users that are simultaneously interested in the same content. However, IP-multicast serves users that belong to the same group *only* and are prepared to receive all of the content that the *group* wants to receive (i.e., not necessarily the parts of it that individual users are interested in).

CCNs, as proposed in [8], constitute the conceptual marriage of the above technologies, *but in the "micro-level".* We classify Web-Caching and CDNs as *"macro-caching"* approaches, since they target caching of entire objects, be it web-pages, or entire files; we consider CCNs as a *"micro-caching"* approach, since caching here is done at the packet-level[1]. Moreover, in the above technologies the setup is fixed and predefined. In contrast, *in CCNs content is cached and multicast "on-the-fly", wherever it is requested or is becoming popular.*

1.2 Single- vs. Multi-cache System Modelling

Web-caching and the properties of the LRU (and other similar) replacement policies have been extensively studied in the past, e.g., [1,10,9]. However, most of these studies focus either on single caches, e.g., [10], or chains of well-known and pre-defined caching points [3]. These studies can model many issues in detail, such as the correlation between subsequent requests, the request frequencies of the various packets in the system and the cache size [11,10,9].

Generally speaking, single-cache modelling has to consider mainly: i) the number of requests for a specific content *at a specific cache*, ii) the number of requests for other contents *at this specific cache*, iii) the size of the cache and iv) correlations between packet requests.

According to the authors in [8], *"CCN caches are the same as IP buffers, but have a different replacement policy".* Trying to model multi-cache systems, where no hierarchy ([3]) exists is not trivial. The problem of modelling the caching system now comprises a multi-dimensional problem, where the above parameters still have to be considered, but they need to be considered *simultaneously for all the routers/caches along the path* from the content client to the content server.

---

[1] In fact, CCN [8] uses the term "message" to refer to both interests (request for data) and "content" (the data itself). Here we are specifically looking at the caching of *addressable* data messages. Whether multiple packets fit in one "message", or multiple "messages" fit in one packet does not affect much our modelling herein.

For example, there is some pre-existing work on multiple cache situations. In [3], the authors develop models that apply to LRU caches which are two levels deep (that is, one cache feeds into a second cache). In [5] and [16] the authors develop a system which can approximate cache miss ratios for hierarchical cache systems using a solution which converges to a solution iteratively. However, the complexity for a single cache is $O(KB)$, where B is the buffer size and K is the number of independent items that have distinct access possibilities. There remains a need for a simple approximate model to estimate loss rates on a system when both B and K may be extremely large. This is where we focus on in this study. We propose a single-cache model, which is simple enough in order to remain valid when extended to multiple caches along the end-to-end path.

The model works from the point of view of estimating how long a particular content packet, the Packet of Interest, *PoI*, remains in *any of the caches* along the path. It outputs the proportion of time that *PoI* is nowhere cached, essentially reflecting the *cache miss ratio for PoI*.

2   Modelling CCN Using Markov Chains

2.1   A Simple Model for a Single Router

Assume that for a single router requests for the *PoI* arrive as a Poisson process with rate λ. Whenever such a request arrives, it moves this packet to the top slot in the cache. Assume that requests which will move the *PoI* further down the cache (either requests for packets not in cache or requests for packets in cache, but further down than the *PoI*) also arrive as a Poisson process with rate μ.

This process can be simply modelled as a continuous time homogeneous Markov chain, where the state of the chain represents the exact slot that the packet currently occupies in the cache. Number the Markov chain states as follows. State 1 is when the *PoI* is at the "top" of the cache (just requested), state N is when our packet is at the bottom of the cache (any request for a packet not already in cache will push our packet out of the cache). State $N + 1$ represents the state where our packet is *not* in the cache. The chain and cache are shown in Fig. 1 (left).

All states (except state 1) have a transition to state 1 with rate λ (another request for the packet arrives and the packet moves to the top of the cache). All states i in $1 \leq i \leq N$ have a transition to state $i+1$ with rate μ (a packet arrives which moves our packet lower in the cache).

Now let $\pi = (\pi_1, \pi_2, \ldots, \pi_{N+1})$ be the equilibrium probabilities of the chain – these have the simply physical interpretation of being the proportion of the time the *PoI* spends at that cache position (or for π_{N+1} the proportion of the time the packet is not in cache). The chain can be trivially shown to be ergodic and hence, these probabilities exist for $\mu, \lambda > 0$ and finite N.

The balance equations can be easily solved to give $\pi_i = \frac{\lambda}{\mu} \left[\frac{\mu}{\mu+\lambda} \right]^i$ for $i < N+1$ and

$$\pi_{N+1} = \left[\frac{\mu}{\mu + \lambda} \right]^N.$$

(1)

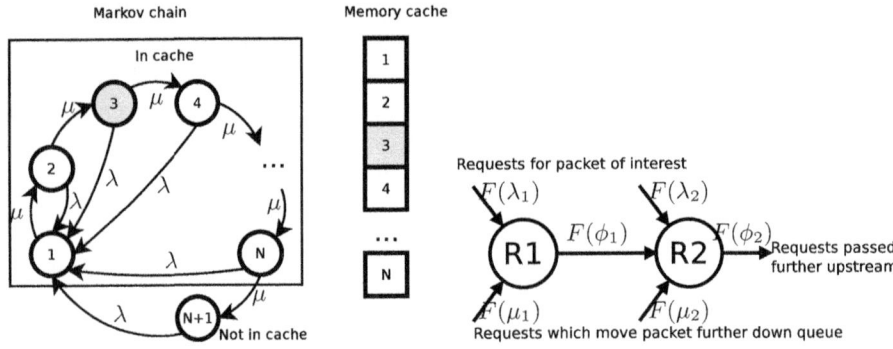

Fig. 1. Models for a single cache (left) and two caches (right)

This equation gives the proportion of the time that the *PoI* is not in cache (and also the proportion of requests for the *PoI* that get a cache miss). Naturally, the proportion of the time our packet gets a "cache hit" is one minus this. Therefore, the mean rate of requests that propagate upstream is: $\lambda \pi_{N+1} = \lambda \left[\frac{\mu}{\mu+\lambda} \right]^N$.

The next question is how can this be extended for the onward process when there are multiple caches in the system. The situation is shown in Fig. 1 (right). Router $R1$ has N_1 memory slots and $R2$ has N_2 memory slots. The notation $F(x)$ means an arrival process (not necessarily Poisson) with a rate x. The processes $F(\lambda_i)$ and $F(\mu_i)$ are Poisson by assumption. The remaining processes $F(\phi_i)$ are those requests for the packet of interest corresponding to cache misses in router Ri. These onward processes can be shown to have a phase distribution (see [12] for a general discussion of the phase distribution) derivable from a simple transform to the chain transition probabilities but it is not analytically tractable.

2.2 Modelling the Multiple Router System as a Markov Chain

Consider again the diagram of two CCN routers in Fig. 1 (right). Note, however, that from (1) we have that the rate $\phi_1 = \lambda_1 \gamma_1^{N_1}$, where $\gamma_1 = \mu_1/(\lambda_1 + \mu_1)$ – the rate of requests passed upstream from router $R1$ for the *PoI* is the rate of incoming requests times the proportion of time spent in the cache miss state. This trivially follows since the Poisson arrival process at rate λ_1 is memoryless, so a proportion $\gamma_1^{N_1}$ of them are passed upstream. The total arrival rate at router 2 is $\lambda_1 \gamma_1^{N_1} + \lambda_2$, call this λ_2'.

Now the question is what proportion of the time is the packet in cache at $R2$. The proportion of time at $R1$ is still given by (1). The problem can again be formulated as a Markov chain, this time with states numbered in $N_1 \times N_2$, where the state number (i, j) refers to the position of the packet of interest in cache at $R1$ and $R2$, respectively. If $i = N_1 + 1$, then the packet is not in cache at router $R1$ and if $j = N_2 + 1$, then the packet is not in cache at router $R2$. Let $\pi_{(i,j)}$ be the equilibrium probability that the *PoI* is in state i at $R1$ and state j at $R2$. Let $\pi_{(i,\bullet)} = \sum_j \pi_{(i,j)}$ be the equilibrium probabilities of the states of $R1$, independent of $R2$. It can be trivially shown that these are exactly those probabilities as for the single router model – that is, the presence of the second router does not affect the first as expected. Calculating $\pi_{(\bullet,j)} = \sum_i \pi_{(i,j)}$ has proved more difficult and the only result obtained so far is that

$$\pi_{(\bullet,1)} = \frac{\lambda_2' - \lambda_2 C/(\mu_2 + \lambda_2)}{\mu_2 + \lambda_2' - C},$$

where λ_2' is the adjusted arrival rate discussed earlier and

$$C = \lambda_1 \left[(\mu_1/(\lambda_1 + \mu_1))^{N_1} - (\mu_1/(\lambda_1 + \lambda_2 + \mu_1 + \mu_2))^{N_1} \right].$$

From this we can see that C is positive and that $\pi_{(\bullet,1)}$ will tend to $\pi_{(\bullet,1)} = \lambda_2'/(\mu_2 + \lambda_2')$ as N_1 increases. This is the equation obtained when the non Poisson nature of the arrival stream is ignored. It can be easily seen that C tends to zero when N_1 increases as both terms inside the brackets are less than one and raised to a positive power.

For the other states, the Markov chain can be explicitly simulated (Monte-Carlo simulation) and the proportion of time spent in each of the states explicitly measured and summed to produce an estimate of the true equilibrium probabilities. One million iterations are performed with results from the first half a million iterations being discarded as "warm up". To ensure convergence, small values for N_1 and N_2 are used. The first simulation was with $N_1 = 6$, $N_2 = 4$ and $\lambda_1 = \lambda_2 = \mu_1 = \mu_2 = 1.0$. The results for $\pi_{(\bullet,j)}$ were in agreement with the Poisson approximation to within ± 0.002. A more rigorous test is provided when the value ϕ_1 is high compared to λ_2. Again with $N_1 = 6$ and $N_2 = 4$ the theory is tested against experiment, but this time with $\lambda_1 = 100.0$, $\mu_1 = 10,000$, $\lambda_2 = 0.1$ and $\mu_2 = 100.0$. For these parameters, the agreement with the Poisson assumption was weaker but all probabilities agreed within ± 0.008. This was the highest level of disagreement found with any of many sets of test parameters tried. We also note that these tests are with N_1 very low compared with the figures which will be used in real runs (small numbers of states were used because Markov chains with fewer states converge better in Monte Carlo simulation).

This section, therefore, provides good evidence that a pair of CCN caches feeding from one to another can be reasonably approximated by the simple single cache model presented before. The low error rate in assuming that the misses from one cache can be used as the input to a second (and the violation of the Poisson assumption ignored) leads us to our final multiple router model. In this model, the miss rates for the *PoI* are calculated by assuming that the input rate to a router is a combination of the rates of requests made directly at that cache and those cache misses passed upstream from previous routers.

2.3 Investigating the μ Process Further

In this section we investigate further the nature of the process μ. It is a subtle but very important point that the *PoI* can be moved down the cache by two related processes. The first possibility is that a packet arrives which is not in cache. Assume requests for all other packets (cached or uncached) arrive at rate ν and that these arrivals are Poisson. However, not all arrivals of the rate ν will move the *PoI* further down the cache. Split ν into ν_c and ν_n for those requests, which are in cache and not in cache respectively – so the rate will sum to $\nu = \nu_c + \nu_n$. Arrivals in ν_n will always move the *PoI* further down (or out of) the cache, if it is in the cache. Arrivals in ν_c will only move the packet further down the cache, if they are requests for a packet which is at a lower cache position. For example, if the *PoI* is at cache position n, a request for any packets in positions $1, \ldots, n-1$ will not move the position of the *PoI* in the cache, but will simply reshuffle those packets above it.

The next step is to assume that packets are equally likely to be found in any position in the cache. This is an unlikely assumption in real life but allows us to calculate an upper bound for the *PoI* cache miss rate. The reason this is an upper bound will be discussed later.

With these assumptions the position of the *PoI* can again be calculated as a Markov chain, but this time, the rate of moving from a state j to a state $j + 1$ is given by $\nu_n + \nu_c(N + 1 - j)/N$, where the first part represents the arrival of a request for a packet not in the cache and the second part represents the arrival of request for a packet which is in the cache, in a position lower than the *PoI* (currently in position j) – impossible if the packet is in position N and always the case for a request for a packet in the cache, if the packet is in position 1. The derivation of equilibrium probabilities π_j can proceed in a similar manner to the previous section, however, while λ remains the same, the μ is replaced by $\nu_n + \nu_c(N + 1 - j)/N$.

With these assumptions then $\pi_{N+1} = \frac{\Gamma[1+N(\nu_c+\nu_n+\lambda)/\nu_c]}{\Gamma[N+N(\nu_c+\nu_n+\lambda)/\nu_c]}$, where Γ is Euler's gamma function. Now, using the well-known formula $\frac{\Gamma[z+a]}{\Gamma[z+b]} = z^{a-b}[1 + O(z^{-1})]$ then

$$\pi_{N+1} = \left[\frac{\nu_c + \nu_n}{\nu_c + \nu_n + \lambda}\right]^N \left[1 + O\left(\frac{\nu_c}{N(\nu_c + \nu_n)}\right)\right]$$
$$\left[1 + O\left(\frac{\nu_c}{N(\nu_c + \nu_n + \lambda)}\right)\right]. \tag{2}$$

The order terms become small as N becomes large, or as ν_c becomes small with respect to ν_n. It should also be noted that the time spent out of cache must be greater than if only the ν_n arrivals were considered. Therefore, the proportion of time the *PoI* is not in cache is $\pi_{N+1} \simeq \left[\frac{\nu_c + \nu_n}{\nu_c + \nu_n + \lambda}\right]^N$ when $\nu_n \gg \nu_c$ (only a small proportion of requests are in cache). In those circumstances, the fact that the μ process in Section 2.1 is a good approximation for the model that takes more careful account of whether arriving packets are in cache, or not and the approximation $\mu = \nu_n + \nu_c$ can be used. In other words, the rate μ is equal to the rate of requests for all packets other than the packet of interest (which is requested at rate λ).

The question remains how (2) alters if the assumption is dropped that arriving non PoI packets are equally likely to be at any position within the cache. This assumption is true only when the non-PoI packets are requested with equal frequency. For non-equal distributions then the term $\nu_n + \nu_c(N + 1 - j)/N$ would always be larger for lower numbered states and π_{N+1} is hence smaller than (2). No lower bound for π_{N+1} exists (indeed π_{N+1} can be arbitrarily close to zero for a given ν_c, ν_n, N and λ). Therefore, (2) is an upper bound which occurs in the case when non-PoI packets are requested with equal frequency and the lower bound is zero. It is worth remembering, however, that the definition of the μ process would, in that case, give the μ process an arrival rate near zero (it is explicitly defined as the arrival rate of packets which move the *PoI* further down the cache).

2.4   Summary of Mathematical Results

The reality of the Poisson assumptions may be questioned. It is well known for example, that Internet traffic is long-range dependent (for a recent review see [4]). However, that is the traffic process as a whole, not the process of requests, which can look very different. Also, it may be argued that nobody knows what traffic statistics in a CCN-based Internet would look like. Many request processes (for example www, ftp and smtp) today can, in fact, be modelled perfectly well with Poisson processes (e.g., Table 3 in [2]).

   Section 2.1 showed a simple closed form solution when requests for the *PoI* are Poisson with rate λ and requests which move the *PoI* further down the cache are Poisson with rate μ. In Section 2.2, it was shown experimentally that certain deviations from Poisson for the requests for the *PoI* make only small differences to the proportion of time spent in cache. In Section 2.3 the assumption of Poisson was relaxed slightly for the μ process by considering requests for packets in cache and requests for packets not in cache separately, however, it was necessary to make strong assumptions about the distribution of packets within the cache. Further work in this direction would require an explicit model of the request frequencies (heavy tailled models would be a good candidate). This would bring the simple model presented here closer to those detailed single cache models discussed in previous sections.

   Finally, therefore, the single cache model of Section 2.1 can be used to model a tree of caches by assuming that the misses from caches propagate towards the original holder of the content and that requests for the *PoI* arriving at a cache are a Poisson process with a total rate which is the sum of the rate of requests directly to that cache and the rate of missed requests to all caches which directly pass on to this cache. This situation is directly amenable to simple simulation even for situations with many caches and large router memories.

3   Simulation Results

3.1   Scenario Setup and Parameter Settings

Two simulation models are evaluated. The first directly encodes the analytical model from Section 2, while the second model is developed on *ns-2* and is presented in Section 3.4.

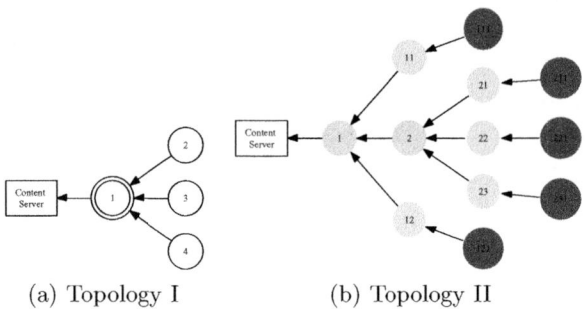

(a) Topology I   (b) Topology II

Fig. 2. Simulation Topologies: Simple and Extended Tree

Our first model takes as input the caching tree topology and, for each cache, the values of λ (request rate for *PoI*), μ (rate at which other requests move the *PoI* down the cache) and N (cache size in packets). For simplicity, results are presented in terms of the ratio λ/μ, referred to as the *Content Popularity Ratio* (R_{CP}). The larger this measure is, the longer the *PoI* will stay in cache. For each router the simulation outputs the *Proportion of Time Not in the Cache* for the *PoI* from Eq. (1) and the proportion of requests that have to be forwarded further on towards the content server, referred to as λ_{Output} and given in Section 2.1. The two topologies tested are given by Fig. 2; here, we are interested in scenarios where end-users request content from professional content-providers, or data-centers, i.e., we are not interested in peer-to-peer scenarios.

It is reasonable to assume that routers which get more requests have more cache and therefore reasonable to assume that μ is proportional to N. For simplicity, the constant of proportionality is set to 1 for most results presented here, unless explicitly stated differently. This reduces the number of parameters to test in the system.

3.2   Scenario 1: Content Popularity and Cache Size

We initially assess the properties of requested content with regard to its popularity, as CCNs were originally proposed to deal with popular content and *flash crowds* [8]. This scenario uses the topology of Fig. 2(a) and experiments with different values for R_{CP}. Values for R_{CP} range from 0.000125 to 0.01 – all of these values represent very popular content but the experiment shows how the popularity affects the time in cache. The buffer size N is set to 200 packets.

Fig. 3 presents the proportion of time the *PoI* is not in any of the caches along the path. Two conclusions can be drawn from this scenario. First, unpopular content spends very little time in the cache. This time is comparable to the life-cycle of an IP packet in the forwarding buffer in the prevailing networking model. This would be the case for most packets. Second, there is a clear difference between caching times for popular content in the core and leaf routers. Caching at the leaf nodes can reverse the expected effect on popular content nearer the server with the data source. That is, more caching at the leaf leads to less caching

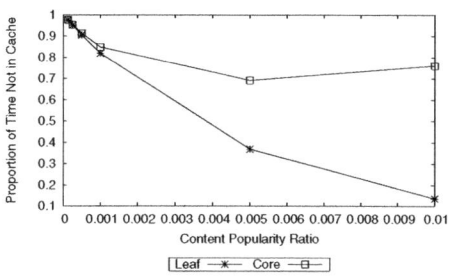

Fig. 3. Scenario 1, Topology I - Increasing Content Popularity

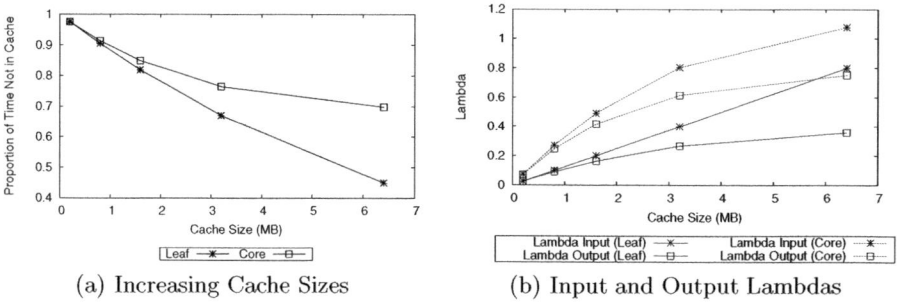

(a) Increasing Cache Sizes   (b) Input and Output Lambdas

Fig. 4. Scenario 1, Topology I - Cache Size

nearer to the data source for very popular content. (Note that the requesters and the data sources are all likely to be "edge" nodes in traditional Internet terms).

We go one step further to investigate the cache-size properties of the CCN paradigm. The buffers in small IP routers can serve the purpose of implementing CCNs, but the gain against today's end-to-end model will be marginal (i.e., even popular content is going to be "forgotten" very quickly). Hence, bigger amounts of memory may have to be considered in order to obtain gains from the paradigm shift. Using the same settings as the previous simulation, and keeping $\mu = N$, the cache size N is varied from 100 to 64,000 packets (rounding packet sizes to 1KB, this is 100KB to 6.4MB).

In Fig. 4(a), we see that for larger caches, even the less popular content has the chance to stay cached for longer times, as would be expected. Above 6MB of cache, even the less popular content is in cache at the leaf node for more than 50% of the time. Again, caching at leaf nodes has reduced the proportion of time spent in cache nearer to the content source.

Fig. 4(b) shows the request rates for the *PoI* at each router (λ_{Input}) as well as those requests passed upstream towards the content server (λ_{Output}). The results agree with Fig. 4(a) and confirm that leaf nodes are passing less traffic upstream when caching levels are higher. This causes this specific content to become less popular nearer the packet source (see Fig. 4(b)).

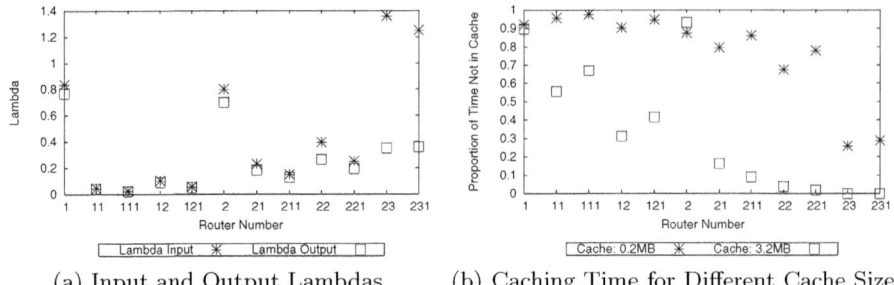

(a) Input and Output Lambdas   (b) Caching Time for Different Cache Sizes

Fig. 5. Scenario 2, Topology II

3.3   Scenario 2: More Complex Tree Topologies

Simulation is now carried out on the *extended tree* topology of Fig. 2(b). In this scenario we apply different λs on each of the leaf routers. The purpose is to evaluate heterogeneity in requests. The request ratios for *PoI*, λ_{Input}, together with the corresponding output λs for each one of the participating routers, are shown in Fig. 5(a). Fig. 5(b) shows the proportion of time that the *PoI* is not in the cache for each of the caches of Fig. 2(b). Results are given for two different cache sizes (0.2MBs and 3.2MBs). The R_{CP} for the content of interest ranges between 0.0001 and 0.005. We observe the following:

1. *Even extremely popular content is "forgotten" quickly by routers near the data source.* Given a tree topology, similar to the one in Fig. 2(b) (a loose reflection of the topology of the Internet), these servers nearer to the data source receive many requests, much of which is for unpopular content. Given that many requests have to be forwarded up the chain to the content server – see Fig. 5(a), routers $R1$ and $R2$) – even popular content gets forgotten quickly – see Fig. 5(b) routers $R1$ and $R2$.
2. *Leaf routers "remember" popular content for longer time.* In Fig. 5(a), we see that $\lambda_{Input} \simeq 0.4$ for router $R22$, while it is more than 0.8 for core router $R2$. Although both routers forward upwards a fairly big percentage of these requests, in Fig. 5(b) we see that the proportion of time not in cache for router $R22$ still drops lower than for core router $R2$.
3. *Larger cache sizes exaggerate point number 2 above, while leave point 1 untouched.* In Fig. 5(b), we see that the proportion of time the *PoI* spends in leaf caches increases with the cache size. The same is not true for routers nearer the source as can be seen in the same Fig. for routers $R1$ and $R2$.

3.4   Scenario 3: Model Simulator vs *ns-2* Simulations

The basic functionality of caching using the CCN paradigm [8] is implemented in *ns-2*. Close approximations to the scenarios from the previous section were tested to estimate the agreement between theoretical modelling and *ns-2* simulation. The full CCN paradigm as stated in [8] cannot easily be implemented in *ns-2*. Some of the reasons why are listed below.

- *CCN Router.* The structure of the CCN router influences several parts of the networking stack. Several designs are possible. It is not clear yet, for example, how traffic from different flows is multiplexed before it enters the outgoing link.
- *Cache Size.* In case of small caches, similar to the IP buffers we have today, the network-wide gain will be limited. On the other hand, larger caches will increase the routers' computational burden and will complicate the collaboration between outgoing interfaces and the cache.
- *Transport Protocol.* The CCN router structure will influence massively the design of the CCN transport entity. The initial proposal focuses on a *receiver-driven* transport protocol [8]. Therefore, flow rate adjustments and retransmission timers will have to be implemented on the receiver side. This constitutes a role- and functionality-swap from the current situation. This will heavily influence simulations and results at the flow level.

The experiments here attempt to minimise the above effects. The topology used is that of Fig. 2(a); the setup includes a simple constant bit rate (CBR) application over UDP that sends requests for 1KB-packet responses. By using UDP and CBR the issues of implementing a CCN-friendly transport protocol (which is yet to be defined) are avoided. Requests arrive as Poisson distributions to the leaf routers $R2$, $R3$ and $R4$. The rate of request for the *PoI* over each leaf node can be set and normalised against the total number of requests. This is similar to the R_{CP} defined in the previous section. However, the exact implementation of μ is hard to achieve in a simulation environment – the effect of the *PoI* being pushed further down the cache is achieved by requesting packets other than the *PoI*, but here, the rate of requests equal to a given value of μ for a given router will be a function of N. The cache size is a fraction of the total number of packets in our pool; this fraction was initially set to 0.5. The simulation time was long enough to guarantee statistically stable results[2]. Simulations were carried out for different values of the cache-size fraction with regard to the number of packets in our pool. Although these results are not presented here, due to space limitations, we report that they follow similar trends.

Initially, $R_{CP} = 0.001$ (for $R2$), $R_{CP} = 0.003$ (for $R3$) and $R_{CP} = 0.01$ (for $R4$) and $N = 200$. Arrivals for other packets are set to a rate of 200 per unit time, to approximate $\mu = 200$. The *ns-2* simulation outputs λ_{Input} and λ_{Output} as in the previous section.

Fig. 6(a) shows λ_{Input} and λ_{Output} for each one of the routers involved. It can be seen that simulation-routers tend to *"forget"* content more readily than model-routers (i.e., NS λ_{Output} is larger than the Model one). Indeed, even for popular content (e.g., router $R4$), the simulation result shows "forgetfulness" of around 30% more than the model result. Differences between the simulator and the model were expected, because of the difficulties of tuning μ. However, it is encouraging the fact that both show results, which are broadly speaking, very similar. The curve shape for both simulations for λ_{Output} follows the same trend.

---

[2] This time varies depending on the R_{CP} and the cache size.

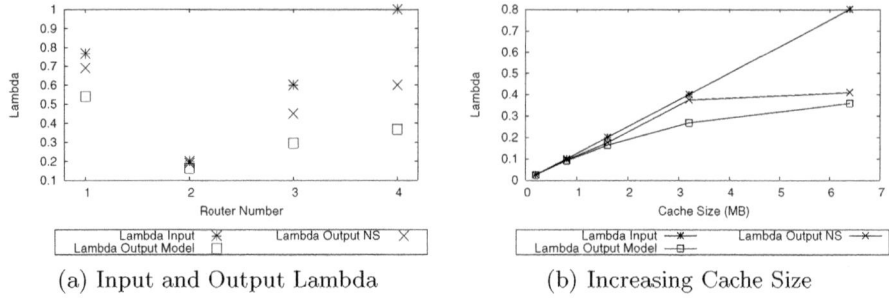

(a) Input and Output Lambda   (b) Increasing Cache Size

Fig. 6. Scenario 3: Model Simulator vs $ns - 2$

In Fig. 6(b), we increase the cache-size and observe the effect on λ_{Output} for the NS and the model simulator, respectively. For small cache sizes, the results are largely in agreement, although there are small differences that cannot be seen here due to the Fig. scale. These differences do not exceed the threshold of 10%. As cache size increases to 3.2MB, the ns-2 λ_{Output} follows the model-simulator input, instead of the model-simulator output (i.e., λ_{Output}). The results are closer to the model-simulator when the cache size increases to 6.4MBs. The disagreement between results in the 3.2MB case could again be attributed to difficulties tuning the μ parameter. In the 6.4MB case, there is better agreement between the results, due to the fact that there are very few cache misses for the *PoI* (i.e., the proportion of time not in the cache for *PoI* is very small, as we have also shown in Fig. 4(a)).

While the results from the two simulators are not directly comparable it is clear they are giving the same indications about the systems being studied.

4   Conclusions

This paper approaches the unexplored area of *Content-Centric Networking*, pro-posed in [8] from the viewpoint of a *packet*, or *message* (i.e., not a flow) and its corresponding caching dynamics. The approach involved a Markov-chain analy-sis, followed by model and *ns-2* simulations. To the best of our knowledge, this is the first study to investigate this new networking paradigm.

The analytical model presented for the caching system presents a simple and tractable model for the time that a packet spends in cache. While the model began with strong assumptions of Poisson behaviour, subsequent analysis shows that many of these assumptions can be weakened and the analytical model re-mains valid. By necessity, the *ns-2* implementation made some simplifying as-sumptions as the detailed protocol of a CCN network is as yet unknown. Results from the model and the *ns-2* simulation were in broad agreement, but with some differences due to the different nature of the approaches.

In summary, the simulation findings indicate that: i) *popular content tends to be cached at the leafs of the network*. Content servers may receive a large number of requests for different content and hence forget more easily than leaf routers. ii)

Sizing CCN caches is not a trivial task, since it depends on the distribution of arrival rates and the flows of requests from upstream. Smaller cache sizes may bring very little gain as only extremely popular content will spend significant time in cache.

Much future work remains to be done. The analytical model can be further developed and in particular the nature of the μ parameter needs to be refined to enable direct and fair comparison with simulation models. More features have to be integrated in both the mathematical model and the simulators when mechanisms for the transport layer are clarified in more detail and the router design is further elaborated. This will allow a rich vein of future investigation for this new networking research field.

References

1. Breslau, L., et al.: Web caching and zipf-like distributions: Evidence and implications. In: INFOCOM, pp. 126–134 (1999)
2. Cairano-Gilfedder, C.D., Clegg, R.G.: A decade of internet research: Advances in models and practices. BT Technology Journal 23(4), 115–128 (2005)
3. Che, H., Wang, Z., Tung, Y.: Analysis and Design of Hierarchical Web Caching Systems. In: INFOCOM, pp. 1416–1424. IEEE, Los Alamitos (2001)
4. Clegg, R.G., Cairano-Gilfedder, C.D., Zhou, S.: A critical look at power law modelling of the Internet. Computer Communications 33(3), 259–268 (2009)
5. Dan, A., Towsley, D.: An approximate analysis of the lru and fifo buffer replacement schemes. In: SIGMETRICS 1990, pp. 143–152 (1990)
6. Floyd, S., et al.: A reliable multicast framework for light-weight sessions and application level framing. IEEE/ACM Trans. Netw. 5, 784–803 (1997), http://dx.doi.org/10.1109/90.650139
7. Fujita, N., Ishikawa, Y., Iwata, A., Izmailov, R.: Coarse-grain replica management strategies for dynamic replication of web contents. Comput. Netw. 45(1) (2004)
8. Jacobson, V., Smetters, D.K., Thornton, J.D., Plass, M.F., Briggs, N.H., Braynard, R.L.: Networking Named Content. In: CoNEXT 2009, pp. 1–12. ACM, New York (2009)
9. Jelenkovic, P.R.: Asymptotic approximation of the move-to-front search cost distribution and least-recently-used caching fault probabilities. The Annals of Applied Probability 9(2) (1999)
10. Jelenković, P.R., Radovanović, A.: Least-recently-used caching with dependent requests. Theor. Comput. Sci. 326, 293–327 (2004)
11. Jelenković, P.R., Radovanović, A., Squillante, M.S.: Critical sizing of lru caches with dependent requests. Journal of Applied Probability 43(4), 1013–1027 (2006)
12. Neuts, M.F.: Probability Distributions of Phase Type. In: Matrix-Geometric Solutions in Stochastic Models: an Algorthmic Approach, ch. 2. Dover Publications Inc., New York (1981)
13. Pallis, G., Vakali, A.: Insight and perspectives for content delivery networks. Commun. ACM 49(1), 101–106 (2006)
14. Ratnasamy, S., Ermolinskiy, A., Shenker, S.: Revisiting IP multicast. In: SIGCOMM 2006, pp. 15–26. ACM, New York (2006)
15. Rosensweig, E.J., Kurose, J.: Breadcrumbs: Efficient, Best-Effort Content Location in Cache Networks. In: INFOCOM, pp. 2631–2635 (2009)
16. Rosensweig, E.J., Kurose, J., Towsley, D.: Approximate models for general cache networks. In: INFOCOM. IEEE, Los Alamitos (2010)

Empirical Evaluation of HTTP Adaptive Streaming under Vehicular Mobility

Jun Yao, Salil S. Kanhere, Imran Hossain, and Mahbub Hassan

School of Computer Science and Engineering
University of New South Wales, Sydney, NSW 2052, Australia
{jyao,salilk,imranh,mahbub}@cse.unsw.edu.au

Abstract. Adaptive streaming is a promising technique for delivering a high-quality video streaming experience. In this technique, the streaming bit-rate is constantly adjusted in accordance to the variations in the underlying network bandwidth conditions. A popular instantiation of this approach is to extend traditional HTTP-based streaming. While several such implementations are widely available, it is unclear how they perform under a typical high-speed vehicular environment, wherein the wireless bandwidth varies significantly and rapidly. In this paper, we seek to provide some insights on this issue through empirical experiments driven by real-world wireless bandwidth traces collected from moving vehicles. Our results suggest that, with appropriate parameter configurations, HTTP adaptive streaming is an effective solution for delivering a high-quality smooth streaming experience even under high-speed vehicular mobility.

Keywords: Adaptive streaming, HTTP streaming, Empirical experiments, Vehicular mobility, Mobile computing.

1 Introduction

With the success of websites, such as YouTube and Vimeo, video streaming over the Internet has become increasingly popular. A recent report [19] estimates that the streaming media business will grow by 27% per year, generating over US$78 billion in revenue in the U.S. alone over the next six years. Given the widespread coverage of high-data rate Wireless Wide Area Networks (WWAN) and the emergence of personal mobile devices with high resolution displays and fast processing speeds (e.g., smartphones and tablets), multimedia streaming is now one of the fastest growing mobile applications.

Till recently, video streaming technologies have mostly adopted the non-adaptive approach, wherein the bit-rate and quality of the video are selected prior to the start of the streaming, and fixed during a streaming session. This is usually sufficient when the viewer is connected via a wired connection, which has high and fairly stable bandwidth capacity. However, in the context of vehicular mobility, where a viewer on a moving vehicle watches the video stream on a mobile device (phone, tablet or laptop) connected to the Internet via a WWAN connection, the non-adaptive solution may not be optimal. This is because, the

J. Domingo-Pascual et al. (Eds.): NETWORKING 2011, Part I, LNCS 6640, pp. 92–105, 2011.

WWAN bandwidth conditions in such an environment fluctuate significantly, even across smaller time scales. This is due to the heterogenous radio characteristics and WWAN network load conditions at different locations as a vehicle makes its way along the route [7]. The ensuing bandwidth fluctuations in such a dynamic environment can seriously compromise the QoS experienced by traditional streaming applications. For example, during a live streaming session, if the WWAN bandwidth suddenly drops significantly below the selected streaming rate, the video viewer can suffer from noticeable "glitches" caused by frame loss or frequent pauses in playback caused by the exhaustion of the playout buffer.

To mitigate the effect of the bandwidth dynamics and achieve the smooth streaming experience, adaptive streaming is emerging as the promising solution [2–4, 9–12, 16, 18, 20]. In adaptive streaming, the streaming servers and/or clients constantly monitor real-time end-to-end network conditions. Based on this information, the video bit-rate and quality are dynamically adjusted to adapt to the changes in underlying network conditions. As a result, adaptive streaming delivers a better quality video stream as compared to conventional non-adaptive approaches. The principle of adaptive streaming has been incorporated with traditional streaming protocols, such as RTSP [11], and evaluated in various prior works [16, 18, 20]. However, due to the popularity of using HTTP in video streaming, the common trend in the industry is to use adaptive HTTP progressive download [2–4, 9–12]. Several such products have even been released by Move Networks, Microsoft, Adobe and Apple. Despite the popularity, a comprehensive empirical evaluation of HTTP adaptive streaming is lacking. In particular, the performance of the HTTP adaptive streaming under high-speed vehicular mobility has not been reported.

Our goal in this paper is to address the above issue and empirically evaluate the performance of HTTP adaptive streaming under vehicular mobility. However, conducting "live" experiments in which the streaming client operates on a mobile device connected to a WWAN network and travelling in a vehicle has several problems. To comprehensively evaluate the effect of varying certain parameters on the streaming performance, we would need to make separate trips, each with a different value of the parameter. However, the WWAN network bandwidth is known to vary from trip to trip [21]. As such, it would be impossible to study the effect of changing parameters in isolation. Further, conducting sufficient tests with statistical reliable results would require making a large number of trips, which is prohibitively expensive in man hours and monetary costs (fuel and WWAN bandwidth subscription). In this paper, we have hence opted to conduct the experiments in a controlled lab environment, but using "real" WWAN bandwidth traces that are empirically collected from a moving vehicle. In our experiments, we report the performance of HTTP adaptive streaming under the effect of various parameters, such as buffer threshold and video chunk size. Our results show that adaptive HTTP streaming is effective in reacting to the widespread fluctuations that are inherent in vehicular mobility and ultimately achieving an improved viewer experience by over *four-fold*. Based on our evaluations, we make recommendations on suitable values of the key parameter settings for achieving optimal performance, which can be of use to industry practitioners.

The rest of this paper is organized as follows. Section 2 discusses the background and related works. In Section 3 we present the vehicular measurement campaign for collecting the bandwidth traces used in this paper. We present the setup of our trace-driven experiments in Section 4. The findings of our experiments are presented in Section 5. Section 6 concludes this paper.

2   Background and Related Works

In this section, we review the state-of-the-art in multimedia delivery in the Internet. We start with the discussion on the traditional non-adaptive streaming approaches. Then we discuss the emerging adaptive techniques with a particular emphasis on HTTP adaptive streaming.

2.1   Non-adaptive Streaming

Most traditional streaming services rely on specialized protocols that are specifically designed for streaming. RTSP (Real Time Streaming Protocol) [17] is a typical example of such protocols. After a streaming session has been established, the media file is sent as a stream of small fixed-sized packets. The server usually sends enough data to fill the client's buffer (typically less than 10 seconds). This traditional streaming approach can be used to stream both on-demand and live video. However, it requires the deployment of dedicated media servers and uses special ports. This may have issues with scalability, caching and penetrating client firewalls [11]. Another popular method for video steaming is *HTTP progressive download*. The idea is similar to a normal file download from an HTTP server. The only difference is that the video can be simultaneously played back while it is being downloaded. This also means that even if the user pauses the media, the server will continue sending data until the whole file is downloaded. Note that, this differs from the first approach explained earlier, where the server stops transmission of packets if the client buffer is filled. The apparent drawback of the HTTP method is that if the user decides to terminate the session, then all stored (but un-viewed) video in the buffer is discarded, thus wasting the bandwidth that was used up in transferring this data. In addition, a significant disadvantage of HTTP progressive download is that it cannot support live streaming, as the size of the entire video is predetermined and fixed during the streaming session [11]. However, the HTTP approach is a cost-effective solution for on-demand video delivery, as it reuses the existing HTTP caches/proxies without the need for specialized servers. Also, the use of HTTP protocol eliminates the issues with firewalls [4, 11]. Nowadays, most popular video sharing websites, e.g., Youtube and Vimeo, exclusively use this approach.

Both of the aforementioned techniques are *non-adaptive*, in that the streaming quality and bit-rate remain unchanged during the entire duration of the streaming. The choice of these parameters is either determined automatically or based on the user's preference at the start of the session. This may serve well in the stationary and wired network environment, where the bandwidth is relatively constant. However, in the context of vehicular mobility (which is the focus of this paper), the wireless bandwidth is known to fluctuate significantly as the vehicle changes its location. Hence the non-adaptive approach can be significantly affected by the bandwidth fluctuations and leads to jerky video playback and ultimately impacts the user viewing experience (see results in Section 5.1).

2.2   Adaptive Streaming

In contrast with the traditional methods, adaptive streaming allows dynamic adjustments in the streaming bit-rate (and quality) to adapt to the varying end-to-end bandwidth conditions, during a live session. Figure 1 illustrates a typical scenario. An adaptive streaming server hosts the videos that have been encoded at various bit-rates. These files are created by encoding the same source video multiple times by varying the quantization and frame rate settings, or using advanced encoding techniques, such as Scalable Video Coding (SVC) [10]. Each of the encoded video streams are then partitioned into a sequence of small "chunks", e.g., Group of Pictures (GoP), which can be decoded independently [10]. Since the video chunks partitioned under different qualities are synchronized, there are various bit-rates available for each video chunk. During streaming, the server can seamlessly switch the streaming quality by using different bit-rates for each video chunk. For example, the server can switch to sending chunks with lower bit-rate when it detects that the available bandwidth reduces, thus gracefully degrading the viewing quality. Note that, the adaptive principle can be readily applied to both streaming-specific protocols and HTTP streaming.

Prior works [16, 18, 20] have proposed mechanisms for implementing adaptive streaming by extending traditional streaming protocols. The basic concept involves the server and/or client relying on some rate control algorithms/protocols to infer the actual bandwidth conditions. Based on this estimate the server selects the next chunk with the appropriate quality from the choices available. Discussion on several such rate adaptation algorithms have been proposed [8, 10]. In general, the network conditions are inferred in response to the changes in observable metrics, such as the occupancy of video playout buffers or certain end-to-end network parameters (i.e. delay, packet loss and throughput).

Due to the advantages mentioned in Section 2.1 of using HTTP in video delivery, HTTP-based adaptive streaming has drawn significant interest in the

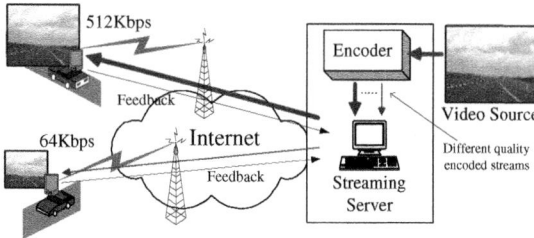

Fig. 1. Adaptive video streaming in a vehicular mobility scenario

industry. In HTTP adaptive streaming, the entire video streaming is split into many small progressive downloads, wherein each HTTP transaction delivers one video chunk to the client. After a session is established, the server returns a manifest file to the client. The manifest file lists the available bit-rates (typically 4-5) for each video chunk [4, 11]. On receiving each chunk, the client measures the throughput to estimate the end-to-end bandwidth conditions. Based on the estimated bandwidth and playout buffer conditions, the client determines the suitable bit-rate and requests the appropriate video chunk through a HTTP GET request. Note that, since each video chunk is individually encoded and transmitted, the streamed video sizes can be unbounded. Thus, unlike its predecessor (HTTP progressive download), HTTP adaptive streaming also fully supports for streaming live events. Recently, Adaptive HTTP Streaming (AHS) has been integrated into the 3GPP Transparent end-to-end Packet-switched Streaming Service (PSS) [2]. 3GPP AHS also has been adopted by Open IPTV Forum (OIPF) as the core component of their adaptive streaming solution [14]. MPEG is also drafting a new standard called Dynamic Adaptive Streaming over HTTP (DASH) [9]. Driven by the popularity, commercial HTTP adaptive streaming products, such as Move Networks [12], Microsoft Smooth Streaming [11], Apple HTTP live streaming [5] and Adobe Dynamic HTTP Streaming [3], have been made available. This technique has been successfully used by NBC in broadcasting the Beijing Olympic Games in 2008 [11].

The performance of the non-HTTP adaptive streaming techniques have been evaluated and studied in previous works [16, 18, 20]. However, evaluations were conducted using simulations with synthetic bandwidth data. Despite the popularity, a detailed empirical evaluation of HTTP adaptive streaming is lacking, to the best of our knowledge. In this paper, we intend to fill this gap and investigate the performance of HTTP adaptive streaming under vehicular mobility.

3   Bandwidth Trace Collection

The goal of this paper is to evaluate adaptive HTTP streaming in a real-world vehicular mobility scenario. The obvious approach is to implement a real client and server and conduct live tests while driving. However, there are several problems in conducting such experiments. To comprehensively evaluate the effect of changing certain parameters (such as playout buffer threshold, video chunk size) on the streaming performance, we would need to make separate trips, each with a different value of the parameter. However, it is known that the WWAN network bandwidth can vary from trip to trip even when the successive trips are made one after the other (see [21] for a detailed investigation on bandwidth variability). As such, it would be impossible to study the effect of changing parameters in isolation. Further, even if the bandwidth remained fairly stable across different trips, conducting all the tests that we have conducted in Section 5 to achieve sufficient statistical significance would require us to make hundreds of trips. Clearly, the amount of man hours and costs (fuel, WWAN subscription) involved are prohibitively high to conduct such tests. Hence, we have opted to conduct trace-driven evaluations in a controlled lab setting. For this, we use the

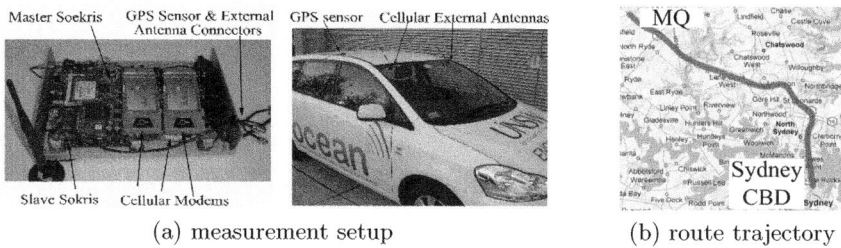

(a) measurement setup   (b) route trajectory

Fig. 2. Bandwidth measurements under vehicular mobility

empirical bandwidth traces collected from our previous wardriving campaign [21] in Sydney. In the following, we briefly review the details of the campaign.

For the measurements, a simple client-server measurement system (further details can be found in [21]) was developed using off-the-shelf hardware. The server was housed in our lab at the University of New South Wales (UNSW). As shown in Fig. 2(a), the client comprised of two Soekris Net4521 boards interconnected via 10 Mbps Ethernet. The boards were enclosed in a protective casing and housed in the boot of a car. Three PCMCIA cellular modems were housed in the system. To enhance the wireless signal reception, the cellular modems were connected to external antennas mounted on the car windshield. We measured two HSDPA [1] networks and one iBurst [6] network. A Garmin GPS sensor was connected to the system for recording the vehicle location.

We developed a lightweight packet-train utility to measure the WWAN bandwidth. We refer readers to [21] for further details and validations. We collected one bandwidth sample for approximately every 200m section of the route. The samples are tagged with location coordinates and time, and stored in a repository. On occasions, some bandwidth samples were missing due to burst packet loss. To mask the effect of missing samples, we use the average bandwidth of all raw samples collected within each 500m segment to represent the bandwidth for the segment. Hence the granularity of the bandwidth traces is 500m.

We have collected bandwidth samples by driving the car along two representative daily commuting routes in the Sydney metropolitan area. During the eight-month measurement period, we have collected 75 traces of WWAN bandwidth along both routes. Fig. 2(b) presents the trajectory of the route that is used in this evaluation study. The chosen route (16.5 Km) runs from Sydney CBD to Macquarie University (MQ). In our experiments, we use the bandwidth traces from all 75 trips for one of the HSDPA networks.

4   Experiment Setup

The goal of this study is to evaluate the performance of HTTP adaptive streaming under real-world bandwidth conditions in high-speed vehicular mobility. We consider the scenario that a passenger in a vehicle watches a streaming video on his mobile device (phone, tablet or laptop) while driving from location A to B. We assume that the viewer watches the video for the entire trip.

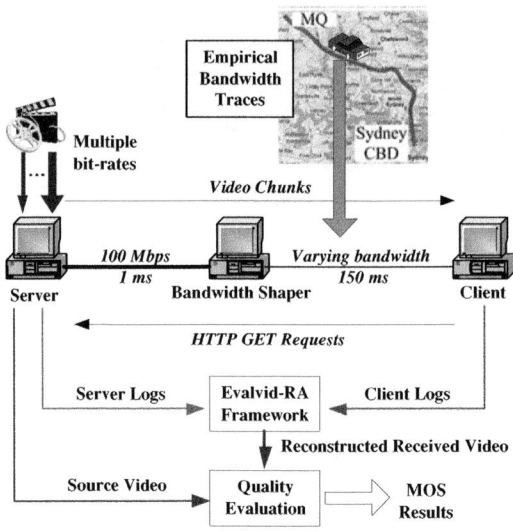

Fig. 3. Experiment setup

We implemented an adaptive HTTP streaming client-server prototype in JAVA and conducted experiments using our empirical bandwidth traces in a controlled lab environment (see discussions in Section 3). Figure 3 presents the experiment setup. The experiment involves 3 Linux (Ubuntu 10.04) machines. The HTTP server and video files are hosted at the server machine. The bandwidth shaper emulates the bandwidth changes of the HSDPA link according to the bandwidth trace files. The client initiates the streaming session and requests video chunks using HTTP from the server as discussed in Section 2.2.

Since the wired links in the Internet and the HSDPA core network have sufficiently high bandwidth and small delays as compared to the last-hop HSDPA link, the server and the bandwidth shaper are connected via a static 100 Mbps Ethernet link with a 10 ms propagation delay to represent the wired Internet. The client and the bandwidth shaper are also physically connected with a 100 Mbps Ethernet link. However, in order to emulate the bandwidth changes as the vehicle travels along its route, we use the system utility *tc* at the bandwidth shaper to throttle the bandwidth of the link between the bandwidth shaper and client. In each experiment run, we emulate one driving trip, wherein the bandwidth shaper varies the bandwidth at each location according to the corresponding empirical bandwidth trace for the trip (collected in Section 3).

For the video clip, we create a looping video (using the medium motion QCIF sequence "Foreman" [10]) that lasts for 40 minutes, which is sufficiently longer than the duration of all trips. Before the experiment starts, the video is pre-encoded at 31 different quantization qualities (corresponds to bit-rates from 80 Kbps to 1 Mbps) using MPEG-4 codec. Note that, commercial products normally use smaller number of bit-rates to save disk space and facilitate file management. The encoded videos are partitioned into a series of equal size video chunks as discussed in Section 2.2. Note that, the size of a chunk is configurable [11]. In Section 5.3, we will investigate the effect of using different chunk sizes.

At the beginning of a trip, the client initiates the session with the server. The server sends the manifest file of the entire video to the client. This file contains all (31, in our experiments) available bit-rates for each video chunk, which can be used during streaming. The client always requests for the lowest bit-rate available for the first chunk. On receiving a chunk from the server, the client measures the instantaneous throughput of the chunk to estimate the bandwidth conditions. The client determines the bit-rate (from those listed in the manifest file) to be used for the next video chunk, based on the estimated bandwidth and the occupancy of the playout buffer. Note that, all received video chunks will be first stored into the playout buffer before being played back. Initially, the buffered video needs to reach an initial playout buffer threshold b_i before the playback starts. Note that, rebuffering will occur whenever the playout buffer is underrun. The playback will be paused until the buffer occupancy reaches b_i. To warrant seamless streaming, the simplest bit-rate selection strategy is to always choose a chunk with a bit-rate that is lower than the current estimated bandwidth. However, this can result in constant increase in the buffer occupancy, if video chunk arrival rate is constantly greater than the playback rate. Excessive buffering is not desired in adaptive streaming, as all un-viewed video in the buffer is discarded, if the user terminates the session. This not only wastes the available bandwidth for transferring the buffered data but also misses the opportunity to deliver better quality video to the client, i.e., by using higher bit-rates. Hence, commercial HTTP adaptive streaming implementations are known to apply some flow control mechanisms to maintain a reasonable buffer occupancy [4, 11]. However, the detailed mechanisms are proprietary. In this paper, we have used a simple threshold-based scheme for this purpose. During streaming, the client consults the buffer occupancy before selecting the bit-rate of the next video chunk. When the buffer occupancy is lower than the threshold b_f, the client selects the highest bit-rate that is lower than the current estimated bandwidth. Otherwise, the client selects the lowest bit-rate that is higher than the current bandwidth. In this case, the arrival rate of the video chunks is expected to decrease, since the selected bit-rate is greater than the available bandwidth. We have assumed the buffer control threshold, $b_f = 1.5 \times b_i$. The coefficient of 1.5 was selected based on our pilot experiments (excluded for reasons of brevity).

For evaluating the video quality, we use the Evalvid-RA framework [10], which is well-accepted for evaluation of the quality of video transmitted over a real or simulated communication network. During streaming, we log the necessary information at the server and client according to the framework. After the experiment is finished, the log files are used to reconstruct the received video sequence. Recall that, during instances of rebuffering, the playback is paused. For those instances, we fill in the paused periods by copying the last played frame. The similar approach is often used by media players to deal with lost frames. We use the tools available from Evalvid-RA to process the source and received videos and generate the Peak Signal-to-Noise Ratio (PSNR) [10] for each frame. To get better insights about the actual viewing experience, we estimate the Mean Opinion Score (MOS) from the PSNR value for each frame. In particular, we employ the empirical model [13] obtained for the specific Foreman sequence used in our experiments,

$$\hat{MOS} = k \times (a - \frac{b}{PSNR}), \tag{1}$$

where $k = 0.56$, $a = 14.2$ and $b = 280.5$. Note that, MOS ranges from 0 to 5. A MOS of 5 suggests an "excellent" viewing experience, while 0 being completely unacceptable. Hence the MOS was set to 0 for the frozen frames during buffering.

5   Experimental Results

In this section, we discuss our findings from the trace-driven experiments. As discussed in Section 4, we calculate MOS to evaluate the viewer streaming experience. It is reported that humans can perceive a drop in the streaming quality, when the MOS remains below 3 consistently for one second or longer [15]. We refer to such an event as a *video glitch*. Clearly, reducing the number of glitches experienced by a viewer directly improves the QoS of a video streaming session. For understanding the effect of glitches on the viewing experience, we define a metric, *glitch duration*, which measures the cumulative time over which glitches occur during a session. Another important aspect of the user viewing experience is the stoppages encountered due to buffering events. As such, we define two metrics to evaluate the impact of buffering. The first metric is the *total number of buffering events* encountered during a viewing session. The second metric is the *buffering duration*, which measures the cumulative time when buffering occurs during a session. Note that, the lower the values of the above metrics (glitch duration, number of buffering events and buffering duration), the better is the viewing experience. In the following, we first compare the performance of the non-adaptive and the adaptive HTTP streaming schemes under vehicular mobility. Further, we study the effect of important streaming parameters, i.e., the buffer threshold and video chunk size, on the adaptive streaming performance.

5.1   Adaptive vs. Non-adaptive

We first investigate how HTTP adaptive streaming can cope with the bandwidth fluctuations under high-speed vehicular mobility. For comparison, we also present the results for non-adaptive streaming scheme. The non-adaptive scheme streams at a constant target bit-rate of 420 Kbps, which is approximately equal to the empirical mean HSDPA bandwidth observed from our traces. For the HTTP adaptive scheme, we set both chunk size and buffer threshold b_i to be 2 seconds, which are the recommended settings in [11].

To gain insights into the behavior of HTTP adaptive streaming, we first present the instantaneous bit-rate, buffer occupancy and MOS results as the vehicle travels along the route during one particular trip in Fig. 4. We observe similar results with the traces from other trips. Fig. 4(a) shows that the HSDPA

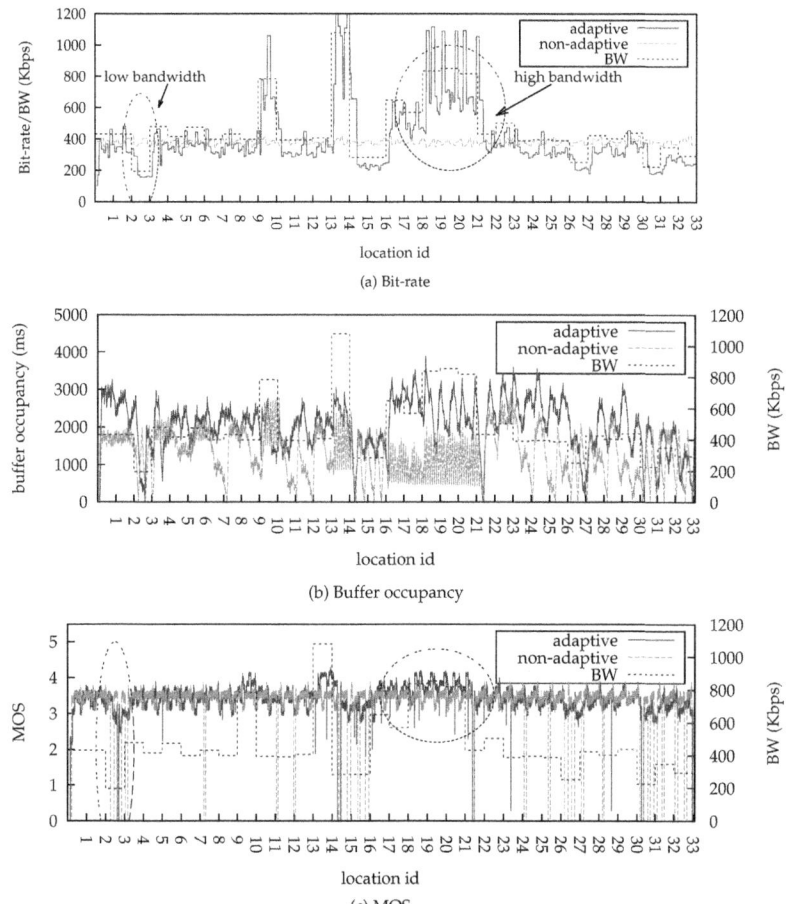

Fig. 4. Microscopic behavior of adaptive and non-adaptive schemes during trip #65

bandwidth keeps fluctuating along the route. For example, the bandwidth drops to 200 Kbps when the vehicle enters location #3, whereas in location #19-#21, the bandwidth increases to 800 Kbps. As is evident, despite the variations in the bandwidth, the non-adaptive scheme keeps sending at near constant bit-rate. At location #3, the steady bit-rate stream overloads the link, which in turn leads to congestion. Fig. 4(b) shows that in this instance, the non-adaptive scheme causes repetitive occurrences of buffer underrun. Hence, a viewer experiences jerky playback as the video pauses to re-buffer frequently. Note that, due to the frozen playback during buffering, the MOS score drops to 0 as shown in Fig. 4(c). In addition, when the HSDPA bandwidth increases between location #19-#21, the non-adaptive scheme is not able to stream the video at a higher bit-rate, as shown in Fig. 4(a). This wastes the opportunities of utilizing the extra bandwidth available to achieve better picture quality. On the other hand, Fig. 4(a) demonstrates that the adaptive scheme varies its bit-rate in accordance to the bandwidth variations. This avoids congestion and draining the playout buffer,

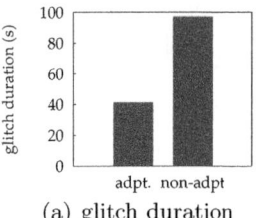
(a) glitch duration

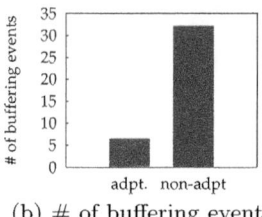
(b) # of buffering events

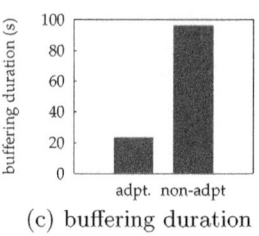
(c) buffering duration

Fig. 5. Comparing adaptive and non-adaptive streaming

when the bandwidth drops, which in turn results in minimal re-buffering. For example, we only observe 1 instance of buffering in location #3, whereas the non-adaptive scheme leads to 3 such instances. When the available bandwidth is high, the adaptive scheme is able to increase its bit-rate to achieve the best quality streaming possible (e.g., in location #19-#21). Fig. 4(c) shows that in these instances, the MOS of the adaptive scheme is significantly higher than that of the non-adaptive. Note that, the higher MOS directly implies better picture quality. Observe that, occasionally the bit-rate of the adaptive scheme exceeds the available bandwidth. This is due to the buffer control mechanism used in our implementation (i.e., increase the bit-rate when the buffer reaches b_f).

The above microscopic analysis of a particular trip reveals that the adaptive approach can effectively reduce the buffering when bandwidth drops and utilize the available bandwidth when bandwidth increases. The mean results over all 75-trip experiments are shown in Fig. 5. Note that, the average duration of a session is about 25 minutes. Fig. 5(a) shows that the adaptive scheme effectively reduces the glitch duration by over 50% as compared to the non-adaptive scheme. More significantly, the number of buffering occasions (Fig. 5(b)) and the buffering duration (Fig. 5(c)) is reduced by 80% and 70%, respectively.

5.2   The Effect of Playout Buffer Threshold

In this set of experiments, we study the effect of playout buffer threshold b_i on HTTP adaptive streaming. Recall that, in the previous experiments (Section 5.1), b_i was set to 2 seconds. Fig. 6 shows the mean results for all three metrics averaged over all 75 trips. Observe that, the glitch duration decreases when a larger b_i is used. This is because a larger b_i allows storage of more video into the buffer, hence better absorbing the bandwidth variations along the route. Fig. 6(b) shows that by setting b_i greater than 6 seconds, the number of buffering events reduces to about only 1 per trip, which is simply due to the initial buffering at the start of the session. This shows that, by using a small buffer of less than 10 seconds, HTTP adaptive streaming can achieve a near un-interrupted streaming experience during a vehicular trip. However, even though the number of buffering event reduces, Fig. 6(c) shows that using a b_i greater than 4 seconds does not further reduce the buffering time. This is due to the fact that by using a larger b_i, a viewer generally spends more time for the initial buffering at the beginning of the trip. This effect is shown in Fig. 7. Further, using a larger

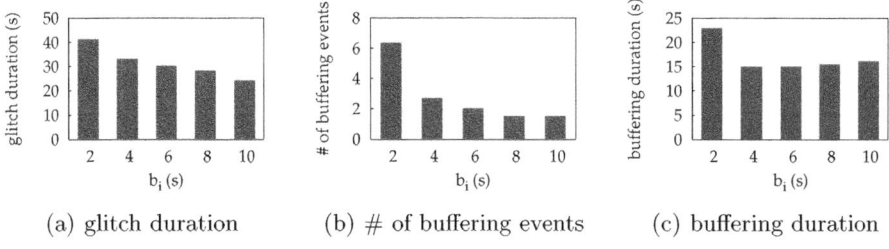

(a) glitch duration   (b) # of buffering events   (c) buffering duration

Fig. 6. The impact of using different buffer threshold

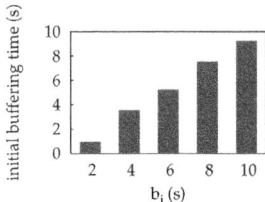

Fig. 7. Initial buffering time under different buffer threshold

playout buffer can also be an issue for memory constrained mobile devices, such as mobile phones. As a result, the buffer threshold needs to be carefully tuned, so that it would not lead to lengthy buffering and memory issues, while effectively smoothing out the bandwidth variations under vehicular mobility.

5.3   The Effect of Video Chunk Size

The video chunk size is also an important parameter in HTTP adaptive streaming. Recall that, each video chunk is stored as an individual file on the HTTP server. Since multiple bit-rates are available for each video chunk, a small chunk size such as 2-second can result in tens of thousands of small files for an hour-long video. This can pose file management issues for video content distributors [11]. Thus, larger chunk sizes are recommended [4]. However, the large chunk size setting may not be sufficient to adapt to the rapid bandwidth fluctuations encountered in vehicular mobility. To understand the effect, Fig. 8 plots the mean results for all the 3 metrics as a function of the video chunk sizes (with a 2-second buffer threshold). Clearly, the glitch duration (as in Fig. 8(a)) increases significantly with the increase in the chunk size. Note that, using a larger chunk size requires more time to transfer each chunk. Since the streaming bit-rate can be only varied once a new chunk is fully received, the larger chunk size reduces the agility of the streaming client in tracking the underlying mobile bandwidth. This further leads to the increase in the number of buffering events and buffering time as shown in Fig. 8(b) and (c). For example, even when a 8 second video chunk is used, both the number of buffering events and the buffering time increase by nearly three-fold, as compared to a 2-second chunk size.

Recall that, in Section 5.2, we have observed that using a larger buffer threshold achieves better streaming performance in the face of frequent bandwidth

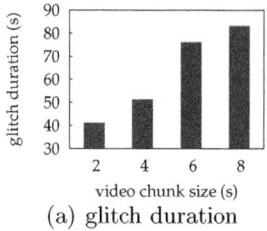 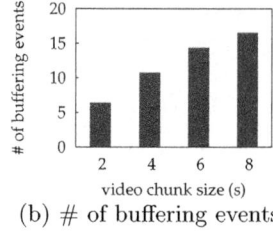 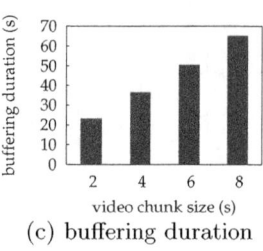

(a) glitch duration   (b) # of buffering events   (c) buffering duration

Fig. 8. Impact of using different video chunk sizes

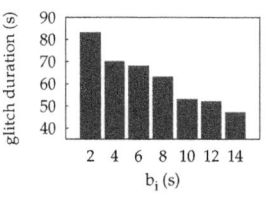 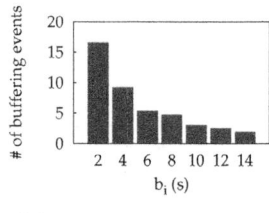 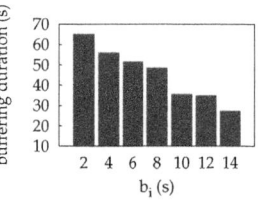

(a) glitch duration   (b) # of buffering events   (c) buffering duration

Fig. 9. Impact of larger chunk size (8 s) with different buffer threshold

fluctuations. Fig. 9 presents the experiment results when incorporating the large video chunk size (8 seconds) with different buffer threshold settings. As is evident, a b_i as large as 14-second is required to effectively reduce glitches and buffering. This highlights that it is important to configure the buffer threshold in accordance with the video chunk size in use, particulary for the scenario under consideration where the bandwidth fluctuates frequently and rapidly. Table 1 lists our recommendations on the buffer threshold for different chunk sizes based on the evaluations.

Table 1. Recommended buffer thresholds for different video chunk sizes

video chunk size (s)	2	4	6	8
recommended b_i (s)	8	10	12	14

6   Conclusion

In this paper, an empirical evaluation of HTTP adaptive streaming under vehicular mobility has been presented. We have implemented a HTTP adaptive streaming prototype and conducted evaluation experiments using "real" WWAN bandwidth traces that are empirically collected from a moving vehicle. We have investigated the performance of HTTP adaptive streaming under the effect of different parameters, such as buffer threshold and video chunk size. Our results have shown that HTTP adaptive streaming is effective in responding to the widespread fluctuations that are inherent in vehicular mobility and ultimately achieving an improved viewer experience. We have highlighted that it is possible to achieve a smooth and high-quality streaming experience, by using appropriate streaming parameter settings.

References

1. 3GPP: High Speed Downlink Packet Access - Overall description - Stage 2
2. 3GPP TS 26.234: Transparent end-to-end packet switched streaming service (PSS); Protocols and codecs (2010)
3. Adobe Systems Incorporated: Dynamic Streaming
4. Akamai: Akamai HD for iPhone Encoding Best Practices - Akamai HD Network
5. Apple Inc.: HTTP Live Streaming Overview
6. Arracomm Inc.: iBurst Broadband Wireless - System Overview
7. Derksen, J., Jansen, R., Maijala, M., Westerberg, E.: HSDPA Performance and Evolution. Ericsson Review (03), 117–120 (2006)
8. Floyd, S., Handley, M., Padhye, J., Widmer, J.: TCP Friendly Rate Control (TFRC): Protocol Specification. RFC5348 (September 2008)
9. ISO/IEC CD 23001-6: Information technology – MPEG systems technologies – Part 6: Dynamic adaptive streaming over HTTP (DASH) (2011)
10. Lie, A., Klaue, J.: Evalvid-RA: Trace Driven Simulation of Rate Adaptive MPEG-4 VBR Video. ACM/Springer Multimedia Systems Journal 14 (November 2007)
11. Microsoft Corporation: IIS Smooth Streaming Technical Overview
12. Move Networks: `http://www.movenetworks.com/`
13. Nemethova, O., Ries, M.: Quality assessment for h.264 coded low-rate and low-resolution video sequences. In: Proc. of Conf. on Internet and Inf. Tech. (2004)
14. OIPF: Specification Volume 2a - HTTP Adaptive Streaming (2010)
15. Orlov, Z.: Network-driven Adaptive Video Streaming in Wireless Environments. In: Proc. of IEEE PIMRC 2008, Cannes, France (September 2008)
16. Schierl, T., Wiegand, T., Kampmann, M.: 3GPP compliant adaptive wireless video streaming using H.264/AVC. In: Proc. of IEEE Intl. Conf. on Image Processing, Rio de Janeiro, Brazil (2005)
17. Schulzrinne, H., Rao, A., Lanphier, R.: Real Time Streaming Protocol (RTSP). RFC2326 (April 1998)
18. Singh, V., Ott, J., Curcio, I.: Rate Adaptation for Conversational 3G Video. In: Proc. of IEEE INFOCOM 2009 Workshop on Mobile Video Delivery, Rio de Janeiro, Brazil (April 2009)
19. The Insight Research Corporation: Streaming Media, IPTV, and Broadband Transport: Telecommunications Carriers and Entertainment Services 2009-2014
20. Wenger, S., Chandra, U., Westerlund, M., Burman, B.: Codec Control Messages in the RTP Audio-Visual Profile with Feedback (AVPF). RFC 5104 (February 2008)
21. Yao, J., Kanhere, S.S., Hassan, M.: An Empirical Study of Bandwidth Predictability in Mobile Computing. In: Proc. of ACM WinTech, SF, CA, USA (September 2008)

MAC Layer Support for Delay Tolerant Video Transport in Disruptive MANETs

Morten Lindeberg, Stein Kristiansen, Vera Goebel,
and Thomas Plagemann

University of Oslo, Department of Informatics,
Postboks 1080 Blindern, 0316 Oslo, Norway
{mglindeb,steikr,goebel,plageman}@ifi.uio.no

Abstract. The overall goal of this work is to improve video delivery
in emergency and rescue scenarios using sparse MANETs that might be
prone to frequent link breaks and network partitions. The core idea of
our approach is to reduce the number of MAC layer retransmissions that
are likely to fail. We do not drop packets that could not be sent after
the final retransmission. Instead we handle them in an overlay for store-
carry-forwarding. The design of the overlay protocol takes the instability
of the network into account, in such a way that each overlay entity works
autonomously and keeps a minimum amount of state. Our experimental
results show that we reduce packet loss seen on broken links, while at the
same time significantly reducing overhead in terms of the total amount
of packets transmitted at the physical layer.

Keywords: Video transmission; MANET; delay tolerant transport;
cross-layer optimization.

1 Introduction

Mobile ad-hoc networks (MANETs) can provide valuable communication services for emergency and rescue (ER) operations in the absence of a working
communication infrastructure. One such service is the transfer of video data
from rescue workers wearing head-mounted video cameras to a remote command and control center (CCC). The dynamicity of MANETs, the possibility
of short and long-term network partitions, and the fact that forwarding nodes
might be small hand held devices introduce many hard research challenges [10].
However, this particular video service is in contrast to classical video streaming more tolerant to delay. Video data that is delayed for seconds, minutes or
even hours can still be very useful in the CCC to understand what happened.
We envisage in the CCC a video client that visualizes the availability of video
sequences on a time axis, and allows play-out of live streams and browsing and
play-out of locally stored video. The visualization will be immediately updated
at the arrival of (delayed) video data. Therefore, video data that passed the
play-out time should not be dropped, but instead delivered as fast as possible.
Any video sequence might be still of importance.

J. Domingo-Pascual et al. (Eds.): NETWORKING 2011, Part I, LNCS 6640, pp. 106–119, 2011.

In order to improve the delivery efficiency, we reduce in this work the number of MAC layer retransmissions that are likely to fail and prevent packet dropping at the MAC layer. Instead of dropping, packets are passed after the maximum number of unsuccessful retransmissions from the MAC layer to an overlay for delay tolerant store-carry-forwarding, called Dts-Overlay. It utilizes information from the routing table and the MAC layer to identify the most promising conditions for forwarding packets.

Packet loss at the MAC layer mainly happens if the link quality, i.e., signal strength, has degraded due to mobility, but the link is still registered in the routing table. In such cases, packets are passed from the transport layer via IP to the MAC layer, which in turn drops the packets after reaching the retransmission limit (typically seven). This problem exists until the routing protocol recognizes that the link is broken, e.g., through a missing hello message in OLSR. The high frequency of link errors between mobile nodes has stimulated cross-layer approaches like CIFLER [1] to use fast link error detection for fast link error recovery. COLLIE [2] is another approach to immediately react to link failures through link layer rate adaptation. Our solution is different; before the overlay sends a packet it checks both the routing table and the queue length at the link layer (*Link Adapt*). Only if both indicate a working link to the next hop it passes the packet to the transport layer. Obviously, if a link is broken it does not make sense to try several retransmissions. Reducing the maximum number of retransmissions results in lower system load, but might lead also to higher packet loss rate. Therefore, the MAC layer returns in our solution after maximum retransmission trials the packet to the Dts-Overlay instead of dropping it (*MAC Return*). This is similar to the approach presented by Voorhaen et al. [3] where the packet is returned to the IP layer, assuming that an alternate route exists through which the packet can be sent instead. However, this assumption is not valid for scenarios with network partitions. In our ER application domain it might be the case that the CCC is only reachable from the location of the accident via message ferries [4]. Another cause of packet loss in MANETs is the separation of neighbourhood discovery in the routing protocol and address resolution in ARP [5]. We study in our system two solutions to avoid this packet loss. First, we avoid the need for ARP by assuming that the devices used during an ER operation are configured and all IP and MAC addresses are exchanged a priori (*Static ARP*). Second, when a new link appears in the routing table, the overlay provokes the address resolution of the connected node and sends only video packets to the node after its address is resolved (*ARP Adapt*).

All core design decisions are based on measurements performed in real-world experiments. Our ns-3 based evaluation shows that the combination of these mechanisms leads to minimal packet loss, reduced overhead, and neglectable increased delay. The remainder of this paper is structured as follows: Section 2 presents system design and implementation, Section 3 the evaluation, related work is discussed in Section 4, and Section 5 summarizes the main achievements and presents future work.

2   Design and Implementation

This section explains the design of our solution which is based on UDP, IP, OLSR [6], and cross-layer enabled IEEE 802.11b (see Figure 1). First, we analyze MAC layer issues and describe the details of our MAC support. Then, we describe Dts-Overlay, and how we address ARP related packet loss. Finally, we present our implementation.

2.1   MAC Layer Issues

IEEE 802.11 minimizes link layer packet loss through positive acknowledgements (ACK frames) and retransmissions if no ACK is received. In current implementations, the default value for the maximum limit of (re-)transmissions is seven, i.e., after seven unsuccessful transmissions the packet is dropped. There are two reasons for packet loss in IEEE 802.11, collisions and link quality. For the first one, more retransmissions will likely result in successful transmissions, while this is not true for links of inadequate quality. In MANETs, link quality might be degrading quickly due to mobility and links might not exist even if they are listed in the routing table, i.e., the routing table is outdated. Obviously, it does not make sense to initiate packet retransmission over a low quality link and especially not over a non-existing link. Reducing in these situations the maximum retransmission limit saves energy and bandwidth, and reduces the probability of collisions. The latter is especially important in highly loaded networks. Results from our experiments with a Nokia N900 as forwarding node in a small MANET [7] reveal a high amount of MAC layer retransmissions when reaching the saturation point in a real wireless ad hoc network setting. The high amount of retransmissions is not only consuming the senders energy and link bandwidth, but also energy and bandwidth of all nodes that are in communication distance to the sender. To investigate this further and identify a better limit for number of retransmissions, we study the overall effect of different retransmission limits in Section 3.

Four cases need to be considered when the MAC layer in the original 802.11 implementation would drop packets: (1) The link towards the destination is down, but there exist another route. (2) The link towards the destination is down, and there exist no alternative route. (3) The receiving node is not able to forward traffic at the rate received from the link [7]. (4) Congestion and collisions lead to

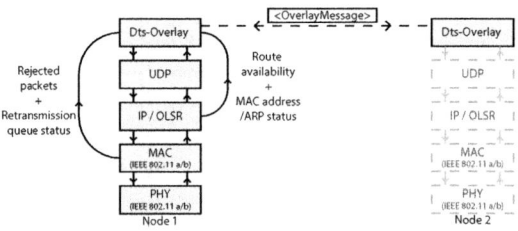

Fig. 1. Cross-layer Information Flow in Dts-Overlay

taildropping in the MAC retransmission queue at the sender side, or rather un-likely, the packet is lost due to collision in all retransmission trials.

For case (1), it is sufficient to hand the packet back to the network layer for re-routing, like in [3]. For case (2), the packet should be handed to the Dts-Overlay for being temporarily stored in the store-carry-forward buffer. For case (3), retransmissions only worsen node contention, thus we either re-route, or temporarily store packets in the forward buffer. Case (4) should only happen in overload situations, thus it is necessary to reduce the load, i.e., less retransmissions and temporary storage. To strive for simplicity and to be minimal invasive with the MAC layer we hand in all cases the packet to the Dts-Overlay. The cost of this is limited. We have measured the costs of passing a packet through an overlay instead of passing it directly to the network layer, on a Nokia 810 running in energy saving mode with low CPU and bus frequency. Results from our two-hop test-bed indicate that throughput is not worsened, and the additional delay is approximately 10 %. In higher CPU frequency settings, this share of additional delay is even reduced.

2.2   Dts-Overlay

The fundamental task of the Dts-Overlay, is to make forwarding decisions, and store transit packets when it is not meaningful to forward them. The Dts-Overlay is formed by autonomously working instances on each node, and packets are forwarded by Dts-Overlay hop-by-hop. Forwarding decisions are based on link status from the MAC layer, and route status from the routing protocol. It is the goal to transport all packets as close as possible to the destination, i.e., the CCC in ER. Dts-Overlay handles packets as follows: (1) If OLSR reports that no route to the destination exists, we check for recent routes in the network topology history, which is kept by Dts-Overlay. (2) If a recent route suggests a next-hop that is within the range of the current node, packets are sent to this next-hop. This is based on the heuristic that the next hop node on the recent route to the destination is probably closer to the receiver. (3) If there are no recent routes, we temporarily suspend transmission by maintaining the packet in a store-carry-forward buffer part of the Dts-Overlay. (4) If the next-hop is identified, we check to see if the MAC layer retransmission queue is filling (representing link status). In that case, we also suspend transmission.

To efficiently leverage message ferries, we currently assume that their IP address is in an a priori defined address range, i.e., message ferries can be identified when their IP address appears in the routing table. Message ferries are used as follows: when no route is found to the destination (CCC), and there exist route(s) to carrier nodes, the packets are forwarded to the closest carrier. To avoid packets looping back, we do not forward packets from carrier to carrier.

In a separate thread of execution, we monitor route updates in the routing table. As route notifications appear, we loop through all suspended packets and see if any of the suspended packets matches the new route(s). If so, they are forwarded to the suggested next-hop. Figure 2 shows the main components of the system, and the main flow of function calls and information exchange between them.

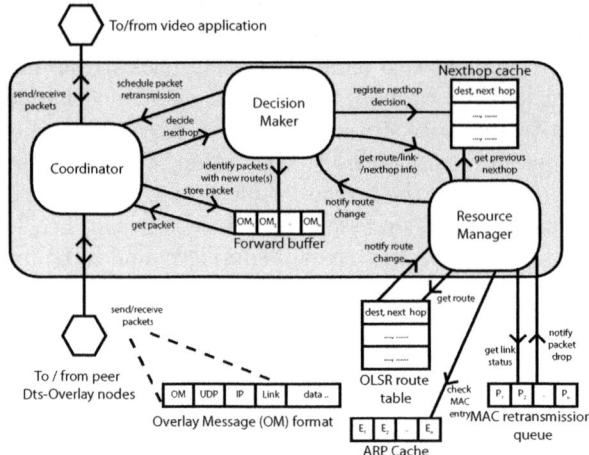

Fig. 2. Dts-Overlay System Design

The **Coordinator** component implements the main interface (send / receive primitives) for the video streaming applications, and holds the UDP socket used for all peer communication. It is responsible for forwarding and receiving packets from other Dts-Overlay peers through the network, and packets scheduled for retransmission from the store-carry-forward buffer.

The main function of the **Decision Maker** is to provide the next-hop IP address for a given packet destination, and control the store-carry-forward buffer. Currently, it implements a simplified and adapted version of the TCP congestion control algorithm to control the send rate in which the feedback is not provided by the receiving node, but instead by the local link status. It is the responsibility of the **Resource Manager** to collect and monitor network state information. Figure 2 illustrates from where it obtains this information. The resource manager serves as information broker, and provides data to the Decision Maker. It also serves as the entry point for packets that reach the final retransmission limit at the MAC layer.

We have implemented the store-carry-forward packet buffer as a drop tail FIFO queue. The size of this queue represents a tradeoff between memory consumption, and delay tolerant packet forwarding. Currently, we have a relatively large queue size (500 MB) to avoid tail dropping. Recent developments in solid-state drives provide relatively cheap, large and fast storage to the mobile devices. Thus, we expect that for low bit rate video streams within relative short network scenarios, it is realistic to expect that the buffer sizes do not limit the store-carry-forward capabilities of nodes. Packet exchange between peers is performed through **overlay messages**. Currently, the header is very simple; it only keeps the IP address of the destination. This information is lost at the IP layer, since overlay messages are only sent hop-by-hop.

2.3   Address Resolution

Carter et al. [3] identified ARP as one cause for packet loss in case several packets are sent immediately after an ARP request before the resolution process has finished, because only one packet could be buffered in ARP. Despite substantially increased buffer space in ARP in the recent ns-3 implementation, we identified a severe packet loss between the Dts-Overlay and the MAC layer. Packets are still silently dropped at the IP layer, when a MAC address for a given IP destination (reported by the routing protocol) has not been discovered. Further investigations revealed that this happens in cases the ARP reply is lost. To avoid these packet drops, we have implemented and evaluated two different approaches. First, we avoid the need for ARP by assuming that all devices in an ER operation are configured a priori with fixed IP addresses and all IP and MAC addresses are stored on all devices. In our second solution, we assure that the address resolution process is successfully finished before sending video packets, by sending an empty dummy packet to the IP destination. This leads to some overhead in terms of transmitted bytes, but our results show that the gain in packet reception at the destination is significant.

2.4   Implementation

We have adapted an already existing UDP based application for transmitting video traces (UdpTraceClient) which is part of the ns-3 code distribution. It generates UDP packets with packet sizes based on a given video trace, and adds a sequence number and timestamp to each packet. Dts-Overlay is implemented in C++ as an ns-3 application. This should make it easily portable to real devices.

We achieve cross-layer parameter exchange through the use of the ns-3 object aggregation system. In essence, ns-3 models network nodes and its protocol instances as objects. The smart pointer system enables access to these objects. This means that our Resource Manager component, realized as an object, can interact with, e.g., the OLSR routing protocol instance through function calls. We have implemented function calls within the ns-3 implementation of the IEEE 802.11 MAC layer, the ARP protocol, and the OLSR routing protocol that enables the Resource Manager to get the necessary state data.

3   Evaluation

In this section, we consider three network topologies, (1) ER scenario, (2) sparse MANETs, and (3) dense MANETs. The simulations are conducted in ns-3[1], version 3.9. For the **ER Scenario**, we have created a mobility model that reflects realistic ER scenarios. It contains two network partitions and a set of designated message ferries moving between them (see Figure 3). On-location nodes move with a speed of 2 m/s (approximately walking speed) 10 s pause time, according to the Random Walk mobility model, within a 500 m x 500 m area. Carrier nodes pause both at the CCC and on location for 60 seconds, allowing data

---

[1] Available at http://www.nsnam.org/

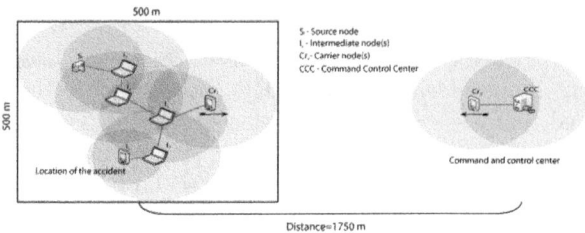

Fig. 3. ER Scenario

exchange. During the carrier phase, they move following a straight line at 10 m/s, the distance is 1750 m. The CCC node is not moving at all.

In addition to the ER scenario, we consider (2) **Sparse MANET** with area size 1250 m x 1250 m, and (3) **Dense MANET** with area size 250 m x 250 m. Each with 20 nodes, Random Walk mobility, node speed of 2 m/s, and 10 s pause time. In all network scenarios, nodes use IEEE 802.11b in ad-hoc mode. The devices simulate direct-sequence spread spectrum (DSSS) modulation, and a constant data transmission mode of 11 Mbps. For modelling the wireless channel, we utilize the constant speed propagation delay model, and the Friis propagation loss model.

3.1   Workload and Metrics

The workload consists of a single unicast stream of the video entitled Foreman, obtained from the Video Traces Research Group[2]. The resolution is CIF (352x288) with a 25 fps frame rate, and the video is pre-encoded using H.264 baseline profile with a target bitrate of 256 kb/s. We packetize the H.264 encoded video into a .mp4 container, with MTU size set to 1400 bytes. The 12-second video stream is repeated 300 times, i.e., we obtain a total duration of 1 hour.

For the dense scenario, our purpose is to identify how MAC support handles packet collisions. To achieve this, four UDP streams are sent between randomly selected intermediate nodes. Each UDP stream comprises 7500 packets per second with a uniform packet size of 1024 bytes. The resulting bitrate is higher than the available bandwidth, thus packet collisions are provoked.

We use four metrics to evaluate our solution: the percentage of video packets that are correctly received at the destination, denoted packet reception Rx_v; number of packets that are stored in intermediate nodes, denoted buffered packets B_{uf}; the percentage of lost packets, denoted packet loss P_l; and the overhead in terms of the total number of bytes of video packets transmitted at the physical layer, denoted Tx_t. Since we support delay tolerant video transfer it does not make sense to use video quality measures based on signal to noise ratio. All numbers are averages from five experiment runs, unless otherwise stated, and we present the standard deviation (σ).

---

[2] http://trace.kom.aau.dk/

3.2   Results

In the MAC Support evaluation, we compare five configurations (**C3** - **C7**) to evaluate the effect of MAC Return (MR) and Link Adapt (LA). The configuration settings are shown in Table 1.

Table 1. MAC Support Evaluation Configurations

	C1	C2	C3	C4	C5	C6	C7
Retransmission Limit	7	7	7	3	3	3	3
MAC Return	no	no	no	no	no	yes	yes
Link Adapt	no	no	no	no	yes	no	yes
ARP setting	default	static	adapt	adapt	adapt	adapt	adapt

Table 2. ARP Effect on Packet Loss

	C1	C2	C3
P_l	39.3 (σ:9.4) %	31.7 % (σ:4.1)	19.5 % (σ:5.1)

Effect of ARP: First, we compare the impact of ARP on packet loss in the default setting **C1** with our solution based on pre-configured IP and MAC addresses (**C2**), and our alternative solution to send packets after the resolution process has successfully finished **C3**. Results are averages from five experiment runs with 30-minute duration, ER scenario. We focus on packet loss P_l. Results indicate that packet loss is heavily affected by our second solution. The packet loss with default settings (**C1**) is 39.3 %, 31.7 % with static ARP (**C2**), and 19.5 % with ARP Adapt (**C3**). In **C1** and **C2**, packets are sent immediately also over links with poor quality. In **C3** packet forwarding is delayed until ARP has successfully finished, which indicates a useful link quality. As a result, we deploy ARP Adapt in remaining experiments.

Retransmission Limit and Link Adapt Limit: To understand and quantify the tradeoffs of reliability and costs related to the maximum number of retransmissions, we measure packet loss and costs for (1) different retransmission limits, and (2) different MAC retransmission queue size limits that prevent the Dts-Overlay to send packets in the LA mechanism. For (1), we show that we reduce overhead by lowering the retransmission limit, at the cost of packet loss. For (2), we see that we actually reduce both packet loss and cost (see Figure 4). The experiments are performed with the default Dts-Overlay configuration in the ER scenario with duration of 30 min. We present the average of five experiment runs.

The highest packet loss occurs with a retransmission limit of one (56.5 %). With a limit of three, packet loss is lower (31 %). A limit of five results in 22.0 %, a limit of seven in 19.5 %, and a limit of eleven in 15.4 % packet loss. At the same time we see an increase in the cost Tx_t of 47 %, from retry limit of one up to eleven. Notice that the curve is almost flat after five retries ($Tx_t = 221$ MB).

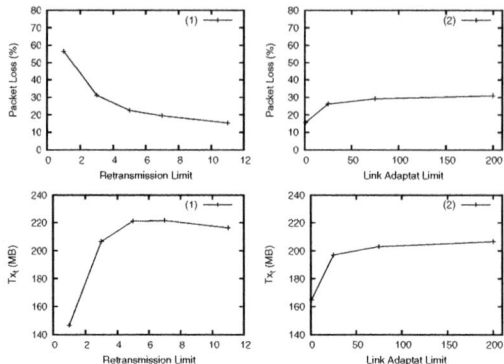

Fig. 4. Effect of Retransmission Limit and Link Adapt Limit Settings

Table 3. Evaluation Results ER Scenario

	C3	C4	C5	C6	C7
Rx_v	77.5 % (σ:1.8)	65.7 % (σ:1.6)	66.3 % (σ:1.0)	93.5 % (σ:0.5)	93.6 % (σ:0.7)
B_{uf}	5.1 % (σ:0.7)	4.5 % (σ:0.6)	4.5 % (σ:0.3)	5.6 % (σ:0.4)	6.1 % (σ:0.6)
P_L	17.4 % (σ:2.1)	29.8 % (σ:1.2)	29.2 % (σ:1.2)	0.9 % (σ:0.3)	0.3 % (σ:0.0)
Tx_t	455 MB (σ:19)	412 MB (σ:7)	411 MB (σ:10)	475 MB (σ:15)	454 MB (σ:28)

In our simulation studies, this behaviour can be attributed to tail dropping in the MAC retransmission queue. The current maximum queue size for this queue in ns-3 is 400 packets, which is very often reached in the experiments.

We have found that MAC retransmission queue size serves as a good indication of the link status. If packets destined on a certain link start piling up, the link is probably down. LA stops all ongoing transfers over the specific link, in case the queue starts filling up. Currently, this adaptation limit is set to 75 for our main experiments. The reason is that the limit is beneficial for MR configurations shown by preliminary experimentation. As a pre-study we investigate its effect on default configurations. Notice that we utilize a fixed retry limit of three. The figure clearly indicates that we get the lowest packet loss with an adaptation limit set to one. At 25, packet reception is worse. Finally at 200, packet loss has almost doubled from the limit set to one, i.e., 31 % of the packets are lost. The gain in reduced packet loss for low limit settings is even combined with reduced cost Tx_t. Such a strict limit means we stop using links in which packets are scheduled for retransmission. For denser networks with more frequent random collisions, we expect such a setting to be too strict, i.e., packets will be kept from being sent over working links.

ER Scenario: Due to the distance between the ER location and the CCC, all packet receptions come via carrier nodes. The numbers in the table show that our MR configurations (**C6** and **C7**) outperform **C3** when it comes to packet reception (16 % improvement), and loss is reduced from 17.4 % to <1 %. Remaining packets are kept in intermediate node buffers. LA (**C5** and **C7**) has

Table 4. Evaluation Results Sparse Manet Scenario

	C3	C4	C5	C6	C7
Rx_v	81.3 % (σ:24.6)	76.6 % (σ:30.7)	76.5 % (σ:30.8)	91.3 % (σ:14.8)	91.4 % (σ:15.0)
B_{uf}	0.0 % (σ:0.0)	0.0 % (σ:0.0)	0.1 % (σ:0.2)	7.3 % (σ:13.6)	8.1 % (σ:14.7)
P_l	18.7 % (σ:24.6)	23.4 % (σ:30.7)	23.4% (σ:30.7)	1.4 % (σ:1.9)	0.5 % (σ:0.6)
Tx_t	397 MB (σ:362)	245 MB (σ:162)	247 MB (σ:165)	381 MB (σ:356)	379 MB (σ:354)

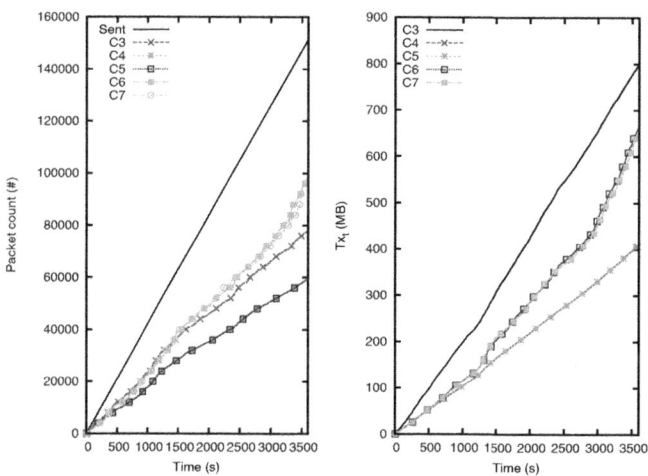

Fig. 5. Sparse MANET single run, Node speed 2m/s

little effect in terms of loss and packet reception. By reducing the retransmission limit for default configuration , packet loss is increased to 29.8 % (**C4**), however, we reduce overhead in terms of meaningless physical layer transmissions.

Most notable, the total transmitted bytes for **C7** is in average 1 MB lower than in **C3**. In comparison, **C6** achieves in average 21 MB more byte transmission than **C7**, thus LA improves our results. To summarize, MAC Support (MR plus LA) increases packet reception, reduces packet loss, and reduces cost in ER scenarios.

Sparse MANET Scenario: In total, we see that packet reception has decreased for the MR configurations, compared to the ER scenario. We achieve with 91 % the highest packet reception in both cases (**C6** and **C7**). Packet reception is better ($\approx$ 10 %) than for **C3**, where packet loss is 18.7 %. Remaining packets are kept in the buffers, although we experience 1 % packet loss for **C6**. The **C4** and **C5** configuration achieve both 77 % packet reception, thus LA does not improve packet reception. Packet loss is 23.4 % in these two cases.

Standard deviation is quite high, also for the total transmission of bytes. The reason for this is that the packet loss is in some runs high and in others rather low, depending on the seed for the generation of random mobility. In the following we only use one seed that creates a scenario with high packet loss.

Table 5. Evaluation Results Dense Manet Scenario

	C3	C4	C5	C6	C7
Rx_v	100.0 % (σ:0.4)	99.0 % (σ:0.6)	99.3 % (σ:0.1)	99.8 % (σ:0.0)	99.8 %(σ:0.1)
Buf	0.0 % (σ:0.0)	0.2 % (σ:0.6)	0.0 % (σ:0.1)	3.5 % (σ:0.1)	3.5 % (σ:0.2)
P_l	0.0 % (σ:0.0)	0.8 % (σ:0.1)	0.6 % (σ:0.1)	-3.4 % (σ:0.1)	-3.4 % (σ:0.1)
Tx_t	149 MB (σ:2)	149 MB (σ:3)	98 MB (σ:6)	151 MB (σ:3)	92 MB (σ:8)

Figure 5 shows Rx_v and Tx_t for one particular experiment run over time. For **C3**, we receive 52 % of the packets. By reducing the retransmission limit to three (**C4**), packet reception is lower, e.g., 39.2 %, and 39.3 % with LA enabled. Packet reception is 64.8 % for **C6**, and 65.0 % for **C7**. This demonstrates that MR is useful in scenarios with high packet loss.

Turning our focus to overhead, we see that overall packet transmission is reduced quite heavily by lowering the retransmission limit from seven to three. For **C3** and **C4**, a reduction of $Tx_t \approx 400$ MB, i.e., a factor of 0.5. Configuration **C7** leads to approximately 63 % more transmitted PHY packets than in **C3**, and close to 65 % more packet reception. It should be noted that those packet that did not reach the destination are kept in forward buffers of intermediate nodes, i.e., no packet loss occurred in **C7**.

Dense MANET Scenario: Nodes are randomly distributed in the relatively small area compared to signal propagation length, i.e., node density is high. As a result, packet loss is mostly caused by collisions, rather than insufficient signal strength. Configuration **C3** handles collisions well, i.e., no packet loss. For configurations **C4** and **C5**, packet collisions cause ≈ 1 % packet loss resulting in ≈ 99 % packet reception. For the MR configurations (**C6** and **C7**), we encompass close to 100 % packet reception and some amount of duplicates, indicated by a negative packet loss of $\approx -3.5\%$. The cause of these duplicates is that our MAC support interferes with the MAC layer functionality to handle frame duplicates due to missing MAC acknowledgements.

We see that **C7** achieves the lowest overhead, from **C3**, a 62 % reduction, which is substantial. In general, a severe reduction applies to all LA configurations, proven highly beneficial in congested and dense MANETs.

4   Related Work

The related work falls into three categories: (1) delay tolerant transport, (2) video delivery over MANETs, and (3) adaptations to MAC or PHY conditions. The typical approach of delay tolerant networking is to provide an alternative to TCP, to achieve some reliability for end-to-end transport. One alternative is to introduce a layer that adds delay tolerance above existing routing protocols, like in SAFT [8]. Reliability is provided through a double control-loop, including end-to-end feedback and TCP hop-by-hop. Applying TCP between each hop substantially increases the number of control packets. Delay-tolerant video transport over IP is achieved in MOMENTUM [9], like in our approach through an overlay. However, only selected nodes, called session nodes, actively transport the video data and require knowledge about all session nodes. Furthermore, adaptations to lower layers are not considered, e.g., to determine the buffer empty rate. Another alternative is to implement a delay tolerant routing protocol, resulting in what is often referred to as space-time routing. This is explored already in [10], the authors suggest that algorithms should take congestion into account, and that performance can still be good without global knowledge. Evaluation results from our autonomous solution support this.

In [11], a detailed survey of research on video streaming over MANETs is given. A popular approach is to combine cross-layered multipath MANET routing with multiple description video coding, often leading to NP-hard optimization problems. It should be noted that all these works assume that one or more routes between source and destination exist.

The effect of MAC layer retransmissions on video quality is studied in [12,13]. Both efforts aim to adapt the retransmission limit, to better meet user video quality demands. COLLIE [2] and CIFLER [1] enable the routing protocol to better react to link status. The reduction of packet loss, caused by lack of exact link status for the OLSR routing protocol is targeted in [3]. In case a packet could not be transferred over a link, it is returned to the IP layer assuming another route exists. Thus, tolerance to path disruptions is neither considered nor supported. Authors show, as we do with our related technique, improvements in terms of reduced packet loss.

5   Conclusion

This work reduces packet loss in disruptive MANETs by MAC layer support, and paired with the Dts-Overlay, achieves tolerance to network disruptions. As part of MAC support, our MAC Return algorithm enables us to drastically reduce packet loss, in turn allowing us to reduce overhead by lowering the MAC layer retransmission limit. In addition, our Link Adapt algorithm allows us to reduce overhead in that we do not try to send packets over links that are down. This has been supported by extensive simulation studies. More precisely, we have shown that our approach achieves both a reduction in physical layer packet transmissions, while avoiding packet loss through the use of intermediate node buffering, paired with a higher packet reception rate in both a sparse MANET setting, and in our special purpose ER scenario. In addition, the usability of our approach is strengthened through evaluation results in a dense MANET setting. With high density, we see improvements of packet reception/packet loss, although at the cost of some packet duplicates.

There are still open issues and remaining challenges. Currently, MAC Return does not support packet fragmentation at the MAC or IP layer. As seen in the dense MANET setting, it can obstruct existing MAC functionality for identifying duplicates. When acknowledgements from the receiver node are lost, we encounter a false negative error. This might lead to the sender retransmitting packets that are already received at the receiver. There are yet parameters to fine-tune. For one, we can improve our rate control algorithm for emptying the store-carry-forward buffer by utilizing physical layer parameters, such as signal strength. We should account for video coding parameters and frame priority in our forwarding decisions. At the network and routing layer, improvements can be done to, e.g., OLSR, by removing links from OLSR data structures known to be down. At the MAC layer, we could benefit from deploying one of the existing rate adaptation algorithms, such as CARA [14]. For future work, we plan to incorporate an energy model in the simulated nodes to enable energy aware forwarding strategies. We aim to implement our work on a real mobile device, such as the Nokia N900. This would strengthen the realism of our performance evaluations. As stated in [7], we are aware that mobile hand helds have lower forwarding capability than what the wireless medium supports. Especially, we hope to investigate how to tackle packet loss due to node contention, as opposed to weak signal strength or packet collisions.

Acknowledgements. This work is funded by the VERDIKT program of the Norwegian Research Council, through the DT-Stream project (no. 183312/S10). The authors would like to thank Sergio Cabrero, Isaías Martínez Yelmo and Daniel Rodríguez-Fernández for valuable insights.

References

1. Yackoski, J., Shen, C.C.: Cross-layer inference-based fast link error recovery for manets. In: Wireless Communications and Networking Conference (WCNC), vol. 2, pp. 715–722. IEEE, Los Alamitos (2006)
2. Rayanchu, S., Mishra, A., Agrawal, D., Saha, S., Banerjee, S.: Diagnosing wireless packet losses in 802.11: Separating collision from weak signal. In: International Conference on Computer Communications (INFOCOM), pp. 735–743. IEEE, Los Alamitos (2008)
3. Voorhaen, M., Blondia, C.: Analyzing the impact of neighbor sensing on the performance of the olsr protocol. In: International Symposium on Modeling and Optimization in Mobile, Ad Hoc and Wireless Networks, pp. 1–6. IEEE, Los Alamitos (2006)
4. Zhao, W., Ammar, M., Zegura, E.: A message ferrying approach for data delivery in sparse mobile ad hoc networks. In: International Symposium on Mobile Ad Hoc Networking and Computing (MobiHoc), pp. 187–198. ACM, New York (2004)
5. Carter, C., Yi, S., Kravets, R.: ARP considered harmful: manycast transactions in ad hoc networks. In: Wireless Communications and Networking (WCNC), vol. 3, pp. 1801–1806. IEEE, Los Alamitos (2003)
6. Jacquet, P., Muhlethaler, P., Clausen, T., Laouiti, A., Qayyum, A., Viennot, L.: Optimized link state routing protocol for ad hoc networks. In: International Multi Topic Conference (INMIC), pp. 62–68. IEEE, Los Alamitos (2001)
7. Kristiansen, S., Lindeberg, M., Rodriguez-Fernandez, D., Plagemann, T.: On the forwarding capability of mobile handhelds for video streaming over manets. In: SIGCOMM Workshop on Networking, Systems, and Applications on Mobile Handhelds (MobiHeld), pp. 33–38. ACM, New York (2010)
8. Heimlicher, S., Baumann, R., May, M., Plattner, B.: SaFT: Reliable transport in mobile networks. In: International Conference on Mobile Adhoc and Sensor Systems (MASS), pp. 477–480. IEEE, Los Alamitos (2006)

9. Cabrero, S., Pañeda, X.G., Plagemann, T., Goebel, V.: Overlay solution for multimedia data over sparse MANETs. In: International Wireless Communications and Mobile Computing Conference, IWCMC (2009)
10. Jain, S., Fall, K., Patra, R.: Routing in a delay tolerant network. SIGCOMM Comput. Commun. Rev. 34, 145–158 (2004)
11. Lindeberg, M., Kristiansen, S., Plagemann, T., Goebel, V.: Challenges and techniques for video streaming over mobile ad hoc networks. Multimedia Systems 17, 51–82 (2011)
12. Chan, A., Lee, S.J., Cheng, X., Banerjee, S., Mohapatra, P.: The impact of link-layer retransmissions on video streaming in wireless mesh networks. In: International Conference on Wireless Internet (WICON), pp. 1–5. ACM, New York (2008)
13. Choudhury, S., Sheriff, I., Gibson, J.D., Belding-Royer, E.M.: Effect of payload length variation and retransmissions on multimedia in 802.11a WLANs. In: International Conference on Wireless Communications and Mobile Computing (IWCMC), pp. 377–382. ACM, New York (2006)
14. Kim, J., Kim, S., Choi, S., Qiao, D.: CARA: Collision-Aware Rate Adaptation for IEEE 802.11 WLANs. In: International Conference on Computer Communications (INFOCOM), pp. 1–11. IEEE, Los Alamitos (2006)

DTN Support for News Dissemination in an Urban Area

Tuan-Minh Pham and Serge Fdida

Laboratoire d'Informatique de Paris 6 (LIP6),
UPMC Sorbonne Universités,
4 place Jussieu, 75005 Paris, France
{tuan-minh.pham,serge.fdida}@lip6.fr

Abstract. We are studying the practicality of news dissemination over a Delay Tolerant Network (DTN) in an urban area. The target application is the distribution of the electronic version of a newspaper in a large city. Therefore, although strict time constraints do not apply, spreading the information should be achieved within a reasonable delay. We consider that mobile users subscribe to their content of interest and expect to receive it within their journey from their home to their office. We provide two contributions. Firstly, we consider a simple DTN environment when content is distributed solely through inter-contact of mobile nodes. We derive analytical expressions for the packet delay in such environments and suggest how to improve effectively the expected message delay in the case of an area with low or high density of mobile nodes. Secondly, if the delay is found to be excessive, we suggest the deployment of some data kiosks in the environment to better support the dissemination of content. Data kiosks are simple devices that receive content directly from the source, usually using wired or cellular networks. We investigate both an upper bound and a lower bound of the number of data kiosks to distribute the content over a geographical area within an expected delay objective. We also show the important property that those bounds scale linearly with the contact rates between a mobile node and a data kiosk. The analytical results are validated through simulation using a number of mobility models.

Keywords: Hybrid DTN, delay, news dissemination, modeling, performance analysis.

1 Introduction

Free daily newspapers have been introduced in many countries worldwide for over 10 years now. Their distribution channel uses point of presence (Kiosk) often located at the entrance to the main transportation systems, such as metro or suburban train stations. They also have a presence on the web and exploit user's contributions and social networks. In addition, the widespread deployment of handheld devices provides the opportunity to use them for content distribution instead of being charged for cellular access. In the context of our research, we

J. Domingo-Pascual et al. (Eds.): NETWORKING 2011, Part I, LNCS 6640, pp. 120–133, 2011.

consider that mobile users will agree to contribute to such an application only for content of their own interest. Information conveyed by free daily newspapers does not need to be instantaneous. However, owners of these newspapers expect that the information will reach the reader within a time window related to the period he/she will spend commuting from their home to their place of work as this is the best time to capture their attention. The contribution of our paper is not restricted to the distribution of news as many other applications will exhibit similar expectations, but it provides a practical use case.

In this paper, we investigate the practicality of news dissemination over a DTN in an urban area. In DTNs, nodes can move freely and exchange content when they are within each other's transmission range. Since an end-to-end path is not available most of the time, a store-carry-forward paradigm is usually used to enable communication. Several questions arose in the above setting in order to design a system that achieved the required service while avoiding deploying too many digital data kiosks. A first step would be to develop a quantitative analysis to investigate the message delay of DTNs. What are the most sensitive parameters: readers' interest in content or contact opportunities? How effective can a DTN be without infrastructure? If the delay is found to be excessive, we suggest deploying some data kiosks in the environment to better support the dissemination of content. Data kiosks are simple devices that receive the content directly from the source, generally using wired or cellular networks. The question is how many data kiosks one has to invest in to satisfy performance constraints and where these data kiosks should be located. We are mostly interested in two performance metrics, the spreading time and the message delay. The message delay is the time needed to transmit content from a mobile node to another node, while the spreading time is the delay required to deliver the content to the last node in a group of nodes, or the time needed for the content to spread over a part of the network. The main question addressed in this paper is to determine the number of data kiosks necessary to disseminate content to a set of mobile nodes, when taking into account the spreading time in a given area and mobility pattern.

Our contribution is to provide explicit solutions to answer the above questions. To the best of our knowledge, our work is the first to obtain explicit expressions for the evaluation of the number of data kiosks needed to reduce the expected spreading time to a target requirement in opportunistic networks. We first investigate both a closed-form formulae and an asymptotic expression of the expected message delay in DTNs where mobile users share their contents only if they share the same interests. We show that the expected message delay is in the order of several hours even if some parameters are optimistic. Second, we consider a hybrid environment and derive closed-form expressions for both an upper and a lower bound of the number of data kiosks needed to satisfy the requirement of the expected spreading time. The results show how the contact rate between a mobile node and a data kiosk influences the quality of service parameters.

The remainder of this paper is structured as follows. In the following section, we describe some related work. In Section 3, we analyze the system formally and derive the main results. In Section 4, we present the analytical model to compute the number of data kiosks. In Section 5, we validate our solution against simulation under three mobility models and give a specific example when applying our results to improve the delay. Finally, Section 6 concludes the paper.

2 Related Work

DTN application to content dissemination has been contemplated with different perspectives. Lenders et al. developed a mobile distribution system where podcasting protocols are extended to the ad hoc domain [11]. Garyfalos and Almeroth designed a system in advertising with a incentive scheme for data sharing through opportunistic contact [5]. In addition, McNamara et al. concentrate on the application of DTN in the context of the movement of people in transit systems due to the fact that people consume time on public transport [13].

The emergence of DTNs has triggered a considerable amount of work to help us to better understand their behavior and improve their performance. Contact opportunities have a strong impact on the performance of DTN and therefore, several contributions have focused on characterizing the inter-contact time distribution. Observing some real world traces, Karagiannis et al. investigated exponential tail behavior of inter-contact times [9]. Studying common mobility models, Groenevelt et al. discovered that the inter-contact time between mobile nodes is almost exponentially distributed [6]. Ibrahim et al., in addition, showed that the inter-contact time between mobiles and stationary nodes is also almost exponential in common mobility models such as the random waypoint or random direction [7].

Based on existing knowledge about contact opportunities, different papers analyzed the performance of DTNs with various assumptions on content distribution schemes including the unrestricted multicopy protocols [6], spray-and-wait [19], k-hop relay [18]. Starting from [6], Hanbali et al. [3] extended the work with lifetime constraints and Zhang et al. [21] studied variations of the epidemic protocol and recovery schemes. Different measures were considered: throughput [2], delay [6,3,21], content age [8].

Recent work was motivated by improving the performance of DTNs. Polat et al. proposed an algorithm to find mobile nodes that can act as message ferries [16]. Another approach is to add some infrastructure nodes. This solution has been used in the context of ad hoc networks where nodes are fixed [12,20,17]. In the DTN's framework, some papers estimated the average message delay in the network with the presence of infrastructure for multi-copy two hop routing protocol and epidemic routing protocol [7,4].

Our work is different as it considers the probable effect of interest for a given content in estimating the expected message delay in DTNs. In addition, it computes the number of infrastructure nodes to meet a performance objective for expected spreading time.

3   System Description and Main Results

We consider a system composed of n mobile nodes and k data kiosks. Mobile nodes are wireless devices carried by people, and data kiosks are stationary nodes in which a content update is downloaded via an infrastructure network (wired or cellular), to be stored and disseminated. We define mobile nodes that subscribe to a given content as subscribers. Subscribers are organized into healthy nodes and infected nodes. An infected (resp. healthy) node is a subscriber that has (resp. does not have) a copy of the content he subscribed to. We assume that every node has the same transmission range, interference from other nodes is negligible, and transmission is always successful and instantaneous in a sparse network.

The dissemination of news relies on opportunistic contacts and a store-carry-and-forward paradigm. Motivated by [6,7], the pairwise inter-contact times for mobile-mobile (resp. mobile-data kiosk) contacts can be approximately represented as exponentially distributed random variables with mean $1/\lambda$, $\lambda > 0$ (resp. $1/\mu$, $\mu > 0$). All these random variables are assumed to be homogeneous and mutually independent. A healthy node will be infected when it meets either a data kiosk or an infected node, and it is interested in the news. We assume that a healthy node can change its interest at any time by subscribing to the content with a probability δ $(0 < \delta \leq 1)$ or unsubscribing with a probability $1 - \delta$, and that only healthy nodes can change their interest when they meet other mobile nodes. One can expect that infected nodes will share their content during their journey as they will mostly read it during that time. We refer to this scheme as a user-preferred content distribution scheme.

3.1   Expected Message Delay

Since a service provider wishes to know whether or not adding some data kiosks is necessary, our first main result addresses the question of performance in a homogeneous environment when no data kiosk exists. The message delay D is our preferred metric and is defined as the time required to deliver information (news/content) from a source to a destination node. It is expected that the destination node will keep its interest until it receives the content. We investigate both a closed-form expression and an asymptotic representation of the expected message delay as a function of $o(\delta)$ when δ is small. We are interested in the influence on the expected message delay of parameters such as the interest of people and contact rates.

Proposition 1. *Under the user-preferred content distribution scheme, the expected message delay is given by*

$$E[D] = \frac{1}{\lambda} \sum_{i=1}^{n-1} \left(P\{N_2 = i\} \sum_{j=1}^{i} \frac{1 - (1 - \delta)^{n-j}}{\delta j \, (n - j)} \right)$$

$$= \frac{1}{\lambda} \left(1 - \frac{n - 2}{4} \delta \right) + o(\delta), \quad \text{for } \delta \to 0$$

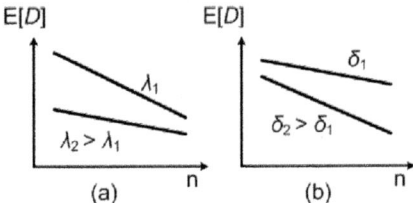

Fig. 1. Asymptotic representation of the expected message delay

where N_2 denote the number of copies of the content at the time it reaches its destination. The distribution of the number of copies $P\{N_2 = i\}$ is given by

$$P\{N_2 = i\} = \frac{1 - (1 - \delta)^{n-i}}{\delta(n-i)} \prod_{j=1}^{i-1} \left(1 - \frac{1 - (1 - \delta)^{n-j}}{\delta(n-j)}\right).$$

The proofs of the proposition can be found in [14] and is derived from [6] adding parameter δ.

The asymptotic representation of the expected message delay suggests that the role of interest is significant in the case of high node density while the role of contact rates becomes more important in the case of low density of nodes (Fig. 1). Consequently, improving the interest is rewarding if the number of mobile nodes is large, while improving contact rates provides better performance if the number of nodes is small. Using the results above, we can decide if the performance is acceptable using only contact opportunities between mobile users or if the addition of data kiosks becomes mandatory.

3.2   Number of Data Kiosks

We consider the situation where a hybrid solution is necessary to satisfy a given performance objective for the application. The performance metric is defined as the spreading time. An upper bound and a lower bound for the number of data kiosks are obtained under the following assumptions:

- In the case of the hybrid solution, we assume that mobile nodes do not change their interest in the lifetime of their journey (of course, they can add new interest),
- the number of data kiosks is greater than λ/μ,
- subscribers do not hold a copy of the content at the beginning of their journey,
- the number of infected nodes is at least equal to two by the spreading time (one usually wants to have a lot of subscribers infected by the spreading time).

The following notations are introduced:

- m is the total number of subscribers at time 0.
- d is a performance objective for the expected spreading time.
- b is the number of infected nodes by the spreading time.

Proposition 2. *If the number of data kiosks is larger than λ/μ, then an upper bound and a lower bound of the number of data kiosks to ensure that b subscribers will receive the content by the expected spreading time d are given by $\lfloor L \rfloor < k < \lceil U \rceil$,*

$$U = \frac{\lambda}{2\mu} + \frac{H + \sqrt{H^2 + 4\lambda d\,(b_\gamma + 1)\,(H+1)}}{2\mu d}, \tag{1}$$

$$L = \frac{-\lambda\,(b_\gamma + 2m)}{4\mu} + \frac{2\varsigma + \sqrt{[\lambda d\,(b_\gamma - 2m) + 2\varsigma]^2 + 8\lambda d\,\left(2\tilde{b}_\gamma + 1\right)}}{4\mu d} \tag{2}$$

where $H = e^{-(0.5+m)\lambda d - \varsigma} - 1$, $\varsigma = \sum_{i=0}^{b-1} \frac{1}{m-i}$, $b_\gamma = b + e^{-\gamma} - 1.5$, $\tilde{b}_\gamma = b - e^{-\gamma}$, γ is Euler-Mascheroni constant $\gamma \approx 0.57721$, e is Euler's number $e \approx 2.71828$. The proof is provided in section 4.

In the formulae, if $b = m$, the computed number of data kiosks ensures that a subscriber will receive the content before d on average. The following corollary is derived directly from Proposition 2.

Corollary 1. *Suppose that the contact rate between a data kiosk and a mobile node in two distinct environments A and B is μ_A and μ_B respectively. Let U_A and L_A (resp. U_B and L_B) be an upper bound and a lower bound of the number of data kiosks in the environment A (resp. B). If we keep the same requirement for the expected spreading time, then a relationship between the bounds of the number of data kiosks and the contact rates in these environments is as follows:*

$$\frac{U_B}{U_A} = \frac{L_B}{L_A} = \frac{\mu_A}{\mu_B}.$$

Corollary 1 shows that the bounds of the number of data kiosks scale linearly with the contact rates between a mobile node and a data kiosk.

4   Analysis of the Number of Data Kiosks

In this section, we focus on the analytical model to compute the number of data kiosks to meet a target for the expected spreading time.

We represent the number of copies of a given content in the network as an absorbing finite state Markov chain comprising $b+1$ states labeled by $\{0, 1, \ldots, b\}$. The chain is in state i when there are i infected nodes in the network. If a current state i is not the absorbing state b, the chain will make a transition into its next state $i+1$ after a time period. The spreading time that we are going to compute is the time until absorption of the chain.

We first compute the expected spreading time. A healthy node is infected the content when it meets either one of k data kiosks or one of i infected nodes. Then, a healthy node is infected with the content at an exponential rate $\lambda i + \mu k$.

The chain leaves state i when one of $m - i$ healthy nodes is infected with the content. The event that a healthy node is infected occurs at an exponential rate $\lambda i + \mu k$, then the event that the chain will make a transition into its next state occurs at an exponential rate $(\lambda i + \mu k)(m - i)$. Let T_i, $0 \leq i \leq b - 1$, be the time interval in which the chain stays in state i. Since T_i has an exponential distribution with parameter $(k\mu + i\lambda)(m - i)$, we have $E[T_i] = \frac{1}{(k\mu+i\lambda)(m-i)}$.

Let T denote the spreading time, i.e. the time until there are b infected nodes in the network. Then, $T = \sum_{i=0}^{b-1} T_i$ and $E[T] = E\left[\sum_{i=0}^{b-1} T_i\right] = \sum_{i=0}^{b-1} E[T_i]$.

We are ready to provide an expression for the expected spreading time d:

$$d = \sum_{i=0}^{b-1} \frac{1}{(k\mu + i\lambda)(m - i)}. \tag{3}$$

We now have to compute the number of data kiosks to meet an objective for the expected spreading time d. Unfortunately, it is not possible to find a closed-form expression for m in the general case and we will therefore provide bounds. Lemma 1 to 4 will give the foundation to derive these bounds. Due to space limits, we only present the proof for Lemma 1. Details of the proofs for Lemma 2 to 4 can be found in [15].

Lemma 1. *An expression of the expected spreading time represented under digamma function is*

$$d = \frac{\psi(u + b) - \psi(u) + \varsigma}{\lambda u + \lambda m} \tag{4}$$

where $\varsigma = \sum_{i=0}^{b-1} \frac{1}{m-i}$, $u = \frac{\mu}{\lambda}k$, $\psi(x)$ *is a digamma function of x.*

Proof. Using an equation 6.3.16 of the digamma function in [1], we have $\psi(x + 1) = -\gamma + \sum_{i=1}^{\infty} \left(\frac{1}{i} - \frac{1}{x+i}\right)$ $(x \neq -1, -2, \ldots)$ with γ is Euler-Mascheroni constant $\gamma \approx 0.57721$. Since $k \geq \lambda/\mu$, we get $u = \mu k/\lambda \geq 1$. Applying the equation of the digamma function, we find $\psi(u + b) = -\gamma + \sum_{i=1}^{\infty} \left(\frac{1}{i} - \frac{1}{u+b-1+i}\right)$ and $\psi(u) = -\gamma + \sum_{i=1}^{\infty} \left(\frac{1}{i} - \frac{1}{u-1+i}\right)$. Subtracting the later equation from the former, we get

$$\psi(u + b) - \psi(u) = \sum_{i=1}^{b} \frac{1}{(u - 1) + i} = \sum_{i=0}^{b-1} \frac{1}{u + i}. \tag{5}$$

From (3), we have $d = \frac{1}{\lambda u + \lambda m} \left(\sum_{i=0}^{b-1} \frac{1}{u+i} + \varsigma \right)$ where $\varsigma = \sum_{i=0}^{b-1} \frac{1}{m-i}$, $u = \frac{\mu}{\lambda}k$.
Substituting (5) into the equation, we get (4) which proves Lemma 1.   □

Lemma 2. *With $u \geq 1$ and $b \geq 2$, inequalities of $\psi(u+b) - \psi(u)$ are*

$$\ln\left(1 + \frac{\tilde{b}_\gamma + 0.5}{u + e^{-\gamma} - 1}\right) < \psi(u+b) - \psi(u) < \ln\left(1 + \frac{b_\gamma + 1}{u - 0.5}\right)$$

where $b_\gamma = b + e^{-\gamma} - 1.5$, $\tilde{b}_\gamma = b - e^{-\gamma}$.

Lemma 3. *One solution of the inequality $\lambda du + \lambda dm - \varsigma < \ln\left(1 + \frac{b_\gamma + 1}{u - 0.5}\right)$ is $\{u \in \mathbf{R} | u \geq 1, u < u_2\}$ where*

$$u_2 = 0.5 + \frac{H + \sqrt{H^2 + 4\lambda d \left(b_\gamma - 0.5\right)\left(H + 1\right)}}{2\lambda d}, H = e^{-(0.5+m)\lambda d - \varsigma} - 1.$$

Lemma 4. *One solution of the inequality $\lambda du + \lambda dm - \varsigma > \ln\left(1 + \frac{\tilde{b}_\gamma + 0.5}{u + e^{-\gamma} - 1}\right)$ is $\{u \in \mathbf{R} | u \geq 1, u > u_4\}$ where*

$$u_4 = \frac{-(b_\gamma + 2m)}{4} + \frac{2\varsigma + \sqrt{\left[\lambda d\left(b_\gamma - 2m\right) + 2\varsigma\right]^2 + 8\lambda d\left(2\tilde{b}_\gamma + 1\right)}}{4\lambda d}.$$

We are almost ready to derive bounds of the number of data kiosks. Using Lemma 1, we find $\lambda du + \lambda dm - \varsigma = \psi(u+b) - \psi(u)$.

Because the number of data kiosks is larger than λ/μ, it follows that $u \geq 1$. Suppose that the number of infected nodes is greater than or equal to two by the spreading time, $b \geq 2$. Hence, using Lemma 2, we get $\ln\left(1 + \frac{\tilde{b}_\gamma + 0.5}{u + e^{-\gamma} - 1}\right) < \lambda du + \lambda dm - \varsigma < \ln\left(1 + \frac{b_\gamma + 1}{u - 0.5}\right)$. From Lemma 3 and Lemma 4, we obtain a solution for the inequalities. Substituting $u = \frac{k\mu}{\lambda}$ into the solution, we find an upper bound and a lower bound of the number of data kiosks $\lfloor L \rfloor < k < \lceil U \rceil$, where U is represented by (1) and L is described by (2), which demonstrates Proposition 2.

5   Validation

In this section, we validate the theoretical results through simulation under three mobility models, and discuss the application of the results. For a given context and a delay objective, we can state if the delay is acceptable to deploy the service. If negative, we provide the number of data kiosks that should be added in a hybrid DTN environment to meet the requirement.

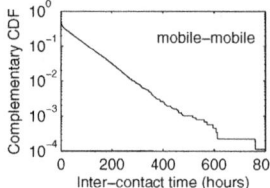 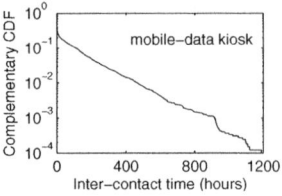

Fig. 2. CCDF of inter-contact time for the random walker mobility model

5.1   Simulation Settings

We simulated the user-preferred content distribution scheme under the random waypoint, random direction and random walker mobility models in the Opportunistic Network Environment [10]. Mobile nodes move according to the mobility models under consideration and data kiosks are uniformly located in a square of size 4 km × 4 km. Radio ranges of nodes are 50 meters.

In the random waypoint mobility model, a node travels to a destination chosen uniformly in an area with a constant speed. The speed of a mobile node (in km/h) is chosen uniformly in $[v_{\min}, v_{\max}] = [4, 10]$. Upon arrival, the node continues its trip by choosing a new destination and a new speed, independently of all previous destinations and speeds.

In the random direction mobility model, a node selects an initial direction, speed and a finite travel time. The node then travels in the direction at the speed for the duration. Once this time expires, it chooses a new direction, speed and travel time which are independent of all its past directions, speeds and travel times. When a node reaches a boundary it is reflected. In our setting, the speed of a mobile node (in km/h) is chosen uniformly in $[v_{\min}, v_{\max}] = [4, 10]$, a direction is uniformly distributed in $[0, 2\pi)$, and travel time is exponentially distributed with mean $1/4$ hour.

In the two-dimensional random walker mobility model, a node moves on a square grid at a constant speed. At crossroads, a node uniformly chooses to go to a next crossroad in the front, in the back, on the left, or on the right. When a node has reached a crossroad at the borders, it will go to the opposite crossroad if the newly chosen crossroad is out of the borders. We set the distance among adjacent crossroads to 80m and a constant speed of 4.8 km/h.

For the random waypoint and random direction mobility model, formulae for λ and μ were introduced [6,7]. Using the parameter settings of our simulation, we computed the values for λ and μ and found that the contact rates are $\lambda \approx 0.0741$ and $\mu \approx 0.0409$ for the random waypoint mobility model, $\lambda \approx 0.0577$ and $\mu \approx 0.0437$ for the random direction mobility model.

A mathematical formula for the exponential tail under the random walker mobility model is, to the best of our knowledge, not known. Therefore we obtained its value through simulation. Fig. 2 plots the complementary cumulative distribution function (CCDF) of the inter-contact time on log-lin scale. Estimates of contact rates for this last mobility model are $\lambda \approx 0.0127$ and $\mu \approx 0.0070$.

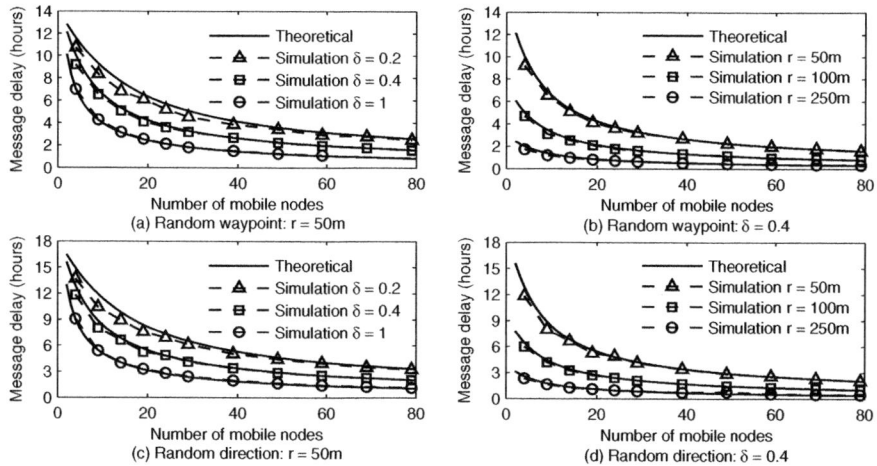

Fig. 3. Message delay vs. number of mobile nodes: Random waypoint, random direction mobility model

5.2   Lower Delay to a Requirement

First, we consider the expected message delay in the situation with zero data kiosk. For the three probability of interest ($\delta = 1, 0.4, 0.2$), and under the three mobility models, we varied the number of nodes between 5 and 80, and computed the average message delay based on 1000 observations obtained from simulation for each configuration. Figs. 3-4 plot the expected message delay as a function of the number of nodes, comparing the theoretical analyses with the simulation results. They confirm the accuracy of the theoretical analysis for these mobility models. Although the inter-contact time is only exponential tail under the random walker mobility model as we see in Fig. 2, the theoretical analysis still can predict the message delay for the different probability of interest and the different contact rates. The error between simulation and theoretical results increases when the probability of interest is small.

Next, in the hybrid case, we run simulations for different number of subscribers ($k = 40, 60$). The number of data kiosks was varied between 2 to 20. For each setting, we ran 1000 observations and computed the expected spreading time by which the content is delivered to 90 percent of subscribers. Then, we used analytical expressions (1) and (2) to calculate an upper bound and a lower bound of the number of data kiosks for a value of the average spreading time that we found from simulation. For each spreading time, we compared the estimated values of the number of data kiosks provided thanks to the theoretical solution or by simulation. Fig. 5 depicts the results under the three mobility models.

We observe that analytical results can approximately predict the number of data kiosks, but the lower bounds deteriorate as the expected spreading time gets very small which is certainly not meaningful. Finally, we give an example for the application of our results with ideal parameters. We first consider the

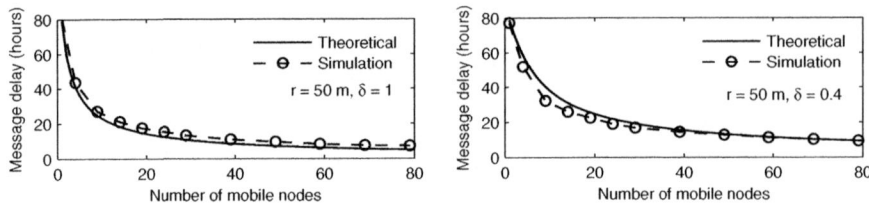

Fig. 4. Message delay vs. number of mobile nodes: Random walker mobility model

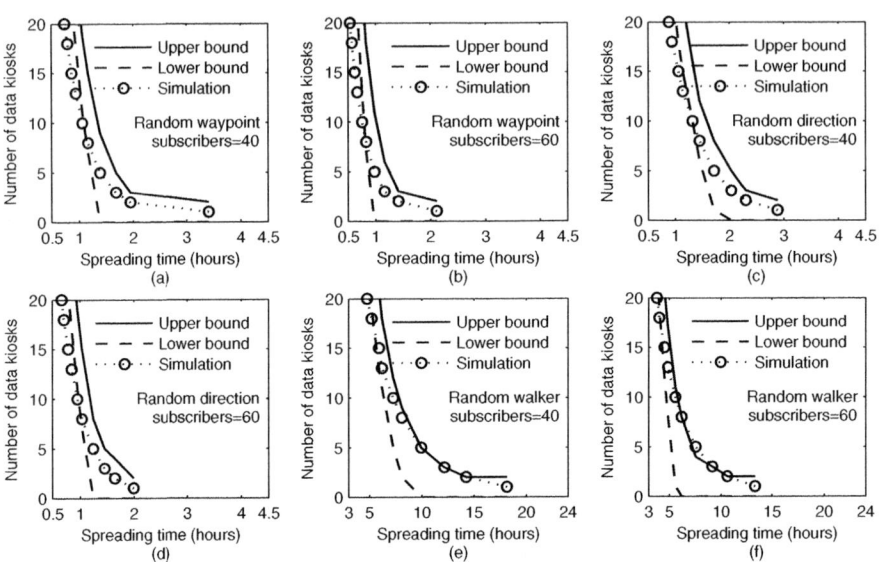

Fig. 5. Number of data kiosks vs. spreading time

scenario without data kiosk. The urban area is a square of size 4 km × 4 km. There are 40 mobile nodes moving under the random waypoint mobility model. The transmission range is 50 meters, the probability of interest is 1, as every node is interested in the content. Under this scenario, the expected message delay is roughly 2 hours (as in Fig. 3(a)). If we wish to spread the content to 90 percent of the subscribers within 1 hour, it appears difficult to satisfy this requirement in a network without data kiosk. We now consider an hybrid environment. We keep these settings and find the number of data kiosks that we should add to meet the requirement. As we see in Fig. 5(a), with 10 data kiosks we can lessen the expected spreading time to 1 hour. We also see that the spreading time is improved effectively with some data kiosks. For example, with 5 data kiosks the expected spreading time can decrease to 2 hours in the random waypoint mobility model (as in Fig. 5(a)), and approximately to less than 8 hours in the random walker mobility model (see Fig. 5(e-f)).

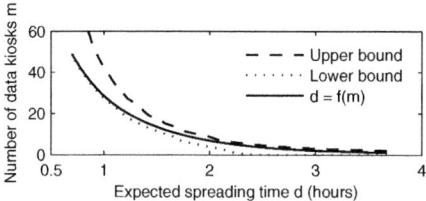

Fig. 6. Bounds of the number of data kiosks and the function of expected spreading time

The simulation results confirm that the analytical models can predict the expected message delay and the number of data kiosks to meet a delay requirement. The lower bound of the number of data kiosks gets close to the simulation results when the expected spreading time is small and the upper bound is accurate when the expected spreading time is large (Fig. 6).

6   Conclusion

In this paper, we investigated the practicality of dissemination of daily news over a DTN in an urban area. When the performance can not fulfill a requirement, a solution is to deploy a hybrid environment with data kiosks to overcome the limit of the performance of DTNs. We obtained both a closed-form and an approximation expression of the expected message delay. The results were used to analyze the influence of user's interest and contact rates on the performance of DTNs. If the interest of people does not vary over the spreading time, we compute the solution for the number of data kiosks that we need to deploy to satisfy an expected spreading time objective. The results show that the message delay in DTNs is in the range of several hours but can be significantly reduced thanks to the addition of data kiosks. One question that we wish to address in our future work is where data kiosks should be located to maximize the contact opportunity between data kiosks and mobile users, or to minimize the delay. The work can also be extended by taking into account interferences, packet sizes and transmission times.

Acknowledgments. This work was supported in part by the Agence nationale de la recherche - ANR - under the CROWD project: http://anr-crowd.lip6.fr/.

References

1. Abramowitz, M., Stegun, I.A.: Handbook of Mathematical Functions: with Formulas, Graphs, and Mathematical Tables. Dover Publications, New York (1965)
2. Al Hanbali, A., Kherani, A.A., Groenevelt, R., Nain, P., Altman, E.: Impact of mobility on the performance of relaying in ad hoc networks - extended version. Comput. Netw. 51(14), 4112–4130 (2007)

3. Al Hanbali, A., Nain, P., Altman, E.: Performance of ad hoc networks with two-hop relay routing and limited packet lifetime (extended version). Perform. Eval. 65(6-7), 463–483 (2008)
4. Banerjee, N., Corner, M.D., Towsley, D., Levine, B.N.: Relays, base stations, and meshes: enhancing mobile networks with infrastructure. In: MobiCom 2008: Proceedings of the 14th ACM International Conference on Mobile Computing and Networking, pp. 81–91. ACM, New York (2008)
5. Garyfalos, A., Almeroth, K.: Coupons: A multilevel incentive scheme for information dissemination in mobile networks. IEEE Transactions on Mobile Computing 7(6), 792–804 (2008)
6. Groenevelt, R., Nain, P., Koole, G.: The message delay in mobile ad hoc networks. Perform. Eval. 62(1-4), 210–228 (2005)
7. Ibrahim, M., Al Hanbali, A., Nain, P.: Delay and resource analysis in manets in presence of throwboxes. Perform. Eval. 64(9-12), 933–947 (2007)
8. Ioannidis, S., Chaintreau, A., Massoulié, L.: Optimal and scalable distribution of content updates over a mobile social network. In: INFOCOM 2009, pp. 1422–1430. IEEE, Los Alamitos (2009)
9. Karagiannis, T., Le Boudec, J.Y., Vojnović, M.: Power law and exponential decay of inter contact times between mobile devices. In: MobiCom 2007: Proceedings of the 13th Annual ACM International Conference on Mobile Computing and Networking, pp. 183–194. ACM, New York (2007)
10. Keränen, A., Ott, J., Kärkkäinen, T.: The ONE simulator for DTN protocol evaluation. In: Simutools 2009: Proceedings of the 2nd International Conference on Simulation Tools and Techniques. ICST (Institute for Computer Sciences, Social-Informatics and Telecommunications Engineering), pp. 1–10. ICST, Brussels (2009)
11. Lenders, V., May, M., Karlsson, G., Wacha, C.: Wireless ad hoc podcasting. SIGMOBILE Mob. Comput. Commun. Rev. 12, 65–67 (2008)
12. Liu, B., Thiran, P., Towsley, D.: Capacity of a wireless ad hoc network with infrastructure. In: MobiHoc 2007: Proceedings of the 8th ACM International Symposium on Mobile Ad Hoc Networking and Computing, pp. 239–246. ACM, New York (2007)
13. McNamara, L., Mascolo, C., Capra, L.: Media sharing based on colocation prediction in urban transport. In: MobiCom 2008: Proceedings of the 14th ACM International Conference on Mobile Computing and Networking, pp. 58–69. ACM, New York (2008)
14. Pham, T.M., Fdida, S.: Delay estimation of a user-preferred content distribution scheme in disruption tolerant networks. In: AINTEC 2009: Asian Internet Engineering Conference, pp. 3–10. ACM, New York (2009)
15. Pham, T.M., Fdida, S.: DTN support for news dissemination in an urban area. Technical report hal-00565307, UPMC (2011), http://hal.upmc.fr/hal-00565307/en/
16. Polat, B.K., Sachdeva, P., Ammar, M.H., Zegura, E.W.: Message ferries as generalized dominating sets in intermittently connected mobile networks. In: MobiOpp 2010: Proceedings of the Second International Workshop on Mobile Opportunistic Networking, pp. 22–31. ACM, New York (2010)
17. Robinson, J., Singh, M., Swaminathan, R., Knightly, E.: Deploying mesh nodes under non-uniform propagation. In: Proceedings of the 29th Conference on Information Communications, INFOCOM 2010, pp. 2142–2150. IEEE Press, Piscataway (2010)

18. Sharma, G., Mazumdar, R., Shroff, N.B.: Delay and capacity trade-offs in mobile ad hoc networks: a global perspective. IEEE/ACM Trans. Netw. 15(5), 981–992 (2007)
19. Spyropoulos, T., Psounis, K., Raghavendra, C.S.: Spray and wait: an efficient routing scheme for intermittently connected mobile networks. In: WDTN 2005: Proceedings of the 2005 ACM SIGCOMM Workshop on Delay-tolerant Networking, pp. 252–259. ACM, New York (2005)
20. Zemlianov, A., Gustavo de, V.: Capacity of ad hoc wireless networks with infrastructure support. IEEE Journal on Selected Areas in Communications 23(3), 657–667 (2005)
21. Zhang, X., Neglia, G., Kurose, J., Towsley, D.: Performance modeling of epidemic routing. Comput. Netw. 51(10), 2867–2891 (2007)

Stochastic Scheduling for Underwater Sensor Networks

Dimitri Marinakis, Kui Wu, and Sue Whitesides

School of Computer Science,
University of Victoria, Canada
dmarinak@kinsolresearch.com, wkui@ieee.org, sue@uvic.ca

Abstract. The context of underwater sensor networks (UWSNs) presents special challenges for data transmission. For that context, we examine the merit of using a simple, stochastic transmission strategy based on the ALOHA protocol. The strategy uses a stochastic scheduling approach in which time is slotted, and each network node broadcasts according to some probability during each time slot. We present a closed-form solution to an objective function that guides the assignment of the broadcast probabilities with respect to overall network reliability. We propose an easily distributed heuristic based on local network density and evaluate our approach using numerical simulations. The evaluation results show that even without using explicit control signalling, our simple stochastic scheduling method performs well for data transmission in UWSNs.

Keywords: underwater senor networks, slotted ALOHA, network reliability, MAC protocols.

1 Introduction

The monitoring and exploration of the ocean is of great importance to the sustainable and environmentally sound development of the Earth. Activities such as oceanographic data collection, offshore exploration, and ocean ecosystem monitoring are facilitated by the deployment of Unmanned or Autonomous Underwater Vehicles (UUVs, AUVs), equipped with underwater sensors, *e.g.*, see Kennedy *et al.* [5]. There is a growing demand for UUVs/AUVs that cooperate to perform monitoring tasks; e.g., a fleet of small, inexpensive underwater AUVs for monitoring underwater waste sites was suggested by Nawaz *et al.* [15]. To cooperate effectively, the nodes must be able to exchange data and control messages with one other.

Underwater communication, however, is itself a challenging area of active research (*e.g.*, [12], [6]). Radio waves propagate underwater only at very low frequencies (e.g., 30-300 HZ) and require high transmission power which generally cannot be afforded on board by UUVs/AUVs. Underwater, optical waves are affected by scattering effects and cannot be used to transmit over long distances. So far, acoustic communication has been the physical layer of choice for underwater communication. Underwater acoustic communication, however, is subject to

J. Domingo-Pascual et al. (Eds.): NETWORKING 2011, Part I, LNCS 6640, pp. 134–146, 2011.

large propagation latency, low bandwidth, high bit error rate (BER), and complex multipath fading. To make the situation worse, there can be large variations in temperature, salinity, and pressure over short distances in the underwater environment, all of which can significantly impact acoustic propagation.

In existing RF communication systems, Medium Access Control (MAC) protocols are used to resolve contention issues in medium access. As a basic requirement, a MAC protocol should be able to find a transmission scheduling scheme that eliminates or minimizes conflicting transmissions. To achieve this, an implicit control mechanism (*e.g.*, Time Division Multiple Access (TDMA)), or explicit control messages (*e.g.*, Request-To-Send (RTS) and Clear-To-Send (CTS) messages in Carrier Sense Multiple Access (CSMA) based protocols), are adopted. Simulation studies have shown that the RTS/CTS based control, which alleviates hidden/exposed terminal problems and improves network throughput [4], actually degrades throughput when the propagation delay becomes large [21].

Interestingly, it has recently been shown that MAC protocols based on the relatively simple ALOHA protocol [1] perform well in an underwater, multi-hop environment in which there are significant propagation delays. Recently, Syed *et al.* [19] modified the slotted ALOHA protocol for underwater, acoustic communication so that ALOHA could achieve a throughput comparable to what it achieves in RF networks. In related work, Petrioli *et al.* [11] evaluated various MAC protocols for underwater sensor networks and found that in multi-hop, underwater acoustic networks, ALOHA variants out-performed protocols in which there were larger overhead costs. Additionally, simulations reported by Zhou *et al.* [21] demonstrated that random ALOHA schemes can provide stable performance in UWSNs.

There is no 'one-size fits all' MAC layer appropriate for all underwater applications. To date, no MAC protocol has been commonly accepted as an industrial standard for UWSNs. For example Partan *et al.* [10] state that medium access is an unresolved problem in underwater acoustic networks. Thus, it is likely that, in specific application areas, lightweight, ALOHA-based MAC protocols will have a place in underwater networking.

Motivated by the role we believe ALOHA based MAC protocols will play in UWSNs, we have examined the merit of a simple, stochastic transmission strategy based on the ALOHA protocol: Time is slotted, and at the beginning of each time slot, each node in the network is assigned a probability for transmission. Such a simple link scheduling method is easy to implement and requires virtually no control overhead. Therefore, we propose that stochastic variants of slotted ALOHA such as the protocol we present here could be used for networking mobile underwater devices. In this context, the network topology may change dynamically, and any energy wasted in colliding transmissions is inconsequential relative to the power requirements of the actuators on the AUVs. Additionally, one key communication requirement is to deliver relatively continuous, but low bandwidth data among proximal nodes for navigation and coordination purposes.

Main Contributions: We lay the groundwork for exploring whether stochastic scheduling for lightweight, ALOHA based MAC variants might provide suitable solutions for underwater network communication challenges. Specifically, for the stochastic variant of slotted ALOHA we proposed above:

- we consider how the transmission probability of a node should be adjusted based on its local (at a given time) communication topology in order to obtain good overall network performance;

- we present a closed form solution to an objective function for assigning the transmission probabilities that is aimed at improving network performance in terms of reliability;

- we show that a heuristic based on the optimizing values of our objective function is easily distributed and shows good performance in simulations.

2   Related Work

2.1   Underwater MAC Protocols

A number of MAC protocols have been proposed to handle the special conditions encountered in underwater multi-hop sensor networks, *e.g.*, [8], [17], [9], [13]. Despite the disadvantages of the inherent propagation delays, a number of modern MAC protocols proposed for underwater communication nevertheless rely on the exchange of handshaking contol messages for medium access. For example, Slotted FAMA, as presented by Molins and Stojanovic [8], is based on carrier sensing. Each network node constantly listens to the channel, but stays idle unless it has permission to transmit, which is granted via an RTS / CTS handshaking mechanism. Collisions are handled through a random back off scheme. Simulations demonstrate that this protocol has promise for underwater mobile networks, although the authors consider an application in which the data packets exchanged are much larger than the control packets used for the handshaking. In this situation, the disadvantage of significant propagation delays when employing handshaking for collision avoidance is somewhat masked. The suitability of this approach for an application in which many small data packets are exchanged on a frequent basis is not clear.

The T-Lohi MAC introduced by Syed *et al.* [17] also employs a synchronized transmission frame with a handshaking scheme for collision avoidance. Unlike Slotted FAMA, however, the protocol allows network nodes to sleep for energy saving purposes. When a node using the T-Lohi protocol is ready to send data, it attempts to reserve the channel by sending a control message (a tone) during a reservation period. If the node does not hear one or more tones from other nodes during this reservation period then it is clear to send; otherwise it backs off and waits. Energy savings through sleeping are achieved by using custom acoustic hardware that triggers the node to wake up when the tone is detected.

Considerable research has demonstrated the promise of underwater MAC layers that incorporate CDMA; *e.g.* the work of Pompili *et al.* [13], the work of Page and Stojanovic [3], and the work of Tan and Seah [20] . The approach is particularly suited for some challenging application areas, such as shallow water operation where multi-path interference is a major factor. In other applications however, *e.g.* where congestion issues dominate, the operational simplicity of ALOHA schemes can be attractive.

In contrast to CDMA based approaches, Slotted FAMA and T-Lohi, the UWAN-MAC protocol presented by Park and Rodoplu [9] does not employ a handshaking mechanism using control messages to reserve channel access. When using UWAN-Mac, each node transmits infrequently but regularly, with a randomly selected offset. The schedule of a node's neighour is learned via synchronization packets sent during an initialization period. The approach achieves energy savings by finding locally synchronized schedules such that network nodes can sleep during idle periods. Although there is no explicit method for avoiding collisions, the collisions are shown to be rare. The approach relies, however, on a static network in which the transmission delays between any pair of nodes remain roughly constant. The UWAN-MAC approach has some similarity in spirit to the stochastic scheduling we consider in this paper, and it should be possible to modify UWAN-MAC to benefit from our analysis, *e.g.*, by adapting the duty cycle of each node based on local network density.

2.2   Stochastic Scheduling

The class of problems related to assigning a slot to each node in a wireless network for the purposes of collision avoidance is referred to as *broadcast scheduling*. Such problems were considered as early as the mid-eighties by Chlamtac and Kutten [2], for example. Later in that decade, Ramaswami and Parhi [14] showed that the problem of finding a minimum length schedule that allows each node to hear from each neighbour is NP-complete and presented effective heuristics.

In previous work [7], Marinakis and Whitesides presented a slotted stochastic transmission strategy which they compared to broadcast scheduling in the context of an alarm network. They addressed the question of how a network of nodes might signal the occurrence of an event capable of disabling the sensors. The approach called for the nodes to exchange messages regularly during normal operation, but to signal the occurrence of an alarm event by ceasing to transmit.

In the work we present in this paper, we consider in detail the merit of such stochastic scheduling approaches for underwater sensor network applications.

3   Network Model

We model the multi-hop communication links available between the network nodes at any instant in time as a directed graph $G = (V, E)$ in which each vertex $v \in V$ represents a network node and each edge $e_{ij} \in E$ denotes a *potential* communication link from node i to node j, *i.e.*, node i can transmit data to node j if and only if $e_{ij} \in E$ in a selected channel.

We make the following assumptions on data communication:

– A node that is transmitting may not receive at the same time. If a node is tuned to receive on a channel m, then a packet can be received if and only if exactly one of its neighbors is transmitting on that channel. This constraint

provides a simple way to model collision issues such as the hidden terminal
problem.

– Time is slotted, and at the beginning of each time slot, a node selects a
  channel m at random (e.g., uniformly). It then transmits on channel m with
  a given probability which is determined according to various performance
  goals. If the node does not transmit, then it tunes its acoustic transceiver to
  receive on channel m.

– All nodes maintain synchronized clocks and may select to time their communications to occur during a particular slot. Note that this assumption is
  common, and there are a number of techniques that could be used to accomplish this task; see Syed and Heidemann [18] for an example of a time
  synchronization method appropriate for acoustic networks, and see Sivrikaya
  and Yener [16] for a more general survey of time synchronization techniques
  in wireless sensor networks.

We will refer to this approach as *Stochastic Scheduling*.

4   Stochastic Scheduling on a Single Channel

In this section, we analyze the single channel case. Rather than handling the collision issue by an assignment of deterministic schedules, instead we propose to
assign to each node in the network a time slot transmission probability. Thus we
want to specify a set of appropriate values $P = \{p_i\}, \forall i \in V$. Since the goal of our
analysis will be to obtain good heuristics that can be used to design simple and
distributed scheduling, we ignore (at first) the propagation delay to ease the analysis; later, in our simulation study, we evaluate the impact of propagation delay.

4.1   Basic Constraints and General Guidelines

As a preliminary, we consider the impact of P on the probability of one node
communicating with another. To this end, we define a throughput graph corresponding to a given network.

Definition 1. *The **throughput graph** of a given network $G = (V, E)$ is a
weighted, directed graph, denoted by $G' = (V, E, R)$, where R denotes the set
of weights r_{ij} on the corresponding edges e_{ij}. The weight r_{ij} of an edge e_{ij}
corresponds to the probability that node j receives a message from a neighbouring
node i during a given time slot.*

We call G' the throughput graph because the weight assigned to the directed edge
e_{ij} is proportional to the amount of data across that link in a long run. Based on
the second assumption in the network model in Section 3, it is easy to obtain:

$$r_{ij} = p_i(1 - p_j) \prod_{k \in N(j), k \neq i} (1 - p_k), e_{ij} \in E \tag{1}$$

where $N(x)$ denotes the neighbours of x in G.

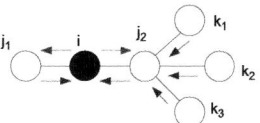

Fig. 1. Example graph showing the influence of p_i. If p_i is adjusted upwards, then the throughput across links (i, j_1) and (i, j_2) will be increased, while that of the links (j_1, i), (j_2, i), (k_1, j_2), (k_2, j_2) and (k_3, j_2) will be decreased.

Equation (1) captures the basic constraint on the value assignment in P. The impact of this constraint is illustrated in Figure 1. In particular, if a node's time slot transmission probability is increased, the long-term throughput from this node to its neighboring nodes will be increased, which may lead to lower transmission opportunities at the neighbours.

4.2   An Objective Function for the Assignment of P Values

We define a natural objective function, Q_r and show how to use it to assign P values.

Definition 2. *We define the **overall reliability** Q_r of network $G = (V, E)$ as a function evaluated over the corresponding throughput graph $G' = (V, E, R)$:*

$$Q_r = \prod_{r_{ij} \in R} r_{ij}. \tag{2}$$

We call Q_r the overall network reliability, because in the long run Q_r represents the chance that a data packet could be successfully routed along an arbitrary path q in $|q|$ time slots, where $|q|$ is the length of the path.

4.3   Maximizing Overall Network Reliability

We will now show how to assign the values in $P = \{p_i\}, \forall i \in V$, to maximize the overall network reliability Q_r. Since Q_r is a function of P, we rewrite it as: $Q_r(P) = \prod_{i,j \in R} r_{ij}$. By taking the log of both sides we get:

$$Q_r'(P) = \sum_{i,j \in R} \log r_{ji},$$

where $Q_r'(P) = \log Q_r(P)$.

Now we would like to find the values of P that maximize Q_r'. We proceed by considering the partial derivatives of Q_r' with respect to the value of $p_i \in P$:

$$\nabla Q_r' = \left(\frac{\partial Q_r'}{\partial p_1}, \frac{\partial Q_r'}{\partial p_2}, \cdots \frac{\partial Q_r'}{\partial p_n} \right)$$

where $n = |V|$. A single partial then becomes:

$$\frac{\partial Q_r'}{\partial p_i} = \sum_{e_{ij} \in E} \frac{1}{r_{ij}} \frac{\partial r_{ij}}{\partial p_i} \quad . \tag{3}$$

The partials for p_i are only non-zero, however, for $r_{ij}, j \in N(i)$ and $r_{ji}, j \in N(i)$ and $r_{kj}, k \in N(j), k \neq i$, based on the basic constraint in Equation (1). We can now consider the partial of a single weight value with respect to p_i. In particular, for the outbound links from i to j:

$$\frac{\partial r_{ij}}{\partial p_i} = \frac{\partial}{\partial p_i} p_i (1 - p_j) \prod_{k \in N(j), k \neq i} (1 - p_k) = \frac{r_{ij}}{p_i} \quad . \tag{4}$$

For the inbound links from j to i where $j \in N(i)$ we have:

$$\frac{\partial r_{ji}}{\partial p_i} = \frac{\partial}{\partial p_i} (1 - p_i) p_j \prod_{k \in N(j), k \neq i} (1 - p_k) = \frac{-r_{ji}}{1 - p_i}, \tag{5}$$

and similarly, for the links from k to j where $k \in N(j), k \neq i$ we have:

$$\frac{\partial r_{kj}}{\partial p_i} = \frac{\partial}{\partial p_i} (1 - p_i)(1 - p_j) \prod_{k \in N(j), k \neq i} (1 - p_k) = \frac{-r_{kj}}{1 - p_i} \quad . \tag{6}$$

Let us denote the set of links for which there exists a non-zero partial with respect to p_i as R_i. This set can be described as all links that have their tail endpoints adjacent either to the vertex i or to one of its neighbours: $R_i = \{r_{kl}\}, k \in \{N(i) \cup i\}$.

We can further categorize the directed links affected by p_i into those with a positive partial derivative: $R_{i+} = \{r_{kl}\}, k \in N(i)$, and those with a negative partial derivative: $R_{i-} = R_i \setminus R_{i+}$. Let δ_i denote the degree of node $i \in V$ and let $\bar{\delta}_i$ denote the sum of the degrees of all the neighbours of i: $\bar{\delta}_i = \sum_{j \in N(i)} \delta_j$. It is easy to see that $|R_{i+}| = \delta_i$ and $|R_{i-}| = \bar{\delta}_i$.

Let us now return to Equation (3) and take the partial derivative of Q'_r with respect to p_i:

$$\frac{\partial Q'_r}{\partial p_i} = \sum_{i,j \in R} \frac{\partial r_{ij}}{\partial p_i} \frac{1}{r_{ij}} = \sum_{i,j \in R_{i+}} \frac{\partial r_{ij}}{\partial p_i} \frac{1}{r_{ij}} + \sum_{i,j \in R_{i-}} \frac{\partial r_{ij}}{\partial p_i} \frac{1}{r_{ij}}$$

$$= \sum_{i,j \in R_{i+}} \frac{r_{ij}}{p_i} \frac{1}{r_{ij}} + \sum_{i,j \in R_{i-}} \frac{-r_{ij}}{1 - p_i} \frac{1}{r_{ij}} = \frac{\delta_i}{p_i} - \frac{\bar{\delta}_i}{1 - p_i} \quad . \tag{7}$$

Since the partial derivative of our objective function Q'_r with respect to a single p_i does not depend on the other elements of P, we can set each partial to zero in order to find the value of each p_i that will maximize Q'_r and of course Q_r as well: $p_i = \delta_i/(\delta_i + \bar{\delta}_i)$. If we now let $\Delta_i = \bar{\delta}_i/\delta_i$ be the *average degree of i's neighbours* we can express the above closed form result as:

$$p_i = \frac{1}{1 + \Delta_i} \quad . \tag{8}$$

5   Stochastic Scheduling on Multiple Channels

In the case where we have $M(M > 1)$ frequency channels, the value for R in G' will take on slightly different values. We can calculate the value of r_{ij} as follows:

$$r_{ij} = \frac{1}{M}p_i\frac{1}{M}(1-p_j) \prod_{k\in N(j),k\neq i} (1 - \frac{1}{M}p_k), e_{ij} \in E \tag{9}$$

where $N(x)$ denotes the neighbours of x in G. This calculation is analogous to Equation (1) for the single frequency variant.

Following the same analysis steps as in the single channel case, we arrive at the following values for P that maximize overall network reliability:

$$p_i = \frac{2\delta_i M + \bar{\delta}_i - \sqrt{\bar{\delta}_i^{\,2} + 4\delta_i M(M-1)}}{2(\delta_i + \bar{\delta}_i)} \ . \tag{10}$$

Due to space limitations we omit the derivation.

6   A Distributed Algorithm for Stochastic Scheduling

The analytical results in the previous sections provide us with heuristics for for designing a simple, distributed algorithm for Stochastic Scheduling. In this section, we present a distributed algorithm that is aimed at maximizing overall network reliability. It is straightforward to assign P values in a distributed manner in order to optimize overall network reliability. Equations (8) and (10) depend only on the knowledge of the local communication topology and the number of frequency channels employed. We will refer to this heuristic for assigning p values as the *Average Neighbourhood Degree Heuristic (ANDH)* .

For the single channel case we present the following distributed algorithm for selecting a suitable value of the transmission probability of an individual node:

1. During initial deployment, a default value for p_i can be assigned to each node given a rough estimate of the typical network density.
2. Each node maintains a neighbour table with one entry for each of its neighbours. Each neighbour table entry includes the unique media access control (MAC) identification of the neighbour, along with the timestamp and the number of neighbours reported by that neighbour.
3. Each node exchanges neighbour count estimates with each of its neighbours, and then updates its transmission probability p_i accordingly.
4. At each time slot, with probability p_i, on a channel selected uniformly at random, a node broadcasts its unique media access control (MAC) identification, the number of entries in its neighbour table, and any data payload.
5. If not transmitting, the node tunes its receiver to a channel selected uniformly at random, and if it receives a message, it adds the appropriate details to its neighbour table and updates its p_i values according to Equation (8).
6. (Optional if the network topology changes, e.g., in a mobile environment) Entries older than a threshold ω may be discarded. In addition, the local p_i value could be smoothed by using, for example, exponential averaging or a similar technique suitable for a low powered platform.

7   Performance Evaluation

We perform simulation studies to evaluate our Stochastic Scheduling approach. In the simulation, we build the network topology using *disk graphs*. The graphs are obtained by selecting points uniformly at random in a region of the plane bounded by a circle of diameter D as the locations of the network nodes. An edge is then assigned between any two vertices if the pair-wise distance between their associated locations is less than a given fraction α of the deployment diameter D. Since the parameter α can be used to control the number of communication links of the network, we call it the *communication ratio*. We assign a delay to an edge proportional to the distance between the pairs (but rounded to a discrete value for ease of simulation). To save space, we only show performance results over a single channel.

7.1   Performance under Varying Load and Propagation Delay

As a general test of network performance we simulated an application in which each device, at random intervals, broadcasts a data packet to each of its neighbours. For this test, we controlled the traffic load by assigning each node a probability that a data packet is passed down to its MAC layer at the beginning of each time slot. We selected T-Lohi [17] for comparison because this protocol represents a typical example of using lightweight control packets (*i.e.* it uses a RTS control packet only). Our implementation of the T-Lohi algorithm is as described in 'Algorithm 1' of [17] using a single time slot as the duration of a contention round.

As shown in Figure 2(a), T-Lohi performs better than our Stochastic Scheduling under light traffic load, because T-Lohi employs control messages for media access. Nevertheless, our Stochastic Scheduling approach performs much better under heavy traffic load. This demonstrates that using control packets may not bring benefits for underwater acoustic communications, because the propagation delay is determined by the pair-wise distance of two communicating nodes, whose locations may be random. Such a randomness offsets the benefit of using control packets.

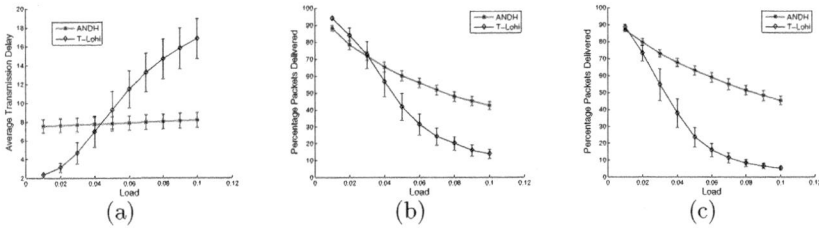

Fig. 2. (a) Average time from queuing a packet until its delivery (T-Lohi *vs.* Stochastic Scheduling using the ANDH); (b) Percentage of successful packet transmissions over *all* links as a function of traffic load; and (c) the same as (b) but under the condition of long propagation delay. Results were averaged over 100 trials of 100-node networks using a communication ratio $\alpha = 0.25$. Error bars depict one standard deviation.

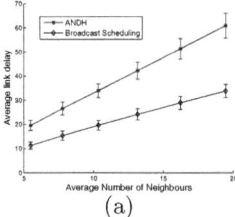 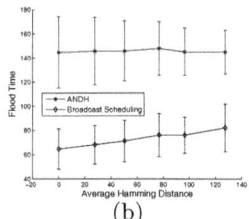 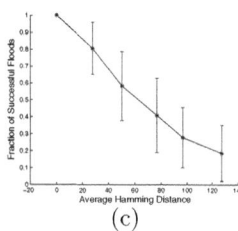

(a)   (b)   (c)

Fig. 3. (a) Average link delay in units of time slots for networks of different node densities. Results were averaged over 100 trials of 100-node networks for each node density. (b) Mean flood time in dynamic networks for *successful floods only* and (c) fraction of broadcast floods that reached all nodes (i.e., succeeded) with broadcast scheduling. Note that all ANDH floods succeeded in all trials. Results are averaged over twenty 100-node networks generated with a communication ratio $\alpha = 0.25$. Error bars depict one standard deviation for all plots.

In Figure 2(a), in order to investigate the average transmission delay, we assumed a very large queue size to avoid buffer overflow. To further study how many packets could be delivered within a given time constraint, we changed the queue size to one so that a packet is considered undelivered, should it be passed down to the MAC layer when the queue is full. Figure 2(b) shows the average percentage of a node's neighbors that receive a packet sent by that node, under different traffic loads. We can also see that the performance of our Stochastic Scheduling with the ANDH outperforms T-Lohi when traffic load becomes heavy.

The results shown in Figures 2(a) and (b) were obtained by setting the propagation delay much smaller than the length of a time slot. We then increased the propagation delay considerably to further investigate its impact. In particular, we set the max propagation delay in the network to three times the length of a time slot. Figure 2(c) shows the results under this set of tests. Comparing Figure 2(b) and Figure 2(c), it can be seen that Stochastic Scheduling is almost unaffected, but the T-Lohi protocol suffers considerably. This is due to the fact that the RTS control packets used by the T-Lohi protocol may fail to reserve the channel due to propagation latency.

7.2   Performance under Dynamic Network Conditions

Although it can be seen that Stochastic Scheduling has some advantages over protocols that rely on handshaking under conditions such as heavy load, we also investigated how it compares to a TDMA based, deterministic scheduling approach.

For this test, we used *Hamming distance* as a metric to measure the dynamic change in the network topology. Assume that the adjacency matrix of graph $G_1 = (V_1, E_1)$ is $A_1 = [a_{ij}]$ and the adjacency matrix of graph $G_2 = (V_2, E_2)$ is $A_2 = [b_{ij}]$, where $V_1 = V_2$ and $|V_1| = |V_2| = n$. The *Hamming distance* of the two graphs is defined as $\sum_{i,j=1}^{n}(a_{ij} - b_{ij})^2$.

For the deterministic scheduling, we implemented the broadcast schedule obtained with the centralized heuristic described in [14]. Simulations suggest that

Stochastic Scheduling results in a poorer average latency than a deterministic transmission schedule when static networks are considered, as shown in plots (a) and (b) of Figure 3.

To test the performance under dynamic networks, we generated networks of 100 nodes, using a communication ratio $\alpha = 0.25$. For each network, the ANDH P values in our Stochastic Scheduling and the deterministic broadcast schedules were determined. Then the edges of the network were changed to make it evolve to a new network, such that the Hamming distance of the two networks was larger than a given value. A global message broadcast was then simulated originating from a single node on this new network, but with transmission schedules based on the old network. We considered how long it would take for a particular piece of information from a single node to broadcast throughout the new network. Plots (b) and (c) of Figure 3 show the results of the flood simulation in which we tested the robustness of stochastic and deterministic approaches to changes in the communication network. It can be seen that although there is a lot of variability in flood times, the Stochastic Scheduling approach continues to function even if the network topology changes. The deterministic approach, on the other hand, depends on a specific topology and fails catastrophically once the underlying communication graph shifts.

To summarize, the stochastic nature of our approach makes it well suited for applications where propagation delay is long and random. For such applications, deterministic scheduling or protocols that rely on control packets to resolve medium contention may not perform well when traffic load becomes heavy.

8   Conclusions and Future Work

We have presented and evaluated the concept of using a lightweight variant of slotted ALOHA in conjunction with a Stochastic Scheduling approach in which network nodes transmit according to some time slot transmission probability. We have obtained a closed-form solution for the assignment of transmission probabilities that maximize overall network reliability and have presented an easily distributed algorithm for assigning these optimal values. Performance results demonstrate that even without using any control signaling, our simple Stochastic Scheduling works well for the context of UWSNs, where propagation delay is not negligible.

In future work we will extend our analysis to incorporate propagation delay (link lengths). This will involve adding a temporal aspect to our analysis beyond the scale of a single communication slot. Additionally, we plan to further investigate our approach though experiments using custom acoustic communication hardware.

Acknowledgements. This research was partially supported by Natural Sciences and Engineering Research Council of Canada (NSERC). We thank the anonymous reviewers for their comments. They have influenced the direction of our future work.

References

1. Abramson, N.: Development of the alohanet. IEEE Transactions on Information Theory 31(2), 119–123 (1985)
2. Chlamtac, I., Kutten, S.: A spatial reuse TDMA/FDMA for mobile multi-hop radio networks. In: INFOCOM (March 1985)
3. Page, E.C., Stojanovic, M.: Efficient channel-estimation-based multi-user detection for underwater acoustic CDMA systems. IEEE Journal of Oceanic Engineering 33(4), 502–512 (2008)
4. Fullmer, C.L., Garcia-Luna-Aceves, J.: Solutions to hidden terminal problems in wireless networks. In: SIGCOMM, pp. 39–49 (1997)
5. Kennedy, J., Gamroth, E., Bradley, C., Proctor, A.A., Heard, G.J.: Decoupled modelling and controller design for the hybrid autonomous underwater vehicle: Maco. Int. Journal of the Society for Underwater Technology 27(1), 11–21 (2007)
6. Liu, L., Zhou, S., Cui, J.: Prospects and problems of wireless communications for underwater sensor networks. Wireless Communications and Mobile Computing 8(8), 977–994 (2008)
7. Marinakis, D., Whitesides, S.: The sentinel problem for a multi-hop sensor network. In: Canadian Conference on Computer and Robot Vision, Ottawa, Canada (May 2010)
8. Molins, M., Stojanovic, M.: Slotted FAMA: A MAC protocol for underwater acoustic networks. In: IEEE Oceans 2006 Conference, Singapore (2006)
9. Park, M.K., Rodoplu, V.: UWAN-MAC: An energy-efficient mac protocol for underwater acoustic wireless sensor networks. IEEE Journal of Oceanic Engineering 32(3), 710–721 (2007)
10. Partan, J., Kurose, J., Levine, B.N.: A survey of practical issues in underwater networks. ACM SIGMOBILE Mobile Computing and Communications Review 11(4), 23–33 (2007)
11. Petrioli, C., Petroccia, R., Stojanovic, M.: A comparative performance evaluation of mac protocols for underwater sensor networks. In: IEEE Oceans 2008 Conference, Quebec City, Canada (September 2008)
12. Pompili, D., Akyildiz, I.F.: Overview of networking protocols for underwater wireless communications. IEEE Communications Magazine (January 2009)
13. Pompili, D., Melodia, T., Akyildiz, I.: A distributed cdma medium access control for underwater acoustic sensor networks. In (Med-Hoc-Net), Corfu, Greece (June 2007)
14. Ramaswami, R., Parhi, K.: Distributed scheduling of broadcasts in a radio network. In: INFOCOM, vol. 2, pp. 497–504 (1989)
15. Nawaz, S., Muzammil Hussain, S.W.N.T., Green, P.N.: An underwater robotic network for monitoring nuclear waste storage pools. In: 1st Int. ICST Conference on Sensor Systems and Software, SCUBE (2009)
16. Sivrikaya, F., Yener, B.: Time synchronization in sensor networks: a survey. IEEE Network 18(4), 45–50 (2004)
17. Syed, A., Ye, W., Heidemann, J.: T-Lohi: A new class of MAC protocols for underwater acoustic sensor networks. In: INFOCOM, Phoenix, Arizona (April 2008)
18. Syed, A., Heidemann, J.: Time synchronization for high latency acoustic networks. In: INFOCOM. IEEE, Barcelona (2006)
19. Syed, A., Ye, W., Krishnamachari, B., Heidemann, J.: Understanding spatio-temporal uncertainty in medium access with ALOHA protocols. In: WUWNet, Montréal, Quebec, Canada, pp. 41–48 (September 2007)

20. Tan, H.X., Seah, W.K.: Distributed cdma-based mac protocol for underwater sensor networks. In: IEEE Conf. on Local Computer Networks, Los Alamitos, CA, USA, pp. 26–36 (2007)
21. Zhou, Z., Peng, Z., Cui, J.: Multi-channel mac protocols for underwater acoustic sensor networks. In: WUWNet (September 2008)

Using SensLAB* as a First Class Scientific Tool for Large Scale Wireless Sensor Network Experiments

Clément Burin des Roziers[2], Guillaume Chelius[2,1,5], Tony Ducrocq[2],
Eric Fleury[1,2,5], Antoine Fraboulet[3,2,5], Antoine Gallais[4], Nathalie Mitton[2],
Thomas Noël[4], and Julien Vandaele[2]

[1] ENS de Lyon. 15 parvis René Descartes - BP 7000 69342 Lyon Cedex 07 - France
[2] INRIA
firstName.lastName@inria.fr
[3] INSA de Lyon
firstName.lastName@insa-lyon.fr
[4] Université de Strasbourg
firstName.lastName@unistra.fr
[5] Université de Lyon

Abstract. This paper presents a description of SensLAB(Very Large
Scale Open Wireless Sensor Network Testbed) that has been developed
and deployed in order to allow the evaluation through experimentations
of scalable wireless sensor network protocols and applications. SensLAB's
main and most important goal is to offer an accurate open access multi-
users scientific tool to support the design, the development tuning, and
the experimentation of real large-scale sensor network applications. The
SensLAB testbed is composed of 1024 nodes over 4 sites. Each site hosts
256 sensor nodes with specific characteristics in order to offer a wide
spectrum of possibilities and heterogeneity. Within a given site, each one
of the 256 nodes is able both to communicate via its radio interface to its
neighbors and to be configured as a sink node to exchange data with any
other "*sink node*". The hardware and software architectures that allow to
reserve, configure, deploy firmwares and gather experimental data and
monitoring information are described. We also present demonstration
examples to illustrate the use of the SensLAB testbed and encourage
researchers to test and benchmark their applications/protocols on a large
scale WSN testbed.

Keywords: Wireless Sensor Network, Testbed, Radio, Monitoring,
Experiments.

1 Introduction

Wireless sensor networks (WSN) have emerged as a premier research topic. In the
industrial domain, wireless sensor networks are opening up machine-to-machine

---

* Supported by the French ANR/VERSO program. http://www.senslab.info

J. Domingo-Pascual et al. (Eds.): NETWORKING 2011, Part I, LNCS 6640, pp. 147–159, 2011.
© IFIP International Federation for Information Processing 2011

(M2M) communications paradigms. However, due to their massively distributed nature, the design, the implementation, and the evaluation of sensor network applications, middleware and communication protocols are tedious and really time-consuming tasks. It appears strategic and crucial to offer to researchers and developers accurate software tools, physical large scale testbeds to benchmark and optimize their applications and services. Simulations remain an important phase during the design and the provisioning step. However, they suffer from several imperfections as simulation makes artificial assumptions on radio propagation, traffic, failure patterns, and topologies [6,8]. As proposed by initiatives in Europe and worldwide, enabling *"open wireless multi-users experiment facility testbeds"*, will foster the emergence of the Future Internet and would be increasingly important for the research community. There is an increasing demand among researchers, industrials and production system architects to access testbed resources they need to conduct their experiments.

In this paper, we describe SensLAB, an open access multi-user WSN testbed (see Figure 1), which has been designed and deployed to answer all these needs (Section 2). SensLAB provides appropriate tools, methods, experimental facilities for testing and managing large scale wireless sensor network applications. As such, it lowers the entry cost to experimentation, often considered as a complex and heavyweight activity, with no extra management burden, accelerating proof-of-concept evaluation and competitiveness. We first describe the hardware and software architectures of the platform (Section 3), then we show how easy and efficient it is to use it through a couple of application examples (Section 4).

2   Context and Design Requirements

One barrier to the widespread use of wireless sensor networks is the difficulty of coaxing reliable information from nodes whose batteries are small, whose wireless medium is sporadic, etc. It is thus very important to conduct *in situ* experiments and researches to better understand the characteristics and compensate for some of these flaws and reach the state of maturity to make them practical. Unfortunately, the development and testing of real experiments quickly become a nightmare if the number of nodes exceeds a few dozens: *(i)* sensors are **small** devices with very **limited capacities** in terms of debugging and friendly programming; *(ii)* software deployment and debugging require the connection of the device yielding to **individual manipulations** of each single node; *(iii)* sensors are generally powered by a **battery** which has limited lifetime, etc.

When the SensLAB project was initiated in 2005/2006, the number of WSN testbeds was limited or did not met the requirements we did have in mind (independence from any specific OS/language, accurate consumption monitoring, reproducible experiments). Nowadays, other WSN testbeds like moteLab [10][1], Kansei [1][2], WASAL[3] or TWIST [4] exist. Our goal is not to compare here all the features of all platforms. Note that it will however really valuable to do so

---

[1] http://motelab.eecs.harvard.edu/

[2] http://ceti.cse.ohio-state.edu/kansei/

[3] http://wasal.epfl.ch/

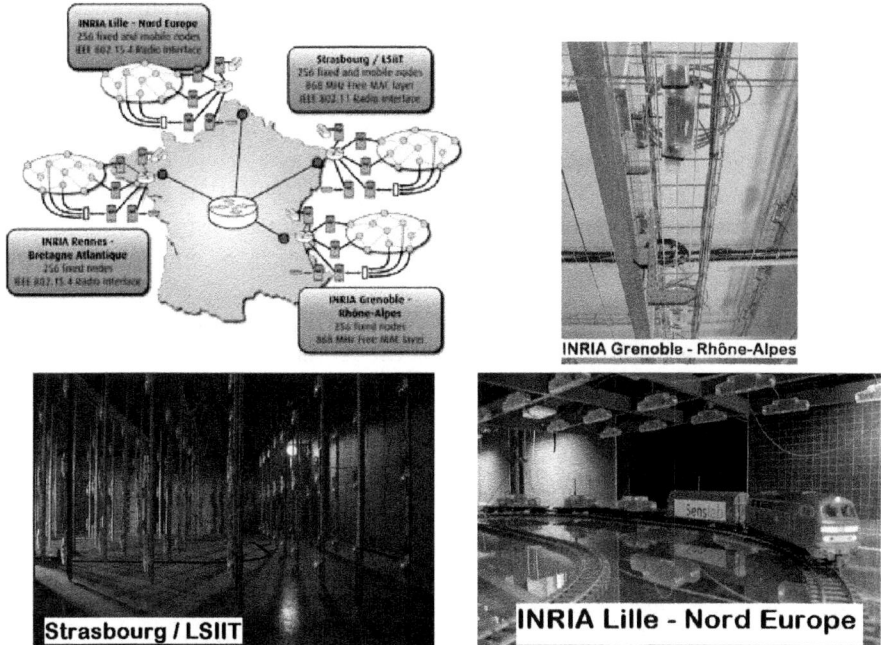

Fig. 1. SensLAB testbed is composed of 4 distributed WSN interconnected by Internet (1024 nodes in total). Mobile nodes are available on Strasbourg and Lille thanks to several toy electric trains remotely controllable and to supply power to the nodes. There is no fundamental limitation on the number of sites or number of nodes per site.

in order to gather all functionalities available in various systems and ease future large scale federations. The Kansei testbed runs 210 Extreme Scale Motes (XSM) hooked individually onto 210 Extreme Scale Stargates (XSS). Kansei provides a testbed infrastructure to conduct experiments with 802.11b networking and XSMs. For some testtbeds specific operating system (*e.g.* TinyOS), MAC layer (*e.g.* IEEE 802.15.4) or/and program language (*e.g.* NesC) are imposed, which drastically limit optimization possibilities. Some platforms do not provide radio or power instrumentation (*e.g.* WASAL) nor noise injection. Recently, the WISEBED[4] project shows the ambition to federate several WSN testbeds in order to establish a European wireless sensor network. It seems that application should be developed by using a specific API dedicated to the WISEBED platform. A great benefit of the WISEBED project outcomes is the release of Wiselib, an algorithm library for sensor networks (localization, routing) since it pursues a common goal of lowering the accessibility investment in terms of time for WSN application development .

We propose to address the problems listed above by operating SensLAB as an *open* research facility for academic and industrial groups. SensLAB provides a research infrastructure for the exploration of sensor network issues in

---

[4] http://www.wisebed.eu/

reproducible experimental conditions. The platform is **generic, open** and **flexible**: it means that a user is able to remotely access (web access) and deploy his/her applications *without any kind of restrictions on the programming language, on the programming model or on the OS that he/she desires to use.* SensLAB provides an easy way to set an experimentation, to let the user choose the number of nodes, sensor & radio characteristics, topology considerations, experimentation time, etc. SensLAB also integrates an efficient reservation tool to schedule different experimentations. During a experiments running, users have an on-line access to his/her nodes (either by the web or by a command line shell) at anytime.

The SensLAB architecture must satisfy several strong requirements in terms of software and hardware: *(i)* **reliable access** to the nodes in order to perform operations such as a reset or firmware flashing whatever is the state of the sensor node or the software it is running. *(ii)* non intrusive and application transparent **real time monitoring** of each sensor node. The external monitoring (*i.e.,* totally independent from the deployed user application code) will include precise and real-time access to fundamental parameters such as energy consumption and radio activity; *(iii)* **real time control** of the experiments by providing a set of commands that may influence an application environment (*e.g.,* turn on/off nodes to mimic crashes, emit radio noise by sending fake data in order to tamper with transmissions, modify the monitoring frequency parameters).

3   Main Elements of SensLAB

Figure 2 gives an overview of the testbed services. The user sets his/her experiment through a webportal. A virtual machine is setup with all the development tools and chains preconfigured (cross compilation chains, OS, drivers, communication libraries). The user can also access and use higher level development and prototyping tools (like cycle accurate hardware platform simulator / radio wireless network simulator [3]). When an experiment is launched, specific SQL tables related to the experiment are created. All monitoring data collected during an experiment are stored in tables to support subsequent analysis. The user is thus able to perform postmortem analysis but the system also provides online data analysis services (OLAP). Each service is replicated on each site in order to be fault tolerant (DNS for users virtual machines, LDAP for authentication...). Fig. 3 details all the software components deployed on each site.

3.1   SensLAB Hardware Components and Infrastructures

The SensLAB hardware infrastructure consists of three main components: *(i)* The **open wireless sensor node** dedicated to the user, *(ii)*; The full **SensLAB node** that encompasses the open node also includes a gateway and a closed wireless node; *(iii)* The global networking backbone that provides power and connectivity to all SensLAB nodes and guaranties the out of band signal network needed for command purposes and monitoring feedback.

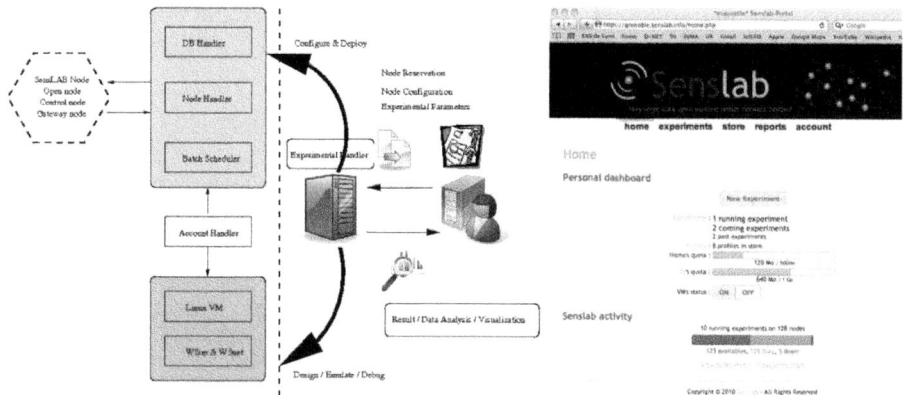

Fig. 2. Simplified view of the platform usage/services (left). SensLAB portal (right).

The **open wireless sensor node** is made available to the user during his/her experimentation. This node is totally open and the user is granted a full access to the memory. The current trend for wireless sensors nodes is geared toward a common architecture based on off the shelf 16-bit micro-controllers. We thus clearly target low power wireless sensor nodes constrained in memory and energy like existing products already on the market[5]. In order to meet with the requirements in terms of energy monitoring, reproducibility, we need to master the architecture and a solution has been to design our own board, named WSN430, in order to include all control signals and thus guarantee a reliable control and feedback[6]. The nodes are based on a low power MSP430-based platform, with a fully functional ISM radio interface[7].

To control the *open* WSN430 node that the user will request and use, we choose to mirror it with another WSN430 whose specific role is to control the open one. In order to link the two WSN430 nodes and also to meet with all mandatory requirements listed previously, we design the SensLAB gateway board in order to insure the control and management of the platform: Automated firmware deployment on open node; Accurate power monitoring of open nodes, both on battery and DC power supply (measure precision is $10\mu A$, and power sampling around 1kHz); RSSI measures and noise injection; Configurable sensor polling on the control node (temperature, light, acoustic activity); Fixed (Ethernet) as well as mobile (Wifi) communication with Node Handler via Connect EM module for the Ethernet version (or a Digi Connect Wi-EM for the Wifi version[8]). These modules embed an ARM7-based module plus Ethernet or Wifi specialized chips; Power over Ethernet support for a standardized and easy power

---

[5] WiEye, Micaz, Tmote-Sky, TinyNodes.

[6] All designs are released under a Creative Commons License.

[7] Two versions are currently available: an open 868MHz radio interface and IEEE 802.15.4 radio interface at 2.4GHz.

[8] These 2 modules are pin-to-pin compatibles, allowing only 1 unique board design, with either an Ethernet or a Wifi module on it.

management; Sink capability for each open node (in/out characters stream redirection). The trade off made was to design our own SensLAB gateway instead of using linux box.

3.2   SensLAB Software Architecture

The SensLAB software architecture is replicated over the four testbed sites, and it is divided in several parts which interact together as shown in Figure 3. We detail below their functionalities:

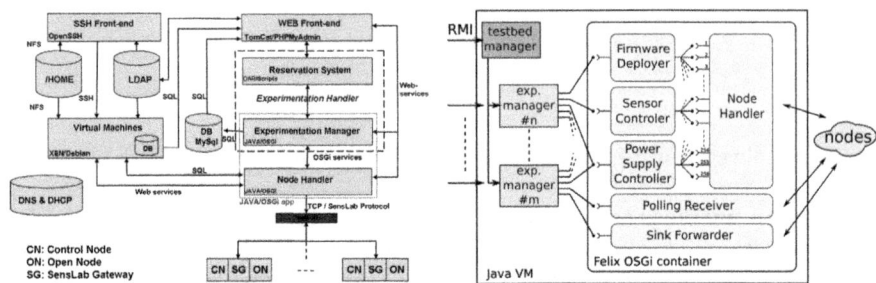

Fig. 3. Software SensLAB architecture and technological choices and the Experiment Handler software structure

Control node software: *i.e.* the firmware running on the control node. It is in charge of powering up/down, resetting and monitoring the open node activity (power consumption, radio activity/RSSI). All these actions can be executed either asynchronously on the user's request, or automatically and periodically. The control node can also be used to set the ADC pins of the open nodes, allowing to send specific stimuli to the open nodes and thus guarantying a totally reproducible environment on the sensing part. By reproducible environment, we mean that it is possible to record a trace of events/values on a ADC sensor device and during another experiment to "*replay*" such values instead of reading real ones. It allows to change application parameters without changing the environment sensed, and thus allows to compare things that are comparable.

Gateway node software: *i.e.* the firmware running on the gateway. It manages the interface between the open and control nodes, and the SensLAB site server over IP communications. It forwards the command frames addressed to the control node, updates the open node's firmware (BSL protocol), and forwards the open node's serial link to the server (sink application).

Experiment handler software: The experiment handler software (Fig. 3) is the server side interaction point with all the 256 nodes of a site. It can execute all methods described above such as firmware update, energy consumption monitoring, polling, etc. It also receives the data coming from the serial links of open nodes relayed by the gateway nodes. It instantiates a 'testbed manager' object and an OSGi container when started. The OSGi container embeds several bundles, responsible for all the interactions with the nodes: the Node Handler bundle sends command frames to the gateway and the control node; the Firmware Deployer bundle provides one service allowing parallel deployment of a firmware on several open nodes; the Sensor Controller bundle allows parallel sensor measurement; the Sink Forwarder bundle provides efficient data redirection between nodes and users' VMs.

Batch scheduler software: Through a web form (or by uploading an xml file), the user configures his/her experiment and specifies requested nodes (either mobile or fixed nodes, with a CC1100 or CC2420 radio chip, outdoor or indoor, location and the number of nodes), experiment's duration and eventually a start date. Those information are transmitted to the batch scheduler software, which is the server-side module allowing optimal experiments scheduling and resources allocation. This module is based on the use of OAR[9], which is a versatile resource manager for large clusters.

User virtual machines: A complete Linux environment is made available to each registered user allowing him/her to build sensor firmwares thanks to the complete set of tools installed. from his/her VM, user interacts with the nodes of his/her running experiment. He/She can also run any dedicated application to handle the nodes' serial link outputs.

4   Conducting Experimentations with SensLAB

When a user wishes to conduct some experiments on the SensLAB platform, he/she first needs to configure his/her experiment through the webportal. As displayed by Fig. 4(a), SensLab offers an easy way to select nodes needed. Through

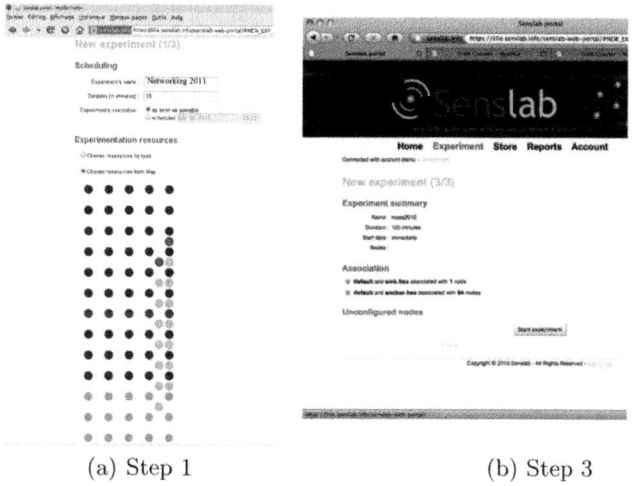

(a) Step 1   (b) Step 3

Fig. 4. The SensLAB web portal for Lille's platform

[9] http://oar.imag.fr

this web portal page: the user is able to choose the experiment name (*Networking 2011* on Fig.), the duration of the experiment (15min) and if the experiment has to start as soon as possible or at a chosen date (*as soon as possible* in our case). Then, nodes on which the experiment will be run have to be chosen, either through their Id or on the map. These nodes are then associated with a specific firmware provided by the user. A summary page is then displayed (Fig. 4(b)). In our case, one node is configured as a sink and 94 as anchors.

Once configuration is done, the experiment is submitted. SensLAB automatically reserves nodes. At starting time, *i.e.* in our case, as soon as chosen nodes are all available, SensLAB configures *control nodes*, deploys firmwares, and resets the experiment's nodes. Every action is performed automatically and in a transparent way for the user.

Once the experimentation is launched, the user is able to interact with the running experiment as we will see in following sections illustrating an experiment. Results chosen to be retrieved (by polling or reading on serial port) are available in the user's virtual machine.

In the remaining of this section, we give some examples of utilization of the platform through several applications examples.

4.1   SensLAB for Testing Geolocalization Protocols

To illustrate the benefits and the use of the SensLAB testbed in designing new algorithms, we focus on an animal tracking application [7]. To geolocalize them, some fixed nodes called *anchors* are spread in the park and receive signals from mobile nodes as soon as they are in range. *Anchor* nodes register the mobile node identifier, the RSSI (Received Signal Strength Indicator) of the signal and the date. Then, these data needs to be routed to a *sink* node. This latter is connected to a computer gathering data and computing *mobile* node location based on these data. Note that the geolocalization application has been simplified as possible since the main purpose here is to highlight SensLAB benefits.

In the setup, the first step is to cover the bounded area by deploying *anchors*. The next step is to set up the routing infrastructure to allow every anchor to send data to the sink. Once again, the routing process used here is simplified. When *anchors* are deployed and powered on, the *sink* is initialized. It then starts to send BEACON and every *anchor* receiving this BEACON attaches itself to the sink. The *sink* becomes its parent. Then every attached *anchor* u forwards the BEACON. Every unattached *anchor* receiving a beacon from u, chooses u as its parent. When every *anchor* has chosen a parent, the whole area is covered and *mobile* messages can be forwarded to the *sink* as follows. When an anchor receives a data message from another anchor or needs to send its own data, it forwards it to its parent. Step by step, the message eventually reaches the *sink*. The *sink* sends data through its serial link and the computer connected to it gathers the different messages and estimates mobile node positions.

Demonstration overview: Demonstration was conducted from the Lille's SensLAB site. Nodes of this site are featured with CC2420 radio chip and 32 among the 256 nodes are mobile, mounted on several toy electric trains. For the need

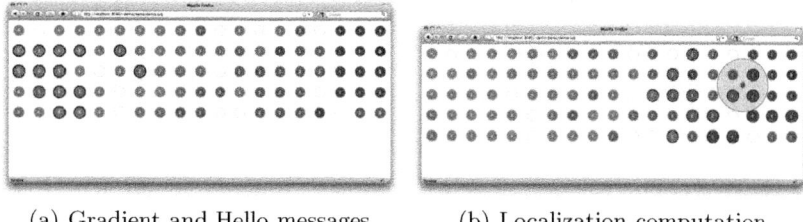

(a) Gradient and Hello messages   (b) Localization computation

Fig. 5. Real time interaction with the running experiment

of the application, we will reserve a grid of 5 × 10 nodes and 2 mobile nodes located on different train paths. Mobile nodes will represent the animal while fixed nodes will stand for anchor nodes. The reservation is be performed through the SensLAB web portal (see Section 3). Three firmwares have to be compiled for the experimentation, respectively for *mobile, anchor* and *sink* nodes. In the experiment only one *profile* will be used tuned with nodes on DC power. Consumption polling is sampled every 50 ms. All fixed nodes except one will be associated with *anchor* firmware and the last node is associated with the *sink* firmware.

Experimentation is launched. During the experiment, monitoring data (power consumption, RSSI, ...) are logged like described in the experiment profile which is the same for all nodes. Once the experimentation is launched, the user is able to interact with the running experiment via his/her VM as shown by Fig. 6.From the VM, and using the UNIX *netcat* tool to create and open a TCP socket on a specific port of the node handler host, which is the serial link redirection of the SensLAB node specially created for the ongoing experiment, the *sink* is ready to be activated and to start sending BEACONS. Results are gathered in real time. A route tree is created by the *anchors* to forward reports to the *sink* (In Fig. 5(a), numbers on nodes give the distance each node estimates itself to the sink at a given time.). *Mobile* nodes are activated through their serial links and start sending Hello Messages received by anchor nodes (Green nodes in Fig. 5(a)). Based on the RSSI of these Hello messages, anchors compute the location of mobile nodes (Fig. 5(b)).

For the experimentation purpose, the VM is hosting an application which collects data from serial links, analyzes them to compute *mobile* node locations and provides a web server to visualize application status in real time (Fig. 6). *Anchors*, routes, messages and estimated *mobile* node positions can then be displayed in a web browser.

4.2   SensLAB for Testing Medium Access Control Protocols

Among the large set of communication protocols that have been studied especially for wireless sensor networks, an important amount of literature focusing on medium access control (MAC) mechanisms has been produced. The MAC layer dictates whenever a node may listen to the channel, transmit and receive

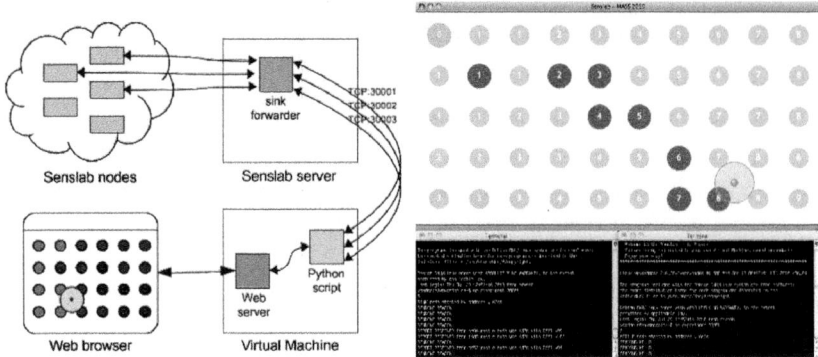

Fig. 6. Demo visualization interface showing routes, messages transmission and mobile nodes location in the upper window, and messages printing on serial links for sink and mobile nodes on lower windows

so that some crucial properties regarding the usage of the shared radio medium (e.g. fairness, reliability) remain guaranteed. In WSN, the MAC layer should also be carefully designed in order to ensure a low energy consumption. The two widespread ideas consist in either allocating time slots to every device (that is allowed to sleep outside of these) or in having regular wake-ups during which any upcoming transmissions should be detectable (i.e. through the use of a preamble sent by the data source). These two main schemes are referred to as synchronized and preamble-sampling protocols, respectively.

Recently, versatile layers, such as B-MAC [9] and X-MAC [2], have gained much attention, especially due to their common tunable low-power listening mechanism that is expected to be suited to any traffic pattern. Still, an overview of today's WSN deployments shows no implementation of any of these protocols [5]. The fact that their performances have never been evaluated under realistic conditions is for sure one of the main reasons why. Indeed, the limited simulation models and the small scale testbeds that were used to evaluate the performances of these solutions have led us to use the SensLAB platform in order to conduct a thorough study of them. We here report some feedbacks concerning such an experiment.

Description of the experiment. The empirical analysis of X-MAC was led over the SensLAB testbed portion installed in Strasbourg. Figure 7 shows the feedback interface that was used for the management and the demonstration of the results. The 3-dimensional grid is composed of 240 static sensor nodes, with an homogeneous step of 1 meter between every pair of devices. Each one of them is equipped with Texas Instruments CC1101 radio interfaces[10].

At the beginning of an experiment, up to 30 sensor nodes were selected as data sources. These nodes implement a time-driven application that transmits a 7 byte sample of data, every 10 seconds toward the sink (located a the center of

---

[10] http://focus.ti.com/docs/prod/folders/print/cc1101.html

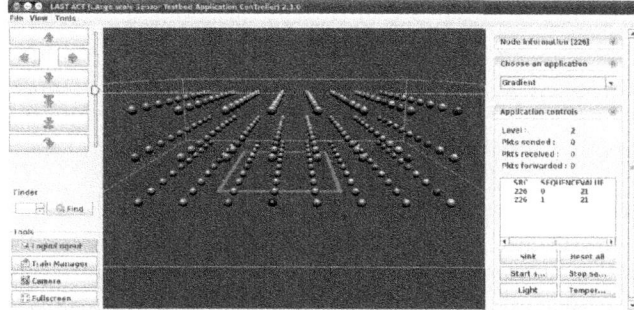

Fig. 7. 3D feedback application

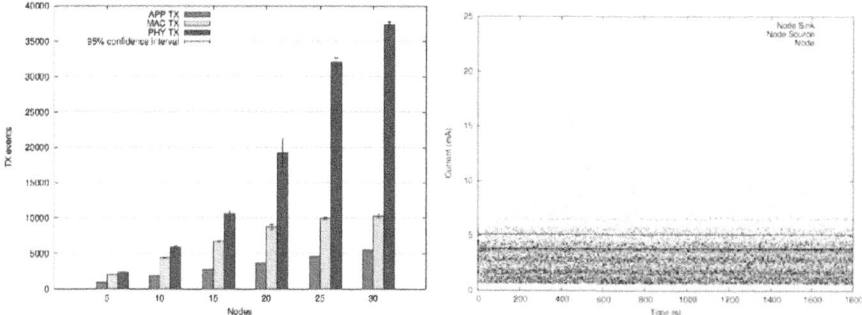

Fig. 8. (left) X-MAC retransmissions. (right) Instantaneous energy consumption among nodes of the SensLAB testbed.

the 3D grid). Multi-hop transmissions were achieved with a basic gradient-based routing protocol. Concerning the X-MAC deployment over the devices, its low-power listening feature was managed by the Wake On Radio (WOR) mechanism embedded in the CC1101 chipset. The sleep period duration was fixed to 101.5ms while the listening time was set to 15ms.

Experimentation results and analysis. For each number of senders, a dozen of experiments were conducted. Each set of experimentations uses the same topology and the same senders to evaluate the MAC layers in a similar environment. The results presented here are an average of the overall collected data.

First, we studied the average overhead induced by MAC layer retransmissions that necessarily occur upon non-reception of an acknowledgment message (e.g. collisions, busy channel during the clear channel assessment). Figure 8 illustrates the average number of retransmissions induced by X-MAC. As expected, we can observe the stress of the medium as more and more data sources are added. While the number of application transmissions linearly increases, both the MAC and the physical layer require much more sent messages.

Energy consumption being one of the crucial points in WSNs, the SensLAB platform allows users to evaluate the instantaneous energy consumption of any node in the network during the experimentations. The INA209 chipset embedded in the nodes was used to retrieve the average value of 128 samples each 68ms. Results obtained during a 30 minute monitoring are exposed on Figure 8.

5   Conclusion

The architecture concepts, the hardware design, the software implementation of SensLAB, a large scale distributed open access sensor network testbed were presented. We have also provided examples showing how to use SensLAB to launch and evaluate experiments, how to configure any nodes as a sink with a serial feed back channel connectable through TCP/IP to any application. The testbed is deployed and operational and researchers are invited to register. Obtaining an account is free and can be done from the main web page: http://www.senslab.info. We are working on the deployment of the OAR-grid version in order to allow fully flexible distributed node reservation. Several research works remain. One extension concerns the use of hybrid simulation within SensLAB by allowing to connect node and or network emulator to the testbed. Another extension is the development of actuator nodes, plugged directly on SensLAB nodes.

Another main direction concerns the study of the federation of research platforms, and more precisely with OneLab. A federation will offer a higher dimension in the spectrum of applications that the research community will design, test, deploy, and tune. But even more important, SensLAB will strongly benefit from the monitoring tools and supervising infrastructure developed and used in OneLab. Thanks to French government's stimulus funds, the FIT (*Future Internet of Things*) project is being awarded 5.8 millions Euros and will enable large scale and diverse experiments with Future Internet technologies, from components to complete systems, and to validate and compare them with existing or evolving solutions. The goal is to develop the tools and the management system required to run a global open federated testing facility.

The work program of the FIT project will consist in establishing the different sites with a set of competitive testbeds including wireless and wired technologies, with a strong emphasis on sensors, networked and embedded objects, multihop and cognitive radio, mobility and overlays. FIT is by nature a distributed facility, involving heterogeneous devices in a set of complementary components situated in adequate and relevant locations. These testbeds are original not only in the cutting edge technologies that they offer, but also in their federated management system, which will provide experimenters a set of common methods to access and control numerous facilities.

References

1. Arora, A., Ertin, E., Ramnath, R., Nesterenko, M., Leal, W.: Kansei: A high-fidelity sensing testbed. IEEE Internet Computing 10, 35–47 (2006)

2. Buettner, M., Yee, G.V., Anderson, E., Han, R.: X-MAC: a short preamble MAC protocol for duty-cycled wireless sensor networks. In: SenSys 2006: 4th International Conference on Embedded Networked Sensor Systems, pp. 307–320. ACM, New York (2006)

3. Chelius, G., Fraboulet, A., Fleury, E.: Worldsens: development and prototyping tools for application specific wireless sensors networks. In: ACM (ed.) International Conference on Information Processing in Sensor Networks, IPSN (2007)

4. Handziski, V., Köpke, A., Willig, A., Wolisz, A.: Twist: A scalable and reconfigurable testbed for wireless indoor experiments with sensor networks. In: RealMAN 2006 (2006)

5. Kuntz, R., Gallais, A., Noël, T.: Medium access control facing the reality of WSN deployments. ACM SIGCOMM Computer Communication Review 39(3), 22–27 (2009)

6. Kurkowski, S., Camp, T., Colagrosso, M.: Manet simulation studies: the incredibles. SIGMOBILE Mob. Comput. Commun. Rev. 9(4), 50–61 (2005)

7. Mitton, N., Razafindralambo, T., Simplot-Ryl, D.: Position-Based Routing in Wireless Ad Hoc and Sensor Networks. In: Theoretical Aspects of Distributed Computing in Sensor Networks. Springer, Heidelberg (2010) (to appear)

8. Pawlikowski, K., Jeong, J.L.R.: On credibility of simulation studies of telecommunication networks. IEEE Communications Magazine (2001)

9. Polastre, J., Hill, J., Culler, D.: Versatile low power media access for wireless sensor networks. In: SenSys 2004: 2nd International Conference on Embedded Networked Sensor Systems, pp. 95–107. ACM, New York (2004)

10. Werner-Allen, G., Swieskowski, P., Welsh, M.: Motelab: A wireless sensor network testbed. In: Special Track on Platform Tools and Design Methods for Network Embedded Sensors, SPOTS (2005)

Using Coordinated Transmission with Energy Efficient Ethernet

Pedro Reviriego[1], Ken Christensen[2], Alfonso Sánchez-Macián[1],
and Juan Antonio Maestro[1]

[1] Universidad Antonio de Nebrija, Calle Pirineos 55, 28040 Madrid, Spain
{previrie,asanchep,jmaestro}@nebrija.es
[2] University of South Florida, 4202 East Fowler Avenue, ENB 118, Tampa, FL 33620, USA
christen@csee.usf.edu

Abstract. IEEE 802.3az Energy Efficient Ethernet (EEE) supports link active
and sleep (idle) modes as a means of reducing the energy consumption of
lightly utilized Ethernet links. A link wakes-up when an interface has packets to
send and returns to idle when there are no packets. In this paper, we show how
Coordinated Transmission (CT) in a 10GBASE-T link can allow for key physical layer (PHY) components to be shutdown to further reduce Ethernet energy
consumption and enable longer cable lengths. CT is estimated to enable an additional 25% energy savings with a trade-off of an added frame latency of up to
40 μs, which is expected to have a negligible impact on most applications. The
effective link capacity is approximately 4 Gb/s for symmetric traffic and close
to 7 Gb/s for asymmetric traffic. This can be sufficient in many situations. Additionally a mechanism to switch to the normal full-duplex mode is proposed to
allow for full link capacity when needed while retaining the additional energy
savings when the link load is low.

Keywords: Ethernet, Energy Efficiency, IEEE 802.3az, 10GBASE-T.

1 Introduction

The reduction of energy consumption in computer networks has been a topic of interest for the networking community in the last decade [1]. This has led to a number of
initiatives aimed at making networks more energy efficient. Among the different
networking technologies, Ethernet was identified as one of the candidates where large
savings could be obtained. The potential savings are in the range of many TWh [2].
To materialize those savings the Energy Efficient Ethernet (EEE) initiative was
launched to standardize mechanisms that would improve the energy efficiency of
Ethernet devices. These efforts have led to the recent approval of the IEEE 802.3az
Energy Efficient Ethernet (EEE) standard [3].

The EEE standard defines two modes of operation for a physical layer (PHY) device, active and sleep (aka Low Power Idle (LPI)). When there are frames to transmit
the device is in the active mode and when there are no pending frames the device can
enter the sleep mode reducing its energy consumption substantially. If frames arrive
while the device is in the sleep mode a transition back into the active mode is

J. Domingo-Pascual et al. (Eds.): NETWORKING 2011, Part I, LNCS 6640, pp. 160–171, 2011.

initiated. The transitions to wake and sleep a device take place in a small time (in the range of microseconds). This minimizes the delay that a frame experiences while it waits for the link to activate. Finally in the sleep mode periodic refresh periods are scheduled during which signals are transmitted to keep the receiver adapted to the current channel's conditions and synchronized with the transmitter. The operation of EEE is illustrated in Fig. 1.

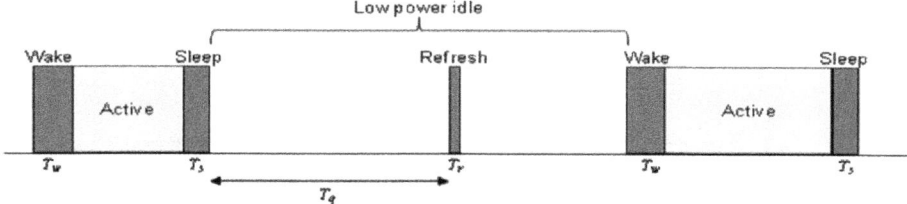

Fig. 1. Transitions between the active and low-power modes in Energy Efficient Ethernet. T_w wake up time, T_s sleep time, T_r refresh time and T_q quiet time.

The current EEE standard covers the most common types of Ethernet PHYs including 100BASE-TX, 1000BASE-T, and 10GBASE-T. These are commonly used in PCs and servers and use Unshielded Twisted Pairs (UTP) as the transmission media. An important point to note is that the PHY power consumption increases significantly with link speed. For example the consumption of a 1000BASE-T PHY can be around 0.5 Watts while for a 10GBASE-T the consumption can be around 4 Watts [4]. Most Network Interface Cards (NICs) support several speeds and when the link is setup the maximum speed supported by both ends is selected [5]. This means that if we were to optimize the energy consumption for one of the speeds, the highest one should be a priority as it is likely to be selected and it is the one that consumes more. For UTP PHYs this speed would be 10 Gb/s and corresponds to the 10GBASE-T standard. 10GBASE-T PHYs are currently used mostly in servers but as in previous UTP PHY generations as the technology matures it is expected that they will be widely adopted in desktop and laptop computers (see [6] for more details). This transition will be accelerated by EEE that significantly reduces power consumption, making it easier to integrate 10GBASE-T PHYs in laptops and other battery run computers. Therefore given its high power consumption and its expected wide adoption in coming years, in the rest of the paper the analysis focuses on 10GBASE-T links.

The 802.3an 10GBASE-T standard [7] supports full duplex communication at 10 Gb/s over up to 100m of UTP cable. Transmission takes place over four cable pairs in parallel. This means that in each pair in addition to the signal from the remote transmitter, echo and both Near End crosstalk (NEXT) and Far End crosstalk (FEXT) signals are received. All those interferences must be cancelled in order to properly receive the remote signal. To that end 10GBASE-T PHYs use advanced signal processing techniques and therefore the circuitry devoted to echo and NEXT cancellation is a significant part of the PHY area and power consumption [8].

EEE has standardized a sleep mode for 10GBASE-T that enables the independent management of each link direction. That is a link can be in sleep mode in one direction while it is actively sending data in the other direction. This is a useful feature as in many cases the traffic on the links is asymmetric. The values of the transition time for wake (T_w), sleep (T_s) and also the duration of the refresh (T_r) and quiet (T_q) periods are shown in Table 1. It can be observed that the duration of the quiet period is much larger than that of the refresh period to maximize energy savings. The transition times are small in terms of their impact on frame delay but large when compared to the frame transmission time. For example a 1500 byte frame takes 1.2 µs to transmit at 10 Gb/s which is significantly less than the wake or sleep transition times. This can lead to large energy overheads due to mode transition when the link is awaken as soon as a frame arrives [9].

Table 1. EEE parameters for 10GBASE-T

T_w	T_s	T_q	T_r
4.48 µs	2.88 µs	39.68 µs	1.28 µs

To minimize the overheads associated with transitions, coalescing of packets can be used such that the link is activated only when a group of frames is ready for transmission. This minimizes the number of mode changes and can significantly reduce energy consumption as discussed in [2].

In this paper, a Coordinated Transmission (CT) scheme is proposed to further reduce the energy consumption of 10GBASE-T PHYs that implement the EEE standard. The objective is to coordinate transmissions at both ends of the links such that a) at a given point in time only one end is transmitting and b) transmissions can only start at some predetermined points in time.

Objective (a) enables the PHYs to save significant energy as the echo and NEXT cancellation circuits could be put into a low power mode. Additional savings can also be obtained in other PHY elements. An additional benefit is that longer cables could be used as there are less interferences in the channel. Objective (b) enables larger savings beyond the PHY as other system elements can be put in low power mode even if their wake up time is larger than that of the PHY.

The proposed CT scheme reuses EEE signalling and therefore minimizes the additional complexity needed to implement it. It also provides mechanisms to signal transitions between CT and normal mode which will be referred to as Uncoordinated Transmission (UCT) mode in the rest of the paper. This enables a link to operate in the traditional UCT mode if the load in both directions is high.

In the rest of the paper the Coordinated Transmission (CT) is described (section 2). Its benefits in terms of energy savings are discussed in section 3. Then performance simulation results of the proposed scheme are reported in section 4. The paper ends with some conclusions in section 5.

2   Coordinated Transmission (CT)

In this section, first the CT scheme is introduced and described in detail. Also some examples are used to illustrate how the scheme operates. Then, a mechanism to switch between CT and UCT modes is discussed.

2.1 Coordinated Transmission Scheme

As described in the introduction the objectives of the CT scheme are a) to ensure that only one link end transmits at a given point in time and b) to provide predetermined time instants at which transmission can take place.

This is achieved by using the EEE refresh periods to coordinate transmissions. The refresh periods are needed to ensure that the receiver is kept aligned to the channel conditions and with the transmitter. Therefore the refresh signal must be, in any case, transmitted each 40.96 μs as per the parameters in Table 1. This means that one link end can at most transmit for 40.96 μs as otherwise its transmission would overlap with the refresh signal in the other direction and objective a) will not be achieved. This leads naturally to using the EEE refresh signal as a means to grant permission to transmit until the next refresh signal. That is a link end can start its transmission when it receives an EEE refresh signal from the other end. Since EEE refreshes are sent periodically a link end is guaranteed to receive a refresh signal each 40.96 μs thus ensuring a low waiting time for transmission. Two examples of this scheme are illustrated in Fig. 2 where obviously the transmission has to be limited to avoid overlapping with the next refresh signal as discussed before. In the first example, link end A has frames to transmit that require less than T_q seconds. In the second, link end A has frames to transmit that require more than T_q seconds and therefore has to activate the link two times to send all the frames.

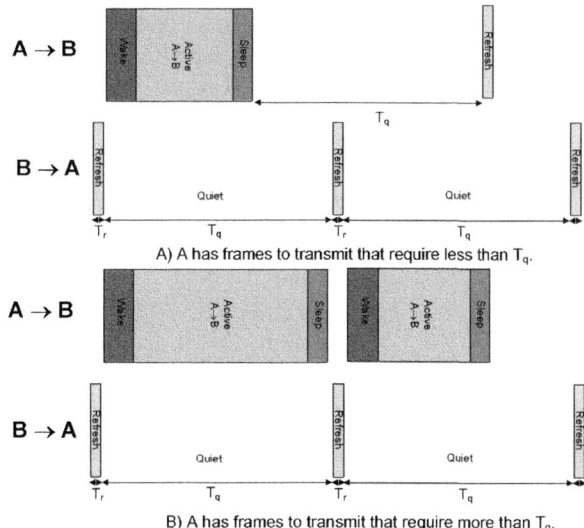

Fig. 2. Examples of transmissions with the proposed scheme

The described scheme assumes that refresh periods occur frequently in both link directions. This is not true when one of the link ends has frames to transmit in every refresh period. In that case that link direction transitions from sleep to active and back to sleep in each period and therefore never spends sufficient time in the sleep mode to send a refresh signal. This issue can be solved by using also sleep transitions in the receive direction to grant permission to transmit in the same way as an EEE refresh signal. In Fig. 3, the case in which link end A has data to transmit in every refresh period and then B also starts to transmit data after a sleep period is illustrated.

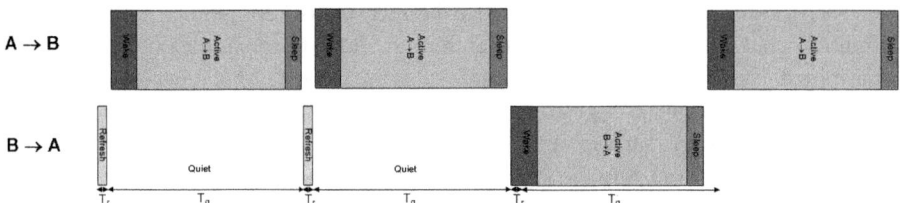

Fig. 3. Example of transmission with the proposed scheme showing alternating transmission between link ends

In the discussion so far, it has been implicitly assumed that the link delay is negligible and therefore the same signal is present at both link ends at the same time. If that is not the case the scheme will not work. For example if the link delay was 50 μs then if one link end sends a refresh at time 0 μs and the other at time 20 μs those would arrive at 50 μs and 70 μs respectively enabling transmissions that can overlap. The delay of 100m of UTP cable is approximately 0.5 μs which is significantly smaller than the refresh period or the transition times. Therefore the scheme will work but care should be taken to account for this extra delay. This implies reducing the transmission time by another 1 μs to avoid overlapping at the remote end with the start of the next refresh signal in the other direction. This is illustrated in Fig. 4. Which shows the detailed timing at both link ends for a complete transmission period starting after a refresh is received at B. T_d corresponds to the link propagation delay and T_{tx} to the actual data transmission time. To avoid the next refresh overlapping with the end of the transmission the following inequality must be met

$$T_r + T_d + T_w + T_{tx} + T_s + T_d < T_r + T_q \tag{1}$$

which can be solved in terms of T_{tx} as

$$T_{tx} < T_q - T_d - T_w - T_s - T_d. \tag{2}$$

Using the values in Table 1 and assuming the worst case link delay of 0.5 μs we obtain a value for the data transition time (T_{tx}) of up to 31.82 μs. In a practical implementation there will be a short time in between the end of the reception of the refresh or sleep signal and the start of the transmission and there should also be some margin in between the end of the transmission and the next refresh period. If we allow for an additional 0.25 μs for each, the transmission time would be up to 31.32 μs, that would enable the transmission of twenty six 1500 byte frames. This means that roughly 75% of the period can be used for actual data transmission. The performance of the proposed scheme will be further discussed in section 4.

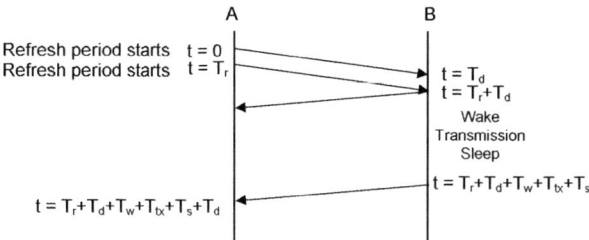

Fig. 4. Detailed transmission time diagram for the proposed scheme

2.2 Signaling Mechanisms for Coordinated Transmission

In the previous subsection the CT scheme was described assuming that both link ends support CT and have somehow agreed to use it. Now the mechanisms that allow link ends to signal support for CT and to switch between CT and UCT modes are described.

To better understand the mechanisms let us first review what exactly they are needed for. When the link is setup we need to convey whether a link end supports CT. This would enable link ends to know if the remote end also supports CT and therefore if it could be used. Then during link operation we need a mechanism that allows to switch between CT and UCT modes. This is needed as CT effectively reduces the link capacity such that when the traffic load is high we may want to switch to UCT even if the energy consumption is larger. As the pattern of use of networks is such that the utilization is low most of the time [10] most links would only need to switch to UCT during a small fraction of time. There is one case where this second mechanism is not needed, that is when CT is used to extend the reach of 10GBASE-T. In that case the link would always operate in CT mode.

The Ethernet standard defines a signalling procedure called Auto-Negotiation that is used when the link is setup to exchange the capabilities of both link ends [11]. During this phase the information is encoded in link pulses so that the exchange can take place even though the link has not been established. Auto-Negotiation is used for example to agree on the speed to use if multiple speeds are supported as discussed before. The support of the CT mode can be easily added to the information exchanged during Auto-negotiation by using a bit in the 10GBASE-T Auto-Negotiation control register defined in the standard. Currently only one bit is used in that register to signal support for fast retrain in 10GBASE-T. Therefore the use of Auto-negotiation is proposed to signal support for the CT mode at link setup.

A 10 Gb/s link in which only one link end can transmit a given time instant can at most carry 10 Gb/s in both directions compared to 20 Gb/s when operating in UCT mode. This means that when the link traffic load is high we should be able to switch to UCT mode operation. This requires coordination between both ends of the link. To ensure that there are no overlapping transmissions during a mode change the scheme shown in Fig. 5 could be used. To switch to UCT mode, the requesting end first enables the echo/NEXT cancellation circuitry and then sends the request. The other end upon reception enables the cancellation circuitry and starts operating in UCT mode. Also a message is sent back to the other end that upon its reception enables the UCT mode of operation. To go back to CT mode the requesting end starts to schedule its sending as if it were already in CT mode and sends a request message to the other end. Upon its reception the other end disables the cancellation circuitry and also starts to schedule sending as in CT mode. Also a message is sent back to the other end that upon its reception can disable the cancellation circuitry. In the transition from UCT to CT we need to be careful to avoid overlapping transmissions and to ensure that the refresh periods in both directions do not take place at the same time. To ensure that, the end that received the request should stop transmitting and go to the sleep mode when it receives the request. Then wait for the other end to either a) stop transmitting or b) send a refresh. At that point wake and start CT operation by sending a message back to the other end confirming the transition to CT mode and any further frames waiting for transmission. This will ensure that the refresh periods are not aligned when entering the CT mode as assumed in the previous subsection.

The message exchange required for the proposed mode switching mechanism can be done above the PHY layer using Ethernet data frames but since the mechanism is heavily related to the PHY it would be better to perform the exchange using PHY layer mechanisms. This can be done using the auxiliary bit defined in the 10GBASE-T standard [7]. This bit provides a channel of approximately 3Mb/s for signalling at the PHY layer so that mode transitions could be done in a short time. The use of this channel for signalling of different advanced PHY features has been proposed by some PHY vendors [12], [13]. In fact since the channel is fully devoted to control signalling, the status of each link ends in terms of the mode of operation (CT or UCT) could be periodically transmitted over this channel to ensure that even in the event of data lost a consistent state would be reached.

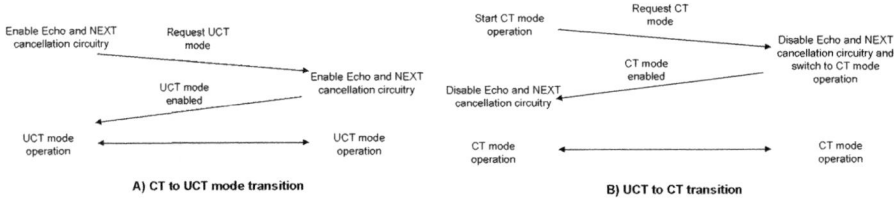

Fig. 5. Message exchanges to change mode of operation

Finally it is worth noticing that the link remains operational during transitions so the effect on frame delay should be small. The decision of when to switch modes can be based on buffer occupancy or on measured load on the link. The study of those algorithms is left for future work.

3   Benefits of Coordinated Transmission

The use of CT provides benefits in terms of additional energy savings and also in extended reach, those are discussed in the rest of this section.

3.1 Energy Savings

The use of CT would only be of interest if it can provide substantial energy savings. To understand the potential savings a basic understanding of how 10GBASE-T works is needed. In 10GBASE-T full duplex transmission takes place over four Unshielded Twisted Pairs (UTP) that are bundled in the same cable. This simultaneous transmission causes that in addition to the remote signal, echo and both near end and far end crosstalk (NEXT and FEXT) are received on each pair. The receiver must remove all those un-wanted signals and for that echo and crosstalk cancellation circuits are used.

In Fig. 6 a block diagram of the circuitry used in each of the four pairs in a 10GBASE-T transceiver [8] is shown. In CT mode the Echo Cancellers and the NEXT cancellers are not needed since there is no transmission when the device is receiving. These elements are an important part of a 10GBASE-T PHY [14]. Echo cancellation is typically done both in the analog domain to reduce the requirements on the Analog to Digital Converter (ADC) and then in the digital domain to eliminate the rest of the echo [8]. This means that in CT mode energy savings would be obtained both in the analog and digital circuitry. NEXT cancellation is typically done in the digital domain. Additional savings maybe obtained in the ADCs as less precision is needed if there is no echo at the input.

Therefore if a PHY operates in CT mode all the circuitry related to echo and NEXT cancellation can be put in low power mode. The energy savings that could be obtained will vary for different implementations but they will be substantial. From the information in [8], savings of 20-30% seem a reasonable estimate.

Additional benefits can be obtained in other PHY elements. For example, since there is no echo, the requirements of the timing recovery are relaxed. This means that some extra power could be saved on the timing recovery circuitry and/or that transitions could be made faster. The main point is that during wake up the timing recovery circuitry must compensate the deviation between the clocks that occurred during the time that the PHY was in low power mode. This is much harder to do if there is echo in addition to the remote signal. Also if echo is present, the adjustment must be done slowly so that the local transmission is not disturbed by the timing recovery and the echo is properly cancelled. Another element that could benefit from the CT mode is the Low Density Parity Check (LDPC) code decoder that could possibly be switched to a simplified mode to save power as the Signal to Noise Ratio (SNR) will probably be larger than needed. This may be easier to do than the reduction of precision in the ADC.

It is expected that since link utilization is typically low, in many cases the EEE-PHYs will stay in the LPI mode most of the time. Therefore it is interesting to also consider the savings that CT enables in LPI mode. When we are in LPI mode the proposed coordination transmission mechanism enables us to know when the other end may start transmitting. This in turn enables us to power down all the receiver circuitry as we are sure that no alert/wake signal will be received. Without CT, some receiver elements must be active at all times to detect the alert/wake signal from the other end as that can start at any point in time. Additionally during the refresh periods some of the energy savings of CT described for the active mode also apply. This means that additional significant savings can be achieved in LPI idle mode when CT is used. A first estimate for the savings in LPI mode can also be in the order of 20% of the LPI mode power consumption.

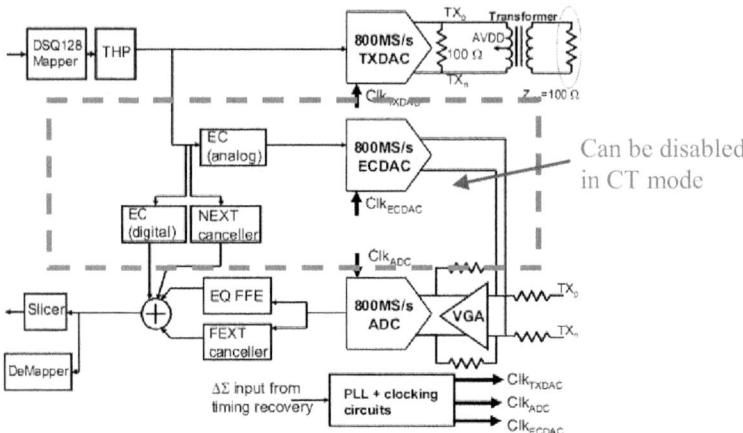

Fig. 6. 10GBASE-T circuitry for each twisted pair (from [8])

Finally, as CT enables a link end to know in advance when the other end can start transmitting, additional savings beyond the PHY could be achieved [15]. For example if link end A schedules a transmission for a complete interval between refresh periods (approximately 40 μs), then no frames will be received during that time. This would enable components with transitions times larger than the PHY to enter the low power mode and still be awake on time to process the data. This could also be done using extended wake timers in the PHY as proposed in [16] but that would increase the PHY power consumption as each EEE mode transition will involve a larger overhead.

3.2 Extended Reach

Another interesting option is to use Coordinated Transmission to extend the reach of a 10GBASE-T link beyond the standard 100 m. The extended reach can be obtained as there is no echo at the receiver such that the signal can be further amplified before the ADC resulting in a better SNR. Also we get rid of the quantization noise from the cancellers and any residual echo/NEXT after cancellation. The number of meters that could be extended will depend on the PHY architecture and is difficult to predict; this requires further work. In this case obviously transitions to UCT mode will not be allowed and the PHYs will always operate in CT mode. This option would provide a capacity in between 1 Gb/s and 10 Gb/s as CT transmission uses some of the capacity for coordination. For symmetric traffic the effective capacity of CT would be approximately 4 Gb/s as discussed in the next section.

4 Performance Evaluation

As mentioned in the introduction, CT introduces an additional frame delay. This is the time that a link end has to wait until it receives a refresh from the other end or the other end stops its transmission if it was transmitting. In the worst case the link end would have to wait for the time between two refresh periods that is approximately 40 μs. This delay is very small compared with the typical delay of Internet connections or with the requirements of multimedia applications such as voice over IP, both in the order of tens of ms. This means that for most users the impact of the added delay will be negligible.

The proposed scheme also reduces link capacity. To evaluate this effect, CT has been implemented in the ns-2 simulator and tested in several experiments. The network configurations used are shown in Fig. 7. The link delays for Link 1 and Link 2 are 0.5 μs that corresponds to the propagation delay of approximately 100m of UTP cable. The intermediate link, models a network that limits the transfer capacity in the 1 Gb/s case and does not in the 10 Gb/s case. The first configuration tries to model a situation in which the bandwidth is limited by the Internet connection having a lower speed than the LAN as it commonly occurs.

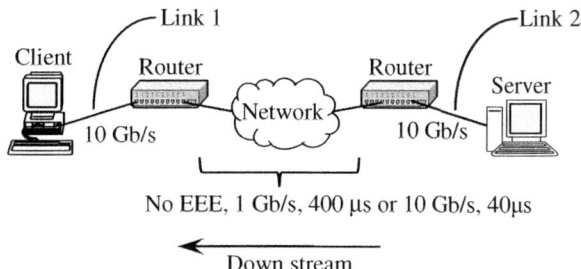

Fig. 7. Configurations for the evaluation experiments

In the first set of experiments a large FTP transfer takes place between the client and the server. The traffic in this case is highly asymmetric with large data frames in one direction and small acknowledgement frames (ACKs) in the other. Therefore the link utilization should be high in the data direction and much smaller in the ACK direction. The results in terms of link utilization are shown in Table 2 (FTP in one direction). When the middle link operates at 10 Gb/s, the link utilization in the data direction is approximately 66% of its capacity and the use of the UCT mode could improve performance by getting close to 100% utilization. Therefore this would be a case in which switching to UCT mode using the proposed mechanism could be desirable to reduce the transfer time for large files. When the middle link operates at 1 Gb/s and a large FTP transfer takes place between the client and the server, the results show how the CT mode is capable of performing the FTP at the same speed as a UCT link (1 Gb/s or 10% utilization). In this case the UCT mode will not give a performance advantage and energy can be saved by using the CT mode. This would be the case when the transfer speed is limited by the network but also when it is limited by the client or server. As discussed in [17] in many cases the transfer speed may be limited by the processor, the operating system, and/or the application software such that the full link speed can not be achieved in any case.

In the second set of experiments a one way 1 Gb/s Constant Bit Rate source is used. The results show that the proposed scheme is able to carry this traffic with no performance lost when the middle link operates at 1 Gb/s and 10 Gb/s. In this case the transmissions are always triggered by the refresh periods in the other direction. This CBR example models the transmission of video or other applications that can only send data as it is generated in the remote end and therefore cannot take advantage of the full link speed.

Table 2. Results from ns-2 experiments

	Config (10 Gb/s, 40 µs delay)		Config (1 Gb/s, 400 µs delay)	
	Link Utilization		**Link Utilization**	
	downstream	upstream	downstream	upstream
FTP in one direction	66%	3.3%	10%	0.5%
CBR source	10%	0%	10%	0%
FTPs in both directions	38.9%	39.1%	10%	10%

Finally, the FTP experiments were re-run with simultaneous FTP transfers in both directions. The results show that when the middle link operates at 10 Gb/s the total link utilization is higher than for the case of one direction only. In this case CT reaches a total utilization of close to 8 Gb/s (4 Gb/s in each direction) compared to around 7 Gb/s in the asymmetrical case. This is so because in this case each time the link is activated transmission takes place for the maximum allowed time in CT (31.2 µs). Therefore the overhead of EEE mode transition is lower than in the asymmetric case in which in one direction the link is activated to send ACKs that require less than 2 µs to be sent. For this configuration the use of UCT transmission would increase the performance and therefore a switch to the UCT mode could be done to reduce transfer time. When the middle link operates at 1 Gb/s, the CT mode is sufficient to perform the FTP file transfers with no performance degradation.

The experiments conducted have served as an initial validation of the proposed scheme. The results show how the CT mode operation can achieve a total link utilization of close to 7 Gb/s for asymmetric traffic and approximately 8 Gb/s (4 Gb/s in each direction) for symmetric traffic. This would be sufficient in many environments where speed is limited by the network connection or by the system hardware and/or software. The same applies to applications that do not require or cannot use the full link speed, like for example multimedia applications. In any case a fast mechanism to switch to UCT mode is provided that enables the use of the complete link capacity when required.

5   Conclusions

In this paper a mechanism to achieve additional savings in 10GBASE-T links that implement the IEEE 802.3az Energy Efficient Ethernet (EEE) standard has been proposed. The mechanism coordinates transmission over the link such that at any given time instant only one link end is transmitting. To do so existing EEE signaling is used. The proposed scheme enables large additional savings in the range of 20-30% in the Physical layer devices (PHYs) as many components such as the echo cancellation circuitry can be put in low power mode. This savings are trade off for additional link delay and reduced link capacity. In the worst case, the proposed approach adds 40 µs of delay for frame transmission that is negligible compared to the typical Internet connection delays and the requirements of most end users applications. The effective capacity of the link is close to 7 Gb/s for asymmetric traffic and approximately 4 Gb/s for symmetric traffic which would be sufficient when the network connection or the computer performance limit the speed. For cases in which the full link speed is needed sporadically, a fast mechanism to switch to normal full-duplex (or UCT) transmission mode has also been described. In this way most of the energy savings would be retained while minimizing the impact on performance.

References

1. Gupta, M., Singh, S.: Greening of the Internet. In: Proc. of ACM SIGCOMM, Karlsruhe, Germany, pp. 19–26 (August 2003)
2. Christensen, K., Reviriego, P., Nordman, B., Bennett, M., Mostowfi, M., Maestro, J.A.: IEEE 802.3az: The Road to Energy Efficient Ethernet. IEEE Communications Magazine 48(11), 50–56 (2010)
3. IEEE Std 802.3az: Energy Efficient Ethernet-2010
4. Aquantia to Deliver World's First Quad 10GBASE-T Solution, press release (May 2009)
5. Gunaratne, C., Christensen, K., Nordman, B., Suen, S.: Reducing the Energy Consumption of Ethernet with Adaptive Link Rate (ALR). IEEE Transactions on Computers 57(4), 448–461 (2008)
6. Van Babel, N., Done, D., Flatman, A., McConnell, M.: Short Haul 10Gbps Ethernet Copper PHY Call for Interest, IEEE 802.3 (November 2005), http://grouper.ieee.org/groups/802/3/cfi/1105_1/10G_Short_Haul_CFI.pdf
7. IEEE Std 802.3an: 10GBASE-T-2006
8. Gupta, S., Tellado, J., Begur, S., Yang, F., Balan, V., Inerfield, M., Dabiri, D., Dring, J., Goel, S., Muthukumaraswamy, K., McCarthy, F., Golden, G., Jiangfeng, W., Arno, S., Kasturia, S.: A 10 Gb/s IEEE 802.3an-Compliant Ethernet Transceiver for 100m UTP Cable in $0.13\mu m$ CMOS. In: IEEE International Solid-State Circuits Conference, pp. 106–599 (2008)
9. Reviriego, P., Hernández, J.A., Larrabeiti, D., Maestro, J.A.: Performance Evaluation of Energy Efficient Ethernet. IEEE Communication Letters 13(9), 697–699 (2009)
10. Odlyzko, A.: Data Networks are Lightly Utilized, and will Stay that Way. Review of Network Economics 2(3), 210–237 (2003)
11. IEEE 802.3 Carrier Sense Multiple Access with Collision Detection (CSMA/CD) access method and physical layer specifications (2008)
12. Diab, W.: Method And System For Asymmetric Transition Handshake. In: An Energy Efficient Ethernet Network US Patent Application 2009/0154467, Filed on October 2, 2008, Published on June 18 (2009)
13. Lee, H.W., McConnell, S.M.: 10GBASE-T link speed arbitration for 30m transceivers, US Patent Application 2008/0219289, Filed on Novemebr 13, 2007, Published on September 11 (2008)
14. Chen, J., Parhi, K.K.: Further cost reduction of adaptive echo and next cancellers for high-speed Ethernet transceivers. In: IEEE Workshop on Signal Processing Systems (SiPS), pp. 227–232 (2008)
15. Dove, D.: Energy Efficient Ethernet: A Switching Perspective, presentation at the IEEE 802.3az, meeting (May 2008), http://www.ieee802.org/3/az/public/may08/dove_02_05_08.pdf
16. Wael, D.: LLDP use in EEE, presentation at the IEEE 802.3az, meeting (January 2009), http://www.ieee802.org/3/az/public/jan09/diab_02_0109.pdf
17. Bencivenni, M., Carbone, A., Fella, A., Galli, D., Marconi, U., Peco, G., Perazzini, S., Vagnoni, V., Zani, S.: High Rate Packet Transmission on 10 Gbit/s Ethernet LAN using Commodity Hardware. In: Proceedings of the 16th IEEE-NPSS Real Time Conference (2009)

Online Job-Migration for Reducing the Electricity Bill in the Cloud

Niv Buchbinder[1], Navendu Jain[2], and Ishai Menache[2]

[1] Open University, Israel
niv.buchbinder@gmail.com
[2] Microsoft Research
{navendu,t-ismena}@microsoft.com

Abstract. Energy costs are becoming the fastest-growing element in datacenter operation costs. One basic approach to reduce these costs is to exploit the spatiotemporal variation in electricity prices by moving computation to datacenters in which energy is available at a cheaper price. However, injudicious job migration between datacenters might increase the overall operation cost due to the bandwidth costs of transferring application state and data over the wide-area network. To address this challenge, we propose novel online algorithms for migrating batch jobs between datacenters, which handle the fundamental tradeoff between *energy* and *bandwidth* costs. A distinctive feature of our algorithms is that they consider not only the current availability and cost of (possibly multiple) energy sources, but also the *future* variability and uncertainty thereof. Using the framework of competitive-analysis, we establish worst-case performance bounds for our basic online algorithm. We then propose a practical, easy-to-implement version of the basic algorithm, and evaluate it through simulations on real electricity pricing and job workload data. The simulation results indicate that our algorithm outperforms plausible greedy algorithms that ignore future outcomes. Notably, the actual performance of our approach is significantly better than the theoretical guarantees, within 6% of the optimal offline solution.

Keywords: cloud computing, energy efficiency, job migration, datacenters.

1 Introduction

Energy costs are becoming a substantial factor in the operation costs of modern datacenters (DCs). It is expected that by 2014, the infrastructure and energy cost (I&E) will become 75% of the total operation cost, while information technology (IT) expenses, i.e., the equipment itself, will induce only 25% of the cost (compared with 20% I&E and 80% IT in the early 90's) [5]. Moreover, the operating cost per server will be double its capital cost (over its amortized lifespan of 3-5 years). According to an EPA report, servers and datacenters consumed 61 billion Kilowatt at a cost of $4.5 billion in 2007, with demand expected to double by 2011 [13]. These costs might further increase due to carbon taxes which are being imposed by several countries to reduce the environmental impact of electricity generation and consumption [11,16].

To reduce the energy costs, research efforts are being made in two main directions: (i) Combine traditional electricity-grid power with *renewable* energy sources such as

J. Domingo-Pascual et al. (Eds.): NETWORKING 2011, Part I, LNCS 6640, pp. 172–185, 2011.

solar, wave, tidal, and wind, motivated by long-term cost reductions and growing unreliability of the electric grid [15,18]; and (ii) Exploit the temporal and geographical variation of electricity prices [19,20]. As shown in Fig 1(a), energy prices can significantly vary over time and location (even within 15 minute windows). Therefore, a natural approach is to shift the computational load to datacenters where the current energy prices are the cheapest. In this paper, we combine these two directions, by proposing *efficient algorithms for dynamic placement (migration) of batch applications (or jobs) in datacenters, based on their energy cost and availability.* Specifically, we focus on executing batch applications such as MapReduce programs, web crawling and index generation for web search, and data analytics, which are typically delay tolerant and can handle the interruption during job migration between datacenters[1].

A key challenge for dynamic job placement is to account for the bandwidth cost of moving the application state and data between source and destination datacenters. The application data and memory footprint depends on the underlying job, and can range from a few KBs to hundreds of MBs [12] [2]. Bandwidth fees are typically charged proportionally to the number of bytes sent and received during the migration process [23]; consequently, bandwidth costs increase linearly with the amount of migrated data.

This paper proposes online algorithms for migrating batch applications within a cloud of multiple datacenters. The input for these algorithms includes the current energy prices and power availability at different datacenters. The algorithms accordingly determine which jobs should be migrated and to which hosting location, while taking into account the bandwidth costs and the current job allocation. Our model accommodates datacenters powered by multiple energy suppliers such as the grid and renewable energy sources. More generally, we assume that the total energy cost in each datacenter is a convex increasing function of utilization. Energy-cost reduction would thus depend on the ability to exploit the instantaneously cheap energy sources to their full availability. A distinctive feature of our approach is that we do *not* settle for an intuitive greedy approach to optimize the overall energy and bandwidth costs, yet take into account future deviations in energy price and power availability. We emphasize that our electricity-pricing model is distribution-free (i.e., we do not make probabilistic assumptions on future electricity prices), as prices can often vary unexpectedly [19].

Related work. The general framework of energy-efficient resource management in cloud datacenters has been an active research area over the last few years. Numerous papers consider the *intra* datacenter optimization, which includes techniques such as switching off idle servers [14] and VM migration and consolidation for load balancing and power management [6]. Motivated by the notable spatiotemporal variation in electricity prices, the networking research community has recently started focusing on *inter* datacenter optimization, which is also the context of our work. Qureshi et al. [19] propose greedy heuristics for redirecting application requests to different datacenters, and evaluate them using simulations on historical electricity prices and Akamai's traffic

---

[1] We note that there are known job migration techniques (e.g., pipelining and incremental snapshots [12]) to minimize the interruption delay.

[2] Note that persistent data such as crawled web pages and click-stream logs, are typically replicated at multiple datacenters and thus do not need to be migrated for individual jobs.

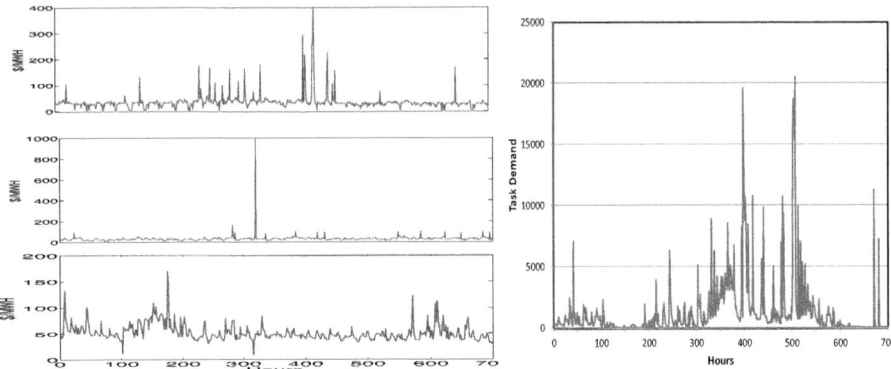

Fig. 1. (a, left) Electricity pricing trends across three US locations (CA (top), TX (middle), and MA (bottom)) over a month in 2009. The graphs indicate significant spatiotemporal variation in electricity prices. (b, right) Hourly task demand data from a cluster running MapReduce jobs. The demand data shows significant variation over time.

data. The authors show that judicious placement can save millions of dollars in energy costs. Rao et al. [20] consider load-balancing of delay sensitive applications with the goal of minimizing the energy cost subject to delay constraints. They use queuing theory and linear programming techniques and report cost savings through simulations with electricity pricing data. Finally, Le et al. [17] propose an optimization-based framework to distribute user requests across datacenters. Their goal is to process requests within a specified cap on energy from non-renewable sources, while meeting the application service-level agreements and minimizing energy costs.

Our paper significantly differs from prior work in three ways. First, we consider the migration of *batch* applications, rather than the redirection of *interactive* applications' requests. Second, our algorithms take into account not only the immediate savings of job migration, but also the future consequences thereof, thereby accounting for future uncertainty in the cost parameters. Third, we incorporate the bandwidth cost into the optimization problem, as it can be a significant cost factor for job migration.

Our solution method lies within the framework of online computation [7], which has been applied for solving problems with uncertainty in networks, finance, mechanism design and other fields. Specifically, we employ online primal-dual techniques [10], including recent ideas from [2,3]. We build upon two well-studied online problems, weighted paging/caching [1,21] and Metrical Task System (MTS) [4,8] (see our technical report [9] for details), and extend them to address two new challenges specific to our problem: (1) optimizing energy costs of executing batch jobs under time-varying electricity prices and job demand, and (2) balancing energy savings with bandwidth costs to minimize the total operation cost. Since our problem is a generalization of the MTS problem [8], we get a lower bound of $\Omega(\log n)$ on the competitiveness of any online algorithm where n is the number of datacenters. A tighter lower bound is subject for future examination.

Contribution and Paper Organization. In this paper, we show, both analytically and empirically, that existing cloud computing infrastructure can perform dynamic application placement based on energy cost and availability and bandwidth costs, for significant economic gains. Specifically,

• We design a basic online algorithm (Section 3) which is $O(\log H_0)$-competitive (compared to the optimal offline), where H_0 is the total number of servers in the cloud.
• To reduce the extensive computation required by the basic algorithm in each iteration, we construct a computationally efficient version of the basic algorithm, which can be easily implemented in practice (Section 4).
• We evaluate our algorithms (Section 5) using real electricity pricing data obtained from 30 US locations over three years, and job-demand data from a large cluster in a datacenter, and show significant improvements over greedy-optimization algorithms. Overall, the cost savings of our online algorithm are typically within 6% of the optimal solution.

The technical proofs of this work are omitted due to lack of space and can be found in an accompanying technical report [9].

2 The Model

In this section we formally define the job migration problem. In Section 2.1 we describe the elements of the problem. In Section 2.2 we formalize the online optimization problem, the solution of which is the main objective of this paper.

2.1 Preliminaries

We consider a large computing facility (henceforth referred to as a "cloud", for simplicity), consisting of n datacenters (DCs). Each DC $i \in \{1, \ldots, n\}$ contains a set $\mathcal{H}_i = \{1, \ldots, H_i\}$ of H_i servers which can be activated simultaneously. We denote by $H_0 = \sum_{i=1}^{n} H_i$ the total number of servers in the cloud, and often refer to H_0 as the *capacity* of the cloud. All servers in the cloud are assumed to have equal resources (in terms of CPU, memory, etc.).

A basic assumption of our model is that *energy (or electricity) costs* may vary in time, yet remain fixed within time intervals of a fixed length of τ (say 15 minutes, or one hour). Accordingly, we shall consider a discrete time system, where each time-step (or *time*) $t \in \{1, \ldots, T\}$ represents a different time interval of length τ. We denote by B_t the total job-load at time t measured in server units, namely the total number of servers which need to be active at time t. For simplicity, we assume that each job requires a single server, thus B_t is also the total number of jobs at time t. We note, however, that the latter assumption can be relaxed by considering a single job per processor or per VM, similar to the virtualization setup. In Section 3 we consider a simplified scenario where $B_t = B$, namely the job-load in the cloud is fixed. We will address the case where B_t changes in time, which corresponds to job arrivals and departures to/from the cloud in Section 4. At each time t, the control decision is to assign jobs to DCs, given the current electricity and bandwidth costs, which are described below.

Energy costs. We assume that the energy cost at every DC is a function of the number of servers that are being utilized. Let $y_{i,t}$ be the number of servers that are utilized in DC i at time t. The (total) energy cost is given by $\tilde{C}_{i_t}(y_{i,t})$ (measured in dollars per hour). Let $C_{i_t} : \mathbb{R}_+ \to \mathbb{R}_+$ be the interpolation of $\tilde{C}_{i_t}(\cdot)$; for convenience, we shall consider this function in the sequel. We assume throughout that $C_{i_t}(y_{i,t})$ is a *non-negative, (weakly) convex increasing function*. Note that the function itself can change with t, which allows for time variation in energy prices. We emphasize that we do not make any probabilistic assumptions on $C_{i_t}(\cdot)$, i.e., this function can change *arbitrarily* over time. The simplest example that complies with the convexity assumption is the linear cost model, in which the total energy cost is linearly proportional to the number of servers utilized. Assuming that servers not utilized are switched off and consume zero power, we have $C_{i_t}(y_{i,t}) = c_{i,t}y_{i,t}$, where $c_{i,t}$ is the cost of using a single server at time t. Further examples can be found in [9].

Bandwidth Costs. The migration of jobs between DCs induces bandwidth costs. In practice, bandwidth fees are paid both for outgoing traffic and incoming traffic. Accordingly, we assume the following cost model. For every DC i, we denote by $d_{i,out}$ the bandwidth cost of transferring a job *out* of DC i, and by $d_{i,in}$ the bandwidth cost of migrating a job into this DC[3]. Thus, the overall cost of migrating a job from DC i to DC j is given by $d_{i,out} + d_{j,in}$. We note that there are also bandwidth costs associated with the arrival of jobs into the cloud (e.g., from a central dispatcher) and leaving the cloud (in case of a job departure). However, these costs are constant and do not depend on the migration control, and are thus ignored in our formulation.

Our goal is to minimize the total operation cost in the cloud, which is the sum of the energy costs at the DC and the bandwidth costs of migrating jobs between DCs. It is assumed that the job migration time is negligible with regard to the time interval τ in which the energy costs are fixed. This means that migrated jobs at time t incur the energy prices at their new DC, starting from time t. We refer to the above minimization problem as the *migration problem*. If all future energy costs were known in advance, then the problem could have optimally be solved as a standard convex program. However, our basic assumption is that electricity prices change in an unpredicted manner. We therefore tackle the migration problem as an *online* optimization problem.

2.2   Problem Formulation

In this subsection we formalize the migration problem. We shall consider a slightly different version of the original problem, which is algorithmically easier to handle. Nonetheless, the solution of the modified version leads to provable performance guarantees for the original problem. Recall that we consider the case of fixed job-load, namely $B_t = B$ for every t. We argue that it is possible to define a modified model in which we charge a migration cost $d_i \stackrel{\triangle}{=} d_{i,in} + d_{i,out}$ whenever the algorithm migrates a job out of DC i, and do not charge for migrating jobs into the DC. The following lemma states that this bandwidth-cost model is equivalent to the original bandwidth-cost model up to an additive constant.

Lemma 1. *The original and modified bandwidth-cost model, in which we charge d_i for migrating jobs out of DC i, are equivalent up to additive constant factors. In particular,*

---

[3] ISPs charge bandwidth to DC operators based on the daily 95th percentile of incoming/outgoing bandwidth over 5 min periods.

a c-competitive algorithm for the former results in a c-competitive algorithm for the latter.

We note that the same result holds for the model which charges d_i for migrating jobs *into* the DC, and nothing for migrating jobs out of the DC.

Using Lemma 1, we first formulate a convex program whose solution provides a lower bound on the optimal solution for the (original) migration problem. Let $z_{i,t}$ be the number of jobs that are migrated from DC i at time t. This variable is multiplied by d_i for the respective bandwidth cost. The optimization problem is defined as

$$\min \quad \sum_{i=1}^{n} \sum_{t=1}^{T} d_i z_{i,j} + \sum_{i=1}^{n} \sum_{t=1}^{T} C_{i_t}(y_{i,t}),$$

subject to the following constraints: $\sum_{i=1}^{n} y_{i,t} \geq B$ for every t (i.e., meet the job demand); $z_{i,t} \geq y_{i,t-1} - y_{i,t}$ for every i, t (i.e., $z_{i,t}$ reflects the number of jobs leaving DC i at time t); $y_{i,t} \leq H_i$ for every i, t (i.e., meet the capacity constraint); $z_{i,t}, y_{i,t} \in \{0, 1, \ldots, H_i\}$ for every i, t. We shall consider a relaxation of the above optimization problem, where the last constraint is replaced with $z_{i,t}, y_{i,t} \geq 0$ for every i, t. A solution to the relaxed problem is by definition a lower bound on the value of the original optimization problem.

Instead of considering the above described convex program (with the relaxation), we construct below an equivalent LP, which will be easier to work with. For every time t, the cost of using the j-th server at DC i is given by $c_{i,j,t}$, where $c_{i,j,t} \in [0, \infty]$ is increasing in j ($1 \leq j \leq H_i$). We assume that $c_{i,j,t}$ may change arbitrarily over time (while keeping monotonicity in j as described above). Note that we allow $c_{i,j,t}$ to be unbounded, to account for the case where the total energy availability in the DC is not sufficient to run all of its servers. Let $0 \leq y_{i,j,t} \leq 1$ be the utilization level of the j-th server of DC i (we allow for "partial" utilization, yet note that at most a single server will be partially utilized in our solution). Accordingly, the energy cost of using this server at time t is given by $c_{i,j,t} y_{i,j,t}$. Define $z_{i,j,t}$ to be the workload that is migrated from the jth server of DC i to a different location. This variable is multiplied by d_i for the respective bandwidth cost. The optimization problem is defined as follows.

(P) : min $\qquad \sum_{i=1}^{n} \sum_{j=1}^{H_i} \sum_{t=1}^{T} d_i z_{i,j,t} + \sum_{i=1}^{n} \sum_{j=1}^{H_i} \sum_{t=1}^{T} c_{i,j,t} \cdot y_{i,j,t}$ (1)

subject to $\qquad \sum_{i=1}^{n} \sum_{j=1}^{H_i} y_{i,j,t} \geq B \quad$ for every t,

$z_{i,j,t} \geq (y_{i,j,t-1} - y_{i,j,t}), \quad 0 \leq y_{i,j,t} \leq 1, \quad z_{i,j,t} \geq 0, \quad$ for every i, j, t.

Observe that the first term in the objective function corresponds to the total bandwidth cost, whereas the second term corresponds to the total energy cost.

One may notice that the above problem formulation might pay the bandwidth cost d_i for migrating data *within* each DC i, which is clearly not consistent with our model assumptions. Nevertheless, since the energy costs $c_{i,j,t}$ in each DC i are non-decreasing in j for every t, an optimal solution will never prefer servers with higher indexes over lower-index servers. Consequently, jobs will not be migrated within a DC. Thus, solving (P) is essentially equivalent to solving the original convex problem. This observation, together with Lemma 1, allows us to consider the optimization problem (P). More formally,

Lemma 2. *The value of (P) is a lower bound on the value of the optimal solution to the migration problem.*

We refer to the optimization problem (1) as our *primal problem* (or just(P)). The dual of (P) is given by

$$(D) : \max \quad \sum_{t=1}^{T} Ba_t - \sum_{i=1}^{n} \sum_{j=1}^{H_i} \sum_{t=1}^{T} s_{i,j,t},$$

$$\text{subject to} \quad -c_{i,j,t} + a_t + b_{i,j,t} - b_{i,j,t+1} - s_{i,j,t} \leq 0,$$

$$0 \leq b_{i,j,t} \leq d_i, \quad \text{and} \quad s_{i,j,t} \geq 0 \quad \text{for every } i, j, t.$$

By Lemma 2 and the weak duality theorem (see, e.g., [22]), an algorithm that is c-competitive with respect to a *feasible* solution of (D), would be c-competitive with respect to the offline optimal solution.

3   An Online Job-Migration Algorithm

In this section we design and analyze an online algorithm for the migration problem (P). The algorithm we design is based on a *primal-dual* approach, meaning that the new primal and dual variables are *simultaneously* updated at every time t. The general idea behind the algorithm is to maintain a feasible dual solution to (D), and to upper-bound the operation cost at time t (consisting of bandwidth cost and the energy cost) as a function of the value of (D) at time t. Since a feasible solution to (D) is a lower bound on the optimal solution, this procedure would immediately lead to a bound on the competitive ratio of the online algorithm.

The online algorithm outputs a *fractional* solution to the variables $y_{i,j,t}$. To obtain the total number of servers that should be activated in DC i, we simply calculate the sum $\sum_{j=1}^{H_i} y_{i,j,t}$ at every time t. Since this sum is generally a fractional number, one can round it to the nearest integer. Because the number of servers in each DC (H_i) is fairly large, the effect of the rounding is negligible and thus ignored in our analysis.

The online algorithm receives as input the energy cost vector $\{c_{i,j,t}\}_{i,j}$ at every time t defining the energy cost at each server (i, j) at time interval t. As a first step, we use a well known reduction that allows us to consider only *elementary cost vectors* [7, Sec. 9.3.1]. Elementary cost vectors are vectors of the form $(0, \ldots, 0, c_{i_t,j_t}, 0, \ldots, 0)$, namely vectors with only a single non-zero entry. This reduction allows us to split any general cost vector into a finite number of elementary cost vectors without changing the value of the optimal solution. Furthermore, we may translate any online migration decisions done for these elementary task vectors to online decisions on the original cost vectors without increasing our cost. Thus, we may consider only elementary cost vectors from now on. Abusing our notations, we use the original time index t to describe these elementary cost vectors. We use c_t instead of c_{i_t,j_t} to describe the (single) non-zero cost at server (i_t, j_t) at time t. Thus, the input for our algorithm consists of i_t, j_t, c_t.

The algorithm we present updates at every time t the (new) dual variables a_t, s_t and $b_{i,j,t+1}$. The values of these variables are determined incrementally, via a continuous update rule (closed-form one-shot updates seems hard to obtain). We summarize in Table 1 the procedure for the update of the dual variables at any time t. We refer to the procedure at time t as the t-th *iteration*. We note that the continuous update rule is

Table 1. The procedure for updating the dual variables of the online problem

---

Input: Datacenter i_t, server j_t, cost c_t (at time t).
Initialization: Set $a_t = 0$. For all i, j: set $b_{i,j,t+1} = b_{i,j,t}$ and $s_{i,j,t} = 0$.
Loop:

- Keep increasing a_t, while **termination condition** is not satisfied.
- Update the other dual variables as follows:
  - For every $i, j \neq i_t, j_t$ such that $y_{i,j,t} < 1$, increase $b_{i,j,t+1}$ with rate $\frac{db_{i,j,t+1}}{da_t} = 1$.
  - For every $i, j \neq i_t, j_t$ such that $y_{i,j,t} = 1$, increase $s_{i,j,t}$ with rate $\frac{ds_{i,j,t}}{da_t} = 1$.
  - For $i, j = i_t, j_t$, decrease $b_{i_t,j_t,t+1}$ so that $\sum_{i,j \neq i_t,j_t} \frac{dy_{i,j,t}}{da_t} = -\frac{dy_{i_t,j_t,t}}{da_t}$.

Termination condition: Exit when either $y_{i_t,j_t,t} = 0$, or $-c_t + a_t + b_{i_t,j_t,t} - b_{i_t,j_t,t+1} = 0$.

---

mathematically more convenient to handle. Nonetheless, it is possible to implement the online algorithm through discrete-time iterations, by searching for the scalar value a_t up to any required precision level.

Each primal variable $y_{i,j,t}$ is continuously updated as well, alongside with the continuous update of the respective dual variable $b_{i_t,j_t,t+1}$. The following relation between $b_{i_t,j_t,t+1}$ and $y_{i,j,t}$ is preserved throughout the iteration:

$$y_{i,j,t} := \frac{1}{H_0} \left(\exp\left(\ln(1 + H_0) \frac{b_{i,j,t+1}}{d_i} \right) - 1 \right). \tag{2}$$

The variable $y_{i_t,j_t,t}$ is updated so that the amount of work that is migrated to other servers is equal to the amount of work leaving j_t (i.e, total work conservation).

The next theorem presents a performance bound for the above described algorithm.

Theorem 1. *The online algorithm is $O(\log(H_0))$-competitive.*

The proof proceeds in a number of steps. We first show that our algorithm preserves a feasible primal solution and a feasible dual solution. We then relate the change in the dual variable $b_{i,j,t+1}$ to the change in the primal variable $y_{i,j,t}$, and the change of the latter to the change in the value of the feasible dual solution. This allows us to upper bound the bandwidth cost at every time t as a function of the change in the value of the dual. Finally, we find a precise relation between the bandwidth cost and the energy cost, which allows us to also bound the latter as a function of the change in the dual value. A detailed proof can be found in [9].

4   An Efficient Online Algorithm

Theorem 1 indicates that the online algorithm is expected to be robust against *any* possible deviation in energy prices. However, its complexity might be too high for an efficient implementation with an adequate level of precision. First, the reduction of [7, Sec. 9.3.1] generally splits each cost vector at any given time t into $O(H_0^2)$ vectors, thereby requiring a large number of iterations per each actual time step t. Second, the

Table 2. The iteration of the EOA. Time indexes are omitted.

Input: (i) Cost vector $\mathbf{c} = (c_1, c_2, \ldots, c_n)$; (ii) $y = (y_1, y_2, \ldots, y_n)$ (the current load vector on the datacenters).

Job Migration Loop: Outer loop: $k = 1$ to n, Inner loop: $i = k + 1$ to n

– Move load from DC i to DC k according to the following migration rule.

$$\min \left\{ y_i, H_k - y_k, s_1 \cdot \frac{c_{i,k} y_i}{d_{i,k}} \cdot (y_k + s_2) \right\}, \tag{3}$$

where $s_1, s_2 > 0$.

---

need for proper discretization of the continuous update rule of the dual variables (see Table 1) might make each iteration computationally expensive. Therefore we design in this section an "easy-to-implement" online algorithm, which inherits the main ideas of our original algorithm, yet decreases significantly the running complexity per time step.

We employ at the heart of the algorithm several ideas and mathematical relations from the original online algorithm. The algorithm can accordingly be viewed as a computationally-efficient variant of the original online algorithm, thus would be henceforth referred to as the *Efficient Online Algorithm* (EOA). While we do not provide theoretical guarantees for the performance of the EOA, we shall demonstrate through real-data simulations its superiority over plausible greedy heuristics (see Section 5). The EOA provides a complete algorithmic solution which also includes the allocation of newly arrived jobs; the way job arrivals are handled is described towards the end of this section.

For simplicity, we describe the EOA for *linear* electricity prices (i.e., the electricity price for every DC i and time t is $c_{i,t}$, regardless of the number of servers that are utilized). In [9], we describe the adjustments needed to support non-linear prices. The input for the algorithm at every time t is thus an n dimensional vector of prices, $\mathbf{c}_t = (c_{1,t}, c_{2,t}, \ldots, c_{n,t})$. For ease of illustration, we reorder the DCs in each iteration according to the present electricity prices, so that $c_1 \leq c_2 \leq \cdots \leq c_n$. For exposition purposes we omit the time index t from all variables (e.g., we write $\mathbf{c} = (c_1, c_2, \ldots, c_n)$ instead of $\mathbf{c}_t = (c_{1,t}, c_{2,t}, \ldots, c_{n,t})$). We denote by $d_{i,k} = d_{i,out} + d_{k,in}$ the bandwidth cost per unit for transferring load from DC i to DC k. We further denote by $c_{i,k}$ the difference in electricity prices between the two DCs, namely $c_{i,k} = c_i - c_k$, and refer to $c_{i,k}$ as the *differential energy cost* between i and k. The iteration of the EOA is described in Table 2.

Before elaborating on the logic behind the EOA, we note that the running complexity of each iteration is $O(n^2)$. Observe that the inner loop makes sure that data is migrated from the currently expensive DCs to cheaper ones (in terms of the corresponding electricity prices). We comment that the fairly simple migration rule (3) could be implemented in a distributed manner, where DCs exchange information on their current load and electricity price; we however do not focus on a distributed implementation in the current paper.

We now proceed to discuss the rational behind the EOA, and in particular the relation between (3) (which will be henceforth referred to as the *migration rule*) and the original online algorithm described earlier. We first motivate the use of the term $c_{i,k}y_i$ in the migration rule. Note first that this term corresponds to the energy cost that could potentially be saved by migrating all jobs in DC i to DC k. The reason that we use differential energy costs $c_{i,k} = c_i - c_k$ is *directly* related to the reduction from general cost vectors to elementary cost vectors (see Section 3). The reduction creates elementary cost vectors by iterating over the differential energy costs $c_i - c_{i-1}$, $i \in [2, n]$.

Further examination of the migration rule (3) reveals that the amount of work that is migrated from DC i to DC k is proportional to y_k (the load at the target DC). The motivation for incorporating y_k in the migration rule follows by differentiating the primal-dual relation (2) with respect to $b_{i,j,t+1}$, which leads to the following relation

$$\frac{dy_{i,j,t}}{db_{i,j,t+1}} = \frac{\ln(1 + H_0)}{d_i} \cdot \left(y_{i,j,t} + \frac{1}{H_0} \right) \tag{4}$$

The above equation implies that the change in the load in DC k should be proportional to the current load. As discussed earlier, this feature essentially encourages the migration of jobs into DCs that were consistently "cheap" in the past, and consequently loaded in the present. Another idea that follows from (4) and is incorporated in the migration rule is that the migrated workload is *inversely* proportional to the bandwidth cost. The last idea which is borrowed from (4) is to include an additive term s_2, which reduces the effect of y_k. This additive term enables the migration of jobs to DCs, even if they are currently empty. Intuitively, the value of s_2 manifests the natural tradeoff between making decisions according to current energy prices (high s_2) or relaying more on usage history (low s_2). The other parameter in the migration rule, s_1, sets the "aggressiveness" of the algorithm. Increasing s_1 makes the algorithm greedier in exploiting the currently cheaper electricity prices (even at the expense of high bandwidth costs). Both s_1 and s_2 can be tuned throughout the execution of the algorithm.

To complete the description of the EOA, we next specify how newly arrived traffic is distributed. We make here the simplifying assumption that jobs arrive to a single global dispatcher, and then allocated to the different DCs. The case of multiple dispatchers is outside the scope of this paper, nonetheless algorithms for this case can be deduced from the rule we describe below. A newly arrived job is assigned through the following probabilistic rule: *At every time t, assign the job to DC i with probability proportional to:* $\frac{1}{d_i}(y_i + s_2)$, *where* $d_i = d_{i,in} + d_{i,out}$. The reasoning behind this rule is the same as elaborated above for the migration rule (3).

5   Simulations

In this section we evaluate the performance of the EOA on real datasets, while comparing it to two plausible greedy algorithms.

Alternative Job-Migration Algorithms. As a first simple benchmark for our algorithm, we use a greedy algorithm referred to as *Move to Cheap* (MTC) that always keeps the load on DCs having the lowest electricity prices (without exceeding their capacity). While its energy costs are the lowest possible by definition, the bandwidth costs turn out to be significant and its overall performance is eventually poor. The second greedy heuristic that we consider is *MimicOPT*. This heuristic tries to emulate the

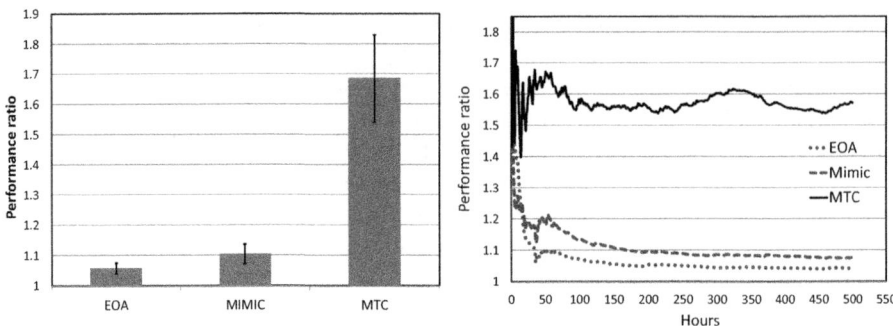

Fig. 2. Performance comparison of the MTC, MimicOPT, and EOA algorithms with respect to OPT: (a, left) Average performance ratios along with standard-deviations, and (b, right) The evolution of the competitive ratio of the algorithms with respect to OPT for a sample run

behavior of the optimal solution i.e., at any given time, MimicOPT solves an LP with the energy costs until the current time t, and migrates jobs so as to mimic the job allocation of the current optimal solution up to time t. As MimicOPT does not have the data on future power costs and does not consider future outcomes, it is obviously not optimal. However, this heuristic is reasonable in non-adversarial scenarios and, indeed, performs fairly well. To make the running time of MimicOPT feasible, we restrict the LP to use a window of the last 80 hours (a further increase in the window size does not seem to significantly change the results). In general, we expect MimicOPT to perform well on inputs in which electricity prices are highly correlated in time. We compare below the performance of the MTC, MimicOPT and EOA algorithms. Our reference is naturally the *offline* solution to the optimization problem, denoted OPT. Obviously, OPT obtains the lowest possible cost.

Electricity Prices and Job-Demand Data. We use in our experiments the historical hourly electricity price data ($/MWh) from publicly available government agencies for 30 US locations, covering January 2006 through March 2009. Due to space constraints, we omit the organization details of the electricity system in the US, see, e.g., [19] for details. In addition to the hourly electricity market, there are also 15-minute real-time markets which exhibit relatively higher volatility with high-frequency variation. Since these markets account for only a small fraction of the total energy transactions (below 10%), we do not consider them in our simulations.

A snapshot of the dynamics of electricity prices in different locations (Figure 1(a)) demonstrates that (i) prices are relatively stable in the long-term, but exhibit high day-to-day volatility, (ii) there are short-term spikes, and (iii) hourly prices are not correlated across different locations.

For our simulations, we use task-demand data obtained from a 10K node cluster running MapReduce jobs for a large online services provider. Fig. 1(b) shows the hourly task demand data exhibiting a significant variation in the number of tasks corresponding to job submissions and different MapReduce phases across running jobs.

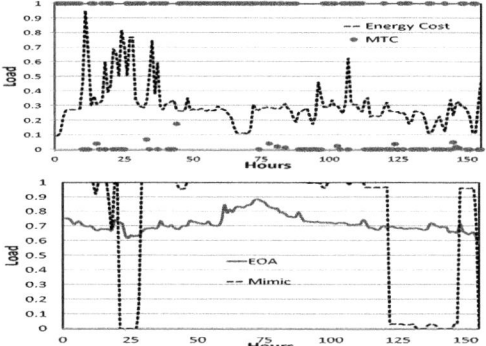
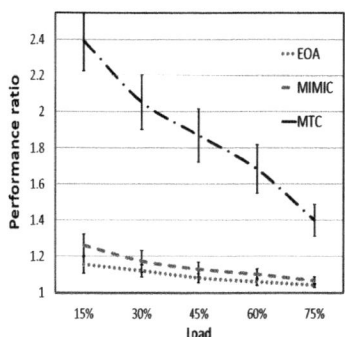

Fig. 3. Job load and normalized electricity price over time at Boston, MA. The top graph shows the electricity price and the load that MTC puts on this DC. The bottom graph shows the load for EOA and MimicOPT.

Fig. 4. The relative performance of EOA, MimicOPT, and MTC with respect to OPT as a function of the DC utilization

Performance Evaluation. We first compare the performance ratio of the total electricity and bandwidth costs incurred by different algorithms over pricing data from 10 locations. We assume that each location has 25K servers, each server consumes 200 Watts, bandwidth price is $0.15 per GB [23], and the task memory footprint is 100 MB (corresponding to intermediate results in MapReduce phases). Based on our discussion with datacenter operators, we compute the total job demand by fixing background demand of 60% (15K servers) per location and a dynamic component from MapReduce traces which runs on the remaining 40% capacity.

(i) Evaluating Total Cost. Fig. 2 compares the total cost (energy + bandwidth) for EOA, MimicOPT and MTC algorithms. Fig. 2(a) shows a ten-run average of the competitive ratio of each algorithm in 20 different executions of 500 hours each. We observe that the EOA performs better than MimicOPT and MTC, with a relatively small standard deviation. EOA performs within 5.7% of the optimum while MimicOPT and MTC perform within 10.4% and 68.5% of the optimum, respectively. Fig. 2(b) shows the competitive ratio of the three algorithms as a function of the number of hours in a 500 hour run. We observe that all algorithms exhibit high variation initially due to a high migration cost, whereas OPT migrates much less and stabilizes quickly, because it knows all future prices in advance.

Fig. 3 shows the workload evolution for the (Boston, MA) location. The top graph shows the normalized electricity costs of the location (exhibiting up to 10x variation), along with the normalized load that MTC puts on the Boston location. Note that MTC always uses the cheapest DCs. Thus, a load of one indicates that Boston is one of the 60% cheaper locations, while a load of zero indicates that it is among the 40% more expensive locations. The bottom figure shows the fraction of servers which are kept active in this location by MimicOPT and EOA. As expected, we observe that the number of active servers for MimicOPT varies significantly with electricity pricing (the DC is either full or empty), indicating that the optimal job assignment varies with time across locations. The EOA algorithm exhibits relatively stable job assignment, as it takes into account the relative pricing of this location, and carefully balances the energy costs with the bandwidth costs of migrating jobs between locations.

(ii) Evaluating Performance as a function of Utilization. Fig. 4 shows the performance ratio as we vary the background utilization of the DCs from 15% to 75%. Each point is the average competitive ratio of ten runs of 500 hours each. We observe that the performance ratio improves (decreases) as the load increases due to the reduced flexibility in the job assignment and EOA performs closest to OPT compared to other algorithms.

Overall, we observe that the job assignment changes significantly with the variation in electricity prices, as all plausible algorithms increase the number of servers at locations with lower prices. Overall, the EOA algorithm provides the best performance ratio and a relatively stable job assignment. While MimicOPT performs close to EOA, the performance gap between the two algorithms becomes larger when subtracting the minimal energy cost that has to be paid by any algorithm. Specifically, the performance-ratio becomes around 1.55 and 1.8 for EOA and MimicOPT, respectively, indicating that the actual cost savings by our algorithm are significant.

6   Concluding Remarks

This paper introduces novel online algorithms for migrating batch applications, with the objective of reducing the total operation cost in a cloud of multiple datacenters. We provide a competitive-ratio bound for the performance of a basic online algorithm, indicating its robustness against any *future* deviation in energy prices. An easy-to-implement version of the basic algorithm is designed, which can be seamlessly integrated as part of existing control mechanisms for datacenter operation. Our simulations on real electricity pricing data and job demand data demonstrate that the actual performance of our algorithm could be very close to the minimal operation cost (obtained through offline optimization). Importantly, our online algorithm outperforms several reasonable greedy heuristics, thereby emphasizing the need for a rigorous online optimization approach.

The algorithms we suggest solve the fundamental energy-cost vs. bandwidth-cost tradeoff associated with migrating batch jobs. Yet, we have certainly not covered herein all possible models and aspects that may need to be considered in the design of future job-migration algorithms. For example, one simplifying assumption we make is that every job can be placed in every DC. In some cases, however, some practical constraints might prevent certain jobs from being run at certain locations. Such constraints could be due to government regulations (e.g., EU data can only be hosted on servers in Europe), delay restrictions (e.g., if a particular DC is too far from the location of the source), security considerations, and other factors. On the other hand, some constraints might require certain jobs to be placed adjacent to others (e.g., to accommodate low inter-component communication delay for interactive applications). Further restrictions might be imposed on the migration paths, e.g., due to the delay overhead of the migration process. Overall, the general framework of online job-migration opens up new algorithmic challenges, as proper consideration of the dynamically evolving electricity prices can contribute to significant cost reductions in cloud environments.

References

1. Bansal, N., Buchbinder, N., Naor, J.: A primal-dual randomized algorithm for weighted paging. In: FOCS, pp. 507–517 (2007)
2. Bansal, N., Buchbinder, N., Naor, J.: Metrical task systems and the k-server problem on HSTs. In: Abramsky, S., Gavoille, C., Kirchner, C., Meyer auf der Heide, F., Spirakis, P.G. (eds.) ICALP 2010. LNCS, vol. 6198, pp. 287–298. Springer, Heidelberg (2010)
3. Bansal, N., Buchbinder, N., Naor, J.: Towards the randomized k-server conjecture: a primal-dual approach. In: SODA (2010)
4. Bartal, Y., Blum, A., Burch, C., Tomkins, A.: A polylog(n)-competitive algorithm for metrical task systems. In: STOC, pp. 711–719 (1997)
5. Belady, C.: In the data center, power and cooling costs more than it equipment it supports. Electronics Cooling Magazine (February 2007)
6. Beloglazov, A., Buyya, R.: Energy efficient resource management in virtualized cloud data centers. In: 10th IEEE/ACM International Conference on Cluster, Cloud and Grid Computing, pp. 826–831 (2010)
7. Borodin, A., El-Yaniv, R.: Online computation and competitive analysis. Cambridge University Press, Cambridge (1998)
8. Borodin, A., Linial, N., Saks, M.E.: An optimal on-line algorithm for metrical task system. Journal of the ACM 39(4), 745–763 (1992)
9. Buchbinder, N., Jain, N., Menache, I.: Online job-migration for reducing the electricity bill in the cloud. Tech. Rep. MSR-TR-2011-20, Microsoft Research (February 2011)
10. Buchbinder, N., Naor, J.: The design of competitive online algorithms via a primal-dual approach. Foundations and Trends in Theoretical Computer Science 3(2-3), 93–263 (2009)
11. Butler, D.: (2009), http://www.nature.com/news/2009/090911/full/news.2009.905.html
12. Clark, C., Fraser, K., Steven, H., Hansen, J.G., Jul, E., Limpach, C., Pratt, I., Warfield, A.: Live migration of virtual machines. In: NSDI (2005)
13. Department of Energy: Carbon dioxide emissions from the generation of electric power in the united states (2000), http://www.eia.doe.gov/cneaf/electricity/page/co2_report/co2emiss.pdf
14. Gandhi, A., Gupta, V., Harchol-Balter, M., Kozuch, M.: Optimality analysis of energy-performance trade-off for server farm management. In: Performance (2010)
15. Kahn, C.: As power demand soars, grid holds up... so far (2010), http://www.news9.com/global/story.asp?s=12762338
16. Komanoff, C.: Carbon tax center (2010), http://www.carbontax.org
17. Le, K., Bilgir, O., Bianchini, R., Martonosi, M., Nguyen, T.D.: Managing the cost, energy consumption, and carbon footprint of internet services. In: Sigmetrics, pp. 357–358 (2010)
18. McGeehan, P.: Heat wave report: 102 degrees in central park (2010), http://cityroom.blogs.nytimes.com/2010/07/06/heatwave-report-the-days-first-power-loss/
19. Qureshi, A., Weber, R., Balakrishnan, H., Guttag, J.V., Maggs, B.V.: Cutting the electric bill for internet-scale systems. In: SIGCOMM, pp. 123–134 (2009)
20. Rao, L., Liu, X., Xie, L., Liu, W.: Minimizing Electricity Cost: Optimization of Distributed Internet Data Centers in a Multi-Electricity-Market Environment. In: INFOCOM (2010)
21. Sleator, D.D., Tarjan, R.E.: Amortized efficiency of list update and paging rules. Communications of the ACM 28(2), 202–208 (1985)
22. Vazirani, V.V.: Approximation Algorithms. Springer, Heidelberg (2001)
23. Windows Azure Platform, http://www.azure.com

Stochastic Traffic Engineering for Live Audio/Video Delivering over Energy-Limited Wireless Access Networks

Nicola Cordeschi*, Tatiana Patriarca, and Enzo Baccarelli

DIET Dept.,"Sapienza" University of Rome
via Eudossiana 18, 00184 Rome, Italy
Tel.: +39 06 44585366; Fax: +39 06 4873300
{cordeschi,patriarca,enzobac}@infocom.uniroma1.it

Abstract. We study the Stochastic Traffic Engineering (STE) problem arising from the support of QoS-demanding live (e.g., real time) audio/video applications over unreliable IP-over-wireless access pipes. First, we recast the problem to be tackled in the form of a suitable nonlinear *stochastic* optimization problem, and then we develop a goodput analysis for the resulting IP-over-wireless pipe that points out the relative effects of fading-induced errors and congestion-induced packet's losses. Second, we present an *optimal* resource-management policy that allows a *joint* scheduling of playin, transmit and playout rates. Salient features of the developed joint scheduling policy are that: *i)* it is *self-adaptive*; and, *ii)* it is able to implement reliable Constant Bit Rate (CBR) connections on the top of unreliable energy-limited wireless pipes.

Keywords: IP-over-wireless connections, real-time streaming, Stochastic Traffic Engineering (STE), self-adaptive rate-control, hard QoS guarantees.

1   Introduction and Related Works

Due to the unreliable randomly time-varying transport quality currently offered by IP-over-wireless connections, providing QoS guarantees to real-time media applications (e.g., live audio/video) over energy-limited congestion-prone wireless IP domains is a (still open) challenging task, that requires an optimized (possibly, self-adaptive) management of the floating resources done available by the underlying pipe [1]. In principle, a suitable means to approach this target is provided by the optimization framework and self-adaptive controlling mechanisms offered by Traffic Engineering (TE). Differently from the (more) usual QoS resource's allocation, the TE solution not only aims to guarantee specified QoS levels, but *also* optimizes an overall *system-wide* performance metric, mainly by performing joint flow-control at the source host, destination host and switching (e.g., serving) nodes. This more conventional TE approach evolves to

---

* Corresponding author.

J. Domingo-Pascual et al. (Eds.): NETWORKING 2011, Part I, LNCS 6640, pp. 186–197, 2011.
© IFIP International Federation for Information Processing 2011

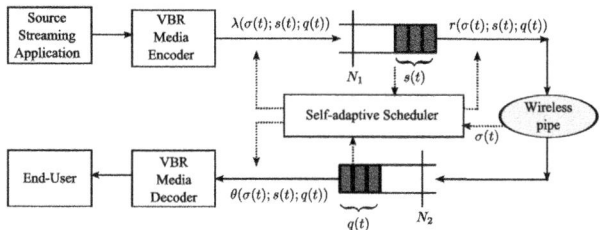

Fig. 1. The considered battery-powered wireless streaming system

the Stochastic TE (STE) one [2],[3], when the underlying connection is affected by some form of uncertainty, as, for example, the random behavior of the state of the IP-over-wireless pipe in the (here considered) streaming scenario of Fig.1. Hence, triggered by these considerations, in this work we resort to the STE principles for designing a *joint self-adaptive* playin-energy-playout scheduler for the streaming system of Fig.1. Specifically, as sketched in Fig.1, we model the (possibly mobile) battery-powered (i.e., energy-limited) source node (e.g., the Source Host (SH)) as a time-slotted fluid G/G/1 queue fed by a Variable Bit Rate (VBR) media encoder, whose output rate (e.g., the playin rate) may be adaptively controlled. Analogously, the receive node (e.g., the Destination Host (DH)) is equipped with a G/G/1 queue, whose output rate (e.g., the playout rate) may be also controlled. In principle, both queues may be considered implemented at the APPlication (APP) layers of the underlying protocol stack, so that the corresponding wireless connection of Fig.1 models the overall resulting peer-to-peer virtual pipe.

The considered system-wide perform metric to be maximized is the resulting transmit rate *averaged* over both the fading and congestion-induced statistics of the state of the wireless pipe, under seven system's constraints. The first one is on the available average energy per slot and it arises from energy limitations typically imposed by the PHY layer of the considered modems. The second and third constraints upper limit the available buffer capacities, while the fourth constraint arises from the APP layer and fixes the maximum instantaneous playin rate. Fifth constraint upper bounds the pre-roll delay [1], while sixth and seventh constraints introduce *hard* (e.g., deterministic) upper and lower bounds on the instantaneous values allowed the playout rate.

Optimized schedulers derived by exploiting the analytical tools of Nonlinear Optimization and Queueing Theory and, thus, implemented via TE-based controlling mechanisms are presented, for example, in [4],[5],[6]. In [4], authors develop an optimized policy for controlling both transmit power and playout rate of the considered streaming system, the target being that to minimize the average power consumption and maximize the rendered average media quality. More recently, some STE aspects related to the *wireless access* are tackled in [5],[6]. However, they do not consider the control of the playout rate, neither account for (possible) limitations on the available average energy-budget. In the application context of multimedia systems and services, Adaptive Media Playout (AMP) policies aiming at minimizing suitable rate-distortion functions are the focus of the contributions in [8],[13]. However, contributions in [8],[13] are not STE-oriented, so that they *do not* consider the adaptive control of the playin rate and/or transmit energy, and (which is the most) *do not* provide hard QoS guarantees to the supported media applications.

2   QoS Real-Time Streaming: System and Traffic Modeling

In the streaming system of Fig.1, time is slotted, with slot-duration of T_s (*sec.*) and the t-th slot spans the (semi-open) interval $[t, (t + 1))$, $t \in \mathbb{N}_0^+$. The APP layers at the source and destination hosts of Fig.1 are equipped with playin and playout buffers of finite capacities N_1 and N_2, respectively. The number of information units (IUs) arriving at the input of the playin buffer *at the end* of slot t is the playin rate $\lambda(t) \in \mathbb{R}_0^+$ (*IU/slot*) at slot t. The underlying wireless physical channel supporting the peer-to-peer pipe of Fig.1 is affected by interference, noise and fading phenomena, we assume constant over each slot (i.e., we assume an ergodic "block fading" physical channel [12]). Hence, the resulting state $\boldsymbol{\sigma}(t) \in \mathbb{R}_0^+$ at the t-th slot of the overall pipe is modelled as a real-valued nonnegative r.v. with (a priori unknown) steady-state pdf $p_\sigma(\sigma)$.[1] Furthermore, the connection state value $\sigma(t)$ is assumed to be known at the SH at the beginning of slot t. The overall streaming system is considered to operate under stationary and ergodic conditions (i.e., it works in the *steady-state*). About backlog dynamics of the playin/playout queues, let $s(t) \in \mathbb{R}_0^+$ be the number of IUs buffered by playin queue of Fig.1 *at the beginning* of slot t (i.e., the backlog of the playin buffer). Thus, after denoting by $r(t)$ (*IU/slot*) the number of IUs to be sent over the wireless connection during t-th slot, the following Lindley's equation:

$$s(t + 1) = [s(t) + \lambda(t) - r(t)]^+, t \geq 0, \tag{1}$$

dictates the evolution of the discrete-time backlog process $\{s(t) \in \mathbb{R}_0^+, t \geq 0\}$ at the playin queue. Let $q(t) \in \mathbb{R}_0^+$ be the number of IUs present in the playout buffer *at the beginning* of slot t (i.e., the backlog of the playout buffer). Hence, after denoting by $\theta(t)$ (*IU/slot*) the number of IUs rendered to the VBR media decoder of Fig.1 *at the end* of slot t, the evolution of the playout backlog is described by the corresponding Lindley's equation:

$$q(t + 1) = [q(t) + r(t) - \theta(t)]^+, t \geq 0. \tag{2}$$

The cost of sending $r(t)$ IUs over slot t is the amount of energy $\mathcal{E}(t)$ (measured in Joule (J)) required for their transmission. Thus, we assume that $r(t)$ depends on *both* $\mathcal{E}(t)$ and $\sigma(t)$ via the rate-function $\mathcal{R}(\cdot; \cdot)$ adopted to measure the (instantaneous) goodput of the considered connection, so that we can write

$$r(t) \triangleq \mathcal{R}(\mathcal{E}(t); \sigma(t)), \quad t \geq 1. \tag{3}$$

---

[1] The meaning of the connection state $\sigma(t)$ is application depending. An example of $\sigma(t)$ modelling the overall behavior of a TCP-friendly connection over an IP-based wireless access network is detailed in Sect.4.

Since $\mathcal{R}(\cdot;\cdot)$ summarizes the goodput-performance of the end-to-end virtual pipe of Fig.1, its actual behavior and structural properties are system depending. Therefore, we limit to introduce few (quite mild) assumptions on $\mathcal{R}(\cdot;\cdot)$, that, by fact, are retained by rate-functions of practical interest [1]. We just assume that it is nondecreasing for $\mathcal{E} \geq 0$ and $\sigma \geq 0$ and, for any assigned $\sigma \neq 0$, the rate-function is strictly concave in the $\mathcal{E}$-variable.

2.1 QoS Support for Real-Time Streaming

The analytical expression assumed by the *distortion-function* $D(\theta) : \mathbb{R}_0^+ \to \mathbb{R}_0^+$ adopted to measure the (subjective) level of satisfaction perceived by the end-user of Fig.1 is strongly application depending, and no general formulae are available for it [1],[13]. Therefore, according to an emerging trend [1], we (only) assume that the adopted distortion-function $D(\theta)$ is a real, nonnegative, *strictly decreasing* function of the actually experimented playout rate θ [1],[13]. This means that, after denoting by $D^{-1}(\cdot)$ the inverse of $D(\cdot)$, any assigned constraint: $D(\theta) \leq D_{max}$ on the subjective QoS level to be guaranteed may be *equivalently* recast in terms of the corresponding constraint: $\theta(t) \geq \theta_{min} \triangleq D^{-1}(D_{max})$ on the minimum rate to be *instantaneously* delivered by the playout buffer.

This constraint guarantees that the maximum (e.g., *worst-case*) queueing delay induced by the playout buffer of Fig.1 equates: (N_2/θ_{min}), so that, for the corresponding playout-delay r.v. $\boldsymbol{T}_{QQ}$ (e.g., the random delay introduced by the playout buffer), the following *deterministic* (e.g., *hard*) upper-bound holds:

$$\boldsymbol{T}_{QQ} \leq (N_2/\theta_{min}) \quad (slot). \tag{4}$$

Passing now to consider the performance limit on the jitter affecting the playout rate, we observe that both the instantaneous jitter and corresponding average squared value become upper-bounded when the following limit on the maximum allowed playout rate is also introduced: $\theta(t) \leq \theta_{max}, \forall t$. In fact, the above constraints guarantee that the resulting playout-jitter is limited (in a *hard* way) as detailed by the following *Proposition 1*, proved in [10].

Proposition 1. *The instantaneous jitter affecting the playout rate is upper-bounded as in* $|\theta(t_1) - \theta(t_2)| \leq (\theta_{max} - \theta_{min}), \forall t_1 \gtrless t_2$. *Furthermore, in the steady-state, the corresponding average playout jitter:* $\sigma_\theta \triangleq \sqrt{E\{(\boldsymbol{\theta}(t) - \bar{\theta})^2\}}$ *(IU/slot) is limited as in* $\sigma_\theta \leq \sqrt{(\theta_{max})^2 - (\theta_{min})^2}$. ∎

To recap, the considered constraints suffice to *guarantee* that the performance limits of the streaming system of Fig.1 are dictated by eqs.(4) and upper-bounds provided by *Proposition 1*.

3 The Tackled STE Problem

In our framework, the overall system's state $\underline{x}(t)$ available at slot t-th for implementing the resulting scheduling action:

$$\underline{x}(t) \triangleq [\sigma(t), s(t), q(t), r(t-1), r(t-2), \ldots, r(t-p))]^T, \tag{5}$$

is composed by the current connection's state $\sigma(t)$ (see Fig.1), the current backlog $s(t)$ of the playin buffer, the current backlog $q(t)$ of the playout buffer and the last $p \geq 0$ transmit rates: $r(t-1), \ldots, r(t-p)$.

Hence, to formally state the resulting STE problem, let: $\mathcal{E}(t) \equiv \varepsilon(\sigma(t); r(t)) \triangleq \mathcal{R}^{-1}(\sigma(t); r(t))$, $t \geq 1$, be the energy to be radiated by the wireless modem present at the PHY layer of the SH of Fig.1 to send $r(t)$ IUs. Therefore, the tackled STE problem may be stated as follows:

$$\max_{[r(\cdot), \lambda(\cdot), \theta(\cdot)]} \mathrm{E}_\sigma\{\mathbf{r}(t)|s(t), q(t)\}, \tag{6}$$

$$s.t. : \mathrm{E}_\sigma\{\varepsilon(\boldsymbol{\sigma}; r(t))|s(t), q(t)\} \leq \mathcal{E}_{ave}, \tag{7}$$

$$0 \leq r(t) \leq s(t), \tag{8}$$

$$0 \leq \lambda(t) \leq \lambda_{max}, \tag{9}$$

$$0 \leq s(t+1) \leq N_1, \tag{10}$$

$$N_0 \leq q(t+1) \leq N_2, \tag{11}$$

$$\theta_{min} \leq \theta(t) \leq \theta_{max}, \tag{12}$$

where the expectations in (6) and (7) are to be carried out with respect to the (a priori unknown) pdf $p_\sigma(\sigma)$ of the connection's state. Hence, from the outset it follows that the playout function $\theta(t)$ to be optimized may be directly expressed as in (see (5)): $\theta(t) \equiv \mathcal{F}(r(t), r(t-1), \ldots, r(t-p))$, $t \geqslant 1$, where $\mathcal{F}(\cdot)$ is a suitable *nonnegative* real-valued function that, in general, may depend on the current transmit rate $r(t)$ and p last ones: $r(t-1), \ldots, r(t-p)$. This points out that p plays the role of *memory-order* of the implemented playout policy.

Before proceeding, we remark that the *stochastic* nature of the considered TE problem arises from the expectations present in eqs.(6),(7). In this regard, eq.(6) points out that we maximize (on a per slot-basis) the *expected* (e.g., average) transmit rate given (i.e., conditioned on) the current backlogs: $(s(t), q(t))$ of the playin and playout buffers. Several contributions (see, for example, [14, Chaps.13,17] and references therein) support the utilization of (6) as an effective metric for characterizing the performance of queuing systems, specially when the supported media traffics exhibit heavy-tailed (e.g., Pareto-like) distributions. Finally, from (11),(12), it follows that the ratio: $\Delta_P \triangleq N_0/\theta_{min}$ covers the role of *allowed pre-roll delay*.

3.1   Algorithmic Solution of the Tackled STE Problem

Let us indicate by $\varepsilon_r(r; \sigma) \triangleq \partial \varepsilon(r; \sigma)/\partial r$ the first-order derivative of the (above introduced) energy-function carried out with respect to the r-argument. Thus, after recognizing that the tackled STE problem is *convex* [10], the resulting optimal solution: $r^{opt}(\cdot; \cdot; \cdot)$, $\lambda^{opt}(\cdot; \cdot; \cdot)$, $\theta^{opt}(\cdot; \cdot; \cdot)$ may be evaluated in *closed-form*, as detailed by the following *Proposition 2* (proved in [10]).

Proposition 2. *Under the above reported assumptions, for the solution of the STE problem of eqs.(6)-(12) the following relationships hold:*

i) *the optimal transmit rate is dictated by*

$$r^{opt}(\sigma(t); s(t); q(t)) = \left[\varepsilon_r^{-1}\left(\sigma; \frac{1}{\mu(s(t); q(t); M_p(t))}\right)\right]_{r_m(M_p(t); s(t); q(t))}^{r_p(M_p(t); s(t); q(t))} \tag{13}$$

where $\varepsilon_r^{-1}(\cdot; \cdot)$ denotes the inverse function of $\varepsilon_r(\cdot; \cdot)$ with respect to the $\mathcal{E}$-variable, while

$$r_m(M_p(t); s(t); q(t)) \triangleq \max\{0; (p+1)\theta_{min} - pM_p(t); (\frac{p+1}{p})(N_0 - q(t)) + M_p(t)\}$$

and

$$r_p(M_p(t); s(t); q(t)) \triangleq \min\{s(t); (\frac{p+1}{p})(N_2 - q(t)) + M_p(t); (p+1)\theta_{max} - pM_p(t)\}$$

are the minimum and peak-rate values allowed at slot t, respectively. Furthermore,

$$M_p(t) \triangleq \frac{1}{p}\sum_{i=t-p}^{t-1} r^{opt}(\sigma(i); s(i); q(i)), \tag{14}$$

represents the short-term average of the (optimal) transmit rates. Finally, $\mu(\cdot; \cdot; \cdot)$ in (13) is the optimal value of the dual variable of the tackled optimization problem, that may be computed by solving the following equation [10]:

$$\int \varepsilon\left(\sigma; \left[\varepsilon_r^{-1}\left(\sigma; \frac{1}{\mu(s(t); q(t); M_p(t))}\right)\right]_{r_m(\cdot;\cdot;\cdot)}^{r_p(\cdot;\cdot;\cdot)}\right) p_\sigma(\sigma) d\sigma = \mathcal{E}_{ave}. \tag{15}$$

ii) *The optimal playin-rate policy is given by the following relationship:*

$$\lambda^{opt}(\sigma(t); s(t); q(t)) = \min\{(N_1 - s(t)) + r^{opt}(\sigma(t); s(t); q(t)); \ \lambda_{max}\}. \tag{16}$$

iii) *The optimal playout-rate policy is dictated by following relationship:*

$$\theta^{opt}(\sigma(t); s(t); q(t)) = r^{opt}(M_p(t); s(t); q(t)) + \frac{p}{p+1}(M_p(t) - r^{opt}(M_p(t); s(t); q(t))). \tag{17}$$

Before proceeding, we stress that a key property of the optimal scheduler of *Proposition 2* is that it utilizes the steady-state pdf $p_\sigma(\sigma)$ of the wireless connection *only* for the computation of the expectation at the left-hand side of (15). Therefore, an approximated version $\widehat{\mu}(.)$ of $\mu(.)$ in (15) may be computed by solving (*on a per slot-basis*) the following sliding-window sample-average equation:

$$\frac{1}{W}\sum_{j=t-W+1}^{t} \varepsilon\left(\sigma(j); \left[\varepsilon_r^{-1}\left(\sigma(j); \frac{1}{\widehat{\mu}(t)}\right)\right]_{r_m(\cdot;\cdot;\cdot)}^{r_p(\cdot;\cdot;\cdot)}\right) = \mathcal{E}_{ave}, \quad t \geq 1, \tag{18}$$

where, due to the (assumed) system's ergodicity, we have that: $\lim_{W \to \infty} \widehat{\mu}(t) = \mu(t)$. Thus, by resorting to (18), the optimal scheduler may be implemented on-the-fly, i.e., without *any* a priori knowledge about actual $p_\sigma(\sigma)$. This implies that the on-the-fly implementation of the optimal scheduler is also capable to *self-track* a priori unknown possible time-variations of the pdf $p_\sigma(\sigma)$ of the connection-state, as those induced, for example, by no stationary fading.

4   Traffic Analysis of IP-over-Wireless Access Connections

Since the access segment is typically the "bottleneck" of current wireless media networks [12], in order to test actual performance of the proposed joint scheduler, in this section we develop the traffic analysis of a wireless access network, where a (possibly nomadic) wireless SH transmits to a (possibly wireless and/or nomadic) DH via a Wireless Access Router (WAR). This last may be simultaneously utilized for the access by other several concurrent Interfering Hosts (IHs), that, in turns, may cause interference and congestion (e.g., buffers' overflow) at the PHY and Network layers of the WAR. The resulting layered structure of the considered protocol stack is detailed in Fig.2. Specifically, FEC-modulation supported by adaptive control of the transmit energy is implemented at the PHY layers of the SH and WAR of Fig.2. The *Data Link* (DL) layer of the SH node of Fig.2 is equipped with a finite-capacity buffer, so to implement a Truncated Selective Repeat ARQ (TSR-ARQ) mechanism [12, Chap.5]. At most, $N_R^{max} \geq 0$ re-transmissions of a same frame may be attempted at the DL layer, before declaring lost the served frame.

Before proceeding to the goodput analysis of the resulting wireless pipe of Fig.2, we detail the considered main operating conditions:

A.i) the forward wireless channel of Fig.2 is frequency-flat and impaired by both i.i.d. Nakagami-distributed fast-fading and log-normal distributed slow-fading [12]. Specifically, the (log-normal distributed) slow-fading phenomena are assumed constant over *at least* $\mathcal{X} \triangleq (N_R^{max} + 1)$ consecutive slot-times, and then may vary according to the Markovian model detailed in the sequel; *A.ii)* when a frame is received incorrectly at the DL layer of DH after FEC/re-transmission recovery, both the corresponding encapsulated datagram and segment are declared loss. Then, triple-duplicate ACKs are generated by the Transport layer of the DH and sent back to the Transport layer of the SH via the ideal backward link of Fig.2. Furthermore, due to the real-time feature of the considered streaming applications, no fragmentation is allowed at the Transport and DL layers of the protocol stack of Fig.2, neither timeout-triggered re-transmissions are assumed feasible at the Transport layer of the SH of Fig.1; *A.iii)* due to implementation complexity considerations [12], the energy $\mathcal{E}(t)$ radiated by the transmitter at the PHY layer of the SH of Fig.2 is assumed to react only to the fluctuations induced by slow-fading. Thus, according to *A.i)*, the transmit energy $\mathcal{E}(t)$ is assumed *constant* over (at least) $\mathcal{X}$ consecutive slot-times.

For flat-fading wireless channels meeting *A.i)*, the channel quality experienced over each slot may be described by the corresponding instantaneous

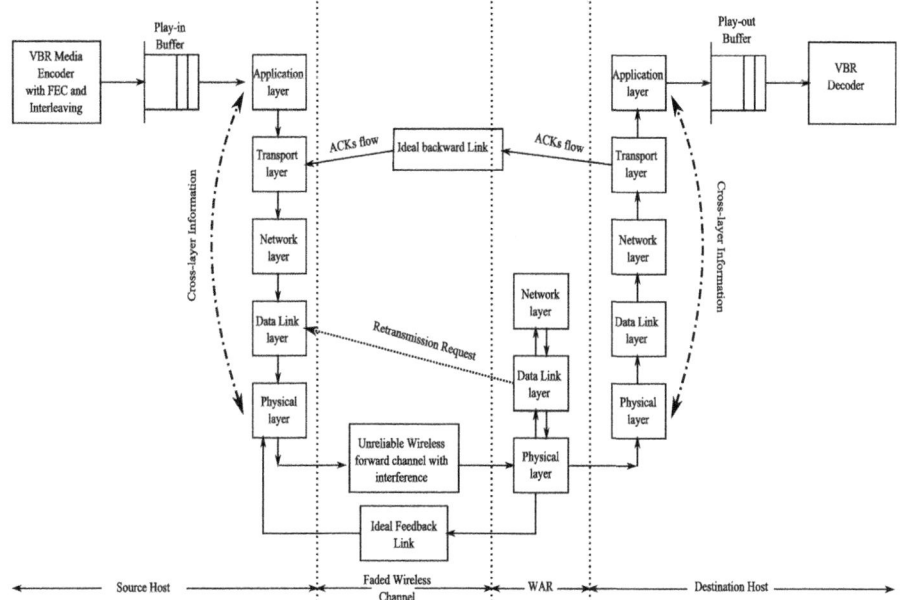

Fig. 2. The layered structure of the considered access network

channel gain $g(t) \in \mathbb{R}_0^+$. In our case, this last is Nakagami-m distributed, so that, for the resulting pdf $p(g)$, we can write [12, Chap.3]

$$p(g) = (m/\overline{g}(t))^m (g^{m-1}/\Gamma(m)) \exp\{-(m/\overline{g}(t))g\}, \quad g \geq 0, \tag{19}$$

where $\Gamma(m)$ is the Gamma function and $m \geq 1/2$ is the Nakagami fading parameter. By definition, $\overline{g}(t)$ in (19) is the instantaneous SINR averaged over (*only*) the fast-fading phenomena. Hence, for log-normal distributed slow-fading, it may be modeled as [12, Chap.3] where the scaling constant: $c_0 \triangleq \exp\{-\frac{(0.1log10)^2}{2}\} \approx 0.9738$ guarantees unit average value for $\overline{g}(t)$, e.g, $\overline{\gamma} \triangleq E\{\overline{g}(t)\} \equiv 1 \ (J^{-1})$. Furthermore, according to [12], $\{z(t), \ t \geq 1\}$ is a zero-mean unit-variance Gaussian-distributed Markovian sequence, that assumes constant value over each frame composed by $\mathcal{X}$ consecutive time-slots (see **A.i**)). For macrocellur land-mobile applications, its correlation coefficient: $h \triangleq E\{z(t)z(t+\mathcal{X})\}$ may be evaluated as in [12]: $h \equiv (0.82)^{T_S v/100}$, where $v(m/sec.)$ is the speed of the SH node of Fig.2.

Thus, according to the performance analysis carried out in [9], the instantaneous Frame Error Rate (FER) measured at the input of the DL layer of the WAR of Fig.2 after FEC-recovery may be well approximated according to [9, eq.(5)]

$$FER(t) \approx \begin{cases} 1, & \text{for} \quad 0 \leq g(t) < C/\mathcal{E}(t), \\ A \exp\{-B\mathcal{E}(t)g(t)\}, & \text{for} \quad g(t) \geq C/\mathcal{E}(t), \end{cases} \tag{20}$$

where the (positive) constants A, B, C in (20) depend on the actually adopted coding/modulation scheme [9].

Therefore, after frame recovery carried out by the Truncated SR-ARQ mechanism implemented at the DL layer, the resulting peer-to-peer IP-based connection available at the Network layer of Fig.2 may be modeled as a *loss-affected* pipe, that may also introduces *random delays* on the conveyed datagrams' flow [7]. Specifically, according to the delay-analysis presented in [7, Sect.II], we model the sequence $\{\Delta(\mathcal{X}t), t \geq 1\}$ of datagram's delays as an i.i.d. exponentially-distributed random series, so that the resulting pdf $p_\Delta(\Delta)$ for the datagram delay reads as (see [7, Sect.II] for the case of a *single* access router): $p_\Delta(\Delta) = \varepsilon_L \delta(\Delta - \infty) + (1 - \varepsilon_L)\omega_R e^{-\omega_R \Delta}$, $\Delta \geq 0$, where $\delta(\Delta - \infty)$ is the Dirac pulse shifted at the infinity, $\omega_R > 0$ is the service rate (in $(slot)^{-1}$) of the WAR of Fig.2, while $\varepsilon_L \in [0, 1)$ is the *overall* datagram loss-rate affecting the wireless pipe of Fig.2.

Finally, under low loss-rate operating conditions, the sending rate $\mathcal{R}(\mathcal{X}t)$ (*byte/slot*) of the TCP-friendly rate-control mechanism implemented at the Transport layer of the SH node of Fig.2 may be well approximated by the following expression [11]:

$$\mathcal{R}(\mathcal{X}t) \equiv [MSS/(\overline{RTT}(\mathcal{X}t)\sqrt{\overline{P_L}(\mathcal{X}t)})]\sqrt{3/4}, \ t \geq 0, \ (byte/slot), \qquad (21)$$

where MSS (*byte*) is the allowed maximum segment size; $\overline{P_L}(\mathcal{X}t)$ is the segment-loss probability *averaged* over the fast-fading statistics of eq.(19); and $\overline{RTT}(\mathcal{X}t)$ is the average segment Round-Trip-Time (RTT) at the $(\mathcal{X}t)$-th slot. According to the Jacobson's formula, this last may be iteratively updated as in $\overline{RTT}(\mathcal{X}t) \equiv 0.75\overline{RTT}(\mathcal{X}(t-1)) + 0.25\Delta(\mathcal{X}t)$, $t \geq 1$.

About the evaluation of the resulting connection goodput, we stress that, in our framework, the reception of triple-duplicate ACKs is *only* utilized for updating the current value of $\overline{P_L}(\cdot)$ in (21). This means that the here considered Transport protocol is, indeed, *UDP based*, but it relies on a suitable *traffic shaper* to work in a TCP-friendly way [11]. Thus, by leveraging on this (key) observation, we are able to arrive at the following *final expression* for the goodput in (21) sustained by the considered connection (see [10] for the derivation):

$$\mathcal{R}(\mathcal{X}t) = K_0 \sigma(\mathcal{X}t)(\mathcal{E}(\mathcal{X}t))^{\mathcal{X}m/2}, \quad t \geq 0, \qquad (22)$$

where $K_0 \triangleq \sqrt{3/4}\, MSS\{(m^m/\Gamma(m))[(C^m/m) + (A\,\Gamma(m; CB))/B^m\,]\}^{-\mathcal{X}/2}$, is a positive constant, while,

$$\sigma(\mathcal{X}t) \triangleq [\overline{g}(\mathcal{X}t)]^{\mathcal{X}m/2}/\overline{RTT}(\mathcal{X}t), \quad t \geq 0, \qquad (23)$$

is the (positive-valued) resulting state of the overall peer-to-peer connection going from the input of the Transport layer at the SH of Fig.2 to the output of the Transport layer at the corresponding DH. Hence, under the above reported assumptions, eq.(22) is able to *fully characterize* the goodput-behavior of the *overall* resulting TCP-friendly connection.

5 Traffic Tests and Final Remarks

According to this conclusion, in order to (numerically) test actual performance of the proposed joint scheduler, we have numerically generated the connection-state sequence $\{\sigma(\mathcal{X}t)\}$ of eq.(23) and, then, we have adopted the expression in (22) for (numerically) measuring the instantaneous goodput $\mathcal{R}(\cdot;\cdot)$ sustained by the overall peer-to-peer pipe. The main simulated parameters are set as $\lambda_{max} = \theta_{max} = 1000$ (*byte/slot*), $MSS = 120$ (*byte*), $N_0 = 0$ (*byte*), $A = 90.2514$, $B = 3.4998$, $C = 1.2865$, $\mathcal{X} \equiv 1$, $b = 2$ and $N_1 = N_2 = 24000$ (*byte*).

5.1 Effects of the Memory-Order of the Playout Policy

Performance of the self-adaptive implementation of the proposed scheduler has been numerically evaluated in terms of average playout rate: $\overline{\theta}^{opt} \triangleq E\{\theta^{opt}(t)\}$ (*byte/slot*) and average playout rate-jitter: $\sigma_\theta^{opt} \triangleq \sqrt{E\{(\theta^{opt} - \overline{\theta}^{opt})^2\}}$ (*byte/slot*). For comparison purpose, we have also implemented a *Benchmark Version* (BV) of the scheduler of Sect.3, where *both* the transmit-energy and playout-rate controls have been removed. The (numerically evaluated) behaviors of the average playout rate $\overline{\theta}_{BV}$ (*byte/slot*) and playout rate-jitter σ_θ^{BV} (*byte/slot*) we obtained by implementing this BV are reported in Figs.3(a),3(b), respectively. For comparison purpose, Figs.3(a),3(b) also report the corresponding performance of the self-adaptive implementation of the optimal scheduler. All the numerical plots of Figs.3(a),3(b) refer to the case of $\omega_R = 1$ (*slot^{-1}*), $\varepsilon_L = 0$, $m = 1$, and $W = 5$ (*slot*).

An examination of these plots allows us to drawn the following insights. First, $\overline{\theta}^{opt}$ is virtually independent from the setting of the memory order p adopted for the playout policy. Second, the gap between $\overline{\theta}^{opt}$ and $\overline{\theta}_{BV}$ tends to decrease

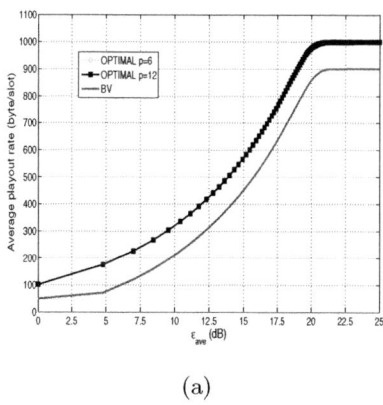

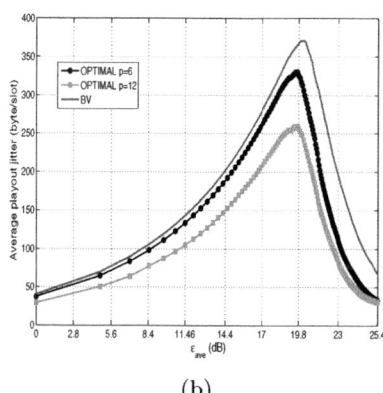

(a)   (b)

Fig. 3. Behaviors of: (a) the average playout rates $\overline{\theta}^{opt}$ (*byte/slot*) and $\overline{\theta}_{BV}$ (*byte/slot*); (b) the average playout jitters σ_θ^{opt} (*byte/slot*) and σ_θ^{BV} (*byte/slot*). Plots of $\overline{\theta}^{opt}$ at $p = 6$ and $p = 12$ are virtually overlapped.

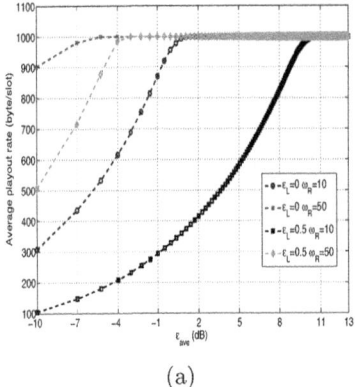
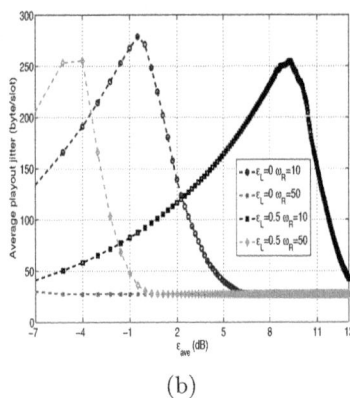

(a)   (b)

Fig. 4. Behaviors of: (a) the average playout rate $\overline{\theta}^{opt}$ ($byte/slot$); and: (b) the average playout jitter σ_{θ}^{opt} ($byte/slot$) of the optimal scheduler under several traffic conditions. The operating scenario of Sect.5.2 is considered.

at both low and high $\mathcal{E}_{ave}$'s, while it is noticeable and of the order of about 45% at $\mathcal{E}_{ave}$'s around 10 (dB). Third, jitter curves of Fig.3(b) point out that, at low/medium $\mathcal{E}_{ave}$'s (we say, at $\mathcal{E}_{ave}$'s limited up to about 19 (dB)), the average jitters affecting the played rates tend to increase for growing $\mathcal{E}_{ave}$'s. However, at high $\mathcal{E}_{ave}$'s (we say, at $\mathcal{E}_{ave}$'s beyond about 20 (dB)), all the implemented transmit policies become able to fully saturate the capacities of the corresponding playout buffers, so that all the resulting jitters tend to decrease (see the right-most parts of the curves of Fig.3(b)). In any case, the practical lesson arising from the plots of Fig.3(b) is that, at medium $\mathcal{E}_{ave}$'s (i.e., at $\mathcal{E}_{ave}$'s around $15-16$ (dB)), the optimal scheduler with $p = 12$ is able to reduce the average jitter affecting the corresponding playout rate of about 33% with respect to the BV.

5.2   Effects of Traffic Load and Traffic Congestion

The average performances of the on-the-fly implementation of the optimal sched-uler under different traffic load and traffic congestion levels are reported in Figs.4(a),4(b) for the case of $m = 1$, $W = 5$ ($slot$), $p = 12$ and $\theta_{min} = 100$ ($byte/slot$). An examination of these curves shows that the relative degra-dations suffered by $\overline{\theta}^{opt}$ and σ_{θ}^{opt} when we pass from: $\varepsilon_L = 0$, $\omega_R = 50$ (i.e., the less congested traffic scenario here considered) to: $\varepsilon_L = 0.5$, $\omega_R = 10$ (i.e., the most congested traffic scenario here considered) are noticeable only at (very) low $\mathcal{E}_{ave}$'s (e.g., at $\mathcal{E}_{ave}$'s below $3 - 4$ (dB)). These degradations quickly decrease, indeed, for $\mathcal{E}_{ave}$'s beyond $4 - 5$ (dB) and, then, virtually vanish at $\mathcal{E}_{ave}$'s over about 9 (dB).

References

1. Van der Schaar, M., Chou, P.A.: Multimedia over IP and Wireless Networks. Aca-demic Press, London (2007)

2. Mitra, D., Wang, Q.: Stochastic Traffic Engineering for Demand Uncertainty and Risk-Aware Network Revenue Management. IEEE/ACM Tr. Networking 13(2), 221–233 (2005)

3. Wang, H., Xie, H., Qui, L., Yang, Y.R., Zhang, Y., Greenberg, A.: Cope: Traffic Engineering in Dynamic Networks. In: Proc. ACM Sigcomm (2006)

4. Yan, L., Markopoulou, A., Bambos, N., Apostolopoulos, J.: Joint power-playout control for media streaming over wireless links. IEEE Tr. on Multimedia 8(4), 830–843 (2006)

5. Sawma, G., Ben-El-Kezadri, R., Aib, I., Bezalel, G., Pujolle, G.: Proactive Traffic Engineering for IEEE 802.11 Mobile Wireless Networks. In: IEEE VTC 2009-Fall (September 2009)

6. Faruque, S.: Traffic Engineering for multi-rate wireless data. In: IEEE International Conference on Electro/Information Technology, pp. 280–283 (May 2008)

7. Chou, P., Miao, Z.: Rate-distortion optimized streaming of packetized media. IEEE Tr. on Multimedia 8(2), 390–404 (2006)

8. Kalman, M., Steinbach, E., Girod, B.: Adaptive media playout for low-delay streaming over error-prone channels. IEEE Tr.on Circuits and Systems for Video Technology 14(6), 841–851 (2004)

9. Liu, Q., Zhou, S., Giannakis, G.B.: Cross-layer combining of adaptive modulation and coding with truncated ARQ over wireless links. IEEE Tr. on Wireless Communications 3(5), 1746–1755 (2004)

10. Cordeschi, N., Patriarca, T., Baccarelli, E.: Energy-Efficient adaptive playin/playout providing hard QoS guarantees to wireless media-streaming applications, DIET Techn. Report, available at the web site, `http://infocom.uniroma1.it/~enzobac/playout_technicalreport.pdf`

11. Jin, S., Matta, I., Bestravos, A.: A Spectrum of TCP-Friendly Window-Based Congestion Control algorithms. IEEE/ACM Tr. on Networking 11(3), 341–355 (2003)

12. Glisic, S.G.: Advanced Wireless Networks-4G Technologies. Wiley, Chichester (2006)

13. Chuang, H.C., Huang, C.Y., Chiang, T.: Content-aware Adaptive Media Playout Controls for wireless video streaming. IEEE Tr. on Multimedia 9(6), 1273–1283 (2007)

14. Park, K., Willinger, W.: Self-Similar Networks Traffic and Performance Evaluation. Wiley, Chichester (2000)

VMFlow: Leveraging VM Mobility to Reduce Network Power Costs in Data Centers

Vijay Mann[1,*], Avinash Kumar[2], Partha Dutta[1],
and Shivkumar Kalyanaraman[1]

[1] IBM Research, India
{vijamann,parthdut,shivkumar-k}@in.ibm.com
[2] Indian Institute of Technology, Delhi
{avinash.ee}@student.iitd.ac.in

Abstract. Networking costs play an important role in the overall costs of a modern data center. Network power, for example, has been estimated at 10-20% of the overall data center power consumption. Traditional power saving techniques in data centers focus on server power reduction through Virtual Machine (VM) migration and server consolidation, without taking into account the network topology and the current network traffic. On the other hand, recent techniques to save network power have not yet utilized the various degrees of freedom that current and future data centers will soon provide. These include VM migration capabilities across the entire data center network, on demand routing through programmable control planes, and high bisection bandwidth networks.

This paper presents VMFlow: a framework for placement and migration of VMs that takes into account both the network topology as well as network traffic demands, to meet the objective of network power reduction while satisfying as many network demands as possible. We present network power aware VM placement and demand routing as an optimization problem. We show that the problem is NP-complete, and present a fast heuristic for the same. Next, we present the design of a simulator that implements this heuristic and simulates its executions over a data center network with a CLOS topology. Our simulation results using real data center traces demonstrate that, by applying an intelligent VM placement heuristic, VMFlow can achieve 15-20% additional savings in network power while satisfying 5-6 times more network demands as compared to recently proposed techniques for saving network power.

Keywords: networking aspects in cloud services, energy and power management, green networking, VM placement.

1   Introduction

In current data centers, networking costs contribute significantly to the overall expenditure. These include capital expenditure (CAPEX) costs such as cost

---

* Corresponding author: Vijay Mann, IBM Research, 4, Block C, Institutional Area, Vasant Kunj, New Delhi - 110070, India.

J. Domingo-Pascual et al. (Eds.): NETWORKING 2011, Part I, LNCS 6640, pp. 198–211, 2011.

of networking equipment which has been estimated at 15% of total costs in a data center [6]. Other networking costs include operational expenditure (OPEX) such as power consumption by networking equipment. It has been estimated that network power comprise of 10-20% of the overall data center power consumption [9]. For example, network power was 3 billion kWh in the US alone in 2006.

Traditionally, data center networks have comprised of single rooted tree network topologies which are ill-suited for large data centers, as the core of the network becomes highly oversubscribed leading to contention [7]. To overcome some of these performance problems of traditional data center networks, such as poor bisection bandwidth and poor performance isolation, recent research in data center networks has proposed new network architectures such as VL2 [7], Portland [16], and BCube [8]. Data centers are also increasingly adopting virtualization and comprise of physical machines with multiple Virtual Machines (VMs) provisioned on each physical machine. These VMs can be migrated at downtime or at runtime (live). Furthermore, emergence of programmable control plane in switches through standardized interfaces such as Openflow [3], has enabled on demand changes in routing paths.

With the above recent advances, opportunities have emerged to save both network CAPEX and OPEX. Network CAPEX is being reduced through several measures such as increased utilization of networking devices, and a move towards cheaper and faster data plane implemented in merchant silicon, while all the intelligence lies in a separate sophisticated control plane [3]. Components of OPEX, such as network power costs, are being reduced through techniques that switch off unutilized network devices as the data center architectures move to higher levels of redundancy in order to achieve higher bisection bandwidth. However, most of the recent techniques to save network power usage do not seem to utilize all the features that current and future data center will soon provide. For instance, recent techniques to save network power in [9] exploit the time-varying property of network traffic and increased redundancy levels in modern network architectures, but do not consider VM migration. On the other hand, current VM placement and migration techniques such as [19] mainly target server power reduction through server consolidation but do not take into account network topology and current network traffic. A recent work by Meng et. al [15] explored network traffic aware VM placement for various network architectures. However, their work did not focus on network power reduction.

There are multiple reasons why saving network power is increasingly becoming more important in modern data centers [13]. Larger data centers and new networking technologies will require more network switches, as well as switches with higher capacities, which in turn will consume more power. In addition, rapid advances in server consolidation are improving server power savings, and hence, network power is becoming a significant fraction of the total data center power. Also, server consolidation results in more idle servers and networking devices, providing more opportunity for turning off these devices. Finally, turning off switches results in less heat, thereby reducing the cooling costs.

In this context, this paper presents VMFlow: a framework for placement and migration of VMs that takes into account both the network topology as well as network traffic demands, to meet the objective of network power reduction while satisfying as many network demands as possible. We make the following contributions:

1. We formulate the VM placement and routing of traffic demands for reducing network power as an optimization problem.
2. We show that a decision version of the problem is NP-complete, and present a fast heuristic for the same.
3. We present the design of a simulator that implements this heuristic and simulates its runs over a data center network with a CLOS topology.
4. We validate our heuristic and compare it to other known techniques through extensive simulation. Our simulation uses real data center traces and the results demonstrate that, by applying an intelligent VM placement heuristic, VMFlow can achieve 15-20% additional savings in network power while satisfying 5-6 times more network demands as compared to recently proposed techniques for saving network power.
5. We extend our technique to handle server consolidation.

The rest of this paper is organized as follows. We give an overview of related research in Section 2. Section 3 presents the technical problem formulation. Section 4 describes our greedy heuristic. We describe the design and implementation of our simulation framework and demonstrate the efficacy of our approach through simulation experiments in Section 5. Section 6 extends our technique to handle server consolidation. Finally, we conclude the paper in Section 7.

2   Background and Related Work

Data center network architectures. Conventional data centers are typically organized in a multi-tiered network hierarchy [1]. Servers are organized in racks, where each rack has a Top-of-Rack (ToR) switch. ToRs connect to one (or two, for redundancy) aggregation switches. Aggregation switches connect to a few core switches in the top-tier. Such a hierarchical design faces multiple problems in scaling-up with number of servers, e..g, the network links in higher tiers become progressively over-subscribed with increasing number of servers. Recently, some new data center network architectures have been proposed to address these issues [7, 8, 16].

VL2 [7] is one such recently proposed data center network architecture (Figure 1), where the aggregation and the core switches are connected using a CLOS topology, i.e., the links between the two tiers form a complete bipartite graph. Network traffic flows are load balanced across the aggregation and core tier using Valiant Load Balancing (VLB), where the aggregation switch randomly chooses a core switch to route a given flow. VL2 architecture provides higher bisection bandwidth and more redundancy in the data center network.

Traffic-aware VM placement. VM placement has been extensively studied for resource provisioning of servers in a data center, including reducing server power usage [19]. Some recent papers have studied VM placement for optimizing network traffic in data centers [12, 15]. These approaches differ from VMFlow in two important ways: (a) the techniques do not optimize for network power, and (b) their solutions do not specify the routing paths for the network demands.

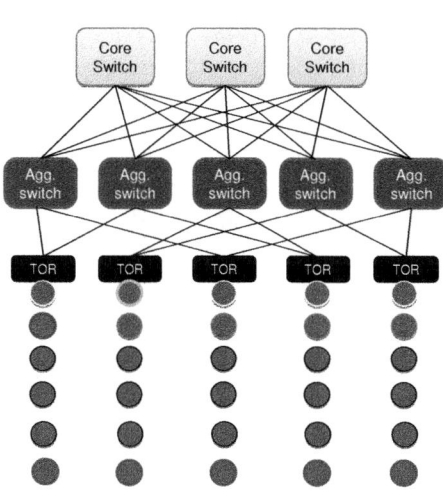

Fig. 1. CLOS network topology

Network power optimization. Most of the recent research on data center energy efficiency has focused on reducing the two major components of data center power usage: servers and cooling [6, 9]. Recently, some papers have focussed on reducing the power consumed by networking elements (which we call, network power) in a data center [9, 13, 17]. In the ElasticTree approach presented in [9], a network power manager dynamically turns on or off the network elements (switches and links), and routes the flows over the active network elements, based on the current network traffic demands. In the primary technique presented in ElasticTree, called greedy bin-packing, every demand is considered sequentially, and the demand is routed using the leftmost path that has sufficient capacity to route the demand. Here, a path is considered to be on the left of another path if, at each switch layer of a structured data center topology, the switch on the first path is either to the left or is identical to the switch on the second path. ElasticTree also proposes an topology-aware approach that improves upon the computation time as compared to greedy bin packing − however, it does not improve the network power in the solution. The VMFlow framework that we present in this paper fundamentally differs from ElasticTree [9] and from the work in [13,17] because we exploit the flexibility of VM placements that is available in the modern data centers. Also, for a given demand, we jointly perform the VM placement and flow routing by greedily selecting the VM placements as well as the routing paths that require minimum additional network power.

3 Network-Aware VM Placement Problem

In this section we present the Network-Aware Virtual Machine Placement (NAVP) Problem. Our problem formulation extends the ElasticTree formulation in [9].

Input. The data center network is composed of network switches and links that connect the hosts (physical servers). The data center network topology is modeled using a weighted directed graph $G = (V, E)$, where V are the set of vertices and the $E \subseteq V \times V$ is the set of directed edges. A link $e = (u, v)$ has capacity (maximum supported rate) of $C(e)$. There are three types of vertices in V: the switches, the hosts, and a special vertex v_E. Vertex v_E models the external clients that communicate with the data center. The edges represent the communication links between the switches, and between the switches and the hosts. (We use edges and links interchangeably in this paper.) Let there be n hosts $H = \{h_1, \ldots, h_n\}$, and q switches $W = \{w_1, \ldots, w_q\}$.

We consider a set of m Virtual Machines (VMs) $M = \{vm_1, \ldots, vm_m\}$, where $m \leq n$, and at most one VM can be placed on a host (we consider the case of multiple VMs per server in Section 6). The network traffic source or destination is one of the m VMs or an external client. We model the traffic to and from any external clients as traffic to and from v_E, respectively. Let M' be $M \cup \{v_E\}$. We are given a set of K demands (rates) among the nodes in M', where the jth demand requires a rate of r_j from source $s_j \in M'$ to destination $d_j \in M'$, and is denoted by (s_j, d_j, r_j).

When a switch w_i or a link e is powered on, let $P(w_i)$ and $P(e)$ denote the power required to operate them, respectively. A *VM placement* is a (one-to-one) mapping $\Pi : M \to H$ that specifies that the mapping of VM vm_i to host $h_{\Pi(vm_i)}$. In addition, we assume that in all VM placements, v_E is mapped to itself.

Constraints. We model the VM placement problem as a variant of the multi-commodity flow problem [4]. A *flow assignment* specifies the amount of traffic flow on every edge for every source-destination demand. We say that a flow assignment on G satisfies a set of demands if the following three standard constraints holds for the flow assignment: edge capacity, flow conservation and demand satisfaction. (Please see [14] for the detailed constraints.)

In addition, we consider another set of constraints that result from the power requirements: (1) a link can be assigned a non-zero flow only if it is powered on, and (2) a link can be powered on only if the switches that are the link's end nodes are powered on. Thus, the *total (network) power* required by a flow assignment is the sum of the power required for powering on all links with non-zero flow assignment, and the power required for powering on all switches that are end nodes of some link that has a non-zero flow assignment.

Due to adverse effect of packet reordering on TCP throughput, it is undesirable to split a traffic flow of a source-destination demand [11]. Therefore, the NAVP problem requires that a demand must be satisfied using a network flow that uses only one path in the network graph (*unsplittable flow constraint*).

The optimization problem. Given the above-mentioned K demands among the nodes in M', we say that a given VM placement Π is *feasible* if there is a flow assignment on G that satisfies the following K demands among the hosts: the jth demands requires a rate of r_j from source host $\Pi(s_j)$ to destination host $\Pi(d_j)$. Then the Network-Aware VM Placement (NAVP) problem is stated as follows: among all possible feasible VM placements, find a placement and an associated flow assignment that has the minimum total power.

Problem Complexity. The following decision version of the NAVP problem is NP-complete: given a constant B, does there exist a feasible VM placement and an associated flow assignment that has total power less than or equal to B. We show the NP-completeness by reduction from the bin packing problem [18].

Theorem 1. *The decision version of the Network-Aware VM Placement (NAVP) problem is NP-complete.*

Please see [14] for the proof.

4   Heuristic Design

We now present a greedy heuristic for the NAVP problem. The algorithm considers the demands one by one, and for each demand, the algorithm greedily chooses a VM placement and a flow assignment (on a path) that needs minimum increase in the total power of the network. We now describe the algorithm in more details. (The pseudocode is presented in Figure 2.)

Primary variables. The algorithm maintains four primary variables: W_{on} and E_{on} which are the set of switches and edges that have been already powered on, respectively, the set H_{free} of hosts that have not yet been occupied by a VM, and a function RC that gives the residual or free capacity of the edges. For each VM v, we assume that initially the placement $\Pi(v)$ is set to $\bot$.

In each iteration of the main loop (in function VMFlow), the algorithm selects a demand in the descending order of the demand rates, and tries to find a feasible VM placement. If a feasible VM placement is found, then the primary variables are updated accordingly; otherwise, the demand is skipped.

Residual graph. To find a feasible VM placement for a demand (s_j, d_j, r_j), the algorithm construct a residual graph G_{res}. The residual graph contains all switches, and all hosts where VMs s_j and d_j can be possibly placed (sets $\Phi(s_j)$ and $\Phi(d_j)$, respectively). G_{res} also contains every edge among these nodes that have at least r_j residual capacity. Therefore, in G_{res}, any path between two hosts has enough capacity to route the demand.

The algorithm next focuses on finding VM placements for s_j and d_j such that there is path between the VMs that requires minimum additional power. To that end, the nodes and edges in the residual graph are assigned weights equal to the amount of additional power required to power them on. Thus, the weight of a switch v is set to 0 if it already powered on, and set to $P(v)$, otherwise. Edges are assigned weights in a similar manner. However, the weights of the hosts are set to ∞ so that they cannot be used as an intermediate node in a routing path. Next, the weight of a path in the residual graph ($pathWt$) is defined as the sum of the weights of all intermediate edges and nodes on the path (and excludes the weights of the end nodes of the path).

VM placement. In the residual graph G_{res}, the algorithm considers all possible pairs of hosts for placing s_j and d_j (where the first element in the pair is from $\Phi(s_j)$ and the second element is from $\Phi(d_j)$). Among all such pairs of hosts, the algorithm selects a pair of hosts that has a minimum weight path. (A minimum

1: **Input:** described in input part of Section 3
2: **function** Initialization
3:     $W_{on} \leftarrow \emptyset; E_{on} \leftarrow \emptyset$ $\{set\ of\ switches\ and\ edges\ currently\ powered\ on\}$
4:     $H_{free} \leftarrow H$ $\{set\ of\ hosts\ currently\ not\ occupied\ by\ a\ VM\}$
5:     $\forall e \in E,\ RC(e) \leftarrow C(e)$ $\{current\ residual\ capacity\ of\ edge\ e\}$

6: **function** VMFlow
7:     select a demand (s_j, d_j, r_j) from the set of demands in the descending order of their rates
8:         **if** $\Pi(s_j) = \bot$ **then** $\Phi(s_j) \leftarrow H_{free}$ **else** $\Phi(s_j) \leftarrow \{\Pi(s_j)\}$
9:         **if** $\Pi(d_j) = \bot$ **then** $\Phi(d_j) \leftarrow H_{free}$ **else** $\Phi(d_j) \leftarrow \{\Pi(d_j)\}$
10:         $V_{res} \leftarrow W \cup \Phi(s_j) \cup \Phi(d_j)$ $\{residual\ nodes\}$
11:         $E_{res} \leftarrow \{(u, v) \in E : (RC(e) \geq r_j) \wedge (u, v \in V_{res})\}$ $\{residual\ edges\}$
12:         $G_{res} \leftarrow (V_{res}, E_{res})$ $\{residual\ graph\}$
13:         $\forall v \in W,$ **if** $v \in W_{on}$ **then** $WT_{res}(v) \leftarrow 0$ **else** $WT_{res}(v) \leftarrow P(v)$ $\{switch\ wt.\ in\ G_{res}\}$
14:         $\forall v \in \Phi(s_j) \cup \Phi(d_j),\ WT_{res}(v) \leftarrow \infty$ $\{host\ wt.\ in\ G_{res}\}$
15:         $\forall e \in E_{res},$ **if** $e \in E_{on}$ **then** $WT_{res}(e) \leftarrow 0$ **else** $WT_{res}(e) \leftarrow P(e)$ $\{edge\ wt.\ in\ G_{res}\}$
16:         $minWt \leftarrow Min\{pathWt(G_{res}, u, v) : (u \in \Phi(s_j)) \wedge (v \in \Phi(d_j)) \wedge (u \neq v)\}$ $\{see\ Sec.\ 4\}$
17:         **if** $(minWt < \infty)$ **then** $\{found\ a\ routing\ for\ this\ demand\}$
18:             $(\Pi(s_j), \Pi(d_j)) \leftarrow$ any (u, v) s.t. $(u \in \Phi(s_j)) \wedge (v \in \Phi(d_j)) \wedge (u \neq v) \wedge$ $(pathWt(G_{res}, u, v) = minWt)$
19:             assign the minimum weight path P from $\Pi(s_j)$ to $\Pi(d_j)$ to the flow for (s_j, d_j, r_j)
20:             for all switches v on path P, $W_{on} \leftarrow W_{on} \cup \{v\}$
21:             for all edges e on path P, $E_{on} \leftarrow E_{on} \cup \{e\}$; $RC(e) \leftarrow RC(e) - r_j$
22:             $H_{free} \leftarrow H_{free} \setminus \{\Pi(s_j), \Pi(d_j)\}$
23:         **else**
24:             skip demand (s_j, d_j, r_j) $\{cannot\ find\ a\ VM\ placement\ for\ this\ demand\}$

Fig. 2. A greedy algorithm for NAVP

weight path between the hosts is found using a variant of Dijkstra's shortest path algorithm whose description is omitted here due to lack of space.) The source and the destination VMs for the demand (s_j and d_j) are placed at the selected hosts, and the flow assignment is done along a minimum weight path. Note that, sometimes G_{res} may be disconnected and each hosts may lie in a distinct component of G_{res}. (For instance, this scenario can occur when the residual edge capacities are such that no path can carry a flow of rate r_j.) In this case, there is no path between any pair of hosts, and the algorithm is unable to find a feasible VM placement for this demand. Depending on the data center's service level objectives, the algorithm can either abort the placement algorithm, or continue by considering subsequent demands.

It is easy to see that each iteration selects a placement and flow routing that require minimum increase in network power. Thus, although the heuristic is sub-optimal, each iteration selects a locally optimal placement and routing.

Lemma 1. *In each iteration, for a given traffic demand, the heuristic selects source and destination VM placements and a flow routing that need minimum increase in network power.*

A practical simplification. We now present a simple observation to reduce the computation time of our heuristic. We observe that in most conventional and modern Data Center network architectures, multiple host are placed under a Top-of-the-Rack (ToR) switch. Thus, for a given demand, if a source (or destination) VM is placed on a host, then the parent ToR of the host needs to be turned on (if the ToR is not already on) to route the demand. Therefore, for VM placement, we first try to place both the source and destination VMs under the same ToR. If such a placement is possible, then only the common parent ToR needs to be turned on to route the demand. If no such ToR is available, then we compute a minimum weight path between all possible pairs of ToRs (where path weights include the end ToR weights), instead of computing minimum weight paths between all possible pairs of hosts. This simplification significantly reduces the computation time because the number of ToRs is much lower than the number of hosts.

5   Experimental Evaluation

We developed a simulator to evaluate the effectiveness of VMFlow algorithm. It simulates a network topology with VMs, switches and links as a discrete event simulation. Each server (also called host) is assumed to host at most one VM (the case of multiple VMs per host is discussed in Section 6). VMs run applications that generate network traffic to other VMs, and VMs can migrate from one node to the other. Switches have predefined power curves − most of the power is static power (i.e. power used to turn on the switch). At each time step, network traffic generated by VMs (denoted by an entry in the input traffic matrix) is read and the corresponding VMs are mapped to hosts as per the algorithm described in Section 4. The corresponding route between the hosts is also calculated based on the consumed network power, the network topology and available link capacities. At each time step of the simulation, we compute the total power consumed by the switches and the fraction of the total number of network demands that can be satisfied. A new input traffic matrix (described below), that represents the network traffic at that time stamp, is used at each time step of the simulation.

Network Topology. The simulator creates a network based on the given topology. Our network topology consisted of 3 layers of switches: the top-of-the-rack (ToR) switches which connect to a layer of aggregate switches which in turn connect to the core switches. Each ToR has 20 servers connected to it. At most one virtual machine (VM) can be mapped to a server (the case of multiple VMs per server is discussed in Section 6). We assume a total of 1000 servers (and VMs). There are 50 ToR switches, and each server connects to a ToR over a 1 Gbps link. Each ToR connects to two aggregate switches over 1 or 10 Gbps links. We assume a CLOS topology between the aggregate and core switches similar to VL2 [7] with 1 or 10 Gbps links as shown in Figure 1. All switches are assumed to have a static power of 100 watts. This is because current networking devices are not energy proportional and even completely idle switches (0% utilization) consume 70-80% of their fully loaded power (100% utilization) [9].

Our simulator has a topology generator component that generates the required links between servers and ToRs, ToRs and aggregate switches, and aggregate switches and core switches. It takes in as input the number of servers and the static power values for each kind of switch. It then generates the number of ToR switches (assuming 20 servers under 1 ToR) and the number of aggregate and

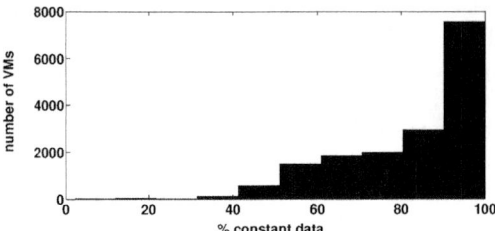

Fig. 3. Histogram of number of VMs with their percentage constant data

core switches using the formulae given in [7] for a CLOS topology. For *(d*d)/4* ToRs, the number of aggregate switches are *d* and the number of core switches are *d/2*. For 1000 servers and 50 ToRs, this results in around 15 aggregate switches and around 8 core switches.

Input Data. To drive the simulator, we needed a traffic matrix (i.e. which VM sends how much data to which VM). Such fine grained data is typically hard to obtain from real data centers [5, 10] because of the required server level instrumentation. We obtained data from a real data center with 17,000 VMs with aggregate incoming and outgoing network traffic from each VM. The data had snapshots of aggregate network traffic over 15 minute intervals spread over a duration of 5 days. To understand the variance in the data, we compared the discrete time differential (Δ) with the standard deviation (σ) of the data. Data that satisfied the following condition was considered constant: $-\sigma/2 \leq \Delta \leq +\sigma/2$. A histogram of percentage of data that was constant for all VMs is given in Figure 3.

It can be observed that most of the VMs have a very large percentage of data that is constant and does not show much variance. A very large variance can be bad for VMFlow, as it will mean too frequent VM migrations, whereas a low variance means that migration frequency can be kept low.

Our first task was to calculate a traffic matrix out of this aggregate per VM data. Given the huge data size, we chose data for a single day. Various methods have been proposed in literature for calculating traffic matrices. We used the simple gravity model [20] to calculate the traffic matrices for all 17,000 VMs at each time stamp on the given day. Simple gravity model uses the following equation to calculate the network traffic from one VM to another: $D_{ij} = (D_i^{out} * D_j^{in})/(\Sigma_k D_k^{in})$. Here D_i^{out} is the total outgoing traffic at VM i, and D_j^{in} is the total incoming traffic at VM j. Although, existing literature [10] points out that traffic matrices generated by simple gravity model tend to be too dense and those generated by sparsity maximization algorithms tend to be too sparse than real data center traffic matrices, simple gravity model is still widely used in literature to generate traffic matrices for data centers [15].

After generating the traffic matrices for each time stamp for the entire data center (17,000 VMs), we used the traffic matrices for the first 1000 VMs in order to reduce the simulation time required. Since gravity model tends to distribute all the traffic data observed over all the VMs in some proportion, it resulted in a large number of very low network demands. In order to make these demands significant, we used a scale-up factor of 50 for all the data (i.e. each entry in all the input traffic matrices was multiplied by 50).

Simulation Results. The simulator compares VMFlow approach with the ElasticTree's greedy bin-packing approach proposed by Heller et. al [9] to save network power. Recall that the ElasticTree's bin-packing approach chooses the leftmost path in a given layer that satisfies the network demand out of all the possible paths in a deterministic left-to-right order. These paths are then used to calculate the total network power at that time instance. For placing VMs for the ElasticTree approach, we followed a strategy that placed the VMs on nodes that had the same id as their VM id.

We assume that all the network power comprise of power for turning on the switches, and the power required for turning on each network link (i.e., for ports on the end switches of the link) is zero. We calculated the total power consumed by the network and the proportion of network demands that were unsatisfied at a given time stamp for both VMFlow and ElasticTree approaches. In the first set of experiments, we simulated an oversubscribed network where the ToR-aggregate switch links and aggregate-core switch links had 1 Gbps capacity. This resulted in a 10:1 oversubscription ratio in the ToR-aggregate switch link layer. The results for the oversubscribed network are shown in Figure 4(a) and Figure 4(b), respectively.

VMFlow outperforms the ElasticTree approach by a factor of around 15% and the baseline (i.e., all switches on) by around 20% in terms of network power at any given time instance. More importantly, one can see the effectiveness of VMFlow in the percentage of network demands that remain unsatisfied. VMFlow saves all the network power while satisfying 5-6 times more network demands as compared to ElasticTree approach.

Since the input data had very little variation over time, we conducted an experiment to compare VMFlow with and without any migration after the first placement was done using the VMFlow algorithm. In order to simulate no migration, VMFlow simulator calculated an optimal placement of VMs based on the first input traffic matrix (representing the traffic for the first time stamp) and only calculated the route at each subsequent time step. Figure 4(a) and Figure 4(b) show the performance of VMFlow approach without any migration.

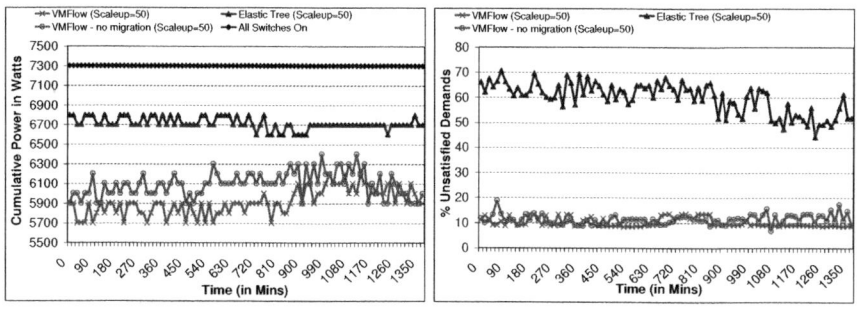

(a) Network power   (b) Unsatisfied network demands

Fig. 4. Comparison at different time steps in the simulation

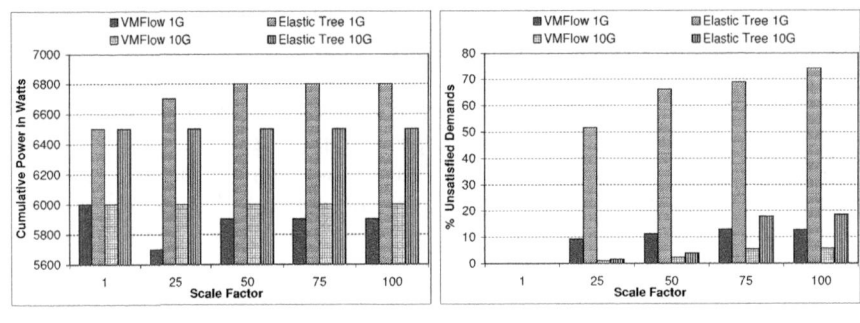

(a) Network power   (b) Unsatisfied network demands

Fig. 5. Comparison at various scale factors in the simulation

One can note that even with low variance input data, VMFlow with migration outperforms VMFlow without migration slightly by a margin of 5% in terms of power. Both the approaches perform roughly the same with respect to percentage of unsatisfied network demands. This indicates that migration frequency has to be properly tuned keeping in mind the variance in network traffic. Each VM migration has a cost associated with it that depends on the size of VM, its current load, the nature of its workload and the Service Level Agreement (SLA) associated with it. We are currently working on a placement framework that will take into account this cost and the potential benefit a migration can achieve in terms of network power and network demand satisfaction.

We also compared the effect of using various scale-up factors on the input network traffic data. In this set of experiments we used a single input traffic matrix (representing the traffic at the first time stamp) and simulated both an oversubscribed network (ToR-aggregate and aggregate-core switch links have 1 Gbps capacity) and a network with no over subscription(ToR-aggregate and aggregate-core switch links have 10 Gbps capacity). The results are shown in Figure 5(a) and Figure 5(b). VMFlow outperforms the ElasticTree approach consistently, both in terms of network power reduction as well as percentage unsatisfied demands in the oversubscribed network (1G). However, in the no over subscription case (10G) the performance difference between VMFlow and ElasticTree is relatively lower. This is along expected lines, since VMFlow is expected to outperform mainly in cases where the top network layers are oversubscribed.

6   Handling Server Consolidation

In the earlier sections, we have assumed that at most one VM can be mapped to a host; i.e., the VM placement mapping Π is one-to-one. It is, however, easy to modify our algorithm to handle the case when multiple VM are mapped to a host, i.e., in case of server consolidation. In modern data centers, server consolidation is often employed to save both on the capital expenditure (e.g., by consolidating several VMs onto a small number of powerful hosts), and operational expenditure (e..g, any unused host can be turned off to save power) [19]. Although the basic idea of server consolidation is simple, deciding which VMs are co-located on a host is a complex exercise that depends on various factors such as fault and performance isolations, and application SLA. Thus, a group of VMs co-located during server consolidation may not be migrated to different hosts during network power optimization. Nevertheless, a group of VMs mapped to a single host can be migrated together to a different host, provided the new host has enough resources for that group of VMs. We now describe, how our greedy algorithm can be easily extended to handle server consolidation.

We assume that we have an initial server consolidation phase which maps zero, one or more VMs to each host. A set of VMs that is mapped to the same host during server consolidation is called a VMset. We make the following two assumptions: (1) while placing VMs on the host, the only resource constraint we need to satisfy is on the compute resource, and (2) a group of VMs consolidated on a host are not separated (i.e., migrated to different hosts) during the network power optimization phase. Now, our greedy algorithm requires two simple modifications to handle server consolidation. First, instead of mapping a VM to a host, the algorithm maps a VMset to a host, and the traffic demand between two VMsets is the sum of the demands of individual VMs comprising the two VMsets. Second, in each iteration for placing the source and destination VMset of a traffic demand, the set of possible hosts ($\Phi(s_j)$ and $\Phi(d_j)$) is defined as the set of free hosts that have equal or more compute resources than the respective VMsets. We omit details of these simple modifications from this paper.

We evaluated the performance of VMFlow in presence of server consolidation using the same simulation framework described in Section 5. Typically server consolidation is done based on the CPU and memory demands of the VMs and the individual host CPU and memory capacities. Since we did not have access to the CPU data for the 1000 VMs that were used in the earlier experiments, we generated synthetic CPU demands that were normally distributed with a given mean and standard deviation. We used the methodology given in [12], to generate mean and standard deviation of the CPU demands. We assumed the servers to be IBM HS22v blades [2], that go into a Bladecenter H chassis (each chassis can host 14 blades). We simulated an oversubscribed network where the server-TOR links, the ToR-aggregate switch links and aggregate-core switch links had 1 Gbps capacity. We assumed 3 different configurations of HS22v blades and created a VM packing scheme based on CPU demands using first fit decreasing (FFD) bin-packing algorithm. Following 3 configurations of HS22v blades (with increasing CPU capacity) were used in the simulation:

1. 1 Intel Xeon E5503 with 2 cores at 2.0 GHz - assuming a mean of 1.6 GHz and std. dev of 0.6, 1000 VMs with normally distributed CPU demands were packed into 612 servers (a packing ratio of 1.63 VMs per server). This required a network with 44 TORs, 14 Aggregate and 7 Core switches.
2. 1 Intel Xeon X5570 with 4 cores at 2.93 GHz - assuming a mean of 2.34 Ghz and std. dev of 0.93, 1000 VMs with normally distributed CPU demands were packed into 265 servers (a packing ratio of 3.76 VMs per server). This required a network with 19 TORs, 9 Aggregate and 5 Core switches.

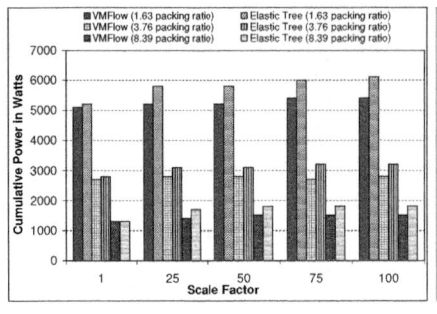   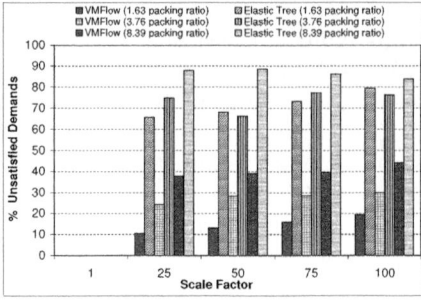

(a) Network power   (b) Unsatisfied network demands

Fig. 6. Comparison at various scale factors and configurations in the server consolidation scenario simulation

3. 2 Intel Xeon X5570 with 4 cores at 2.93 GHz - assuming a mean of 2.34 Ghz and std. dev of 0.93, 1000 VMs with normally distributed CPU demands were packed into 119 servers (a packing ratio of 8.39 VMs per server). This required a network with 9 TORs, 6 Aggregate and 3 Core switches.

The input network traffic matrix was modified such that input and output network demands for VMs that were packed together to create a VMset, were added. VMsets resulting from the VM packing step were then used by the simulator to compare VMFlow and Elastic tree at various network traffic scale factors. Results for network power and percentage unsatisfied demands are shown in Figures 6(a) and 6(b),respectively. VMFlow continues to outperform Elastic tree even in presence of server consolidation. While the network power savings are consistently in the range of 13-20% at higher scale factors, the gains for percentage unsatisfied demands decrease (5x at 1.63 packing ratio, 3x at 3.76 packing ratio, and 2x at 8.39 packing ratio) as more VMs are packed into a single server and network traffic matrix becomes denser.

7   Future Work

This paper presented VMFlow, a framework for reducing power used by network elements in data centers. VMFlow uses the flexibility provided by VM placement and programmable flow-based routing, that are available in modern data centers, to reduce network power while satisfying a large fraction of the network traffic demands. We formulated the Network-Aware VM Placement as an optimization problem, proved that it is NP-Complete, and proposed a greedy heuristic for the problem. Our technique showed reasonable improvement (15-20%) in network power and significant improvement (5-6 times) in the fraction of satisfied demands as compared to previous network power optimization techniques.

Each VM migration in a data center has a cost associated with it. Similarly, the benefits that can be achieved by a VM migration in terms of network power reduction and meeting more network demands, depends on the variance in network traffic over time. In future, we plan to work on a placement framework that takes into account this cost and the potential benefit a migration can achieve in terms of network power and network demand satisfaction. We also plan to apply our technique to other network topologies and evaluate its benefits.

References

1. Cisco: Data center: Load balancing data center services (2004)
2. IBM BladeCenter HS22V, http://www-03.ibm.com/systems/xbc/cog/bc_hs22v_7871/bc_hs22v_7871aag.html
3. The OpenFlow Switch, http://www.openflowswitch.org
4. Ahuja, R., Magnanti, T., Orlin, J.: Network Flows - Theory, Algorithms and Applications. Prentice-Hall, Englewood Cliffs (1993)
5. Benson, T., Akella, A.A.A., Zhang, M.: Understanding Data Center Traffic Characteristics. In: ACM SIGCOMM WREN Workshop (2009)
6. Greenberg, A., Hamilton, J., Maltz, D., Patel, P.: The Cost of a Cloud: Research Problems in Data Center Networks. In: ACM SIGCOMM CCR (January 2009)
7. Greenberg, A., Jain, N., Kandula, S., Kim, C., Lahiri, P., Maltz, D., Patel, P., Sengupta, S.: VL2: A Scalable and Flexible Data Center Network. In: ACM SIGCOMM (August 2009)
8. Guo, C., Lu, G., Li, D., Wu, H., Zhang, X., Shi, Y., Tian, C., Zhang, Y., Lu, S.: BCube: A High Performance, Server-centric Network Architecture for Modular Data Centers. In: ACM SIGCOMM (August 2009)
9. Heller, B., Seetharaman, S., Mahadevan, P., Yiakoumis, Y., Sharma, P., Banerjee, S., McKeown, N.: ElasticTree: Saving Energy in Data Center Networks. In: USENIX NSDI (April 2010)
10. Kandula, S., Sengupta, S., Greenberg, A., Patel, P., Chaiken, R.: The Nature of Data Center Traffic: Measurements and Analysis. In: ACM SIGCOMM Internet Measurement Conference, IMC (2009)
11. Kandula, S., Katabi, D., Sinha, S., Berger, A.: Dynamic load balancing without packet reordering. Computer Communication Review 37(2) (2007)
12. Korupolu, M.R., Singh, A., Bamba, B.: Coupled placement in modern data centers. In: IEEE IPDPS (2009)
13. Mahadevan, P., Banerjee, S., Sharma, P.: Energy proportionality of an enterprise network. In: 1st ACM SIGCOMM Workshop on Green Networking (2010)
14. Mann, V., Kumar, A., Dutta, P., Kalyanaraman, S.: VMFlow: Leveraging VM Mobility to Reduce Network Power Costs in Data Centers. Tech. rep. (2010)
15. Meng, X., Pappas, V., Zhang, L.: Improving the Scalability of Data Center Networks with Traffic-aware Virtual Machine Placement. In: IEEE INFOCOM (2010)
16. Mysore, R., Pamboris, A., Farrington, N., Huang, N., Miri, P., Radhakrishnan, S., Subramanya, V., Vahdat, A.: PortLand: A Scalable Fault-Tolerant Layer 2 Data Center Network Fabric. In: ACM SIGCOMM (August 2009)
17. Shang, Y., Li, D., Xu, M.: Energy-aware routing in data center network. In: 1st ACM SIGCOMM Workshop on Green Networking (2010)
18. Vazirani, V.: Approximation Algorithms. Addison-Wesley, Reading (2001)
19. Verma, A., Ahuja, P., Neogi, A.: pMapper: Power and migration cost aware application placement in virtualized systems. In: ACM/IFIP/USENIX Middleware (2008)
20. Zhang, Y., Roughan, M., Duffield, N., Greeberg, A.: Fast accurate computation of large-scale IP traffic matrices from link loads. In: ACM SIGMETRICS (2003)

A Collaborative AAA Architecture to Enable Secure Real-World Network Mobility

Panagiotis Georgopoulos, Ben McCarthy, and Christopher Edwards

School of Computing and Communications,
Lancaster University, Lancaster, LA1 4WA, UK
{panos,b.mccarthy,ce}@comp.lancs.ac.uk

Abstract. Mobile Networks are emerging in the real world in various scenarios, from networks in public transportation to personal networks in consumer electronics. The NEMO BS protocol provides constant network connectivity and reachability for the nodes of these Mobile Networks in a seamless manner despite their roaming. However, NEMO BS has yet to show its advantages in real world deployment because it lacks trouble-free and secure network access for the whole network, and secure data transmission for the nodes it provides connectivity for. On the other hand, Access Networks provide connectivity for Mobile Networks, but currently lack a robust AAA service which would enable network mobility support in a fast, trouble-free, but also secure and authenticated manner. Our paper describes a collaborative Unified Architecture that satisfies the requirements of both Mobile Networks and Access Networks, and our evaluation proves its efficiency and applicability for real world deployment in today's Internet infrastructure.

Keywords: Network Mobility, NEMO BS, AAA, RADIUS, Security.

1 Introduction

Mobile IP (both Version 4 [11] and Version 6 [8]) is designed to support the mobility of individual mobile nodes, such as laptops, netbooks or smartphones that are roaming from one network to another, maintaining session continuity for their applications. However, there is a strong requirement nowadays to support the mobility of an entire group of devices (termed a Mobile Network (MN)) that usually moves as a whole, whereas its individual devices (termed as Mobile Network Nodes (MNNs)) remain relatively immobile in relation to one another. For example, networks in public transportation, such as in trains, buses, coaches and airplanes that offer connectivity to passengers' devices as they move can be considered as MNs, whilst the end-devices are MNNs. Another example of a MN is a PAN in a consumer electronics scenario, where connectivity for all the different devices a user may carry as he moves, is maintained by a lightweight personal mobile router. For example, a mountain rescuer may have such a PAN that could consist of a wireless IP camera, a PDA with a live mapping application, IP sensor devices on his body to monitor his health condition and a VoIP application for communication with fellow rescuers during a rescue mission.

J. Domingo-Pascual et al. (Eds.): NETWORKING 2011, Part I, LNCS 6640, pp. 212–226, 2011.

The NEMO BS protocol [4] is responsible for offering constant connectivity and seamless session continuity to the MNNs of a MN without them having to be aware of their mobility, even though the MN might change it point of attachment from one Access Network (AN) to another. However, for real-world deployment, in order for the MR to run NEMO BS and support the mobility of the network, it has to be able to obtain secure and authenticated access from an AN in a fast and trouble-free manner. In addition, in such wireless mobile scenarios secure transmission of data is paramount, both in the range of the wireless hotspot the MR offers, but also as packets travel beyond the MN to the Internet via the AN. Furthermore, the AN has to have an efficient and scalable AAA infrastructure in order to authenticate, authorize and account the MN's connectivity.

Building on our previous work [6], this paper describes a Unified Architecture (UA) that combines the strengths of Network Mobility and AAA services to satisfy in a secure manner the requirements of both MNs and ANs. The rest of the paper is organized as follows. Section 2 discusses the motivation behind our research and Section 3 gives some background information on its cornerstones. Section 4 provides the design of our UA and Section 5 evaluates our approach qualitatively and quantitatively. Finally, Section 6 concludes this body of work and highlights its benefits.

2   Motivation

In order to describe the motivation behind this work, let us discuss a real-life example that reveals the current problems and explains the reasoning and motivation behind our research. Let us consider a scenario where a bus company decides to offer Internet connectivity to passengers that board its buses every day. In this scenario, as the bus does its every-day route through town, its MR is responsible to obtain Internet connectivity from various publically available AN's Access Points (APs) being sporadically located around town, and share it to the passengers' devices by projecting a wireless hotspot in the bus.

During this real-life scenario, the MN of the bus requires:

1. Seamless mobility and uninterrupted connectivity for all its MNNs, despite the fact that the MR has to change its IP address whilst roaming from one AN to another, thus causing the MNNs' applications sessions to break.
2. Quick, effortless and secure network access to each AN it connects to. The MR should avoid requiring to be configured with the different type of authentication credentials and protocols each AN it encounters require.
3. Dynamic trust establishment with the AP it roams to, to avoid being connected to available but deceitful APs that would try to sniff its authentication credentials and hack into the MNNs' transmitted packets.
4. Avoid revealing its identity to each AN it roams to for privacy purposes.
5. Secure data transmission both locally, in the vicinity of the hotspot it provides to its MNNS, but also globally, as its MNNs' data are transmitted from the MN to the Internet via the AN.

At the same time, AN providers have their own requirements to satisfy in order to provide connectivity for MNs. To be specific, ANs require :

1. Financial benefit for setting up and administering wireless networks possible scattered around a large area, for the MNs to connect to.
2. A robust and well configured service to Authenticate, Authorize and Account the MNs' connectivity in a practical and scalable manner. It is unrealistic to expect that each AN should know in advance each MR requesting network access on behalf of a MN.
3. Avoid compromising their security policies and disallow unauthorized access to their networks.

3   Background

This section introduces the cornerstones of this research work; NEMO BS, IPsec, Radius, TLS based authentication methods and wireless security protocols.

3.1   NEMO BS Protocol

The NEMO BS protocol [4] grants session continuity to the nodes of a MN that is roaming across different ANs, changing its point of attachment to the Internet. NEMO BS takes advantage of the large IPv6 address pool that guarantees global reachability of every MNN and also, manages to keep the mobility of the network transparent to its MNNs, by removing completely the need for the MNNs to run any extra protocols themselves to support their mobility.

According to NEMO BS, when a MR moves away from its Home Network (HN) and finds an AN to connect to, it sends a Binding Update (BU) to its Home Agent (HA) containing its new topologically correct IPv6 Care-of-Address (CoA), optionally along with one or more Mobile Network Prefixes (MNPs). These MNPs represent the networks that the MR serves and advertises to its MNNs in Router Advertisements, so that each MNN can configure an IPv6 address itself and be globally reachable. When the HA receives a valid BU from a MR, it updates its binding cache for that MR and its MNPs, and then acknowledges it with a Binding Acknowledgment (BA). From this moment a bidirectional tunnel between the HA and the MR is instantiated and the HA intercepts packets destined for all the MNPs that the MR has registered, and forwards them to the MR's CoA to ensure reception of packets by the MNNs.

3.2   IPsec

IPsec [9] is a protocol suite for establishing secure IP communications between peers (hosts or gateways) over an insecure network. IPsec uses a combination of protocols to provide mutual authentication, data confidentiality, data integrity, non-repudiation and anti-replay protection on a per packet basis without any regard to the communication path between parties. IPsec mainly uses three cryptographic protocols; IKE, AH and ESP. IKE is used to exchange cryptographic keys among peers, establish the Security Associations (SA) among them for inbound and outbound traffic per peer and negotiate all the cryptographic algorithms of the secure communication channel that IPsec will operate upon. AH and ESP are used in two different modes based on the requirements of the application scenario; transport mode, where protection is provided from the Transport layer and higher, and tunnel mode, where protection is provided for the whole packet. Use of IPsec is mandatory in mobility scenarios [2],[4] to avoid attacks such as man-in-the middle, passive wiretapping or impersonation. Our UA complies with the aforementioned references and uses ESP in transport mode to protect control traffic between the MR and the HA in both directions, and ESP in tunnel mode to protect all the application traffic the MNNs generate.

3.3   RADIUS AAA Protocol

The RADIUS AAA protocol [12] is the most well known and widely deployed AAA protocol worldwide. Its functionality is built on the generic AAA framework defined in [10] and mainly involves three entities; the supplicant (end-device), the Network Access Server (NAS) and the AAA server.

The process of performing a AAA service for a wireless device using RADIUS is as follows. When a device requests network access it uses a Layer 2 protocol (such as PPP or EAP) to communicate with the NAS and, among other information, send to it its authentication credentials. The chosen authentication method (e.g. EAP-TLS, PEAP etc.) will define the number of packets that should be exchanged for the authentication of the client, in addition to the type and format of the authentication credentials (e.g. a hashed password, a certificate etc.). Since the NAS has no appropriate means to authenticate the supplicant itself, it will initially collect all the information the supplicant sends, then use its AAA client implementation to encapsulate it in appropriate AAA packets, and then encrypt those with a strong key it shares with the AAA server before finally sending them to the server. When the AAA server receives these packets, it authenticates the supplicant usually with the aid of other resources, such as local databases or a PKI. If authentication is successful, the RADIUS server tries to authorize the user by checking its authorization policies which are, usually, ISP specific. When the AAA server reaches a decision whether the user should be granted or denied access to the network, it replies using AAA packets to the NAS, which is then responsible for relaying the reply to the supplicant over Layer 2 frames. When this phase is completed, the user is granted (or denied) access with a defined authorization level and the AAA server starts collecting accounting information for the supplicant's network usage from the NAS using specific accounting messages, that update the AAA server at regular intervals.

3.4   TLS Based Authentication Methods

There are more than 40 EAP based authentication methods that can be used in conjunction with a AAA protocol, which will encapsulate the data found in EAP frames into appropriate AAA messages and transfer them from the NAS to the AAA server. However, EAP authentication methods that are based on a Transport Layer Security Tunnel, such as EAP-TLS, EAP-TTLS, PEAP or

EAP-FAST and others, appeal more to our research not only because they are more secure, but also because they bring significant advantages to roaming users. During Phase 1 of such an authentication method, the AAA server is authenticated to the supplicant, and then a secure TLS tunnel is created between them, which is then used from the client to submit its own authentication credentials. This process allows the supplicant to maintain its privacy as it does not disclose its authentication credentials to an intermediate AP or AN, but ensures that these are only transmitted to the AAA server over the tunnel they have established. Of particular importance for our research are EAP-TLS [13] and EAP-TTLS [5]. During Phase 2, where the AAA server authenticates the client, EAP-TLS uses a client's certificate, whereas EAP-TTLS uses a password in a hashed format, usually dictated by another authentication mechanism such as CHAP. One important advantage that EAP-TLS and EAP-TTLS offer to roaming users, is the ability to skip Phase 2 of the authentication process if the user has been authenticated with the AAA server before. This feature is called Session Resumption (SR) and when enabled, allows the AAA server to keep a unique cache entry for a client that gets successfully authenticated once. Therefore, when the client roams to another AP requesting the same authentication procedure to occur, after completion of Phase 1, the AAA server realizes that it has a cached entry for it from its previous authentication, and thus skips Phase 2, even in scenarios where the client has not been connected to that AP before.

3.5   Wireless Security

The need for securing wireless communication is much higher for publicly available APs. The WiFi Alliance has designed WPA and WPA2 protocols to secure wireless networks by offering packet encryption, message integrity, protection against replay attacks and authorized network access with the use of cryptographic algorithms. WPA and WPA2 protocols support two authentication modes; Personal mode and Enterprise mode. When Personal mode is used, wireless clients can connect to an AP using a preshared key (PSK), whereas when Enterprise mode is used authentication is performed with the aid of a AAA backend server. One of the big advantages of the Enterprise mode of WPA2, is that after successful authentication, the wireless client and the AAA server are the only owners of a Master Key, which they then use to derive a Pairwise Master Key (PMK). The PMK is then sent from the AAA server to the AP in a secure AAA message, and is being used as a symmetric key, bound to the session of the AP and the client. With the knowledge of the PMK, a subsequent 4-way handshake is performed between the AP and the wireless node, that derives, binds and verifies additional operational keys that are used in the future communication of the node with the AP. This significant feature of WPA2 in Enterprise mode means that session keys are being securely derived and negotiated in a way that security is enhanced and preconfiguration is avoided. One additional advantage that the aforementioned key derivation procedure brings to mobile users, is that when a roaming node connects to a WPA2 AP, it firstly checks if it has a collection of keys, called PMK Security Association (PMKSA), that can be used with this AP. If this information has been cached from a previous association, then there is no need for a full authentication procedure with a AAA server, but only the 4-way handshake has to be performed locally between the AP and wireless client. PMKSA caching significantly speeds up the authentication procedure of the roaming client without compromising the security of the network.

4   Design

We devised a Unified Architecture (UA) with the goal of bridging the gap between Network Mobility and AAA services in a secure manner and satisfying the requirements of all the parties involved. Fig 1 illustrates our design where we overlay the AAA model in its extended form for roaming scenarios [10] over NEMO BS's architecture and integrate these services in a unified way. According to our design, the MN's HN now consists of a HA and a AAA Home Server (AAAHS) that are responsible for providing the Mobility and AAA services for the MN respectively, whilst it is away from its HN. Conceptually, in our bus scenario the HN could be represented by servers located at the IT department of the company or provided by an ISP offering these services. The AN depicted in Fig. 1 represents any network that can provide Internet connectivity via wireless hotspots in the area the MN roams. The AN (also known as Foreign Network) consists of its own AAA Foreign Server (AAAFS) and many APs that act as NASs and are able to exchange packets with its AAAFS. Conceptually the AN could be a wireless ISP's town network or a municipal network, offering publicly available WiFi connectivity to MNs.

It is important to emphasize that our UA does neither augment nor alter the design of the AN itself, making it ready to be used in the current Internet infrastructure. What our UA requires though, is that the AN has a Service Level Agreement (SLA) with the HN of a MN, through which, the AAAFS has the ability to relay the authentication process to the MN's AAAHS. The AAAHS has evidently more appropriate information to authenticat the MR, and permit it to offer connectivity to its MNNs. This collaborative design provides important benefits to all parties involved, as the AN does not require to know the MR in advance, neither has to have any preconfigured information about it or its MNNs. In addition, if a TLS based authentication method is being used the MR avoids revealing its identity and credentials to the AN itself, as its authentication data are forwarded securely to its HN over a tunnel, after

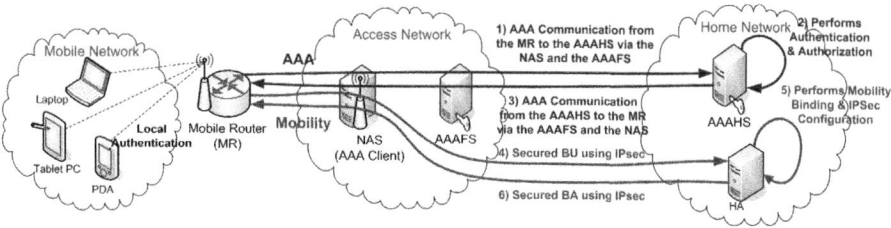

Fig. 1. The devised Unified Architecture

the initial authentication of the AAAHS to the MR (as described in Subsection 3.4). If the authentication procedure of the MR is successful, this means that the AAAFS has a secure partnership with the AAAHS dictated by the SLA and confirmed from the knowledge of the shared secret they use to secure the AAA packets they exchange, thus, the MR can trust the AN's AP. ANs have financial incentives to get SLAs with HNs because these would allow them to serve MNs and in turn bill their HNs appropriately for the provided service. HNs are also interested in getting SLAs with as many ANs as possible, because the latter will serve their MNs when they are away from "home", inducing financial benefits to both network providers and bigger connectivity coverage for the end users. Our UA does not oblige the establishment of SLAs between all small-scale networks, on the contrary, it can easily facilitate a hierarchical model where only big ISPs have SLAs between each other and through them accommodate the smaller networks they provide connectivity for. This model is simply fitted into our UA by introducing a chain of intermediate AAA servers of the involved ISPs in the path between the AAAFS of the AN and the AAAHS of the HN. According to this model, each AAAFS will play the role of the proxy AAA server and forward packets to the next AAA server in the chain until data reaches an ISP that has a partnership with the MN's HN and is able to finally route these packets to the MN's AAAHS, leading to a model that scales for the real world.

Following the principle of NEMO BS, where the MR is responsible to provide the mobility service for all its MNNs without them having to run any mobility protocols themselves, our UA dictates that the MR should carry out the AAA service on behalf of the whole network as well in a similar fashion. Therefore, when the MR connects to an AN it is responsible to authenticate itself and the entire network to its AAAHS, and include the MNP of the network it is serving in its BU. This important process ensures that only MNNs of a certain IPv6 address pool are authenticated and authorized for Internet connectivity. However, since the MR is carrying the authentication procedure on behalf of the whole network, it should also perform local authentication of the MNNs connected to it, using for example WPA2/PSK or WPA2/EAP-TLS. In our bus scenario this can easily be facilitated if, for example, regular commuters subscribe to the Internet-on-the-bus service and obtain a username and password to authenticate to the bus' MR, or if non-regular commuters could get a WPA2/PSK scratch card when they purchase their ticket. The two distinct authentication procedures that our UA defines (MNNs authenticated to the MR and the MR authenticated to its AAAHS) provide important advantages in roaming scenarios. Although the MR might be changing its point of attachment from one AN to another roaming, the MNNs will experience only a slight connectivity disruption whilst the roaming take place, but none of them would have to reauthenticate after this roaming. In addition, the MNNs could join or leave the MN whenever they want, without having to "inform" a distant server about it, reducing significantly the number of packets that have to travel from the MN to the Internet (and vice versa) saving bandwidth and minimizing processing delays.

Let us now consider the phases a MR has to go through to provide Mobility, Security and AAA for all its locally authenticated MNNs, from the time it starts roaming to a new AN until it obtains full Internet connectivity. According to our design, in order for the MR to become fully operational it has to perform its Layer 2 handover, its AAA communication as required by RADIUS and the chosen TLS based authentication method, its mobility tasks as required by NEMO BS and its security related configuration, as required by the local AP it connects to and the use of IPsec for the traffic generated by or destined to its MNNs. These occur sequentially in the following three distinct phases :

• **Phase 1 - Layer 2 Association:** The MR performs the Layer 2 association with the AP of the AN it roams to.

• **Phase 2 - Layer 2 & 3 AAA Communication and WiFi Security Configuration:** During this phase the MR will request network access to the AN and will perform the AAA authentication with its HN using the RADIUS protocol (described in Subsection 3.3). Since the MR does not have an IP address yet, its communication with the AP occurs using EAP frames that carry all the required information using Layer 2 (MAC) addresses. The AAA client implementation of the AP, encapsulates the data from the EAP frames to IP AAA packets and sends them to its AAAFS. However, since we are using a TLS based EAP authentication method in our UA, the initial packets that are sent from the MN should eventually reach its AAAHS server in order to establish a TLS tunnel. In order to accomplish this, in these initial packets the MR presents its "identity" to the AP, essentially revealing only its domain name in the form of "anonymous@homenetwork.com" by complying to the standardized Network Access Identifier (NAI) (defined in [1]), and thus enabling the local AAAFS to identify where it should forward all the AAA packets to. When the AAAFS relays the initial authentication data to the MN's AAAHS, the latter forms a tunnel with the MR and carries out the full authentication process by exchanging packets back and forth according to the chosen authentication method. When the authentication process finishes, the AAAHS replies to the AAAFS of the AN, and then the later according to the authentication reply either grants network access to the MN or denies it. Authorization usually occurs after authentication and is related to the actual policies the AN and HN have in place for roaming networks. When authentication and authorization finish, and if the MN has been granted access, the MN derives secure session keys for its communication with the AP as described in Subsection 3.5, and the AAAFS starts the accounting procedure for the MN and updates the AAAHS with billing records.

• **Phase 3 - Layer 3 Mobility & IPsec configuration:** If the MR successfully finishes Phase 2 and is granted access, it obtains a topologically correct IPv6 address either by contacting a DHCPv6 server or using IPv6 autoconfiguration. When the MR has an IPv6 address, its first task is to perform its mobility binding with its HA and at the same time to configure its security associations and apply its IPsec policies. According to [4] and [2], at this moment, the MR ensures that its control traffic to and from its HA will be secured by ESP in transport mode and all the subsequent traffic to and from its MNNs will be secured by ESP in tunnel mode. Therefore, to successfully finish Phase 3 the MR sends its secured BU with its MNPs to the HA and waits for the matching BA that denotes a successful binding and a fully operational MN.

5   Evaluation

This section describes a qualitative and quantitative evaluation of our UA.

5.1   Qualitative

Our UA brings significant benefits to both the roaming MN and the AN that serves it. Section 2 discussed the motivation behind our research. Here we revisit the presented motivation and detail it in requirements that are satisfied with our approach. Our UA satisfies the following requirements of the MN :

1. Secure, unobtrusive and trouble-free network access :
   a) The MR does not need to be configured with different types of credentials each visiting AN requires, since the actual AAA procedure is performed with the MN's AAAHS and this configuration is known to the MR in advance.
   b) The MR does not reveal its identity to each AN it is visiting, thus keeping its privacy whilst roaming.
   c) The MR is able to establish trust dynamically with the AN it roams to, by relaying this task to its HN during its authentication. If its AAAHS does not trust the AAAFS when the latter forwards packets to the former, the AAA process fails and thus the MR should not trust the AN's AP.
2. Secure transmission of data locally, in the range of the AP using WPA/ WPA2, and globally, as data leave the AN and travel to the Internet using IPsec.
3. Constant and reliable connectivity is provided with the use of NEMO BS, which in conjunction with the trouble-free network access that is provided using the AAA service, leads to seamless and quick roaming for the MN.

   Our UA satisfies the following requirements of the AN:
1. Authentication of the MR without requiring to have information about it in advance. The AN relays the authentication procedure to the MN's HN that has more appropriate information to authenticate the MR.
2. Authorization of the MR according to its local policies.
3. Accounting of the MR for its MNNs' network use in order to bill its HN for the provided service appropriately.
4. All the previous transactions are performed without compromising the security policies of the AN and by bringing financial benefits to it.

5.2   Quantitative

In order to evaluate the true applicability and efficiency of our approach, we carried out a series of performance Tests on our experimental testbed. In this Section we describe the hardware and software setup of our testbed, the tests that we carried out and finally, we analyze and discuss the results observed.

Hardware and Software Testbed Setup : To evaluate the capabilities and performance of our UA we configured the testbed illustrated in Fig. 2. Our testbed consists of three Access Networks (AN1, AN2 and AN3), a MN, and the HN the MN originates from. All PCs on our testbed have a P4 2.8GHz CPU, 2GB RAM and a 80 GB hard drive and run Ubuntu 10.04 LTS. Each AN consists of two PCs, one of them acting as an AP by projecting a 802.11g wireless hotspot using the ath5k driver and the hostapd deamon (version 0.7.3), and the other acting as the AN's AAAFS. The AP in AN1 is being configured in WPA1-Enterprise mode, whereas the APs in AN2 and AN3 are being configured in WPA2-Enterprise mode, allowing us to experiment with different wireless AP configurations. The HN consists also two PCs, one of them acting as a HA by running the NEMO BS protocol stack from [7] in HA configuration, and the other acting as a AAAHS. All the AAA Servers on our testbed run the FreeRadius AAA Server (version 2.1.10) and have a certificate which we issued by creating a Certificate Authority. Furthermore, each AAAFS has a shared secret with the AAAHS in order to communicate with it securely. All the equipment on our testbed is IPv6 enabled. AN1 and AN2 are connected with the HN over Ethernet using native IPv6 addresses, whereas AN3 is connected to the HN over the Internet using an IPv6 Tunneling service from HE Tunnel Broker [3]. This IPv6 tunneling approach introduces approximately a 315 ms delay (630ms roundtrip) and routes all the packets from AN3 to HN and vice versa via the Internet, using global IPv6 addresses. This technique ensures that our tests are being carried out not only on a local basis, but also over a long distance route over the Internet that presents real-time traffic characteristics in the communication. The MR used in our tests runs the NEMO BS stack from [7] being configured in MR mode, with the appropriate MNPs and IPsec configuration that matches the one at its HA. Finally, the MR runs WPA_supplicant (version 0.7.3) to allow it to connect to the ANs' APs, and also runs the hostapd deamon to create a wireless hotspot for three MNNs (two laptops and an HTC Touch 2 PDA) and use its internal RADIUS functionality to authenticate the MNNs.

Testing Sets : The aforementioned testbed setup allows testing to take place via APs that have different configuration and over different routes, mimicking how communication would take place in an actual deployment scenario where MNNs are connected to a MR, and the MR roams from one AN to another. Following the design of our UA, we perform two separate Testing Regimes, one to test the performance of local authentication of the MNNs to the MR, and another one to evaluate the roaming of the MR from one AN to another. We repeat each Test of each Testing Regime 50 times, with the focus on how quickly the MNNs or the MR become fully operational using different authentication methods and configuration over our UA.

In our first Testing Regime we use three different MNNs (two laptops and a PDA) and record how long it takes them to connect to the MR's AP (Layer 2 association), to authenticate to it, and finally, to obtain an IPv6 address and become fully operational (Table 1). The PDA and one laptop are using WPA2-PSK for authentication, whereas the other laptop is using EAP-TTLS.

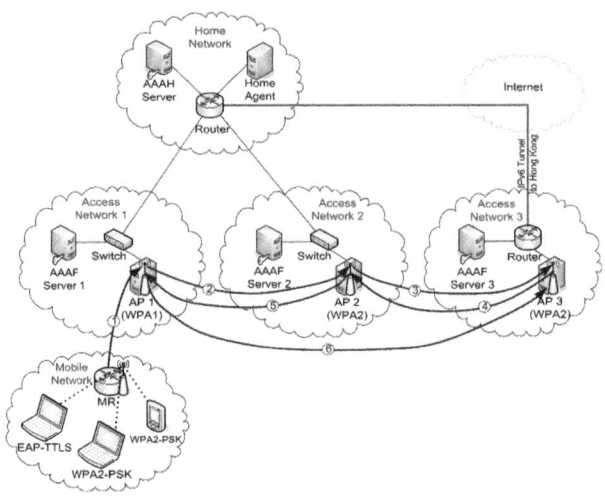

Fig. 2. Experimental Testbed

Our second Testing Regime, includes four different Tests where the MR authenticates to the ANs using EAP-TLS without SR (Test 1, Table 3), EAP-TLS with SR (Test 2, Table 4), EAP-TTLS without SR (Test 3, Table 5) and EAP-TTLS with SR (Test 4, Table 4) respectively. Each Test includes six roaming movements (Stages) of the MR from one AP to another which are presented in Fig. 2 in blue arrows, associating the MR with AP1, AP2, AP3, AP2, AP1 and finally AP3. We decided to perform the aforementioned roaming movements because they demonstrate realistic scenarios, where a MR might connect repeatedly to the same AP, or swap from one AP to another repeatedly.

Results : Table 1 details the results from our first Testing Regime where our three MNNs connected to the MR's AP, authenticated to it using WPA2-PSK or WPA2-EAP-TTLS and obtained an IPv6 address from the MNP the MR advertised. Although the results from this Testing Regime seem very reasonable, we could note the significant difference on the average time it takes for the PDA to do its WPA2-PSK authentication (3.690 sec.) compared to the laptop's average time (0.05 seconds), which is attributed to the big difference on the resources the two devices have. All three MNNs remained connected constantly to the MR and were transmitting packets while we were performing the roaming Tests of the second Testing Regime that follow.

For our second Testing Regime we performed four different Tests where we roamed the MR of the MN, and essentially the whole MN from one AP to another in Six Stages using four different authentication methods and configuration. Before focusing on the results of the different authentication methods (noted as Phase 2 on Tables 3, 4, 5 and 6), we could note that Phase 1 results (Layer 2 Association) were similar in all Tests, varying from 0.883 seconds to 1.499 seconds. Similarly, Phase 3 results in all Tests, where the MR configures an IPv6 CoA, performs its binding registration with its HA and applies its IPsec policies

Table 1. Authenticating MNNs to the MR

	WPA2-PSK PDA	WPA2-PSK LAPTOP	WPA2-EAP-TTLS LAPTOP
Layer 2 Assoc. (sec.)	2.585	2.300	3.670
Layer 3 Auth. (sec.)	3.690	0.050	0.320
Layer 3 IP (sec.)	3.167	2.390	1.905
Total (sec.) :	9.442	4.740	5.895

Table 2. Local/Global number of Packets for Phase 2

# of Packets	TEST 1	TEST 2	TEST 3	TEST 4
Stage 1	7/14	7/16	7/16	7/16
Stage 2	5/14	5/6	5/16	5/6
Stage 3	5/14	5/6	5/16	5/6
Stage 4	4/0	4/0	4/0	4/0
Stage 5	7/14	7/6	7/16	7/6
Stage 6	4/0	4/0	4/0	4/0

for all the traffic the MNNs might send, did not present notable differences, varying from 1.372 seconds to 2.574 seconds.

Focusing on the results of the different authentication methods and the route they take place via, we concentrate on Phase 2 of each Test of the second Testing Regime. During Test 1 of our second Testing Regime, the MR used EAP-TLS without SR and roamed from one AN to another as presented in Fig. 2. As Table 3 illustrates, in Stage 1 the MR connected to AP1 and carried out its EAP-TLS authentication in approximately 0.780 seconds, which is remarkably low, since this phase requires 21 packets to be exchanged in total. As Table 2 presents, 7 out of the 21 packets are transmitted locally between the MR and the AP, and 14 are transmitted "globally", following a 3 hop route from the MR to the AN's AP, then to the AAAFS and finally, to the MR's AAAHS. In Stage 2 of Test 1, the MR roamed to the WPA2 AP2, and did its EAP-TLS authentication faster and transmitted two less local packets. Stage 3 consists of the MR being connected to AP3, where all the Phase 2 packets are routed to the AAAHS over the IPv6 tunnel to Hong Kong. The tunnel overhead increased Phase 2 timing to an average of 4.549 seconds, almost 10 times more compared to the previous Stages. This is an expected delay as 14 packets have now to travel over

Table 3. EAP-TLS Results without Session Resumption

Phases \ Stages	ST.1	ST.2	ST.3	ST.4	ST.5	ST.6
Phase 1 (sec.)	1.208	1.045	1.084	0.959	0.969	1.016
Phase 2 (sec.)	0.780	0.208	4.549	0.014	0.690	0.009
Phase 3 (sec.)	1.630	1.585	2.310	1.833	2.123	2.433
Total (sec.) :	3.618	2.838	7.943	2.806	3.782	3.458

Table 4. EAP-TLS Results with Session Resumption

Stages Phases	ST.1	ST.2	ST.3	ST.4	ST.5	ST.6
Phase 1 (sec.)	1.110	0.985	0.883	1.068	0.887	1.081
Phase 2 (sec.)	0.345	0.090	1.906	0.038	0.126	0.045
Phase 3 (sec.)	1.770	1.372	1.754	1.686	2.487	2.290
Total (sec.) :	3.225	2.447	4.543	2.792	3.500	3.416

the IPv6 tunnel which adds significant delay. The results from Stage 4, where MR roams to AP2 where it has been connected before in Stage 2, illustrate the benefits of PMKSA caching of WPA2, as the MR does not perform a full EAP-TLS authentication, but just a 4-way handshake. With PMKSA caching Phase 2 of Stage 4 completes in just 0.014 seconds, almost 15 times less compared to Phase 2 of Stage 2, when the MR connected to AP2 for the first time. Stage 5 is where the MR connects to AP1, where, although it has been connected to before, as it is a WPA1 AP, it does not support PMKSA caching and thus records similar timings with Stage 1. Stage 6 further affirms the advantages of PMKSA caching, as Phase 2 completes only in 0.009 seconds despite the tunnel setup to Hong Kong, since only the local 4-way handshake is required. Due to PMKSA caching, 15 packets less are now being exchanged in Stage 6 compared to Stage 3, which also leads to a reduction of approximately 4.5 seconds of the total time it takes for Stage 6 to complete.

Test 2 of the second Testing Regime, repeats Test 1 but with the SR feature enabled at the AAAH Server. During Phase 2 of Stage 2 of this Test, the AAAHS realized that the MR had performed a successful authentication with it some seconds ago (during Stage 1) and thus skips the second part of the EAP-TLS procedure. As Table 2 presents, only 6 packets compared to 14 are now exchanged in Phase 2, which completes in only 0.090 seconds (Table 4), more than twice as fast compared to Test 1 when SR was disabled. Further demonstration of the benefit of SR is illustrated in Stage 3, where again only 6 packets are transmitted over the tunnel, reducing the time of this Phase to just 1.906 seconds compared to 4.549 seconds of Test 1. Stages 4 and 6 present similar results with Test 1, as PMKSA caching takes effect. During Stage 5, SR affirms its advantages again, with a significant reduction of Phase 2 timing down to 0.126 seconds compared to 0.690 seconds of Stage 5 of Test 1, since PMKSA caching is not applicable as AP1 is in WPA configuration. Overall, it has to be noted that the SR feature, where applicable (Stages 2, 3 and 5) has demonstrated significant advantages and reduction in Phase 2 timings compared to Test 1 and further improved the overall timings of the Stages where PMKSA caching was not applicable.

To further evaluate our UA we repeated Tests 1 and 2 using a different authentication method, namely EAP-TTLS, that uses a username/password pair to authenticate the MR instead of a certificate. As Table 5 shows Phase 2 timings in this Test, are in similar levels with Test 1 although the number of "global" packets now required for EAP-TTLS are now 16, compared to 14 in EAP-TLS (Table 2). During Stage 3 we observed an expected increase of the timing of

Table 5. EAP-TTLS Results without Session Resumption

Stages / Phases	ST.1	ST.2	ST.3	ST.4	ST.5	ST.6
Phase 1 (sec.)	1.299	1.207	0.932	0.961	1.491	1.090
Phase 2 (sec.)	0.335	0.242	5.124	0.010	0.335	0.015
Phase 3 (sec.)	1.876	2.269	2.574	2.320	1.763	2.340
Total (sec.) :	3.510	3.718	8.630	3.291	3.589	3.445

Table 6. EAP-TTLS Results with Session Resumption

Stages / Phases	ST.1	ST.2	ST.3	ST.4	ST.5	ST.6
Phase 1 (sec.)	1.603	2.086	0.977	0.981	0.845	1.090
Phase 2 (sec.)	0.291	0.055	1.888	0.022	0.080	0.032
Phase 3 (sec.)	1.554	1.544	2.660	2.222	2.620	2.220
Total (sec.) :	3.448	3.685	5.525	3.225	3.545	3.342

Phase 2, as now more packets have to travel over the IPv6 Tunnel and thus the additional delay is reflected in the results. However, once again, Phase 2 timings of Stage 4 and Stage 6 are remarkably low, thanks to PMKSA caching which prohibits the need for any "global" packets exchange. Generally, the overall timings of this Test remain at the same level compared to Test 1, with slight differences due to the difference of the authentication method being used.

Finally, Test 4 repeats Test 3 with SR being enabled at the AAAHS. All the overall timings of this Test (Table 6) are decreased compared to those of Test 3 (Table 5), as both Session Resumption and PMKSA caching are triggered where applicable. In particular, Phase 2 timings for Stages 2, 3 and 5 are remarkably low (0.055 seconds, 1.888 seconds and 0.080 seconds respectively), because SR reduces the number of "global" packets that needed to be exchanged from 16 down to 6 (Table 2). Phase 2 of Stages 4 and 6 of this Test, required only 4 local packets to be exchanged, compared to 21 in total for a full EAP-TTLS authentication with a WPA2 AP, because again PMKSA caching was enabled and ensured that only the 4-way handshake was performed.

6 Conclusion

In this paper we presented a UA that combines the strengths of NEMO BS and AAA services in a secure way and satisfies the requirements of both MNs and ANs. Our UA enables roaming MNs to experience constant Internet connectivity with trouble-free but secure network access, and secure transmission of their data despite their frequent roaming. Using our UA, ANs are able to provide efficient and secure AAA services in a profitable fashion. Our qualitative evaluation discussed the merits of our approach and how it satisfies the requirements of all the parties involved. The results from our thorough quantitative evaluation with different authentication methods and configuration, demonstrated the performance and applicability of our approach for a real world deployment.

References

1. Aboba, B., Beadles, M., Arkko, J., Eronen, P.: The Network Access Identifier. IETF RFC 4282 (December 2005)
2. Arkko, J., Devarapalli, V., Dupont, F.: Using IPsec to Protect Mobile IPv6 Signaling Between Mobile Nodes and Home Agents. IETF RFC 3776 (June 2004)
3. Hurricane Eectric IPv6 Tunnel Broker, http://www.tunnelbroker.net/
4. Devarapalli, V., Wakikawa, R., Petrescu, A., Thubert, P.: NEMO Basic Support Protocol. IETF RFC 3963 (January 2005)
5. Funk, P., Blake-Wilson, S.: Extensible Authentication Protocol Tunneled Transport Layer Security (EAP-TTLSv0). IETF RFC 5281 (August 2008)
6. Georgopoulos, P., McCarthy, B., Edwards, C.: Towards a Secure and Seamless Host Mobility for the real world. In: 8th International Conference on Wireless On-demand Network Systems and Services (WONS 2011), Italy (January 2011)
7. Umip Mobile IPv6 Stack, http://umip.org/
8. Johnson, D., Perkins, C., Arkko, J.: Mobility Support for IPv6. IETF RFC 3775 (June 2004)
9. Kent, S., Seo, K.: Security Architecture for the Internet Protocol. IETF RFC 4301 (December 2005)
10. de Laat, C., Gross, G., Gommans, L., Vollbrecht, J., Spence, D.: Generic AAA architecture. IETF RFC 2903 (August 2000)
11. Perkins, C.: IP Mobility Support. IETF RFC 2002 (October 1996)
12. Rigney, C., Willens, S., Rubens, A., Simpson, W.: Remote Authentication Dia. In: User Service (RADIUS). IETF RFC 2865 (June 2000)
13. Simon, D., Aboba, B., Hurst, R.: The EAP-TLS Authentication Protocol. IETF RFC 5216 (March 2008)

Markov Modulated Bi-variate Gaussian Processes for Mobility Modeling and Location Prediction

Paulo Salvador and António Nogueira

Department of Electronics, Telecommunications and Informatics
Institute of Telecommunications
University of Aveiro
{salvador,nogueira}@ua.pt

Abstract. A general-purpose and useful mobility model must be able to describe complex movement dynamics, correlate movement dynamics with the nodes geographic position and be sufficiently generic to map the characteristics of the movement dynamics to general geographic regions. Moreover, it should also be possible to infer the mobility model parameters from empirical data and predict the location of any node based on known positions. Having the ability to model movement and predict future positioning of mobile nodes in complex environments can be very important to several operational and management tasks. Existing mobility modeling approaches are based on simple and limited models, specifically designed for particular application scenarios or requiring the complete knowledge of the mobility environment. These features make them unusable in complex scenarios with no (or partial) knowledge of the environment. This paper presents a discrete time Markov Modulated Bi-variate Gaussian Process that is able to characterize the position and mobility of any mobile node, assuming that the position within a generic sub-region can be described by a bi-variate Gaussian distribution and the transition between sub-regions can be described by an underlying (homogeneous) Markov chain. With this approach, it is possible to describe the mobile node movement within and between a set of geographic regions determined by the model itself and, due to the Markovian nature of the model, it is also possible to capture complex dynamics and calculate the future probabilistic position of a mobile node. The proposed approach can be applied to scenarios where the possible pathways are unknown or too complex to consider in a real model that must make a prediction in a very short time. The results obtained by applying the proposed model to real and publicly available data demonstrate the accuracy and utility of this approach: the model was able to efficiently describe the movement patterns of mobile nodes and predict their future position. Besides, the model has also revealed higher performances when compared to other modeling approaches.

Keywords: mobility modeling, location prediction, Markov model, modulated bi-variate Gaussian process.

J. Domingo-Pascual et al. (Eds.): NETWORKING 2011, Part I, LNCS 6640, pp. 227–240, 2011.

1   Introduction

Being able to describe the movement and predict future positioning of mobile nodes in complex environments can be very important to different operational and management tasks: evaluation of different system design alternatives and implementation costs, simulation and study of new routing protocols for Mobile and Delay Tolerant Networks, knowledge of the nodes distribution for performance evaluation and message delivery optimization, prediction of the nodes preferred locations for connectivity issues, etc. Other research topics, like mobile computing, can also greatly benefit from the ability to track and predict the location of mobile devices: an accurate location predictor can significantly improve the performance or reliability of many applications in pervasive computing.

There are two different strands of the mobile modeling problem: the ability to describe or track the movement of mobile nodes, which is mainly used for simulation purposes, and the ability to predict the mobile nodes position, which can be useful to several management tasks. An efficient mobility model must be able to describe complex movement dynamics, correlate the movement dynamics with the nodes position and be sufficiently generic to map movement's dynamics characteristics to general geographic regions where the road/path topology is unknown or imprecise. The parameters of the mobility model should be easily inferred from real mobility data and the model should provide an almost real-time prediction of the nodes position.

This paper proposes a discrete time Markov Modulated Bi-variate Gaussian Process to characterize the position and mobility of any mobile node, assuming that the position can be described by a bi-variate Gaussian distribution and the transition between regions can be described by an underlying (homogeneous) Markov chain. This model is able to describe the mobile node movement within and between a set of geographic regions determined by the model itself. Due to the Markov nature of the model, it is possible to determine the future probabilistic position of a mobile node.

The results obtained by applying this model to real mobility data prove that this approach can accurately predict the future position of mobile nodes presenting, at the same time, higher performance than other prediction models that were selected for comparison.

The paper is organized as follows: section 2 presents some of the most relevant works related to the proposed model; section 3 presents the mathematical background on the multidimensional discrete time Markov Modulated Bi-variate Gaussian Process (dMMBVGP) and the description of the proposed model; section 4 presents the algorithmic details of the inference procedure; section 5 presents the results obtained in a proof of concept and a discussion of those results; finally, section 6 presents the main conclusions.

2   Related Work

Multiple proposals for modeling movement of mobile nodes have already been published [5]. The proposed models can be mapped into four categories: (i) purely random; (ii) with temporal dependency, when the movement of a mobile node is likely to be affected by its movement history; (iii) with spatial dependency between subjects, when mobile nodes tend to travel in a correlated manner, and (iv) with geographical dependency/restriction, when the movement of nodes is bounded by streets, freeways or obstacles.

In purely random mobility models, mobile nodes move randomly and freely without restrictions, that is, the destination, speed and direction are all chosen randomly and independently of other nodes. The Random Waypoint model [4] and two of its variants, named Random Walk model [5] and Random Direction model [18], are the most important examples of this type of models. Obviously, these models fail to represent some mobility characteristics that are likely to exist in mobile ad hoc networks: the temporal and spatial dependency of velocity and the geographic restrictions of movement. Since, in general, the velocities of a single node at different time slots are "correlated", other models were proposed to capture this temporal dependency behavior: the Gauss-Markov mobility model [11], where the velocity of a mobile node is assumed to be correlated over time and modeled as a Gauss-Markov stochastic process and the Smooth Random Mobility Model [2], where the speed and direction of the node movement is incrementally and smoothly changed in order to generate more realistic movement behaviors. In [7], the authors proposed a time-variant community mobility model that realistically captures spatial and temporal correlations. Synthetic traces that match the characteristics of different mobility traces, including wireless LAN traces, vehicular mobility traces and human encounter traces, were generated, demonstrating the flexibility of the proposed model. Several Markovian approaches have also been proposed to address the mobility problem: reference [8] proposed the Markovian Waypoint model, where mobile nodes move along a straight line segment from one waypoint to another at a certain speed. In this approach, waypoints are obtained from a discrete time Markov process having a subset of $\mathbb{R}^2$ as the state space. Other mobility models, such as [2], are able to produce more smoothly turning trajectories.

In most real life applications, a node's movement is subject to the environment. Therefore, nodes may move in a pseudo-random way on predefined pathways in the environment field. Mobility models with geographical restrictions have been proposed to describe this kind of environments: in the Pathway Mobility Model [22] the node movement is restricted to pathways in a map, while in the Obstacle Mobility Model [9] the effects of obstacles on node mobility and radio propagation are taken into account. All these classes of models are unable to constrain/define the movement dynamics as a function of the mobile node positioning.

There has also been a significant interest in mobility models specially designed for partitioned networks, where end-to-end connectivity rarely or never exists. In [3], authors proposed the Area Graph Based Mobility Model that considers the inhomogeneity of spatial node distribution by using a directed graph with areas of different densities; in [6] similar approaches, called Weighted Waypoint Mobility and Clustered Mobility Models, respectively, were proposed to capture node preferences in choosing possible destinations; the Community Based Mobility Model, proposed in [21], was designed to recreate the characteristics of mobility in sparse networks. All these models exhibit a skewed distribution of location visits, which could result in forming islands of connectivity at popular locations and being unable to stabilize. The Heterogeneous Random Walk model, proposed in [17], is a very parsimonious model where clusters emerge because of low node speed in popular areas. Musolesi et al. [13] presented a mobility model that incorporates sociological relationships between mobile nodes, resulting in a quite complex model. Recently, some studies have been devoted to model the realistic social dimension of human mobility. A human mobility model based on heterogeneous centrality and overlapping community structure in social networks was proposed in [23], being able to match characteristics observed from real human motion traces, especially heterogeneous human mobility popularity. This type of models requires pre-established division of the network geographic environment and are not able to adapt themselves to unknown generic scenarios.

Several approaches have been proposed to predict the future location of mobile entities. In [12], authors proposed a method recording a sequence of cell identifiers to predict future location of users in wireless ATM networks, while [1] use logical information for location prediction. Other approaches use road topology information to predict user location [19]. In [10], a semi-Markov model is used to describe the state transitions between stationary behavior and movement in a WLAN cell, where all model parameters can be obtained from movement monitoring. Most of the existing methods are restricted to a specific purpose, leading to straightforward implementation but limited effectiveness in generic network management tasks. Designing a robust flexible mobility and location model that can be used in generic environments is the motivation for this paper.

3   Model and Positioning Prediction

3.1   Markov Modulated Bi-variate Gaussian Processes Model

The proposed discrete time Markov Modulated Bi-variate Gaussian Process (dMMBVGP) model characterizes position and mobility of a subject based on the following assumptions: (i) within a restricted generic geographic region the position may be described by a bi-variate Gaussian distribution, (ii) the geographical space can be divided in N distinct regions (non-determined *a priori*) and (iii) the transition between regions can be described by an underlying (homogeneous) Markov chain, where each state maps the movement characteristics of a specific region. The dMMBVGP can then be described as a random process (Z) with a bi-variate Gaussian distribution, which determines a position in a two-dimensional environment, whose parameters are a function of the state (J) of the modulator Markov chain (Z, J) with N states. More precisely, the (homogeneous) Markov chain $(Z, J) = \{(Z_k, J_k), k = 0, 1, \ldots\}$ with state space $\mathbb{R}^2 \times U$, with $U = \{1, 2, \ldots, N\}$, is a dMMBVGP if and only if for $k = 0, 1, \ldots,$

$$P(Z_{k+1} = \mathbf{z}, \ J_{k+1} = n | J_k = m) = p_{mn} \Gamma_n(\mathbf{z}) \tag{1}$$

where p_{mn} represents the probability of a transition from state m to state n of the underlying Markov chain in time interval $[k, k+1]$, and $\Gamma_n(\mathbf{z}) = \frac{1}{2\pi|\Sigma_n|^{1/2}} e^{-\frac{1}{2}(\mathbf{z}-\Lambda_n)^T \Sigma_n^{-1}(\mathbf{z}-\Lambda_n)}$ is the bi-variate Gaussian distribution in region n centered in Λ_n and having covariance matrix Σ_n.

The dMMBVGP may be also regarded as a Markov random walk [14] where the two-dimensional positions increments (Δ) in each instant k have a distribution given by the difference of two bi-variate Gaussian distributions whose parameters are a function of the state of the modulator Markov chain

$$P(Z_{k+1} = \mathbf{z} + \Delta, J_{k+1} = n | Z_k = \mathbf{z}, J_k = m) =$$

$$= p_{mn} \frac{1}{2\pi |\Sigma_n + \Sigma_m|^{1/2}} e^{-\frac{1}{2}(\Delta - \Lambda_n + \Lambda_m)^T (\Sigma_n + \Sigma_m)^{-1}(\Delta - \Lambda_n + \Lambda_m)}$$

Whenever (1) holds, we say that (Z, J) is a dMMBVGP with a set of modulating states with size N and parameter matrices $\mathbf{P}$, $\mathbf{S}$, and $\mathbf{L}$. The matrix $\mathbf{P}$ is the transition probability matrix of the modulating Markov chain J,

$$\mathbf{P} = \begin{bmatrix} p_{11} & p_{12} & \cdots & p_{1N} \\ p_{21} & p_{22} & \cdots & p_{2N} \\ \cdots & \cdots & \cdots & \cdots \\ p_{N1} & p_{N2} & \cdots & p_{NN} \end{bmatrix} \tag{2}$$

matrix $\mathbf{L}$ defines the centers of each region,

$$\mathbf{L} = \begin{bmatrix} \Lambda_1 & \Lambda_2 & \cdots & \Lambda_N \end{bmatrix}, \Lambda_n = \begin{bmatrix} \lambda_n^{(x)} \\ \lambda_n^{(y)} \end{bmatrix} \tag{3}$$

and matrix $\mathbf{S}$ contains the covariance (sub-)matrices of each region:

$$\mathbf{S} = \begin{bmatrix} \Sigma_1 & \Sigma_2 & \cdots & \Sigma_N \end{bmatrix}, \Sigma_n = \begin{bmatrix} (\sigma_n^{(x)})^2 & \rho_n \sigma_n^{(x)} \sigma_n^{(y)} \\ \rho_n \sigma_n^{(x)} \sigma_n^{(y)} & (\sigma_n^{(y)})^2 \end{bmatrix}. \tag{4}$$

Moreover, we denote by $\mathbf{\Pi} = [\pi_1, \pi_2, \ldots, \pi_N]$ the stationary distribution of the underlying Markov chain.

Specifying $\mathbf{z} = \begin{bmatrix} x & y \end{bmatrix}^T$, where x and y represent longitude and latitude, respectively, and defining the general (not weighted) bi-variate Gaussian distribution within a region n by

$$\Gamma_n(x, y) = \frac{1}{2\pi \sigma_n^{(x)} \sigma_n^{(y)} \sqrt{1 - \rho_n^2}} e^{Q_n} \tag{5}$$

with $Q_n = -\frac{1}{2(1-\rho_n^2)} \left(\frac{(x-\lambda_n^{(x)})^2}{(\sigma_n^{(x)})^2} + \frac{(y-\lambda_n^{(y)})^2}{(\sigma_n^{(y)})^2} - \frac{2\rho_n (x-\lambda_n^{(x)})(y-\lambda_n^{(x)})}{\sigma_n^{(x)} \sigma_n^{(y)}} \right)$, the stationary probability density map of position (x, y) can be determined by a weighted sum of N bi-variate Gaussian distributions:

$$f^N(x, y) = \sum_{n=1}^{N} \pi_n \Gamma_n(x, y) \tag{6}$$

obtaining a position probability density map.

3.2   Positioning Prediction

Defining $\mathbf{q}_k = \{q_{k,1}, q_{k,2}, \ldots, q_{k,N}\}, k = 0, 1, \ldots, K$, where $q_{k,n}$ is the probability that, at time interval k (within a total of K observations), the mobile node is in region n (i.e. the probability of the underlying Markov chain of the mobility model is in state n at time instant k) and based on equation (6) we can define the predicted position probability map at time instant k as:

$$\bar{f}_k(x, y) = \sum_{n=1}^{N} q_{k,n} \Gamma_n(x, y) \tag{7}$$

with

$$\mathbf{q}_k = \mathbf{q}_0 \mathbf{P}^k \tag{8}$$

where $\mathbf{q}_0$ represents the probability vector of the last known position (x_0, y_0) belonging to the different regions, and can be inferred as

$$q_{0,n} = \frac{\pi_n \Gamma_n(x_0, y_0)}{f(x_0, y_0)} \tag{9}$$

4   dMMBVGP Inference Procedure

Hereafter, we will refer to the empirical position probability density map as $f^e(x, y)$ and $\epsilon_{n-1}(x, y) = f^e(x, y) - f^n(x, y)$ will represent the difference function between the empirical and fitted position probability maps with n bi-variate Gaussian distributions (f^n).

The inference procedure can be divided in two steps: (i) inference of matrices $\mathbf{L}$ and $\mathbf{S}$ and vector $\mathbf{\Pi}$ by adjusting $f^e(x, y)$ by a weighted sum of bi-variate Gaussian distributions and (ii) inference of matrix $\mathbf{P}$ by analyzing the transitions between the underlying Markov chain inferred in step (i).

The first step of the inference procedure relies on the minimization process defined by

$$\min |f^n(x, y) - f^e(x, y)|$$

where $f^n(x, y)$, as defined in (6), is a weighted sum of n bi-variate Gaussian distributions and n is determined as part of the inference process and does not need to be defined a priori.

The iterative fitting procedure starts with $n = 1$, finds the global maximum of $\epsilon_0(x, y) = f^e(x, y)$ and defines that point as the center of the first mobility region. Each mobility region (n) will have associated a bi-variate Gaussian distribution with parameters $\lambda_n^{(x)}, \lambda_n^{(y)}, \sigma_n^{(x)}, \sigma_n^{(y)}, \rho_n$ and weight π_n. The n-th mobility region center (i.e. the bi-variate Gaussian distribution mean) is determined by:

$$(\lambda_n^{(x)}, \lambda_n^{(y)}) = \epsilon_{n-1}^{-1} \left(\max(\epsilon_{n-1}(x, y)) \right) \quad, n \geq 1. \tag{10}$$

The inference of the remaining parameters of the bi-variate Gaussian distributions are estimated by analyzing the bi-dimensional local decay behavior of

$f^e(x, y)$ in the vicinity of the local maximum points along an orthogonal plane. Therefore, based on the bi-variate Gaussian distribution function (5), it is possible to determine the ratio between the partial derivative of $\pi_n \Gamma_n(x, y)$ with respect to the variable x in the plane defined by $y = \lambda_n^{(y)}$ and $\pi_n \Gamma_n(x, \lambda_n^{(y)})$ as

$$\frac{\pi_n \frac{\partial \Gamma_n(x, \lambda_n^{(y)})}{\partial x}}{\pi_n \Gamma_n(x, \lambda_n^{(y)})} = -\frac{(x - \lambda_n^{(x)})}{(1 - \rho_n^2)(\sigma_n^{(x)})^2} \Leftrightarrow (\sigma_n^{(x)})^2 = \frac{\gamma_n^{(x)}}{1 - \rho_n^2} \quad (11)$$

and the ratio between the partial derivative of $\pi_n \Gamma_n(x, y)$ with respect to the variable y in the plane defined by $x = \lambda_n^{(x)}$ and $\pi_n \Gamma_n(\lambda_n^{(x)}, y)$ as

$$\frac{\pi_n \frac{\partial \Gamma_n(\lambda_n^{(x)}, y)}{\partial y}}{\pi_n \Gamma_n(\lambda_n^{(x)}, y)} = -\frac{(y - \lambda_n^{(y)})}{(1 - \rho_n^2)(\sigma_n^{(y)})^2} \Leftrightarrow (\sigma_n^{(y)})^2 = \frac{\gamma_n^{(y)}}{1 - \rho_n^2} \quad (12)$$

where

$$\gamma_n^{(x)} = -\frac{(x - \lambda_n^{(x)})\Gamma_n(x, \lambda_n^{(y)})}{\frac{\partial \Gamma_n(x, \lambda_n^{(y)})}{\partial x}} \quad \text{and} \quad \gamma_n^{(y)} = -\frac{(y - \lambda_n^{(y)})\Gamma_n(\lambda_n^{(x)}, y)}{\frac{\partial \Gamma_n(\lambda_n^{(x)}, y)}{\partial y}}. \quad (13)$$

Based on definitions (5) and (13) it is possible to redefine the n-th weighted bi-variate Gaussian distribution function

$$\Gamma_n(x, y) = \Gamma_n(\lambda_n^{(x)}, \lambda_n^{(y)})e^{Q'_n} \quad (14)$$

with $\Gamma_n(\lambda_n^{(x)}, \lambda_n^{(y)}) = \pi_n \frac{\sqrt{1 - \rho_n^2}}{2\pi\sqrt{\gamma_n^{(x)}\gamma_n^{(y)}}}$ and

$$Q'_n = -\frac{1}{2}\left(\frac{(x - \lambda_n^{(x)})^2}{\gamma_n^{(x)}} + \frac{(y - \lambda_n^{(y)})^2}{\gamma_n^{(y)}} - \frac{2\rho_n(x - \lambda_n^{(x)})(y - \lambda_n^{(x)})}{\sqrt{\gamma_n^{(x)}\gamma_n^{(y)}}}\right).$$

¿From equation (14), it is possible to define the n-th bi-variate Gaussian distribution correlation:

$$\rho_n = \frac{\sqrt{\gamma_n^{(x)}\gamma_n^{(y)}}}{(x - \lambda_n^{(x)})(y - \lambda_n^{(y)})}r_n(x, y) \quad (15)$$

where
$r_n(x, y) = \ln\left(\frac{\Gamma_n(x, y)}{\Gamma_n(\lambda_n^{(x)}, \lambda_n^{(y)})}\right) + \frac{(x - \lambda_n^{(x)})^2}{2\gamma_n^{(x)}} + \frac{(y - \lambda_n^{(y)})^2}{2\gamma_n^{(y)}}$, and the n-th bi-variate Gaussian distribution weight (i.e. the stationary probability of the underlying Markov chain n-th state) is given by:

$$\pi_n = \frac{2\pi\Gamma_n(\lambda_n^{(x)}, \lambda_n^{(y)})\sqrt{\gamma_n^{(x)}\gamma_n^{(y)}}}{\sqrt{1 - \rho_n^2}} \quad (16)$$

It is assumed that within a square region with width $2\Delta_n^{(x)}$, height $2\Delta_n^{(y)}$ and center at the n-th local maximum of $f^e(x, y)$ (equation (10)), the position is

modeled exclusively by the n-th weighted bi-variate Gaussian distribution, i.e.,
$\Gamma_n(x, y) \approx \epsilon_{n-1}(x, y), x \in [\lambda_n^{(x)} - \Delta^{(x)}, \lambda_n^{(x)} + \Delta^{(x)}] \wedge y \in [\lambda_n^{(y)} - \Delta^{(y)}, \lambda_n^{(y)} + \Delta^{(y)}]$.
Therefore, it is possible to find estimators for parameters $\gamma_n^{(x)}$ and $\gamma_n^{(y)}$ based
on the average value of equations (13) calculated in the vicinities of the $f^e(x, y)$
local maximums:

$$\bar{\gamma}_n^{(x)} = -\frac{1}{2\Delta^{(x)}} \int_{\lambda_n^{(x)} - \Delta^{(x)}}^{\lambda_n^{(x)} + \Delta^{(x)}} \frac{(x - \lambda_n^{(x)})\epsilon_{n-1}(x, \lambda_n^{(y)})}{\frac{\partial \epsilon_{n-1}(x, \lambda_n^{(y)})}{\partial x}} \, dx \tag{17}$$

and

$$\bar{\gamma}_n^{(y)} = -\frac{1}{2\Delta^{(y)}} \int_{\lambda_n^{(y)} - \Delta^{(y)}}^{\lambda_n^{(y)} + \Delta^{(y)}} \frac{(y - \lambda_n^{(y)})\epsilon_{n-1}(\lambda_n^{(x)}, y)}{\frac{\partial \epsilon_{n-1}(\lambda_n^{(x)}, y)}{\partial y}} \, dy. \tag{18}$$

The estimator for the n-th bi-variate Gaussian distribution correlation is defined
as the average value of (15) calculated over a square region in the vicinity of the
$f^e(x, y)$ local maximums:

$$\bar{\rho}_n = \frac{\sqrt{\bar{\gamma}_n^{(x)} \bar{\gamma}_n^{(y)}}}{4\Delta^{(x)}\Delta^{(y)}} \int_{\lambda_n^{(x)} - \Delta^{(x)}}^{\lambda_n^{(x)} + \Delta^{(x)}} \int_{\lambda_n^{(y)} - \Delta^{(y)}}^{\lambda_n^{(y)} + \Delta^{(y)}} \frac{\bar{r}_n(x, y)}{(x - \lambda_n^{(x)})(y - \lambda_n^{(y)})} \, dy dx \tag{19}$$

where $\bar{r}_n(x, y) = ln \left(\frac{\epsilon_{n-1}(x, y)}{\epsilon_{n-1}(\lambda_n^{(x)}, \lambda_n^{(y)})} \right) + \frac{(x - \lambda_n^{(x)})^2}{2\gamma_n^{(x)}} + \frac{(y - \lambda_n^{(y)})^2}{2\gamma_n^{(y)}}$.

The weight of the n-th bi-variate Gaussian distribution can be calculated
using the estimators obtained with equations (17), (18) and (19):

$$\bar{\pi}_n = \frac{2\pi\epsilon_{n-1}(\lambda_n^{(x)}, \lambda_n^{(y)})\sqrt{\bar{\gamma}_n^{(x)}\bar{\gamma}_n^{(y)}}}{\sqrt{1 - \bar{\rho}_n^2}} \tag{20}$$

The n-th bi-variate Gaussian distribution variances can be obtained based on
equations (11), (12), (17), (18) and (19):

$$(\bar{\sigma}_n^{(x)})^2 = \frac{\bar{\gamma}_n^{(x)}}{(1 - \bar{\rho}_n^2)} \quad \text{and} \quad (\bar{\sigma}_n^{(y)})^2 = \frac{\bar{\gamma}_n^{(y)}}{(1 - \bar{\rho}_n^2)} \tag{21}$$

The final iteration step is to re-calculate the difference function and update
$n = n + 1$:

$$\epsilon_n(x, y) = \epsilon_{n-1}(x, y) - \bar{\pi}_n \Gamma_n(x, y). \tag{22}$$

The iterative process finishes after a predefined number of iterations or when
$\min |f^n(x, y) - f^e(x, y)|$ reaches a predefined fitting error; if that goal is not
achieved, the value n is increased by one and the iterative process continues
recomputing equations (10) to (22). At the end of the iteration process it is
possible to determine the model matrices **L** and **S**.

The final step of the inference procedure is to infer matrix $\mathbf{P}$, i.e. the transition probabilities between the regions/states defined in the first step. The task is achieved by probabilistically mapping each known position $(x_k, y_k), k = 0, 1, \ldots, K$ to one region and then average the probabilistic transitions between states.

The position in time interval k (x_k, y_k) will be probabilistically mapped to one of the N regions according to a probability vector $\mathbf{q}_k = \{q_{k,1}, q_{k,2}, \ldots, q_{k,N}\}, k = 0, 1, \ldots, K$ defined by:

$$q_{k,n} = \frac{\pi_n \Gamma_n(x_k, y_k)}{f(x_k, y_k)} \tag{23}$$

Since the underlying Markov chain at instant k is in state m with probability $P(J_k = m)$, it is possible to define a partial estimator for p_{mn} at all k time instants as (taking equations 1 and 2 into account):

$$\begin{aligned} \bar{p}_{mn,k} &= P(J_k = m)p_{mn} = P(J_k = m)P(J_{k+1} = n | J_k = m) = \\ &= P(J_k = m)P(J_{k+1} = n) = q_{k,m}q_{k+1,n} \end{aligned} \tag{24}$$

Therefore, the overall estimator for p_{mn} can then be inferred by averaging over k the partial estimators defined in (24),

$$\bar{p}_{mn} = \frac{1}{K-1} \sum_{k=0}^{K-1} \bar{p}_{mn,k}, \quad m, n = 1, \ldots, N \tag{25}$$

Rewriting (25) in matrix form, it is possible to obtain matrix $\mathbf{P}$:

$$\mathbf{P} = \frac{1}{K-1} \sum_{k=0}^{K-1} \mathbf{q}_k^T \times \mathbf{q}_{k+1} \tag{26}$$

5   Results

As proof of concept, we applied the proposed model to mobility data of taxicabs in San Francisco, USA. Data was provided by Exploratorium, the San Francisco museum of science, art and human perception, through the cabspotting project (http://cabspotting.org). A pre-processed data set is freely available [16] and contains GPS coordinates, sampled at approximately 1 per minute, of 500 taxis collected over 30 days of May 2008 in the San Francisco Bay Area. Fig. 1 depicts a sample trajectory of one taxi in the San Francisco Bay area. In this paper, in order to infer and test the proposed model we have only considered the movements within a rectangular region in the San Francisco downtown, which is depicted in Fig. 1 within the white rectangle, with longitude limits between -122.447° and -122.388° and latitude limits between 37.77° and 37.81°. The data subset was randomly divided in two sub-sets (of 250 cars each), one to infer the model parameters and the other to test the model.

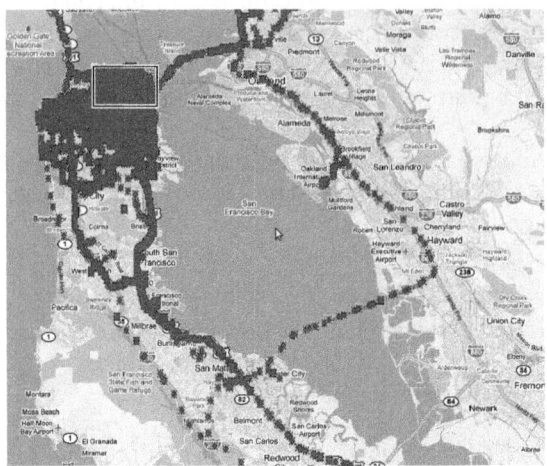

Fig. 1. San Francisco map with one sample taxi movement (base map source: GoogleMaps)

Applying the proposed mobility model to the test data sub-set, a model with 64 states (sub-regions) was obtained. The model inference time interval was less than 30 minutes with an octave/Matlab implementation in a generic PC. Fig. 2 shows the position probability distribution map for a taxi in San Francisco downtown and Fig. 3 shows the stationary position probability distribution map of the dMMBVGP inferred model. It is possible to observe that the dMM-BVGP model was able to incorporate and model the complexity of the cars movement/location in San Francisco Downtown. With a more detailed analysis, it is possible to observe that the inference algorithm centered the sub-regions in the more visited/crowded San Francisco's landmarks.

In order to evaluate the location prediction performance of the model, we used the test sub-set to predict the node future position based on a known position, using different time-lags (1 to 20 timeslots, approximately 1 to 20 minutes). The performance metric was the (average) distance (in meters) between the predicted and real future node position. After inferring the model parameters, position prediction using the dMMBVGP, equation (7), is performed in hundreds of microseconds (average value $\sim 921\mu s$), using also an octave/Matlab implementation in a generic PC. Note that the prediction time interval depends on the number of future time slots for which the prediction is made.

As a complementary evaluation of the proposed model performance, we also evaluate other four mobility and location prediction models in the same scenario:

- Complete random model, where node location follows a uniform distribution over the complete geographic region under study.
- Grid random model, where node location is probabilistically determined in two phases; firstly, the location sub-region is determined according to the empirical location distribution map inferred over a uniform grid and, secondly,

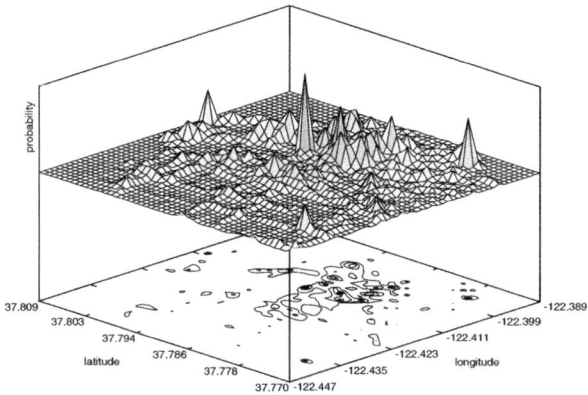

Fig. 2. Real position probability map, San Francisco downtown only

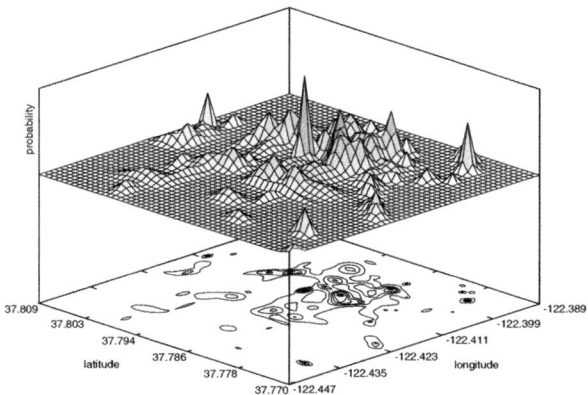

Fig. 3. Fitted dMMBVGP position probability map, San Francisco downtown only

the exact position within the chosen sub-region is determined according to a uniform distribution.

– Random walk model [5] with reflection at the region borders [15], where the node direction of movement is uniformly distributed in $[0, 2\pi]$ and the speed has a Gaussian distribution with mean and standard deviation inferred from empirical data.
– Markov grid model, which is a sub-region based Markov family model [20] where location is determined by an underlying Markov chain that determines the transition between sub-regions. We choose to define the sub-regions as an uniform grid. Within a sub-region, node location has an uniform distribution.

The complete random model has no geographic dependency and has no information of the node movement history. The model location prediction does not depend on the current node position. This model is used as the worst case reference value. The grid random model prediction does not depend also on the current node position, however it incorporates some geographic dependency

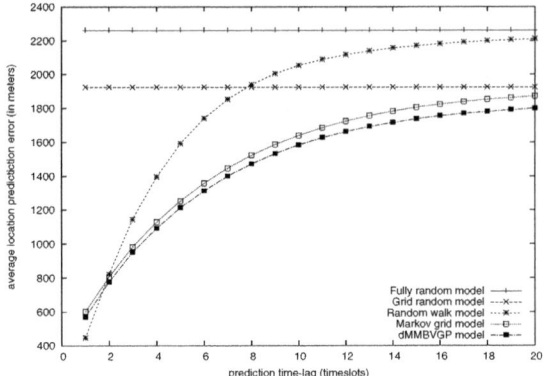

Fig. 4. Models relative location prediction performance

inferred from empirical historic data. A 8x8 grid (64 equal sub-regions) was used to infer the empirical location distribution map. The random walk model has no geographic dependency but the prediction process utilizes the current node position and extrapolates the future position based on a known constant speed and direction. The model speed was modeled by a Gaussian distribution with mean 302.1m/minute ($\approx$ 18.1Km/h or $\approx$ 11.2mph) and standard deviation of 346.7m/minute. For the inference of the Markov grid model, a 8x8 grid (64 equal sub-regions) was used in order to obtain a model with the same level of complexity as the dMMBVGP. Note that the Markov grid model differs from the dMMBGP model in the way regions are defined and position is modeled within each region; the time dependability (based on an underlying Markov chain) is modeled and inferred in a similar way in both models.

Figure 4 depicts the models prediction performance. It is possible to observe that the dMMBGP outperforms all the other models and it is possible to obtain relatively small prediction errors, in a region so large and complex as the one under study, even in long range predictions (20 timeslots $\sim$ 20 minutes). As expected, the worst performing models are the ones that do not receive any real-time input at the time of the prediction, and therefore the prediction error is independent of the time-lag. The random walk model is able to surpass the others in very short time-lag predictions ($\leq$ 2 minutes); however the model is not able to capture the complex dynamics of the taxis movements and the prediction error greatly increases when the prediction range is larger than 1 slot and rapidly approximates the error of a completely random location choice. The Markov grid model, due to its nature, is able to capture the complexity and dynamism of the movements; however, the definition of generic regions restricts the model performance. Therefore, the dMMBVGP model is able to model the complexity of the movements due to its Markovian nature, while its adaptable definition of sub-regions allows it to maintain the model complexity at low levels and achieve very good location prediction results.

6   Conclusions

Being able to model movement and predict future positioning of mobile nodes in complex environments can be very important to several operational and management tasks, like evaluation of different system design alternatives, simulation and study of new routing protocols, message delivery and nodes connectivity optimization or mobile computing. Since existing mobility modeling approaches are based on simple and limited models that were specifically designed for particular application scenarios or require the complete knowledge of the mobility environment, this paper presented a discrete time Markov Modulated Bi-variate Gaussian Process that is able to characterize the position and mobility of any mobile node by assuming that the position within a generic sub-region can be described by a bi-variate Gaussian distribution and the transition between sub-regions can be described by an underlying (homogeneous) Markov chain. This approach can be applied to scenarios where the possible pathways are unknown or too complex to consider in a real model that must make a prediction in a very short time. The results obtained by applying the proposed model to real and publicly available mobility data of San Francisco taxi-cabs demonstrate the accuracy and utility of this approach: the model was able to efficiently describe the movement patterns of the mobile nodes and predict their future position. Besides, the model has also revealed higher performances when compared to other modeling approaches.

References

1. Akyildiz, I., Wang, W.: The predictive user mobility profile framework for wireless multimedia networks. IEEE/ACM Transactions 12(6), 1021–1035 (2004)
2. Bettstetter, C.: Mobility modeling in wireless networks: Categorization, smooth movement and border effects. ACM Mobile Computing and Communications Review 5(3), 55–67 (2001)
3. Bittner, S., Raffel, W.U., Scholz, M.: The area graph-based mobility model and its impact on data dissemination. In: Proceedings of the Third IEEE International Conference on Pervasive Computing and Communications Workshops (2005)
4. Broch, J., Maltz, D.A., Johnson, D.B., Hu, Y.C., Jetcheva, J.: A performance comparison of multi-hop wireless ad hoc network routing protocols. In: Proceedings of the Fourth Annual ACM/IEEE International Conference on Mobile Computing and Networking, Mobicom 1998 (October 1998)
5. Camp, T., Boleng, J., Davies, V.: A survey of mobility models for ad hoc network research. Wireless Communications and Mobile Computing - Special Issue in Mobile Ad Hoc Networking: Research, Trends and Applications 2(5), 483–502 (2002)
6. Hsu, W., Merchant, K., Shu, H., Hsu, C., Helmy, A.: Weighted waypoint mobility model and its impact on ad hoc networks. SIGMOBILE Mobile Computing Communication Review 9(1), 59–63 (2005)
7. Hsu, W.J., Spyropoulos, T., Psounis, K., Helmy, A.: Modeling spatial and temporal dependencies of user mobility in wireless mobile networks. IEEE/ACM Transactions on Networking 17(5), 1564–1577 (2009)

8. Hyytia, E., Lassila, P., Virtamo, J.: A markovian waypoint mobilitymodel with application to hotspot modeling. In: Proceedings of the IEEE ICC (2006)

9. Johansson, P., Larsson, T., Hedman, N., Mielczarek, B., Degermark, M.: Scenario-based performance analysis of routing protocols for mobile ad-hoc networks. In: Proc. of the International Conference on Mobile Computing and Networking, MobiCom 1999 (1999)

10. Lee, J.K., Hou, J.C.: Modeling steady-state and transient behaviors of user mobility: formulation, analysis, and application. In: MobiHoc 2006: Proceedings of the 7th ACM International Symposium on Mobile Ad hoc Networking and Computing, pp. 85–96. ACM, New York (2006)

11. Liang, B., Haas, Z.J.: Predictive distance-based mobility management for PCS networks. In: Proceedings of IEEE Information Communications Conference, INFOCOM 1999 (April 1999)

12. Liu, T., Bahl, P., Chlamtac, I.: Mobility modeling, location tracking and trajectory prediction in wireless networks. IEEE Journal on Special Areas in Communications 16(6), 922–936 (1998)

13. Musolesi, M., Mascolo, C.: Designing mobility models based on social network theory. SIGMOBILE Mobile Computing Communication Review 11(3), 59–70 (2007)

14. Pacheco, A., Tang, L.C., Prabhu, N.U.: Markov-modulated processes & semiregenerative phenomena. World Scientific Publishing Co. Pte. Ltd., Hackensack (2009)

15. Papoulis, A.: Probability, random variables, and stochastic processes. McGraw-Hill, New York (1984)

16. Piorkowski, M., Sarafijanovic-Djukic, N., Grossglauser, M.: CRAWDAD data set epfl/mobility (v. 2009-02-24) (February 2009), Downloaded from http://crawdad.cs.dartmouth.edu/epfl/mobility

17. Piorkowski, M., Sarafijanovoc-Djukic, N., Grossglauser, M.: A Parsimonious Model of Mobile Partitioned Networks with Clustering. In: The First International Conference on COMmunication Systems and NETworkS (COMSNETS) (January 2009), http://www.comsnets.org

18. Royer, E.M., Melliar-Smith, P.M., Moser, L.E.: An analysis of the optimum node density for ad hoc mobile networks. In: Proceedings of the IEEE International Conference on Communications, ICC (June 2001)

19. Soh, W., Kim, H.: Dynamic bandwidth reservation in cellular networks using road topology based mobility predictions. Proceedings of the IEEE Infocom 4, 2766–2777 (2004)

20. Song, L., Kotz, D., Jain, R., He, X.: Evaluating location predictors with extensive wi-fi mobility data. In: INFOCOM 2004, Twenty-third Annual Joint Conference of the IEEE Computer and Communications Societies, vol. 2, pp. 1414–1424 (7-11, 2004)

21. Spyropoulos, T., Psounis, K., Raghavendra, C.S.: Performance analysis of mobility-assisted routing. In: Proceedings of the 7th ACM International Symposium on Mobile Ad Hoc Networking and Computing (2006)

22. Tian, J., Hahner, J., Becker, C., Stepanov, I., Rothermel, K.: Graph-based mobility model for mobile ad hoc network simulation. In: Proceedings of 35th Annual Simulation Symposium, in Cooperation with the IEEE Computer Society and ACM (April 2002)

23. Yang, S., Yang, X., Zhang, C., Spyrou, E.: Using social network theory for modeling human mobility. Network 24(5), 6–13 (2010)

Mobility Prediction Based Neighborhood Discovery in Mobile Ad Hoc Networks

Xu Li, Nathalie Mitton, and David Simplot-Ryl

INRIA Lille - Nord Europe, Univ Lille Nord de France,
USTL, CNRS UMR 8022, LIFL, France
{Xu.Li,Nathalie.Mitton,David.Simplot-Ryl}@inria.fr

Abstract. Hello protocol is the basic technique for neighborhood discovery in wireless ad hoc networks. It requires nodes to claim their existence/aliveness by periodic 'hello' messages. Central to a hello protocol is the determination of 'hello' message transmission rate. No fixed optimal rate exists in the presence of node mobility. The rate should in fact adapt to it, high for high mobility and low for low mobility. In this paper, we propose a novel mobility prediction based hello protocol, named ARH (*Autoregressive Hello protocol*). Each node predicts its own position by an ever-updated autoregression-based mobility model, and neighboring nodes predict its position by the same model. The node transmits 'hello' message (for location update) only when the predicted location is too different from the true location (causing topology distortion), triggering mobility model correction on both itself and each of its neighbors. ARH evolves along with network dynamics, and seamlessly tunes itself to the optimal configuration on the fly using local knowledge only. Through simulation, we demonstrate the effectiveness and efficiency of ARH, in comparison with the only competitive protocol TAP (Turnover based Adaptive hello Protocol) [9]. With a small model order, ARH achieves the same high neighborhood discovery performance as TAP, with dramatically reduced message overhead (about 50% lower 'hello' rate).

Keywords: Neighborhood Discovery, Hello Protocol, MANET.

1 Introduction

In mobile ad hoc networks (MANET), a fundamental issue for many network operations is *neighborhood discovery*, where each node finds out which other nodes are within its communication range (i.e., neighboring it). Having up-to-date neighborhood knowledge, a node is able to make proper networking decisions. The basic technique for neighborhood discovery is *hello protocol*. The first hello protocol was described in the Open Shortest Path First (OSPF) routing algorithm [13] for IP networks. In OSPF, nodes exchange 'hello' message carrying required information periodically at fixed frequency. When node a receives 'hello' message from node b, it creates an entry for b, or update the existing entry of b, in its neighbor table depending on whether or not b is a new neighbor. If a does not receive 'hello' message from b for a predefined amount of time, it will

J. Domingo-Pascual et al. (Eds.): NETWORKING 2011, Part I, LNCS 6640, pp. 241–253, 2011.
© IFIP International Federation for Information Processing 2011

consider b has left its neighborhood and remove b's entry from the table. The protocol enables nodes to maintain neighborhood information in the presence of node mobility and dynamic node addition and removal.

1.1 Motivation

The usefulness of a hello protocol highly depends on the transmission rate of 'hello' message [3]. It is not a trivial task to choose proper rate. If the rate is too high with respect to node mobility, precious communication bandwidth and energy will be wasted for unnecessarily frequent transmissions. On the other hand, if it is too low, neighbor tables will quickly become out-of-date, leading to failure in other network operations and thus bandwidth waste and energy loss in those other operations. An optimal hello protocol maintains accurate neighborhood information at minimal 'hello' transmission rate. Unfortunately, no constant rate can always remain optimal in dynamic MANET. The rate should evolve together with the network along time for the best performance.

In literature, a majority of MANET protocols adopt hello protocol in one form or another as a building block. But the impact of hello protocol on the network performance has not been studied until recently in [11, 14]. Existing hello protocols all have noticeable limitations and weaknesses. They are inferior for possible applications, compared to the protocol proposed in this paper, for a variety of reasons, e.g., assumption of static networks, use of fixed 'hello' frequency, requirement of extra input parameters. A survey of these previous work will be presented later, in Sec. 2. The importance of the topic and the incompleteness and insufficiency of relevant research motivate our work presented here.

1.2 Contributions

We address the problem of neighborhood discovery in MANET by proposing a novel mobility prediction based hello protocol, named ARH (Autoregressive Hello protocol). This protocol adaptively adjusts 'hello' message rate to the optimal value according to time-varying node mobility. The idea is to let each node constantly estimate its neighbors' position using past location reports and transmit 'hello' message (reporting its current position) when its own location estimated by a neighbor is not accurate enough. For ease of presentation, terms 'predict' and 'estimate' are used interchangeably in the sequel.

More specifically, each node n samples its position at regular intervals and considers the position samples as a time series of data. From this series, it computes two associated time series, respectively for its moving direction and velocity. It applies autoregressive (AR) modeling on the two series and obtains a mobility model for itself. Each neighboring node m builds and maintains an identical mobility model for n using n's previous location reports (carried by 'hello' message) and estimates n's position. In the meantime, n predicts its own mobility (moving direction and velocity) and position using the AR-based mobility model so that it has the same location estimates for itself as its neighbors. It transmits a 'hello' message carrying its current position only when the predicted position leads to false topological change (which means its location estimates by neighboring nodes are no longer accurate). Further, AR-based 'hello' frequency prediction is suggested to detect neighborhood change caused by node removal/departure and improve the algorithm performance.

Through extensive simulation, we study the performance of ARH using real mobility trace data, in comparison with the best known competitive hello protocol TAP (Turnover based Adaptive hello Protocol) [9]. Our simulation results indicate that both protocols require a short learning curve to stabilize. Their learning curves have roughly equal length, $20 - 30$ seconds. Once passing the learning curve, they stabilize to the same high neighborhood discovery performance, about 96% accurately reflecting true neighborhood situation. In particular, ARH results in dramatically lower 'hello' frequency than TAP, i.e., around 50% less message overhead, therefore saving both bandwidth and energy.

The rest of the paper is organized as follows. We review related work in Sec. 2 and introduce autoregressive modeling in Sec. 3. We present ARH in Sec. 4 and report our simulation study in Sec. 5, followed by the closing remarks in Sec. 6.

2   Related Work

In [10], the authors considered a static network whose size is known a priori, and they aimed to reduce the overall energy consumption for communication. Time is slotted. At the beginning of each time slot, a node chooses with a predefined probability p_{state} to enter one of the three states: transmitting ('hello' message), listening, and sleeping. The sum of the probabilities for the three states is 1. The authors studied the optimal values for p_{state}. In dynamic networks such as MANET, where node mobility and node addition/removal are present, finding optimal p_{state} remains to be an open problem.

Similar to [10], three protocols RP, LP, and SP are presented for static networks in [1, 15] rather than MANET. A node can be either in talking state or in listening state, and it can stay in a state for a random period. In RP, a node at each time slot enters talking state with probability p and listening state with probability $1-p$. In LP, if current state is talking, the next state will be listening; otherwise, it will make a random decision as in RP. In SP, if the current and previous states are different, the node will back off for a while. The backoff period is modeled as a uniform random variable with values in a predefined range. After this period, the node sends a message back to the original sender.

In [8], a two-state hello protocol is presented. In this protocol, there are two different 'hello' message frequencies, 1s for low-dynamic network (default frequency) and 0.2s for high-dynamic network, whose selection is however not justified. Which frequency to use depends on two factors: Time to Link Failure (TLF) and Time Without link Changes (TWC). Each node estimates TLF and TWC among its neighborhood based on past neighborhood change. If TLF is smaller than a threshold, the system switches to the High dynamics state. When TWC becomes greater than another threshold, the mechanism switches back to the Low dynamics state. This protocol alternates only between two fixed transmission rates. Its adaptation to network dynamics is obviously very limited.

In [7], three hello protocols are presented. In an adaptive protocol, a node transmits 'hello' message if its travel distance is beyond a threshold since last transmission. Transmission frequency is additionally subject to predefined a minimum value (enabling static nodes to transmit) and a maximum value (preventing 'hello' message storm). This protocol may cause unnecessary transmissions, e.g., when the node moves along a small circle. In a reactive protocol, a node starts, before sending a data packet, neighborhood discovery where it transmits 'hello' message and expects a reply from each neighbor. If no reply is received within a predefined period of time, the process is repeated, up to a maximum number of times. This protocol brings large delay into data communication and is vulnerable to high mobility. In an event-based protocol, fixed 'hello' rate is used. However, a transmission may be skipped if no communication activity is observed in previous 'hello' interval. As such, it is possible that some nodes moving from 'quiet' area to 'quiet' area are never discovered.

In [9], the authors defined turnover ratio as the ratio of the number of new neighbors to the total number of neighbors during a time period Δt. They studied optimal (expected) turnover r_{opt}, and concluded that node velocity does not have any impact on r_{opt} and that r_{opt} is related only to 'hello' frequency and communication radius. They then suggested to adjust 'hello' rate toward the optimal value for obtaining the optimal turnover (i.e., expecting to discover all new neighbors). Based on this idea, a protocol named TAP (Turnover based Adaptive hello Protocol) is presented. In the protocol, nodes initially transmit 'hello' message at a default frequency. Turnover is checked periodically, everytime when 'hello' message is transmitted; the 'hello' rate is immediately adjusted by certain formula that takes current turnover and optimal turnover as input.

TAP is to our knowledge the only adaptive protocol comparable to our proposed ARH here. Other protocols have various obvious weaknesses as summarized above. As we will see in the sequel, ARH assumes location-awareness on each node while TAP does not. This assumption limits ARH to the scenarios where location information is available. But nevertheless, by making good and reasonable use of it ARH outperforms TAP to a great extent.

3   Preliminaries

The ARH protocol to be proposed later in the paper uses autoregressive (AR) model for mobility prediction. For a better understanding of ARH, in this section we briefly introduce AR modeling.

3.1   Autoregressive Model

Autoregressive model (AR) [2] is a tool for understanding and predicting a time series of data. It estimates the current term z_k of the series by a linear weighted sum of previous p terms (i.e., observations) in the series. The model order p is generally less than the length of the series. Formally, AR(p) is written as

$$z_k = c + \sum_{i=1}^{p} \phi_i z_{k-i} + \epsilon_k \quad , \tag{1}$$

where c is a constant standing for the mean of the series, φ_i autoregression coefficients, and ϵ_k zero-mean Gaussian white noise error term. For simplicity, the constant c is often omitted. Deriving $\mathrm{AR}(p)$ involves determining the coefficients φ_i for $i \in [1..p]$ that give a good prediction. It can be done through computation-intensive complex calculus, for example, by the Yule-Walker equations [5] or Burg method [4]. The model can be updated continuously as new samples arrive so as to ensure accuracy, or it may be recomputed only when the predication is too far from the true measurement (beyond certain threshold).

3.2   Simplified Autoregressive Model

In [6], a simplified AR model is proposed. This model requires to be fed with its own estimates as samples to generate new estimates. Although its order p may be larger than 1, it can be updated using only a single sample (in contrast to the whole sample set in the traditional model). Model update needs only trivial calculus, greatly reducing the requirement on the computational power on the nodes that implement the model. Thus it becomes embeddable on tiny and computationally weak devices. We elaborate on this simplified model below.

Recall that p is the model order. A single sample is needed for initializing the model. At initialization, i.e., at time $k = 0$, we set $\phi_i = \frac{1}{p}$ for all $i \in [0..p]$ and $z_i = z_0$ for all $i < 1$. Let the estimate error be $e_k = z_k - \hat{z}_k$ at time $k > 0$. When e_k is too big, the ϕ coefficients need to be adjusted.

Assume that e_k results from using estimates as samples. We spread the error evenly over the p samples and the current estimate z_k. Each sample has error $\frac{1}{p+1}e_k$. Then the estimate with maximum error will be $\hat{z}_k = z_k - e_k + \frac{1}{p+1}e_k = z_k - \frac{p}{p+1}e_k$. We update ϕ coefficients by enforcing the following equation:

$$\sum_{i=1}^{p} \phi_i z_{k-i} = z_k - \frac{p}{p+1}e_k \quad . \tag{2}$$

We update ϕ coefficients iteratively. We first take ϕ_p as variable and the others as constant. We compute new ϕ_p (denoted by ϕ'_p) by solving the above equation. Then we plug ϕ'_p into the equation. Meanwhile, we need to increase the right side by $\frac{1}{p+1}e_k$ because the use of ϕ'_p will bring that much more error. After that, we compute a new ϕ_{p-1}, and so on. We update the coefficients one at a time in this way from ϕ_p downto ϕ_1. We compute in this order because the farther away a sample is in time, the less important its coefficient. The update sequence is shown below:

$$\phi'_p = \frac{1}{z_{k-p}}\left(z_k - \frac{p}{p+1}e_k - \sum_{i=1}^{p-1}\phi'_i z_{k-i} - c\right) \quad ;$$

$$\vdots$$

$$\phi'_j = \frac{1}{z_{k-j}}\left(z_k - \frac{j}{p+1}e_k - \sum_{i=1}^{j-1}\phi'_i z_{k-i} - \sum_{i=j+1}^{p}\phi_i z_{k-i} - c\right) \quad ;$$

$$\vdots$$

$$\phi_1' = \frac{1}{z_{k-1}}(z_k - \frac{1}{p+1}e_k - \sum_{i=2}^{p}\phi_i z_{k-i} - c) \quad .$$

At last, value c can be updated such as $z_k = \hat{z_k}$:

$$c = z_k - \sum_{i=1}^{p}\phi_i' z_{k-1}$$

4   Autoregressive Hello Protocol

In this section, we present a novel *Autoregressive Hello protocol* (ARH) for mobile ad hoc networks. The protocol records historical location information and applies AR modeling to predict node mobility (moving direction and velocity). It determines when to transmit 'hello' message according to prediction accuracy. We start with our assumptions. Then we give an overview of the algorithm, followed by elaboration on individual algorithmic building blocks.

4.1   Assumptions

Nodes have locomotion by being attached to, for example, human being or animal, and are free to move. They may be constrained computing devices. They are aware of their own geographic location $X = [x, y]^T$ by equipped GPS devices or other localization means. For simplicity, we assume that time is synchronized. However, this time synchronization assumption is by no means necessary and can be easily relaxed, as explained at the end of Sec. 4.2.

Nodes independently divide the time domain into slots of equal length λ. The global parameter λ is a positive real number, the same for all the nodes. It defines position sampling rate. That is, each node samples its position $X_i = [x_i, y_i]^T$ at the beginning of every time slot i. Node movement is described by direction θ_i and velocity s_i. We assume that λ is selected small enough such that a node's moving direction does not change more than 2π in each time slot. Nodes have equal communication radius r_c. Each of them is associated with a unique identifier (ID) such as MAC address or manufacturer serial number by which it can be distinguished from others.

4.2   Algorithm Overview

After being added into the network, each node n samples its position X at the beginning of every time slot. The position samples X_i, $i = 0, 1, 2, \cdots$ constitute a time series. n applies AR modeling to this time series to predict its own mobility and future position. Considering potential computational weakness of nodes, here we choose to use the simplified AR model [6], as detailed in Sec. 4.3. At time slot k, node n estimates its position at next time slot $k+1$, and the position estimate is denoted by $\hat{X}$. As soon as n obtains the ground truth X_{i+1}, it checks position estimate error. If the error is acceptable, i.e., $\hat{X_{i+1}}$ is close enough to X_{i+1}, it will update the AR-based mobility model by feeding $\hat{X}_{k+1}$ back to it as sample (this is the feature of simplified AR modeling); otherwise, it will update the mobility model using the real sample X_{k+1}. In the latter case, n also transmits 'hello' message that carries X_{k+1} and its ID. Section 4.4 explains how to evaluate position estimate error in detail.

Upon receiving 'hello' message from n, each neighboring node m initializes an AR-based mobility model for n using the enclosed position sample of n if it has not yet done so, and updates the model otherwise. Between two successive 'hello' messages, m uses this locally maintained mobility model to estimate n's position and keep updating the model using position estimates. As such, m makes the same position estimates for n as n does. This justifies why n transmits 'hello' message according to the accuracy of position estimate. Note that n must transmits 'hello' message right after getting its first position sample in order to enable parallel model initialization and identical position estimation by m.

In Sec. 4.1, we assumed time synchronization for simplicity. The algorithm however is not dependent of this assumption. When time is not synchronized, m and n may not have exactly the same AR model, thus identical location estimates, for n. To overcome this problem, n can simply encapsulate its latest AR mode (all the coefficients and the entire sample set) rather than just the last position sample into each 'hello' message. Then m easily synchronizes (replaces) its locally maintained AR model for n with the one maintained by n itself.

4.3   Mobility Modeling and Position Estimation

At each node n, the position samples $X_0, X_1, X_2, \cdots$ constitutes a time series. From it, two associated time series are computed, one for direction θ and the other for velocity s, describing the mobility of n over time. At time slot $i > 1$, direction is computed as

$$\theta_i = \begin{cases} \arctan(\frac{y_i - y_{i-1}}{x_i - x_{i-1}}) & \text{if } x_i > x_{i-1} \\ \arctan(\frac{y_i - y_{i-1}}{x_i - x_{i-1}}) + \pi & \text{if } x_i < x_{i-1} \\ \frac{1}{2}\pi & \text{if } x_i = x_{i-1} \text{ and } y_i > y_{i-1} \\ -\frac{1}{2}\pi & \text{if } x_i = x_{i-1} \text{ and } y_i \leq y_{i-1} \end{cases} \tag{3}$$

and speed

$$s_i = \sqrt{(x_i - x_{i-1})^2 + (y_i - y_{i-1})^2} \quad . \tag{4}$$

Future direction $\hat{\theta}_{i+1}$ and speed $\hat{s}_{i+1}$ are predicted in order to deduce a position estimate of

$$\hat{X}_{i+1} = \begin{bmatrix} \hat{x}_{i+1} \\ \hat{y}_{i+1} \end{bmatrix} = \hat{X}_i + \begin{bmatrix} cos(\hat{\theta}_{i+1}) \\ sin(\hat{\theta}_{i+1}) \end{bmatrix} \times \hat{s}_{i+1} \quad . \tag{5}$$

The s series can be directly applied in Eqn. 1; whereas, θ is a cyclic value in $[0, 2\pi]$ and needs an adaptation. We thus translate the θ series into $[0, \infty[$ by introducing a t series computed as follows:

$$t_i = \begin{cases} \theta_1 & \text{if } i = 1 \\ \theta_i + 2k\pi & \text{otherwise, with } k \in \mathbb{N} \text{ s.t.} \\ & |t_i - t_{i-1}| \text{ is minimum.} \end{cases} \tag{6}$$

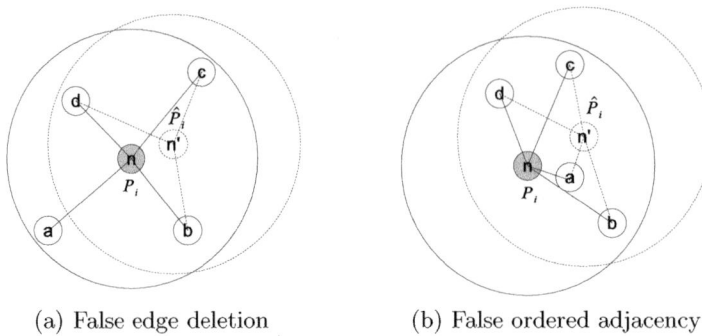

(a) False edge deletion   (b) False ordered adjacency

Fig. 1. Evaluating location estimate

Note that the sampling rate $1/\lambda$ is chosen such that between two successive samples, the moving direction of node n can not vary more than 2π. Equation 1 is thus applied on the t series for estimating $\hat{t}_{i+1}$. Finally, $\hat{\theta}_{i+1}$ is obtained from $\hat{t}_{i+1}$ as follows:

$$\hat{\theta}_{i+1} = \hat{t}_{i+1} - \lfloor \frac{\hat{t}_{i+1}}{2\pi} \rfloor \times 2\pi \quad . \tag{7}$$

Summarizing, the mobility model of node n is composed of two simplified AR models, one for s series and one for t series. The two series are processed independently. Their corresponding AR models evolve along time by being input with estimates produced by themselves and are periodically corrected by being fed with new position samples, as described in Sec. 3.2.

4.4   Evaluating Position Estimates

When evaluating position estimates, a naive way is to use a predefined error threshold δ. That is, node n transmits 'hello' message at time slot i if $|\hat{X}_i - X_i| > \delta$. This method introduces additional parameter δ, and renders the algorithm performance subject to the selected parameter value. If δ is too small, 'hello' message frequency may unnecessarily increase; if it is too large, each node could have a stale or ineffective neighborhood map, which degrades the performance of other networking protocols, for example, routing protocols. The best value of δ depends on global network conditions, which are usually beyond the knowledge available to each node. To enable ARH to work in all mobility scenarios and adjust itself seamlessly to one of optimal protocols for any particular mobility, we propose a parameterless evaluation method.

Specifically, at each time slot i, node n maintains two neighborhood subgraphs G_n and $\hat{G}_n$ of its neighbors and itself using its true location X_i and location estimate $\hat{X}_i$, respectively. In both graphs, neighbors' locations are estimates. If an edge between n and neighbor m in G_n disappears in G'_n, we say there is a *false edge deletion* in G'_n. Node n constructs a circular order among its incidental edges according to their appearance sequence in certain direction, either clockwise or counter-clockwise. If the order is different in the two graphs, we say there is a *false ordered adjacency* in G'_n. Node n transmits 'hello' message if G'_n contains either false edge deletion or false ordered adjacency. This topology-based evaluation method does not require any pre-set parameter and provides a great degree of flexibility.

Figure 1 illustrates the two false situations. Node n has 4 neighbors. Big circles indicate its communication range. G_n is represented by solid links, while G'_n is shown by dashed links. In Fig. 1(a), false edge deletion happens to edge an; in Fig. 1(b), false ordered adjacency involves edges an and bn.

4.5   Detecting Neighborhood Change

If a node does not transmit 'hello' message, its neighbors will consider their location estimates for that node are correct. They are not able to distinguish this situation from node failure, where the node is malfunctioning and will never transmit 'hello' message. If a new neighbor arrives without sending 'hello' message (it is possible when the current set of neighbors of that node all have acceptable location estimates), the node will not be able to know it or update its neighborhood. Additional mechanism is needed for detecting neighborhood change and improving the algorithm performance.

A straightforward method is to use constant low frequency 'hello' message, regardless of node mobility. While 'hello' message loss implies node departure, forced 'hello' transmission increases the chance of new neighbor discovery. However, this method reserves our efforts so far and brings us back to the original problem – how to determine the best 'hello' frequency. It is not a solution.

Here we propose a 'hello'frequency predication based solution, similar to the main body of ARH. We let each node n use an AR model to predict the inter-arrival time of 'hello' message of each neighbor m. If 'hello' message does not arrive on time for more than a threshold number α of times, then n considers that m has left its neighborhood. Node m builds an AR model for itself to monitor the inter-departure time of its own 'hello' message. The model is approximately equal to the model that n builds for it. So, m will follow this model for 'hello' message transmission. It transmits next 'hello' message within α number of successive predicated intervals, even if it does not need to do so by the main algorithm.

With some additional computation overhead on each node, this solution adapts to node mobility and ensures detection of leaving neighbors. It still does not guarantee new neighbor detection since the new neighbor may possibly not transmit any message before moving away. But nevertheless, this is a problem for any 'hello' protocol. In fact, it is not possible to guarantee new neighbor detection because 'hello' message is transmitted at discrete time instants.

5   Performance Evaluation

In this section, we evaluate our new hello protocol ARH through simulation, in comparison with TAP [9] that is to date the most efficient adaptive hello protocol known. In contrast to ARH, TAP supposes that nodes are not equipped with GPS-like devices and thus are not aware of their position.

As the objective of a hello protocol is neighborhood discovery, the protocol must be able to keep the consistency of neighborhood tables among nodes at minimal 'hello' frequency (i.e., message overhead). Thus in addition to 'hello' frequency, we use two evaluation metrics: *neighborhood accuracy* and *neighborhood error*. Assuming that $N(u)$ is the set of *actual* neighbors of a node u, and $N'(n)$ the set of neighbors *known* to u (*i.e.* whose identifier is present in its neighborhood table), these two metrics are defined below. From their definition, notice that $acc(u) + err(u)$ is not necessarily equal to 1.

Definition 1. *Neighborhood accuracy $acc(u)$ is the proportion of actual neighbors of node u that have been indeed detected by u.*

$$acc(u) = \frac{|N(u) \cap N'(u)|}{|N'(u)|} \times 100.$$

Definition 2. *Neighborhood error $err(u)$ measures both how many neighbors of node u have not been detected, and how many "false neighbors" remain in its neighborhood table (*i.e. old neighbors that have not been removed*).*

$$err(u) = \frac{|N(u) \backslash N'(u)| + |N'(u) \backslash N(u)|}{|N(u)|} \times 100.$$

We implemented ARH and TAP using WSNet/Worldsens [16] event-driven simulator, with IEEE 802.11 DCF being implemented at MAC layer and free space propagation model at physical layer. Packet collision and contention were also implemented. In ARH, we set AR model order $p = 5$, position sampling interval $\lambda = 2s$ and $\alpha = 5$ (the order of the AR model for 'hello' frequency predication; see Sec. 4.5). We generated nodal mobility trace based on logs obtained from real experiments on pedestrian runners. Node moving speed was thus spread around a mean value of 1 meter per second. Mobility trace data can be found in [12]. A varying number of nodes (from 100 to 600) were uniformly randomly deployed in a 1000×1000 square region. These nodes have the same transmission range $r_c = 100m$. For each setting, we conducted 50 simulation runs to obtain average results.

5.1   Performance in Relation with Time

We first evaluate the performance of ARH and TAP along time with a fixed number (100) of nodes. Figure 2(a) plots the message overhead, i.e., 'hello' frequency, of both protocols. In ARH, each node first sends 'hello' message with higher frequency so as to train its mobility model on neighboring nodes. After the training period (20 − 30 seconds in our simulation), the number of 'hello' messages sent greatly decreases to stabilize at about the half of messages sent by nodes with TAP. This indicates that ARH is much less costly than TAP in both bandwidth usage and energy consumption.

Figures 2(b) and 2(c) compare neighborhood discovery performance for ARH and TAP. Starting from a low performance point, TAP achieves increasingly

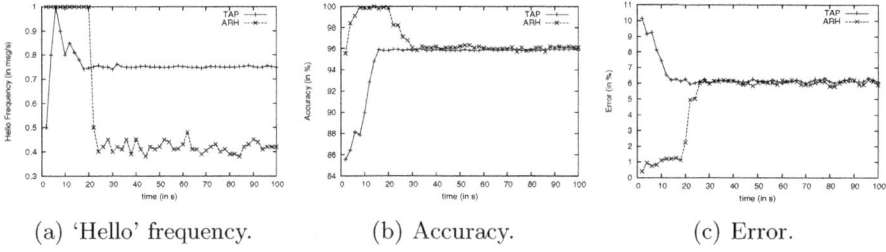

(a) 'Hello' frequency.   (b) Accuracy.   (c) Error.

Fig. 2. Performance in relation with time

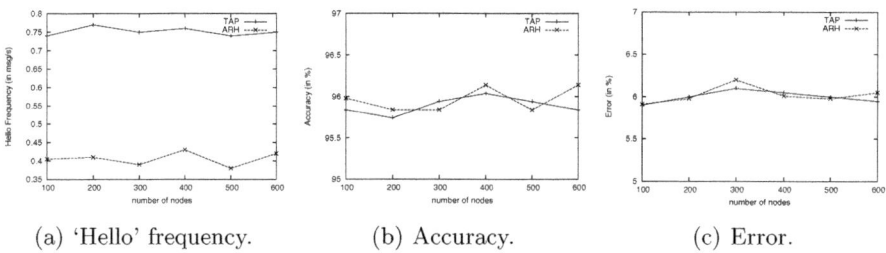

(a) 'Hello' frequency.   (b) Accuracy.   (c) Error.

Fig. 3. Performance in relation with number of nodes

better accuracy (resp. lower error) since it adapts 'hello' frequency till achieving a stabilized turnover and better performance. Contrarily, in ARH each node at first sends a lot of 'hello' messages for mobility model training. Frequent 'hello' transmission leads to very good neighborhood accuracy. After the mobility model is fully trained, satisfactory location prediction can be computed, and the number of 'hello' messages sent decreases, resulting in a slight decrease in accuracy and a slight increase of error. After the initial short training period, both protocols stabilize with the same excellent performance, around 96% of accuracy and 6% error in neighborhood tables.

5.2 Performance in Relation with Number of Nodes

Figure 3 plots the simulation results obtained with regards to different number of nodes after the protocols ARH and TAP pass their initial learning curve. We observe consistent performance as the number of nodes changes. That is to say, the number of nodes has no impact on the protocol performance. Indeed, in TAP, the adaptation of 'hello' rate is related to the mean speed of nodes and thus is not impacted by the node density. In ARH, 'hello' rate depends on the error that a node detects between its position estimate and its real position, both independent from the number of neighbors and the total number of nodes.

6   Conclusions

We proposed a novel *Autoregressive Hello protocol* (ARH) for neighborhood discovery in mobile ad hoc networks (MANET). Each node predicts its neighbors mobility and position by autoregressive (AR) modeling, based on historical location reports; it also predicts its own mobility and position using position samples in the same way. The node updates its location among neighbors when the its own location estimate leads to false topology change in its neighborhood. Each location update corresponds to a 'hello' message transmission. Simulation results indicate that ARH achieves as high neighborhood discovery performance as the best-known algorithm TAP [9], at dramatically reduced 'hello' rate (about 50% smaller). This is a great advantage in wireless communications since more message transmissions indicate more bandwidth usage and more energy consumption. It is at the cost of an additional requirement for location-awareness on each node. We conclude that ARH is a highly-efficient alternative to TAP when location information is readily available.

In our current work, the AR model order p was set to a small value 5. It is possible that a smaller p (thus yet smaller storage overhead at individual nodes) leads to similar neighborhood discovery performance. On the other hand, if a larger-valued p (thus increased model complexity and accuracy) is used, ARH would possibly have higher neighborhood discovery performance. In addition, the selection of p could be subject to the mobility nature of nodes. For example, it is probably necessary to select p differently when nodes are vehicles rather than human. In the future, we will study the trade-off between model complexity and algorithm performance in different mobility scenarios.

Acknowledgements

This work was partially supported by CPER Nord-Pas-de-Calais/FEDER Campus Intelligence Ambiante and the ANR SensLAB project.

References

1. Alonso, G., Kranakis, E., Wattenhofer, R., Widmayer, P.: Probabilistic protocols for node discovery in ad hoc, single broadcast channel networks. In: Proc. IEEE IPDPS (2003)
2. Box, G., Jenkins, G.M., Reinsel, G.C.: Time Series Analysis: Forecasting and Control, 4th edn. Wiley, Chichester (2008)
3. Chakeres, I., Belding-Royer, E.: The utility of hello messages for determining link connectivity. In: Proc. WPMC, pp. 504–508 (2002)
4. Collomb., C.: Burg's Method, Algorithm and Recursion. Internet resource
5. Eshel, G.: The Yule Walker Equations for the AR Coefficients. Internet resource
6. Ghaddar, A., Razafindralambo, T., Simplot-Ryl, I., Simplot-Ryl, D., Tawbi, S.: Towards Energy-Efficient Algorithm-Based Estimation in Wireless Sensor Networks. In: Proc. MSN (2010)
7. Giruka, V., Singhal, M.: Hello protocols for ad hoc networks: Overhead and accuracy trade-offs. In: Proc. IEEE WoWMoM, pp. 354–361 (2005)

8. Gomez, C., Cuevas, A., Paradells, J.: A two-state adaptive mechanism for link connectivity maintenance in AODV. In: Proc. REALMAN, pp. 98–100 (2006)

9. Ingelrest, F., Mitton, N., Simplot-Ryl, D.: A Turnover based Adaptive HELLO Protocol for Mobile Ad Hoc and Sensor Networks. In: Proc. IEEE MASCOTS, pp. 9–14 (2007)

10. McGlynn, M., Borbash, S.: Birthday protocols for low energy deployment and flexible neighbor discovery in ad hoc networks. In: Proc. ACM MobiHoc, pp. 137–145 (2001)

11. Medina, A., Bohacek, S.: A performance model of neighbor discovery in proactive routing protocols. In: Prof. ACM PE-WASUN, pp. 66–70 (2010)

12. MobilityLog, http://researchers.lille.inria.fr/~mitton/mobilitylog.html

13. Moy, J.: OSPF - Open Shortest Path First. RFC 1583 (March 1994)

14. Razafindralambo, T., Mitton, N.: Analysis of the Impact of Hello Protocol Parameters over a Wireless Network Self-Organization. In: Proc. ACM PE-WASUN, pp. 46–53 (2007)

15. Sawchuk, G., Alonso, G., Kranakis, E., Widmayer, P.: Randomized protocols for node discovery in ad hoc multichannel networks. In: Pierre, S., Barbeau, M., An, H.-C. (eds.) ADHOC-NOW 2003. LNCS, vol. 2865. Springer, Heidelberg (2003)

16. WSNet/Worldsens Simulator, http://wsnet.gforge.inria.fr/

STEPS - An Approach for Human Mobility Modeling

Anh Dung Nguyen[1,2], Patrick Sénac[1,2], Victor Ramiro[3], and Michel Diaz[2]

[1] ISAE/University of Toulouse, Toulouse, France
[2] LAAS/CNRS, Toulouse, France
[3] NIC Chile Research Labs, Santiago, Chile
{anh-dung.nguyen,patrick.senac}@isae.fr, vramiro@niclabs.cl,
michel.diaz@laas.fr

Abstract. In this paper we introduce Spatio-TEmporal Parametric Stepping (STEPS) - a simple parametric mobility model which can cover a large spectrum of human mobility patterns. STEPS makes abstraction of spatio-temporal preferences in human mobility by using a power law to rule the nodes movement. Nodes in STEPS have preferential attachment to favorite locations where they spend most of their time. Via simulations, we show that STEPS is able, not only to express the peer to peer properties such as inter-contact/contact time and to reflect accurately realistic routing performance, but also to express the structural properties of the underlying interaction graph such as small-world phenomenon. Moreover, STEPS is easy to implement, flexible to configure and also theoretically tractable.

1 Introduction

Human mobility is known to have a significant impact on performance of networks created by wireless portable devices e.g. MANET, DTN. Unfortunately, there is no model that is able to capture all the characteristics of human mobility due to its high complexity. In this paper we introduce Spatio-TEmporal Parametric Stepping (STEPS) - a new powerful formal model for human mobility or mobility inside social/interaction networks. The introduction of this new model is justified by the lack of modeling and expressive power, in the currently used models, for the spatio-temporal correlation usually observed in human mobility.

We show that preferential location attachment and location attractors are invariants properties, at the origin of the spatio-temporal correlation of mobility. Indeed, as observed in several real mobility traces, while few people have a highly nomadic mobility behavior the majority has a more sedentary one.

In this paper, we assess the expressive and modeling power of STEPS by showing that this model successes in expressing easily several fundamental human mobility properties observed in real traces of dynamic network:

1. The distribution of human traveled distance follows a truncated power law.
2. The distribution of pause time between travels follows a truncated power law.

J. Domingo-Pascual et al. (Eds.): NETWORKING 2011, Part I, LNCS 6640, pp. 254–265, 2011.
© IFIP International Federation for Information Processing 2011

3. The distribution of inter-contact/contact time follow a truncated power law.
4. The underlying dynamic graph can emerge a small-world structure.

The rest of this paper is structured as follows. After an overview of the state of the art in Section 2, we present the major idea behind the model in Section 3. Section 4 formally introduces STEPS as well as some implementation issues. Section 5 shows capacity of STEPS to capture salient features observed in real dynamic networks, going from inter-contact/contact time to epidemic routing performance and small-world phenomenon. Finally we conclude the paper in Section 6.

2 Related Works

Human mobility has attracted a lot of attention of not only computer scientists but also epidemiologists, physicists, etc because its deep understanding may lead to many other important discoveries in different fields. The lack of large scale real mobility traces made that research is initially based on simple abstract models e.g. Random Waypoint, Random Walk (see [3] for a survey). These models whose parameters are usually drawn from an uniform distribution, although are good for simulation, can not reflect the reality and even are considered harmful for research in some cases [14]. In these model, there is no notion of spatio-temporal preferences.

Recently, available real data allows researchers to understand deeper the nature of human mobility. The power law distribution of the traveled distance was initially reported in [6] and [1] in which the authors study the spatial distribution of human movement based on mobile phone and bank note traces. The power law distribution of the inter-contact time was initially studied by Chaintreau et al. in [4]. In [8], Karagiannis et al. confirm this and also suggest that the inter-contact time follows a power law up to a characteristic time (about 12 hours) and then cut-off by an exponential decay.

Based on these findings, some more sophisticated models have been proposed. In [5], the authors have proposed an universal model being able to capture many characteristics of human daily mobility by combining different sub-models. With a lot of parameters to configure, the complexity of this type of model make them hard to use.

In [9], the authors propose SLAW - a Random Direction liked model, except that the traveled distance and the pause time distributions are ruled by a power law. An algorithm for trajectory planning was added to mimic the human behavior of always choosing the optimal path. Although being able to capture statistical characteristics like inter-contact, contact time distribution, the notion of spatio-temporal preferential attachment is not expressed. Moreover, the routing protocol performance results have not been compared with real traces.

Another modeling stream is to integrating social behaviors in the model. In [10], a community based model was proposed in which the movement depends on the relationship between nodes. The network area is divided in zones and the social attractivity of a zone is based on the number of friends in the same zone. The comparisons of this model with real traces show a difference for the contact time distribution. Moreover, the routing performance has not been shown.

Time Varying Community [7] is another interesting model in which the authors try to model the spatio-temporal preferences of human mobility by creating community zones. Nodes have different probabilities to jump in different communities to capture the spatial preferences. Time structure was build on the basis of night/day and days in a week to capture temporal preferences.

A recent research shows that some mobility model (including [7]), despite of the capacity of capturing the spatio-temporal characteristics, deviate significantly the routing performances compared to ones obtained with real traces [12]. This aspect that has not always been considered in existing models is indeed really important because it shows how a model can confront the real dynamic networks.

3   Characterizing Human Mobility

STEPS is inspired by observable characteristics of the human mobility behaviour, specifically the spatio-temporal correlation. Indeed, people share their daily time between some specific locations at some specific time (e.g. home/office, night/day). This spatio-temporal pattern repeats at different scales and has been recently observed on real traces [7].

On a short time basis (i.e. a day, a week), we can assume that one have a finite space of locations. We define two mobility principles :

- Preferential attachment: the probability for a node to move in a location is inverse proportional to the distance from his preferential location.
- Attractor: when a node is outside of his preferential location, he has a higher probability to move closer to this location than moving farther.

From this point of view, the human mobility can be modeled as a finite state space Markov chain in which the transition probability distribution express a movement pattern. In the next section, we answer the question what exactly this probability distribution. Figure 1(a) illustrates a Markov chain of 4 states which corresponds to 4 locations: (A) House, (B) Office, (C) Shop and (D) Other places.

4   Model Description

In STEPS, a location is modeled as a zone in which a node can move freely according to a random mobility model such as Random Waypoint. The displacement between zones and the staying duration in a zone are both drawn from a power law distribution whose the exponent value expresses the more or less localized mobility. By simply tuning the power law exponent, we can cover a large spectrum of mobility patterns from purely random ones to highly localized ones. Moreover, complex heterogeneity can be described by combining nodes with

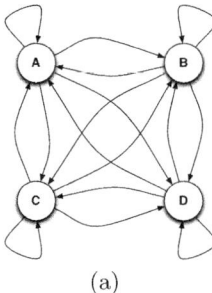
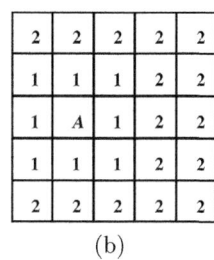

(a)   (b)

Fig. 1. (a) Human mobility modeling under a markovian view: States represent different localities e.g. House, Office, Shop and Other places, and transitions represent the mobility pattern. (b) 5×5 torus representing the distances from a location A to the other locations.

different mobility patterns as defined by their preferential zones and the related attraction power. Group mobility is also supported on our implementation.

4.1   The Model

Assume that the network area is a square torus divided in $N \times N$ square zones. The distance between theses zones is defined according to a metric (here we use Chebyshev distance). Figure 1(b) illustrates an example of a 5×5 torus with the distances from the zone A. One can imagine a zone as a geographic location (e.g. building, school, supermarket) or a logical location (i.e. a topic of interest such as football, music, philosophy, etc). Therefore, we can use the model to study human geographic mobility or human social behaviours. In this paper, we deal only with the first case.

In the so structured space, each node is associated to a preferential zone Z_0. For the sake of simplicity we assume is this paper that each node is attached to one zone, however this model can be extended by associating several preferential zones to each node. The movement between zones is driven by a power law satisfying two mobility principles described above. The pdf of this power law is given by

$$P\left[D = d\right] = \frac{\beta}{(1 + d)^\alpha}, \tag{1}$$

where d is the distance from Z_0, α is the power law exponent that represents the attractor power and β is a normalizing constant.

From (1) we can see that : the farther a zone is from the preferential zone, the less probability the node to move in (i.e. principle of preferential attachment). On the other hand, when a node is outside of its preferred zone he has a higher probability to move closer to this one than moving farther (i.e. principle of attraction).

Algorithm 1: STEPS algorithm

Input: Initial zone $\leftarrow Z_0$

repeat

> - Select randomly a distance d from the probability distribution (1);
> - Select randomly a zone Z_i among all zones that are d distance units away from Z_0;
> - Select randomly a point in Z_i;
> - Go linearly to this point with a speed randomly chosen from $[v_{min}, v_{max}], 0 < v_{min} \leq v_{max} < +\infty$;
> - Select randomly a staying time t from the probability distribution (2);
> **while** t *has not elapsed* **do**
> > | Perform Random Waypoint movement in Z_i ;
> **end**

until *End of simulation*;

The staying time in a zone is also driven by a power law

$$P[T = t] = \frac{\omega}{t^\tau}, \tag{2}$$

where τ is the temporal preference degree of node and ω is a normalizing constant.

From this small set of modeling parameters the model can cover a full spectrum of mobility behaviours. Indeed, according to the value of the α exponent a node has a more or less nomadic behavior. For instance,

- when $\alpha < 0$, nodes have a higher probability to choose a long distance than a short one and so the preferential zone plays the repulsion role instead of a attraction one,
- when $\alpha > 0$, nodes are more localized,
- when $\alpha = 0$, nodes move randomly towards any zone with a uniform probability.

We summarize the description of STEPS in Algorithm 1.

4.2 Markov Chain Modeling

In this section, we introduce the Markovian model behind STEPS. This analytical analysis makes it possible to derive routing performance bounds when using the model. From a formal point of view STEPS can be modeled by a discrete-time Markov chain of N states where each state corresponds to one zone. Note that the torus structure gives nodes the same spatial distribution wherever their preferential zones (i.e. they have the same number of zones with an equal distances to their preferential zones). More specifically, for a distance d, we have $8 \times d$ zones with equal distances from Z_0. Consequently, the probability to choose one among these zones is

$$P[Z_i | d_{Z_i Z_0} = l] = \frac{1}{8l} P[D = l] = \frac{\beta}{8l(1 + l)^\alpha}. \tag{3}$$

Because the probabilities for a node to jump to any zone do not depend on the residing zone but only on the distance from Z_0, the transition probabilities of the Markov chain is defined by a stochastic matrix with similar lines. Therefore the resulting stochastic matrix is a idempotent matrix (i.e. the product of the matrix by itself gives the same matrix)

$$
P = \begin{bmatrix} p(Z_0)\ p(Z_1)\ \dots\ p(Z_{n-1}) \\ p(Z_0)\ p(Z_1)\ \dots\ p(Z_{n-1}) \\ \vdots\ \ \ \vdots\ \ \ \ddots\ \ \ \vdots \\ p(Z_0)\ p(Z_1)\ \dots\ p(Z_{n-1}) \end{bmatrix}.
$$

Hence it is straight-forward to deduce the stationary state of the Markov chain

$$
\Pi = \big(p(Z_0)\ p(Z_1)\ \dots\ p(Z_{n-1}) \big). \tag{4}
$$

From this result, it is interesting to characterize the inter-contact time (i.e. the delay between two consecutive contacts of the same node pair) of STEPS because this characteristic is well known to have a great impact on routing in dynamic networks. To simplify the problem, we assume that a contact occurs if and only if two nodes are in the same zone. Let two nodes A and B move according to the underlying STEPS Markov chain and initially start from the same zone. Let assume that the movement of A and B are independent. For each instant, the probability that the two nodes are in the same zone is

$$
p_{contact} = P_A(Z_0)P_B(Z_0) + \dots + P_A(Z_{n-1})P_B(Z_{n-1})
$$
$$
= \sum_{i=0}^{n-1} P(Z_i)^2 = \sum_{i=0}^{dMax} \frac{1}{8i} \left[\frac{\beta}{(i+1)^\alpha} \right]^2 ,
$$

where $dMax = \lfloor \frac{\sqrt{N}}{2} \rfloor$ is the maximum distance a node can attain in the torus.

Let ICT be the discrete random variable which represents the number of instants elapsed before A and B are in contact again. One can consider that ICT follows a geometric distribution with the parameter $p_{contact}$, i.e. the number of trials before the first success of an event with probability of success $p_{contact}$. Hence, the pdf of ICT is given by

$$
P[ICT = t] = (1 - p_{contact})^{t-1} p_{contact}. \tag{5}
$$

It is well known that the continuous analog of a geometric distribution is an exponential distribution. Therefore, the inter-contact time distribution for i.i.d. nodes can be approximated by an exponential distribution. This is true when the attractor power α is equals to 0 (i.e. nodes move uniformly) because there is no spatio-temporal correlation between nodes. But when $\alpha \neq 0$, there is a higher correlation in their movement and in consequence the exponential distribution is not a good approximation. Indeed, [2] reports this feature for the Correlated Random Walk model where the correlation of nodes induces the emergence of a power law in the inter-contact time distribution. A generalized closed formula

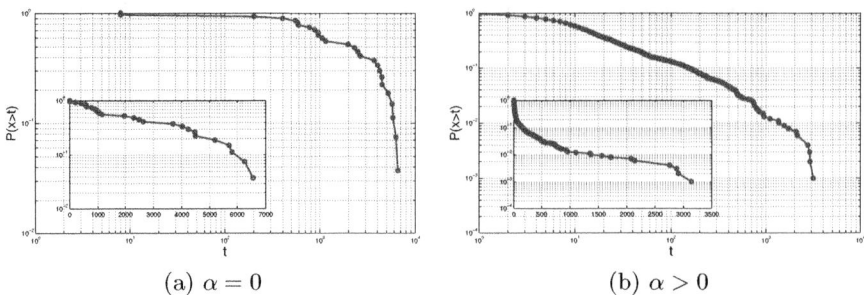

Fig. 2. Theoretical Inter-Contact Time Distribution of STEPS

for STEPS inter-contact time is an on-going work. We provide here simulation results related to this feature.

Figure 2 gives the linear-log plot of the complementary cumulative distribution function (CCDF) of the inter-contact time that results from a simulation of the Markov chain described above when $\alpha = 0$ and $\alpha > 0$. In the first case, the inter-contact time distribution fits an exponential distribution (i.e. is represented by a linear function the linear-log plot) while in the second case it fits a power law distribution (i.e. is represented by a linear function in the log-log plot) with an exponential decay tail. This result confirms the relationship between the spatio-temporal correlation of nodes and the emergence of a power law in inter-contact time distribution.

5   Model Properties

It is worth mentioning that a mobility model should express the fundamental properties observed in real dynamic networks. In this section, we show that STEPS can really capture the seminal characteristics of human mobility and that when used for testing routing performance STEPS deliver the same performances as the ones observed on top of real traces (this features is too often neglected when introducing a new mobility model).

5.1   Inter-Contact Time vs. Contact Time Distributions

Inter-Contact Time Distribution. The inter-contact time is defined as the delay between two encounters of a pair of nodes. Real trace analysis suggest that the distribution of inter-contact time can be approximated by a power law up to a characteristic time (i.e. about 12 hours) followed by an exponential decay [8]. In the following we will use the set of traces presented by Chaintreau et al. in [4] as base of comparison with STEPS mobility simulations. Figure 3(a) shows the aggregate CCDF of the inter-contact time (i.e. the CCDF of inter-contact time samples over all distinct pairs of nodes) for different traces. In order to demonstrate the capacity of STEPS to reproduce this feature, we configured STEPS to exhibit the results observed in the Infocom 2006 conference trace. Table 1 summarizes the characteristics of this trace.

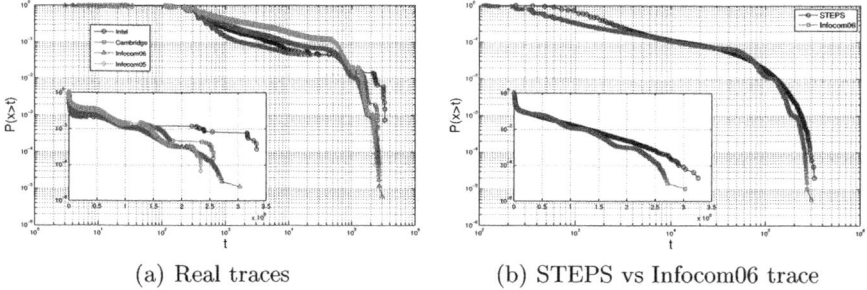

(a) Real traces   (b) STEPS vs Infocom06 trace

Fig. 3. CCDF of Inter-Contact Time

Table 1. Infocom 2006 trace

Number of nodes	98
Duration	4 days
Technology	Bluetooth
Average inter-contact time	1.9 hours
Average contact duration	6 minutes

To simulate the conference environment, we create a 10×10 torus of size $120 \times 120m^2$ that mimics rooms in the conference. The radio range is set to $10m$ which corresponds to Bluetooth technology. Figure 3(b) shows the CCDF of inter-contact time in log-log and lin-log plots. We observe that the resulting inter- contact time distribution as given by the STEPS simulations fits with the one given by the real trace.

Contact Time Distribution. Because of the potential diversity of nodes behavior it is more complicated to reproduce the contact duration given by real traces. Indeed, the average time spent for each contact depends on the person (e.g. some people spend a lot of time to talk while the others just check hands). To the best of our knowledge, the abstract modeling of social behavior has not been studied precisely yet. We measured the average contact duration and the celebrity (i.e. the global number of neighbor nodes) of the Infocom06 nodes and ranked them according to their average contact duration. The result is plotted in Figure 4(a). According to this classification, it appears that the more/less popular the person is, the less/more time he spends for each contact. Because the contact duration of STEPS depends principally on the pause time of the movement inside zone (i.e. the pause time of RWP model), to mimic this behavior we divided nodes in four groups. Each group corresponds to a category of mobility behavior: highly mobile nodes, mobile nodes, slightly mobile and rarely mobile. The pause time for each groups is summarized in Table 2.

With this configuration, we aim to mimic the behavior observed in Infocom06 trace where a large percentage of nodes have short contacts and a few nodes have long to very long contacts. The CCDFs of contact time of STEPS

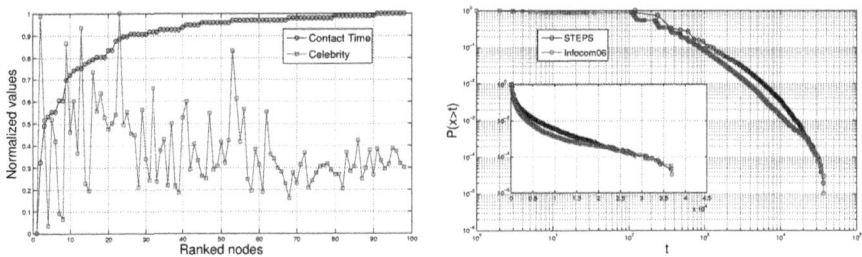

(a) Social Characteristic Observed in In-   (b) CCDF of Contact Time of STEPS vs
focom06   Infocom06 Trace

Fig. 4. Contact Time Behavior of STEPS vs Real Trace

Table 2. Group categories

Dynamicity categories	RWP pause time range (s)	Number of nodes
Very high	$[0, 60]$	65
High	$[60, 900]$	15
Low	$[900, 3600]$	10
Very low	$[3600, 43200]$	8

and Infocom06 trace as shown in Figure 4(b) show that STEPS can also capture with high accuracy this mobility behavior.

5.2   Epidemic Routing Performance

A mobility model has not only to capture the salient features observed in real traces but must also reproduce the performances given by routing protocols on top of real traces. In order to assess the capacity of STEPS to offer this important property we ran Epidemic routing on STEPS and Infocom06 trace and compared the respective routing delays. For each trace, the average delay to spread a message to all the nodes is measured in function of the number of nodes who received the message. Figure 5 shows that STEPS is able to reflect at the simulation level the performance of Epidemic routing when applied on real traces.

5.3   Human Mobility Structure Studied with STEPS

A mobility model should allow not only to express faithfully peer-to-peer interactions properties such as inter-contact time and contact duration, but should be able also to reproduce the fundamental structure of the underlying interaction graph as modeled by a temporal graph, i.e. graphs with time varying edges. The structural properties of static interaction graphs have been studied leading to the observation of numerous instances of real interaction graph with a high clustering and a low shortest path length [13]. Such a structure of graph is called small-world. With respect to routing, the small world structure induces fast message spreading in the underlying network.

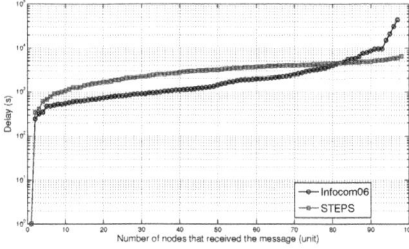 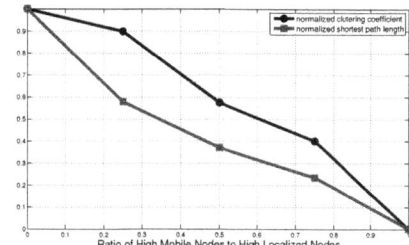

Fig. 5. Epidemic Routing Delay of STEPS vs Infocom06 Trace

Fig. 6. Small-World Structure in STEPS

We extended the notions of clustering coefficient and shortest path length introduced in [13] for dynamic graph. In this paper, we present only the most important definitions due to the lack of space.

Let $\mathcal{G}(t) = (\mathcal{V}(t), \mathcal{E}(t))$ be a temporal graph with a time varying set $\mathcal{V}(t)$ of vertexes and with a time varying set $\mathcal{E}(t)$ of edges.

1. Temporal Clustering Coefficient : Let $\mathcal{N}(w, i)$ be the set of neighbors of node i in a time window w. The temporal clustering coefficient is defined as the ratio of the actual number of connections between neighbors of i to the theoretical number of connections between them during a time window w. Intuitively, it represents the cliquishness of a time varying friendship circle. Formally, that is

$$C_i = \frac{2 \sum_j |\mathcal{N}(w, j) \cap \mathcal{N}(w, i)|}{|\mathcal{N}(w, i)| \left(|\mathcal{N}(w, i)| - 1 \right)}, \quad (6)$$

where j is a neighbor of i and $|X|$ denotes the cardinal of X.

2. Temporal Shortest Path Length : Let $\mathcal{R}_{ij}(t) \in \{0, 1\}$ denotes a direct or indirect (i.e. via multiple connections at different times) connection between node i and node j at time t. The shortest path length between i and j is defined as the earliest instant when there is a connection between them. That is

$$L_{ij} = \inf\{t | \mathcal{R}_{ij}(t) = 1\}. \quad (7)$$

To visualize the phenomenon in STEPS, we create scenarios where there are two categories of nodes with different attractor power. In the first category, nodes have a high mobile behaviour (i.e. α is small) while in the second one, they are more localized (i.e. α is large). The idea is to "rewiring" a high clustered dynamic graph (i.e. with the population is in the second category) by introducing high mobile nodes into the population. By tuning the ratio of number of nodes between these two categories, we measure the metrics defined in (6) and (7). Figure 6 shows that between two extrema, with certain value of the ratio, the network have a structure where the clustering coefficient is high while the shortest path length is low. This result suggests the existence of the small-world phenomenon in dynamic graph.

6   Conclusion

In this paper, we introduce STEPS, a generic and simple mobility model which abstracts the spatio-temporal correlation of human mobility. Based on the principles of preferential attachment and location attractor, this model can cover a large spectrum of human mobility patterns by tuning a small set of parameters. Via simulations, the model is shown to be able to capture different characteristics observed on top of real mobility traces. On the other hand, the model can also reflect accurately realistic routing performances, one of important aspect often neglected in proposed mobility models. Moreover, STEPS can reflect the structural features of the underlying dynamic graph as well as the peer-to-peer ones. Finally, the underlying Markovian basis make it possible to derive analytical results from this model.

References

1. Brockmann, D., Hufnagel, L., Geisel, T.: The scaling laws of human travel. Nature 439(7075), 462–465 (2006)
2. Cai, H., Eun, D.: Toward stochastic anatomy of inter-meeting time distribution under general mobility models. In: Proceedings of the 9th ACM International Symposium on Mobile Ad hoc Networking and Computing, MobiHoc 2008, pp. 273–282. ACM, New York (2008)
3. Camp, T., Boleng, J., Davies, V.: A survey of mobility models for ad hoc network research. Wireless Communications and Mobile Computing 2(5), 483–502 (2002)
4. Chaintreau, A., Hui, P., Crowcroft, J., Diot, C., Gass, R., Scott, J.: Impact of human mobility on opportunistic forwarding algorithms. IEEE Transactions on Mobile Computing 6(6), 606–620 (2007)
5. Ekman, F., Keränen, A., Karvo, J., Ott, J.: Working day movement model. In: Proceeding of the 1st ACM SIGMOBILE Workshop on Mobility Models, pp. 33–40. ACM, New York (2008)
6. González, M., Hidalgo, C., Barabási, A.: Understanding individual human mobility patterns. Nature 453(7196), 779–782 (2008)
7. Hsu, W., Spyropoulos, T., Psounis, K., Helmy, A.: Modeling spatial and temporal dependencies of user mobility in wireless mobile networks. IEEE/ACM Transactions on Networking 17(5), 1564–1577 (2009)
8. Karagiannis, T., Le Boudec, J., Vojnovic, M.: Power law and exponential decay of inter contact times between mobile devices. IEEE Transactions on Mobile Computing, 183–194 (2010)
9. Lee, K., Hong, S., Kim, S., Rhee, I., Chong, S.: Slaw: A mobility model for human walks. dspace.kaist.ac.kr (2009)
10. Musolesi, M., Mascolo, C.: A community based mobility model for ad hoc network research. In: Proceedings of the 2nd International Workshop on Multi-hop Ad hoc Networks: from Theory to Reality, pp. 31–38. ACM, New York (2006)
11. Tang, J., Musolesi, M., Mascolo, C., Latora, V.: Characterising temporal distance and reachability in mobile and online social networks. ACM SIGCOMM Computer Communication Review 40(1), 118–124 (2010)

12. Thakur, G.S., Kumar, U., Helmy, A., Hsu, W.-J.: Analysis of Spatio-Temporal Preferences and Encounter Statistics for DTN Performance (July 2010)
13. Watts, D., Strogatz, S.: Collective dynamics of small-world networks. Nature 393(6684), 440–442 (1998)
14. Yoon, J., Liu, M., Noble, B.: Random waypoint considered harmful. In: IEEE Societies INFOCOM 2003, Twenty-Second Annual Joint Conference of the IEEE Computer and Communications, vol. 2 (2003)

Epidemic Spread in Mobile Ad Hoc Networks: Determining the Tipping Point

Nicholas C. Valler[1,*], B. Aditya Prakash[2], Hanghang Tong[3],
Michalis Faloutsos[1], and Christos Faloutsos[2]

[1] Dept. of Computer Science and Engineering, University of California, Riverside
{nvaller,michalis}@cs.ucr.edu
[2] Computer Science Dept., Carnegie Mellon University
{badityap,christos}@cs.cmu.edu
[3] IBM T.J. Watson Research Center
htong@us.ibm.com

Abstract. Short-range, point-to-point communications for mobile users enjoy increasing popularity, particularly with the rise in Bluetooth -equipped mobile devices. Unfortunately, virus writers have begun exploiting lax security in many mobile devices and subsequently developed malware exploiting proximity-based propagation mechanisms (e.g. Cabir or CommWarrior). So, if given an ad-hoc network of such mobile users, will a proximity-spreading virus survive or die out; that is, can we determine the "tipping point" between survival and die out? What effect does the average user velocity have on such spread? We answer the initial questions and more. Our contributions in this paper are: (a) we present a framework for analyzing epidemic spreading processes on mobile ad hoc networks, (b) using our framework, we are the first to derive the epidemic threshold for any mobility model under the SIS model, and (c) we show that the node velocity in mobility models does not affect the epidemic threshold. Additionally, we introduce a periodic mobility model and provide evaluation via our framework. We validate our theoretical predictions using a combination of simulated and synthetic mobility data, showing ultimately, our predictions accurately estimate the epidemic threshold of such systems.

1 Introduction

The prevalence and increased functionality of mobile phones present an unique opportunity for malicious software writers. According to popular reports, 45.4

* This material is based upon work supported by the Army Research Laboratory under Cooperative Agreement No. W911NF-09-2-0053, the National Science Foundation under Grants No. IIS-1017415, CNS-0721736 and CNS-0721889, NETS-0721889 and NECO-0832069 and a Sprint gift. Any opinions, findings, and conclusions or recommendations in this material are those of the authors and should not be interpreted as representing the official policies, either expressed or implied, of the Army Research Laboratory, the U.S. Government, the National Science Foundation, or other funding parties. The U.S. Government is authorized to reproduce and distribute reprints for Government purposes notwithstanding any copyright notation here on.

J. Domingo-Pascual et al. (Eds.): NETWORKING 2011, Part I, LNCS 6640, pp. 266–280, 2011.

million U.S. citizens own smartphones in 2010 [12], roughly 15% of the total population. These devices are becoming more and more essential for the daily lives of end-users, especially as smartphones offer more and more capabilities. At the same time, mobile devices are equipped with device-to-device (a.k.a. point-to-point) communication technologies (such as Bluetooth), where the communication does not use the phone's service provider's infrastructure. There are estimated over 920 million Bluetooth-equipped devices worldwide shipped in 2008, making Bluetooth the most common point-to-point communication protocol in today's smartphones. To date, at least two smartphone worms have been found in the wild, Cabir and CommWarrior, which spread using Bluetooth.

Given a system of mobile agents, such as smartphones, what can we say about the propagation of a virus[1] within the system? A key question is to identify the tipping point, known as the system's *epidemic threshold*, or *take-off point*, below which a virus is guaranteed to "die out." For the epidemic models, we focus on one of the most popular models, the flu-like one susceptible-infected-susceptible - SIS (see Section 5 where we handle other models). There, agents maintain no immunity, and become susceptible, immediately after they heal. Our key contributions are as follows:

1. *Framework and Formula*: We present a framework for estimating the epidemic threshold on *any, arbitrary* mobile ad hoc network model. To the best of our knowledge, ours is the first theoretical study for the general mobility case. The idea is to derive a sequence of adjacency matrices, and then compute the first eigenvalue of the so-called *system matrix* (see Theorem 1). There, we show that the epidemic threshold depends only on this first eigenvalue, and nothing else.
2. *Closed Formulas*: We show how to use our framework to derive simple, approximate (but accurate) formulas for several, popular special cases (Random Walk model, Levy flight model).
3. *Insensitivity to Velocity*: To our surprise, our results show that the epidemic threshold does *not* depend on the node velocity ($v > 0$). Our experiments confirm the accuracy of our approximations, as well as our 'insensitivity' observation.

Jumping ahead, Figure 1 showcases the accuracy of our results (Lemma 3) of the epidemic threshold for the SIS (=flu-like) model on the so-called 'Levy Flight' mobility model. See section 4 for more details - but the point is that our prediction for the take-off point (= *epidemic threshold*, indicated with a black arrow) is exactly where all curves take off.

We have two additional contributions: through extensive simulation experiments, we show that similar insensitivity results hold for other popular mobility models like Levy flight, Random Waypoint etc.; and, moreover, we introduce

---

[1] We focus on proximity propagation, which in mobile networks is affected by mobility. However, virus for smartphones can also propagate through email, mms, or direct access to the web. We do not consider these cases, sine they are not directly affected by mobility.

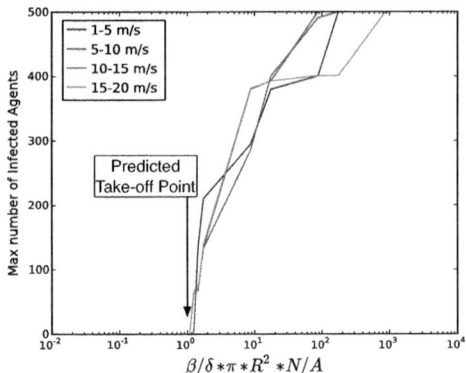

Fig. 1. Accuracy of our results (Lemma 3), for the 'Levy-Flight' model. Take-off plot, plotting the max number of infections vs. strength of the virus. Notice that our predicted take-off point (black arrow) agrees with the simulations, for several node velocities.

the *periodic mobility* model, which is very popular in biological virus epidemiology [1,4], and show how to use our framework to estimate its epidemic threshold.

The rest of this paper has the typical organization: background (Section 2, proposed framework and theorems (Section 3), experiments (Section 4), additional observations (Section 5), and finally, related work and conclusions. (Sections 6 and 7).

2   Preliminaries

In this section, we present a general background on proximity-based epidemic spreading models and formulate our problem statement.

2.1   Epidemic Model: SIS (Flu-Like)

The SIS epidemic model resembles a flu-like virus, where nodes have no immunity. Healthy ('S' = susceptible) nodes become sick ('I' = infected) stochastically from their infected neighbors with a probability β. Alternatively, a sick node becomes healthy (and open to re-infection), with a probability δ. These two parameters are also referred to as the *birth* rate (β) and *death* rate (δ) of the virus.

The tipping point τ, or *epidemic threshold*, of an SIS epidemic model is the condition under which an infection will die out exponentially quickly irrespective of initial infection, as opposed to spreading out, causing and epidemic (technically, a pandemic). For a survey on SIS and numerous other epidemic models, see Hethcote [13], or [7,11].

2.2   Problem Formulation

Using the background discussed above, we now formulate our problem statement. See Table 1 for definitions of various symbols. In this paper, we consider an epidemic on a mobile network, which provides an underlying contact structure for the virus to use as it propagates. By doing so, at any point in time, the system is non-homogenous, as nodes may only transmit the virus to its neighbors.

Given:

1. Mobile ad hoc network mobility models, described below.
2. The SIS model parameters, i.e., the virus birth and death probabilities β and δ.

Find:

The epidemic threshold τ or *tipping point* for the system such that for $\tau < 1$ an infection will die out quickly, irrespective of initial conditions.

Our problem naturally leads to other issues like the effect of node velocity in models on the threshold, giving approximations in specific cases etc. We elaborate on them in the upcoming sections.

Table 1. Terminology

Symbol	Definition and Description
General Terms:	
$\mathbf{A}, \mathbf{B}, \ldots$	matrices (bold upper case)
$\mathbf{A}(i,j)$	element at the ith row and j^{th} column of $\mathbf{A}$
$\mathbf{A}(i,:)$	i^{th} row of matrix $\mathbf{A}$
$\mathbf{A}(:,j)$	j^{th} column of matrix $\mathbf{A}$
$\mathbf{I}$	standard $n \times n$ identity matrix
$\mathbf{a}, \mathbf{b}, \ldots$	column vectors
$\mathcal{I}, \mathcal{J}, \ldots$	sets (calligraphic)
$\lambda_{\mathbf{B}}$	first eigenvalue (in absolute value) of a matrix $\mathbf{B}$
Mobility Terms:	
M	mobility model
$P_{i,t}$	position of node i at time t
N	number of nodes
A	simulation area
ρ	node density (N/A)
ΔT	Time step
T	number of different alternating behaviors
$\mathbf{A}_1, \ldots, \mathbf{A}_T$	T corresponding size ($n \times n$) symmetric alternating adjacency matrices
Epidemic Terms:	
β	virus transmission probability in the SIS model
δ	virus death probability in the SIS model
τ	epidemic threshold
Acronyms and Terms:	
USS	Uniform Steady-State Approximation
EAAM	Eigenvalue of Average Adjacency Matrix Approximation
Take-off Plot	Max number of infected agents vs. Epidemic Threshold τ

3 Framework

In this section, we detail our framework for analyzing mobility models and then move on to specific approximations and questions arising out of the framework. We will present extensive simulations demonstrating the results later in Section 4. Also please see Section 4.2 for a description of the mobility models.

Note that node-to-node contacts at a particular time can be represented by an *adjacency matrix* $\mathbf{A}$. We next provide a general theorem expressing the epidemic threshold for mobility models.

3.1 Epidemic Thresholds on Mobility Models

Theorem 1 (Mobility model threshold). *If a mobility model can be represented as a sequence of connectivity graphs $L = \{\mathbf{A}_1, \mathbf{A}_2, \ldots, \mathbf{A}_T\}$, one adjacency matrix $\mathbf{A}_t$ for each time step $t \in \{1..T\}$, then the epidemic threshold is:*

$$\tau = \lambda_S \tag{1}$$

where $\lambda_\mathbf{S}$ is the first eigenvalue of matrix $\mathbf{S}$ and $\mathbf{S} = \prod_i \mathbf{S}_i$ and $\forall_i \in \{1..T\}$ $\mathbf{S}_i = (1 - \delta)\mathbf{I} + \beta\mathbf{A}_i$ ($\mathbf{I}$ is the standard $N \times N$ identity matrix).

Proof. If the mobility model can be represented as a sequence of graphs, then the epidemic threshold depends on the first eigenvalue of the system matrix [25]. Hence, $\tau = \lambda_{\prod_i((1-\delta)\mathbf{I}+\beta\mathbf{A}_i)}$. □

We can now give a simpler closed-form approximation for the threshold in Equation 1 in the following lemma:

Lemma 1 (EAAM Approximation for Threshold). *Under the same conditions as in Theorem 1, the following is an approximation for the epidemic threshold:*

$$\tau \approx \frac{\beta}{\delta} \times \lambda_{\mathbf{A}_{avg}} \tag{2}$$

where $\mathbf{A}_{avg} = \sum_i \mathbf{A}_i/T$ is the average adjacency matrix.

Proof. Note that,

$$\mathbf{S} = \prod_i ((1 - \delta)\mathbf{I} + \beta\mathbf{A}_i)$$

$$= (1 - \delta)^T\mathbf{I} + \beta\sum_i \mathbf{A}_i + O(\beta^2) + O(\beta * \delta) + O(\delta^2)$$

$$\approx (1 - T\delta)\mathbf{I} + T\beta\mathbf{A}_{avg} \tag{3}$$

where we neglected second or lower order terms involving β and δ. Hence, we find that $\mathbf{B} = (1 - T\delta)\mathbf{I} + T\beta\mathbf{A}_{avg}$ is a first order approximation for the $\mathbf{S} = \prod_i \mathbf{S}_i$ matrix. Hence from Theorem 1 we want $\lambda_B < 1$ which implies Equation 2. □

We will refer to the above approximation as the 'Eigenvalue of the Average Adjacency Matrix' (EAAM) approximation.

3.2   Specific Approximations

Lemma 2 (Random-Walk Threshold). *In the random-walk mobility model and under the SIS model, the following is an approximate epidemic threshold:*

$$\tau \approx \beta/\delta \times \pi R^2 \times N/A \qquad (4)$$

where R is the radius of influence of each node.

Proof. Under the random-walk model, at every point of time, each node is at a random (x, y) position, uniformly distributed on the field of interest. Each node has a radius of possible connections (like the BlueTooth radius) R. Consequently each node has $d = \pi R^2 \times N/A$ neighbors on average (ignoring boundary effects). The connectivity graph at each time step is roughly a random graph with average degree d. Hence it has first eigenvalue $\lambda_1 = d$ on average. Hence this is approximately equivalent to having a static graph under the SIS model where the epidemic threshold [7] is $\tau = \beta/\delta \times \lambda_1$. We now obtain the lemma after obvious substitutions. $\square$

In fact, we can go further and generalize this to *any* mobility model where the geographic steady state distribution is uniform.

Lemma 3 (Uniformly-Distributed Steady State (USS) Threshold). *For any mobility model where the geographic distribution of nodes at the steady state is uniform over the area of interest and under the SIS model, the following is an approximate epidemic threshold:*

$$\tau \approx \beta/\delta \times \pi R^2 \times N/A \qquad (5)$$

where R is the radius of influence of each node.

Proof. The proof for Lemma 2 goes through even here precisely because of the geographically uniformly distributed nature of the steady state. Each node has roughly the same number of connections and hence the adjacency graph is approximately a homogenous graph with constant first eigenvalue. The result follows as before. $\square$

We will refer to the above approximation as the '**U**niform **S**teady-**S**tate' (USS) approximation. Mobility models like Levy Flight and Random-Walk are examples of models with a geographically uniform distribution of the nodes at the steady state. Lemma 3 allows us to quickly estimate the threshold for these and many other models.

3.3   Insensitivity to Node Velocity

As there is no factor depending on the node velocity in Lemma 3, we conclude the following surprising implication:

Corollary 1 (Node velocity and threshold). *The node velocity ($v > 0$) does not affect the epidemic threshold in mobility models where the steady state has a geographically uniform steady state distribution like Random Walk, Levy Flight etc.*

We conjecture that the velocity does not affect the threshold even for non-geographically uniformly distributed steady state mobility models like Random Waypoint. We provide empirical results supporting this claim later in Section 4.

Conjecture 1 (Effect of velocity). The node velocity ($v > 0$) does not affect the epidemic threshold in the Random Waypoint mobility model.

Does Velocity have an impact at all? The above discussion raises the point whether the node velocity has any effect at all on the dynamics of the epidemic spreading. We expect that the velocity of motion does have an effect, when we are above threshold. Furthermore, simulations resulted in a non-obvious observation. The velocity had an impact on the steady-state number of infected agents in the system. We elaborate more on these issues in Section 5.

4   Simulation Methodology and Results

4.1   Experimental Setup

To facilitate the simulation, we wrote a custom Python2.6 simulation program using the NumPy/SciPy python libraries. All simulations were conducted on a 4 core Intel(R) Xeon(R) CPU operating at 2.53 GHz and 72 GB of memory running CentOS-5.5 (Linux kernel 2.6). We varied the number of agents (nodes) N between 250 and 1500 within a simulation field of area $A = 40,000 m^2$ (200m by 200m). Thus, node density ρ, commonly defined a N/A, was between 0.125 and .125 nodes per m^2. All nodes had a transmission range of 5.0 meters. We did not account for signal attenuation, reflection nor other wireless phenomena. Prior to the beginning of the simulation, nodes were distributed on the simulation field in a uniform fashion. Simulations were generally run for a period of 100s with time intervals of $\Delta T = 0.1$ seconds.

We studied three mobility models common to mobile ad hoc networking: Random Walk, Levy Flight and Random Waypoint. In the following sections, we provide detail on each model as well as simulation results. The position $P_{i,t}$ of each node in the system at time t is a function of mobility model and previous position and time step ΔT, such that $P_{i,t+1} = M(P_{i,t}, \Delta T)$, where M is the mobility model.

The purpose of our simulations was to determine the role of the mobility model in the spread of malware in a point-to-point contact network loosely describing Bluetooth communication technology.

4.2   Mobility Models

Random Walk. The Random Walk (RW) mobility model (also referred to as *Brownian Motion*) was originally formulated to describe the seemingly random motion of particles. Numerous variations exists, here we describe our implementation.

Each node i in the system is parameterized by *speed (V_i)* and *angle (θ_i)*. Both V_i and θ_i are drawn uniformly from systemwide predefined ranges, $[v_{min}, v_{max}]$ and $[0, 2\pi)$, respectively. Clearly, such a system is memoryless. The model we employ varies from the simple RW model by introducing a *flight time* for each node, T_i. Flight time is drawn uniformly from a range $[\tau_{min}, \tau_{max}]$. The spatial distribution of the RW mobility model is uniform over the simulation field. According to our framework, we predict the RW mobility model will follow Lemma 2.

Levy Flight. Levy Flight mobility models have recently attracted attention due to their statistical similarities with human mobility [27]. At the beginning of each flight, each node selects an angle uniformly from within $(0, 2*\pi]$, a flight time drawn from some distribution, a flight length and a pause time. Flight length and pause time are drawn from Levy distributions $p(t) \propto |t|^{-(1+\alpha)}$ and $\psi(t) \propto t^{-(1+\beta)}$, where time $t > 0$, respectively. When $\alpha = 2$ and $\beta = 2$ the result is a special case of the Levy distribution resulting in a Gaussian distribution. As with the Random Walk, the spatial distribution of the Levy Flight mobility model is uniform over the simulation field.

Random Waypoint. The Random Waypoint (RWP) mobility model is often cited as the *de facto* mobility model in ad hoc networks. As originally proposed by Johnson et. al [14], the RWP mobility model each node i is described by three parameters: *current location (P_i)*, *speed (V_i)*, *waypoint (W_i)*, and *pause time (ρ)*. In general, the RWP mobility model operates as follows: Initially, a node is stationary. After a pause time ρ, the node selects a waypoint uniformly from the simulation field, then, travels along the shortest path to its waypoint P at a velocity V_i drawn uniformly from $(v_{min}, v_{max}]$. Upon arrival at their waypoint, each node will pause for a time $t = \rho$. After the pause period is done, each node will repeat the process. The spatial distribution of the RWP mobility model is bell-shaped [5].

4.3   Summary of Results

Accuracy of Approximations, Random Walk. We present a series of simulation studies of the Random Walk mobility model in Figure 2. In these studies, we varied the birth (β) and death (α) parameters of the SIS infection model. We refer to the resulting plots as *"take-off plots,"* which show the maximum number of infected agents seen in our simulation against our approximated epidemic threshold. For each plot, we labeled the estimated take-off point according to the specific threshold approximation.

Figure 2(A), we approximated the epidemic threshold using the first eigenvalue of the average adjacency matrix, $\mathbf{A}_{avg}$ (the EAAM threshold approximation of Lemma 1). We indicate the predicted threshold value at $\tau = 1$. As expected, no epidemic was present at values of $\tau < 1$. At values of $\tau > 1$, we see explosive growth in the max number of infected agents.

In Figure 2(B), we plot the the USS approximation of the epidemic threshold (Lemma 3), i.e. $\beta/\delta \times \pi R^2 \times N/A$. As in the EAAM threshold, USS behaves

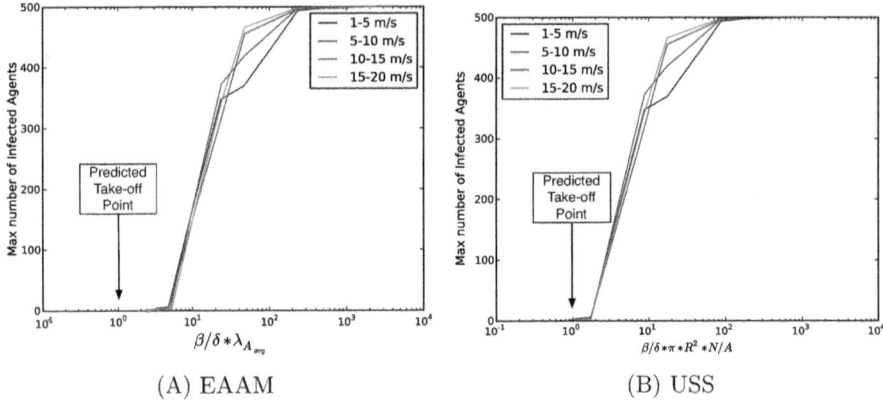

Fig. 2. Accuracy of Framework Approximations, Random Walk Take-off plots for EAAM and USS. Notice that our predictions (black arrows) are accurate.

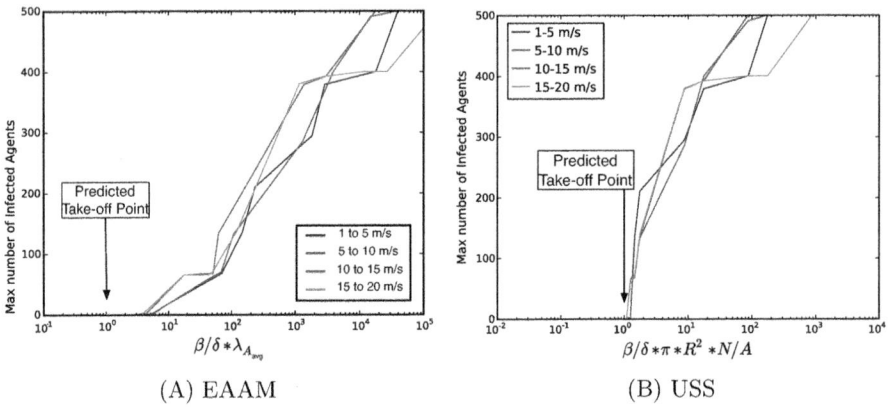

Fig. 3. Accuracy of Framework Approximations, Levy Flight Take-off plots for EAAM and USS. Again our predictions (black arrows) prove accurate.

as expected (i.e. no epidemic below $\tau = 1$). In fact, in each of these figures, we note that no infected agents (aside from patient zero) are present for epidemic threshold values below 1. Compared to EAAM, this threshold value takes off at values closer to $\tau = 1$.

Accuracy of Approximations, Levy Flight. As with RW, the spatial distribution of nodes following Levy Flight mobility model is uniformly distributed on the simulation field. Thus, we expect Levy flight to perform similar to RW. Figure 3 presents a take-off plots for Levy Flight simulations. Again, we note that no infected agents exist below either threshold approximations. Furthermore, we find that USS performs better than EAAM.

Accuracy of Approximations, Random Waypoint. The next series of simulations were conducted using the popular Random Waypoint mobility model. We also selected this model specifically because it does not result in a uniform

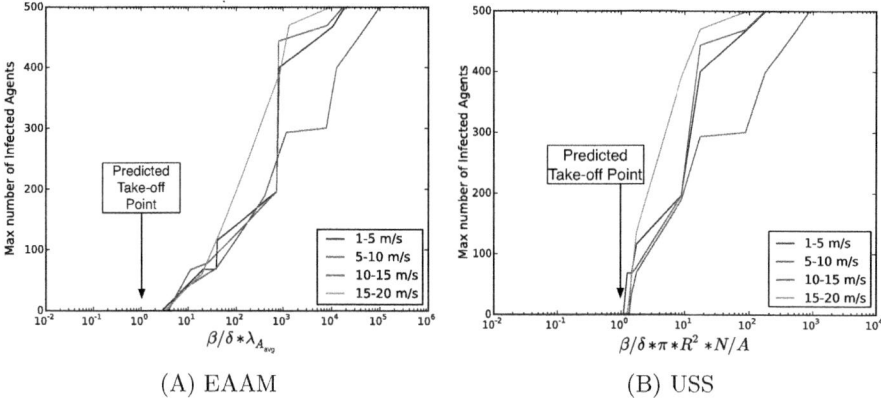

Fig. 4. Accuracy of Framework Approximations. Random Waypoint Take-off plots for EAAM and USS.

spatial distribution of nodes, therefore does not fall under USS. The results of these simulations are presented in Figure 4. Surprisingly, both threshold approximations perform well against the RWP mobility model. As the earlier mobility models exemplified, USS performs better than EAAM.

Insensitivity to Velocity. In order to illustrate Corollary 1 and validate Conjecture 1, we performed a series of simulation in which we varied the node velocity, the results of which are illustrated in Figures 2, 3, 4.

As expected, for both the RW and Levy Flight mobility models, the take-off points were not greatly affected by increasing the nodes velocity. Furthermore, Figure 4 shows that the take-off point for the RWP mobility model was not affected by velocity, affirming Conjecture 1.

5 Discussion

We elaborate here on the effect of node velocity on the dynamics of epidemic spreading. We also introduce the periodic mobility model and present an analysis of it via our framework. In addition, we touch upon other epidemic models as well.

5.1 More on Impact of Node Velocity

As discussed previously in Section 3, node velocity does not seem to effect the threshold in many models. We now ask whether the velocity affects the epidemic at all?

For an "above threshold" system, two more parameters are of interest: (a) *steady-state maximum*, and (b) *warm-up period*. The steady-state maximum is the maximum number of infected agents in the system till steady state, whereas the warm-up period is the time necessary to reach steady state. We expect that the velocity of motion does have an effect, when we are above threshold. Clearly,

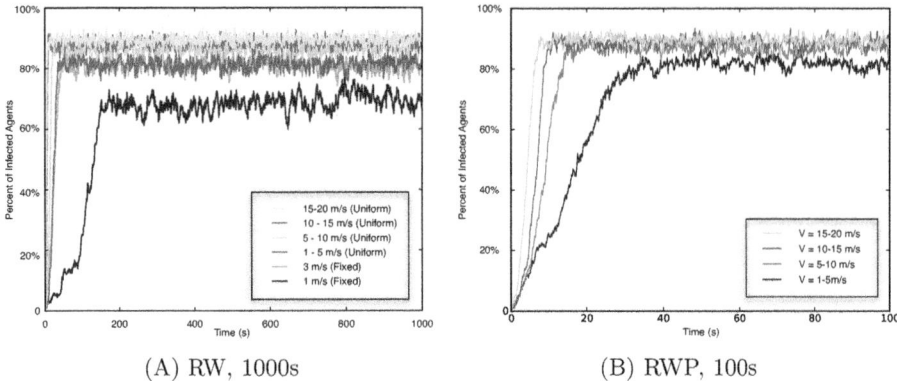

Fig. 5. Number of Infected Agents vs. Time on the Random Walk and Random Waypoint. Node velocities were varied as indicated in the legends. Steady-State number of infected agents increases with node velocity, while warm-up period shrinks (best viewed in color).

speed will effect the speed of propagation of the virus and thus the warm-up period. Higher velocity means better mixing of agents, and thus faster convergence to the steady state. This observation is also demonstrated through simulations.

Figure 5 (best viewed in color) shows the number of infected agents per unit time (in seconds) for both the Random Walk and Random Waypoint mobility models. The velocity varied between a fixed 1 m/s fixed and an uniform selected 15 − 20 m/s, as indicated in the legends. We performed a longer simulation in order for the systems to settle in a steady state.

Less intuitive is that velocity appears to affect the number of infected agents at steady-state. For example, in Figure 5(A), the line corresponding to 1 m/s appears to reach a steady-state of approximately 65 − 70% infected agents, whereas, at 15 − 20 m/s, the steady-state is roughly 90%. The steady-state for velocities between these two extremes lay in-between. We suspect the degree of mixing, influenced by node velocity, is the root cause of the above observation.

5.2   The Periodic Mobility Model

As we indicated in Section 3, our framework predicts the epidemic threshold of mobility models that can be represented as a series of adjacency matrices. The *Periodic Mobility Model* is a special case of such a series, where a set of k adjacency matrices $\{\mathbf{A}_1, \mathbf{A}_2, \cdots, \mathbf{A}_k\}$ are repeated periodically.

This is a typical model used in biological virus studies [4] to model general movements of a population. As an example, let $\mathbf{A}_1$ be an adjacency matrix of people during the day (say, at the office). Let $\mathbf{A}_2$ be an adjacency matrix representing contacts/interactions during the evening (say, at home). So the series formed by repeating $\{\mathbf{A}_1, \mathbf{A}_2\}$ represents the daily, repeated interactions of our population. The periodic model offers a realistic, yet general model of mobility, capturing general patterns rather than specific movements of the system.

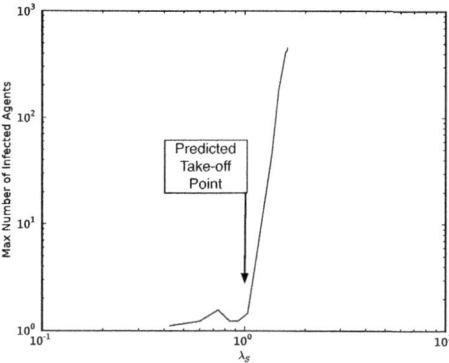

Fig. 6. Take-off Plot for System Matrix Eigenvalue of a Periodic Mobility Model. A_1 and A_2 had $N = 500$ nodes spread uniformly across a 200×200 simulation field. Notice the accurate our prediction (black arrow).

Lemma 4 (Periodic Model Threshold). *Under the periodic mobility model with k alternating behaviors repeating periodically, the epidemic threshold is given by $\tau = \lambda_S$ where $\mathbf{S} = \prod_{i=1}^{k} \mathbf{S}_i$ and, as before, $\mathbf{S}_i = (1 - \delta)\mathbf{I} + \beta \mathbf{A}_i$.*

Proof. Omitted for brevity, similar to Theorem 1.

As an example, Figure 6 shows the take-off plot of a periodic mobility models where $N = 500$ nodes, $k = 2$ adjacency matrices. As predicted by Lemma 4, the max number of infected agents over the simulation period takes off at $\lambda_S = 1$.

5.3   Other Epidemic Models

Given recent results on epidemic thresholds on static networks [24], we believe that our results will carry through for many other epidemic models as well e.g. SIR (mumps-like), SIRS, SEIR, MSEIR etc [13] which capture differences between the way various diseases spread.

Conjecture 2 (Other Epidemic models). Our results for all the mobility models discussed in this paper hold for all the epidemic models covered in [24] as well.

6   Related Work

Here, we review the related work. It is worth pointing out that while most of existing studies about epidemic spread on mobile networks focus on (1) some particular types of network structures, and/or (2) one specific mobility model; our framework is very general and it applies to *arbitrary* network structure, and all the three popular mobility model.

General Epidemic Modeling. Bailey provides the canonical text on epidemic modeling [2]. A more recent survey is provided by Hethcote in [13]. Kephart and White [15,16] were among the first to propose epidemiology-based models to analyze the spread of computer viruses. The model they suggest provides a good approximation of virus propagation in networks where contact among individuals is essentially homogeneous. Recent discoveries suggest real networks (including social networks [9], router and AS networks [10], and Gnutella overlay graphs [28]) follow a power-law structure instead, prompting a re-evaluation of the homogeneity assumption common in the works above.

Epidemics on Static Networks. Observation suggests that real networks are not homogeneous, rather, overwhelming evidence suggests real networks follow a power law structure instead. By introducing an underlying structure for a disease to spread, such as a static network, removes the original homogeneous assumption pioneered by those reference above. Newman [21] studied the epidemic thresholds for multiple competing viruses on special, *random* graphs. Pastor-Satorras and Vespignani studied viral propagation for such power-law networks [20,22,23]. They developed an analytic model for the Barabási-Albert (BA) power-law topology [3]. However, their derivation depends on some assumptions which does not hold for many real networks [17,10]. Pastor-Satorras et al. [23] also proposed an epidemic threshold condition, but this uses the "mean-field" approach, where all graphs with a given degree distribution are considered equal. There is no particular reason why all such graphs should behave similarly in terms of viral propagation. Chakrabarti et. al. [7] observe that epidemic threshold of an arbitrary graph can be captured in a single parameter, the first eigenvalue of the adjacency matrix $\lambda_{1,A}$. Their observation was rigorously confirmed in [8] and independently by [30]. We again leverage the above observations to formulate our solution in Sect. 3

Epidemics on Mobile Networks. Prompted by the emergence of mobile devices, such as Bluetooth-equipped smartphones, researchers introduced mobility to epidemic spread. Mickens et. al. were among the first to examine device-to-device spreading of malicious software in mobile ad hoc networks [19,18]. In their work, they present a *queue*-based technique for the RWP model to overcome the limitations of the earlier homogeneous models of Kephart and White. In a similar work, Yan et. al. extend the observations of Mickens et. al. by examining additional mobility models and their effect on epidemic spreading of a SIS virus [31]. Their work is unique in that it models virus propagation, in detail, a Cabir-like bluetooth worm, including the Bluetooth stack and unique worm properties.

Mobility Models. The mobility models used in Sect. 4 are fairly common, with significant literature devoted to the subject. For an overview on mobility models, we refer our readers to the following surveys [6]. The RWP mobility model has been extensively used, despite well know flaws. For a discussion of the merits of RWP, refer to [32]. The Levy mobility model was first described in [29], yet has been used extensively to model human and animal movements [27,26].

7   Conclusions

To conclude, recent malware in the wild, using device-to-device virus propagation schemes, prompted our study of the epidemic threshold in mobile ad hoc networks. Our contributions in this paper are:

1. *Framework:* We present a framework for the determining the epidemic threshold (for the SIS model) on *any* mobility model which can be converted into a series of adjacency matrices and give a formula for it (Theorem 1).
2. *Closed Formulas:* We also give a closed-form approximation for the SIS epidemic threshold on general mobility models (Lemma 1).
3. *Insensitivity to Velocity:* We analyze the impact of velocity in popular mobility models like Random walk, Levy Flight, Random waypoint etc. and find that it unexpectedly does not affect the threshold (Lemmas 2, 3 and Conjecture 1).

In addition, we introduced the "periodic mobility model," popular in other fields like epidemiology [1,4], to the networking community and solved it using our framework (Lemma 4). Finally we presented extensive simulations to demonstrate our analysis and results. Future work may concentrate on providing theoretical analysis on the effect of velocities on the steady state behavior of an epidemic on various mobility models.

References

1. Anderson, R.M., May, R.M.: Infectious Diseases of Humans. Oxford University Press, Oxford (1991)
2. Bailey, N.T.J.: The Mathematical Theory of Infectious Diseases and its Applications. Charles Griffin and Company Ltd., London (1975)
3. Barabasi, A.L., Albert, R.: Emergence of Scaling in Random Networks. Science 286(5439), 509–512 (1999)
4. Barrett, C.L., Bisset, K.R., Eubank, S.G., Feng, X., Marathe, M.V.: EpiSimdemics: an Efficient Algorithm for Simulating the Spread of Infectious Disease over Large Realistic Social Networks. In: ACM/IEEE Conf. on Supercomputing (2008)
5. Bettstetter, C., Resta, G., Santi, P.: The Node Distribution of the Random Waypoint Mobility Model for Wireless Ad Hoc Networks. IEEE Trans. on Mobile Computing 2(3) (2003)
6. Camp, T., Boleng, J., Davies, V.: A Survey of Mobility Models for Ad Hoc Network Research. Wireless Communication and Mobile Computing, 483–502 (2002)
7. Chakrabarti, D., Wang, Y., Wang, C., Leskovec, J., Faloutsos, C.: Epidemic thresholds in real networks. ACM TISSEC 10(4) (2008)
8. Chakrabarti, D., Leskovec, J., Faloutsos, C., Madden, S., Guestrin, C., Faloutsos, M.: Information Survival Threshold in Sensor and P2P Networks. In: IEEE ICC (2007)
9. Domingos, P., Richardson, M.: Mining the Network Value of Customers. In: ACM SIGKDD 2001, pp. 57–66. ACM Press, New York (2001)
10. Faloutsos, M., Faloutsos, P., Faloutsos, C.: On power-law relationships of the Internet topology. In: SIGCOMM, pp. 251–262 (August-September 1999)
11. Ganesh, A., Massoulie, L., Towsley, D.: The effect of network topology in spread of epidemics. IEEE INFOCOM (2005)
12. Gonsalves, A.: Android Phones Steal Market Share. Information Week (04, 2010), http://www.informationweek.com/news/mobility/smart_phones/showArticle.jhtml?articleID=224201881

13. Hethcote, H.W.: The mathematics of infectious diseases. SIAM Rev. 42(4), 599–653 (2000)
14. Johnson, D.B., Maltz, D.A.: Dynamic Source Routing in Ad Hoc Wireless Networks. In: Mobile Computing (1996)
15. Kephart, J.O., White, S.R.: Directed-Graph Epidemiological Models of Computer Viruses. In: IEEE Security and Privacy, pp. 343–359 (May 1991)
16. Kephart, J.O., White, S.R.: Measuring and Modeling Computer Virus Prevalence. In: IEEE Security and Privacy 1993, pp. 2–15 (May 1993)
17. Kumar, S., Raghavan, P., Rajagopalan, S., Tomkins, A.: Trawling the Web for Emerging Cyber-Communities. Computer Networks 31(11-16), 1481–1493 (1999)
18. Mickens, J.W., Noble, B.D.: Modeling Epidemic Spreading in Mobile Environments. In: ACM WiSe 2005, pp. 77–86. ACM, New York (2005)
19. Mickens, J.W., Noble, B.D.: Analytical models for epidemics in mobile networks. In: IEEE WiMob 2007, p. 77 (2007)
20. Moreno, Y., Pastor-Satorras, R., Vespignani, A.: Epidemic Outbreaks in Complex Heterogeneous Networks. The European Physical Journal B 26, 521–529 (2002)
21. Newman, M.: Threshold Effects for Two Pathogens Spreading on a Network. Phys. Rev. E (2005)
22. Pastor-Satorras, R., Vespignani, A.: Epidemic Dynamics and Endemic States in Complex Networks. Physical Review E 63, 066117 (2001)
23. Pastor-Satorras, R., Vespignani, A.: Epidemic Dynamics in Finite Size Scale-Free Networks. Physical Review E 65, 035108 (2002)
24. Prakash, B.A., Chakrabarti, D., Faloutsos, M., Valler, N., Faloutsos, C.: Got the Flu (or Mumps)? Check the Eigenvalue! arXiv:1004.0060v1 [physics.soc-ph] (2010)
25. Prakash, B.A., Tong, H., Valler, N., Faloutsos, M., Faloutsos, C.: Virus Propagation on Time-Varying Networks: Theory and Immunization Algorithms. In: Balcázar, J.L., Bonchi, F., Gionis, A., Sebag, M. (eds.) ECML PKDD 2010. LNCS, vol. 6323, pp. 99–114. Springer, Heidelberg (2010)
26. Rhee, I., Shin, M., Hong, S., Lee, K., Chong, S.: Human Mobility Patterns and their Impact on Delay Tolerant Networks. In: ACM HotNets IV (2007)
27. Rhee, I., Shin, M., Hong, S., Lee, K., Chong, S.: On the Levy-Walk Nature of Human Mobility. In: IEEE INFOCOM 2008, pp. 924–932. IEEE, Los Alamitos (2008)
28. Ripeanu, M., Foster, I., Iamnitchi, A.: Mapping The Gnutella Network: Properties of Large-Scale Peer-to-Peer Systems And Implications For System Design. IEEE Internet Computing Journal 6(1) (2002)
29. Shlesinger, M.F., Klafter, J., Wong, Y.M.: Random Walks with Infinite Spatial and Temporal Moments. J. Stat. Phys. 27, 499–512 (1982)
30. Van Mieghem, P., Omic, J., Kooij, R.: Virus Spread in Networks. IEEE/ACM Trans. Netw. 17(1), 1–14 (2009)
31. Yan, G., Flores, H.D., Cuellar, L., Hengartner, N., Eidenbenz, S., Vu, V.: Bluetooth Worm Propagation: Mobility Pattern Matters! In: ACM ASIACCS 2007, New York, NY, USA, pp. 32–44 (2007)
32. Yoon, J., Liu, M., Noble, B.: Random Waypoint Considered Harmful. In: INFOCOM 2003, vol. 2, pp. 1312–1321 (2003)

Small Worlds and Rapid Mixing with a Little More Randomness on Random Geometric Graphs

Gunes Ercal

University of Kansas
gunes@ku.edu

Abstract. We theoretically and experimentally analyze the process of adding sparse random links to a random wireless networks modeled as a random geometric graph. While this process has been previously proposed, we are the first to (i) theoretically consider sparse new-wiring (with asymptotically less than constant fraction of nodes allowed very small constant new wired edges), (ii) prove bounds for any new-wiring process upon random geometric graphs, and (iii) consider the effect of such sparse new-wiring upon the spectral gap of the resultant normalized Laplacian. In particular, we consider the following models of adding new wired edges: Divide the normalized space into bins of length $k\frac{r}{2\sqrt{2}} \times k\frac{r}{2\sqrt{2}}$, given that the radius is on the order required to guarantee asymptotic connectivity. For each bin, choose a bin-leader. Let the G_1 new wiring be such that we form a random cubic graph amongst the bin-leaders and superimpose this upon the random geometric graph. Let the G_2 new wiring be such that we form a random 1-out graph amongst the bin-leaders and superimpose this upon the random geometric graph. We prove that the diameter for G_1 is $O(k + log(n))$ with high probability, and the diameter for G_2 is $O(klog(n))$ with high probability, both of which exponentially improve the $\Theta(\sqrt{\frac{n}{\log n}})$ diameter of the random geometric graph, thus also inducing small-world characteristics as the high clustering remains unchanged. Further, we theoretically motivate and experimentally demonstrate that the spectral gap for both G_1 and G_2 are significantly greater than that of the original random geometric graph. These results further motivate future hybrid networks and advances in the use of directional antennas.

Keywords: small world, wireless networks, spectral gap, theory.

1 Introduction and Related Work

Ever since the first observation of "six degrees of separation" by Stanley Milgram [1], small-world phenomenon have been noted in numerous diverse network domains, from the World Wide Web to scientific co-author graphs [2]. The pleasant aspect of the small-world observations is that, despite the high clustering characteristic of relationships with "locality", these various real world networks

J. Domingo-Pascual et al. (Eds.): NETWORKING 2011, Part I, LNCS 6640, pp. 281–293, 2011.
© IFIP International Federation for Information Processing 2011

nonetheless also exhibit short average path lengths as well. This is surprising because purely localized graphs such as low dimensional lattices have very high average path lengths and diameter, whereas purely non-localized graphs such as random edge graph models of Erdos and Renyi [3] exhibit very low clustering coefficient. With intuition consolidating these two extremal graph types, the first theoretical and generative model of small world networks was proposed by Watts and Strogatz [4]: Start with a one dimensional k-lattice, and re-wire every edge to a new uniformly at random neighbor with a small constant probability. They showed that even for a very small but constant re-wiring probability, the resulting graph has small average path lengths while still retaining significant clustering.

Despite the prevalence of small world phenomenon in many real-world networks, wireless networks, in particular ad-hoc and sensor networks, do *not* exhibit the small average path lengths required of small-world networks despite the evident locality arising from the connectivity of geographically nearby nodes. Although taking a high enough broadcast radius r clearly can generate a completely connected graph of diameter one, this is a non-realistic scenario because energy and interference also grow with r. Rather, from a network design and optimization perspective, one must take the smallest reasonable radius from which routing is still guaranteed. To discuss such a radius in the first place, we must employ a formalization which is common in all theoretical work on wireless networks [5,6], namely we fix the *random geometric graph* model of wireless networks. Given parameters n, the number of nodes, and r, the broadcast radius, the random geometric graph $\mathcal{G}(n,r)$ is formed by uniformly at random dispersing the n nodes into the unit square (which is a normalized view of the actual space in which the nodes reside), and then connecting any two nodes iff they are within distance r of each other. Note that due to the normalization of the space, r is naturally viewed as a function of n. Given such a model, it is a seminal result of Gupta and Kumar [6] that the *connectivity* property exhibits a sharp threshold for $\mathcal{G}(n,r)$ at critical radius $r_{con} = \sqrt{\frac{\log n}{\pi n}}$, which also corresponds to an average degree of $\log n$. As connectivity is a minimal requirement for routing, r_{con} is the reference point to take for analysis of $\mathcal{G}(n,r)$, and yet, as we shall see, such a radius still yields average path lengths of $\Theta(\sqrt{\frac{\pi n}{\log n}})$ with high probability.

This serves as a first motivation for the question: In the spirit of small-world generative models [4] that procured short average path lengths from a geographically defined lattice by adding random "long" edges, can we obtain significant reduction in path lengths by adding random "short cut" wired links to a wireless network? The first to ask this question in the wireless context was Ahmed Helmy [7] who experimentally observed that even using a small amount of wires (in comparison to network size n) that are of length at most a quarter of the physical diameter of the network yields significant average path lengths reduction. Another seminal work on this question is that of Cavalcanti et al. [8] which showed that introducing a fraction f of special nodes equipped with two radios, one for short-range transmission and the other for long-range transmission, improves the connectivity of the network, where this property is seen to exhibit a sharp threshold dependent on both the fraction f and the radius r. Other work yet include an optimization approach with a specified sink, in which the placement of wired links is calculated to decrease average path lengths in the resulting topology [9]. The existing body of literature authored by practitioners in the field of wireless networks on inducing small-world characteristics (particularly shortened average path lengths) into wireless networks by either introducing wired links or nodes with special long range radios or directional antennas yields that such hybrid scenarios are eminently reasonable to consider for real networks. Nonetheless, no theoretical bounds have yet been proven for such processes of long-range link addition on random geometric graphs. We establish such bounds in this work for models of sparse random wired link additions, with the sparsity controlled by a parameter k.

In particular, we consider the following models of adding new wired edges: Divide the normalized space into bins of length $k\frac{r}{2\sqrt{2}} \times k\frac{r}{2\sqrt{2}}$, given that the radius is on the order required to guarantee asymptotic connectivity. For each bin, choose a bin-leader. Let the G_1 new wiring be such that we form a random cubic graph amongst the bin-leaders and superimpose this upon the random geometric graph. Let the G_2 new wiring be such that we form a random 1-out graph amongst the bin-leaders and superimpose this upon the random geometric graph. We prove that the diameter for G_1 is $O(k + log(n))$ with high probability, and the diameter for G_2 is $O(klog(n))$ with high probability, both of which exponentially improve the $\Theta(\sqrt{\frac{n}{\log n}})$ diameter of the random geometric graph, thus also inducing small-world characteristics as the high clustering remains unchanged. Our results on resulting average path lengths are also stable in comparison to using a constant fraction of wire lengths, as that in the work of Ahmed Helmy [7]. To see this note that, for example, using a maximum wire length of one-quarter the maximum distance can be simulated by subdividing the unit square into 16 parts and applying results to the parts separately, then combining into a maximum average path length that is still at most 16 of that within each part.

In addition to establishing bounds on resultant path lengths, we are the first to consider the resultant *spectral gap*, namely the difference between the first and second eigenvalues of the normalized adjacency matrix, in comparison to that of the random geometric graph. It is well established that the spectral gap is intrinsically related to the node expansion, edge expansion [10], connectivity, and random walk properties of the given graph [11,12]. Although short average path lengths is necessary for a graph to exhibit optimal random walk sampling properties, it is far from sufficient (the barbell graph being a notable counterexample). As a strange omission, the small world literature thus far has primarily ignored spectral gap as a measure in their analyses and generative models despite the known expansion of random edge graph models.

Our motivation is as follows: Yet another limitation of random geometric graphs in comparison to random edge graph models that is especially problematic for oblivious routing, sampling, and gossiping applications[13,14,15] is that, whereas sparse random regular graphs as well as random connected Erdos-Renyi graphs have a large spectral gap [16,17], connected random geometric graphs $\mathcal{G}(n, \Theta(r_{con}))$ have asymptotically quickly diminishing spectral gap [18,19,20]. In general, additional edges need not improve the spectral gap of the resulting graph. However, the author notes that the *conductance argument* used to bound the spectral gap of random geometric graphs in [19], would no longer immediately yield bad bounds using the same counterexample of dividing down the middle if random edges are added to the graph with any sparseness so long as the number of such edges grows asymptotically in network size. Motivated by this observation, we experimentally calculate the graph spectra, however using the normalized Laplacian which is guaranteed to be symmetric and similarly is intrinsically related to random walk sampling properties [21], and observe that indeed there is an asymptotic improvement in the spectral gap of G_1 and G_2 in comparison to the starting random geometric graph.

Finally, we must note the work [22] of Abraham Flaxman, which is an excellent related work in which the spectra of randomly *perturbed* graphs have been considered in a generality that already encompasses major small-world models thus far. [22] demonstrates that, no matter what is the starting graph G_0, adding random 1-out edges at every node of G_0 will result in a graph with constant spectral gap (the best possible asymptotically). The work also presents a condition in which a random Erdos-Renyi graph superimposed upon the nodes of G_0 would yield good expansion, whereas without that condition the resulting graph may have poor expansion. Despite the apparent generality of the work in terms of the arbitrariness of the underlying graph that is considered, unfortunately the results do not generalize to situations in which not all nodes are involved in a wired linkage. Notably, the small world models thus far also require such a high probability of new random links. In contrast, in this work, we focus on adding sparse random wires. In particular, the fraction of nodes involved in a wired link will be no more than $O(\frac{1}{\log n})$, and in general shall be $O(\frac{1}{k^2 \log n})$, both of which are asymptotically diminishing fractions.

2   Theoretical Results

2.1   Random Geometric Graph Preliminaries

Our analytical results for modifications of random geometric graphs are based on certain "smooth lattice-like" properties that those graphs have. These properties include the uniformity of nodes distribution and the regularity of node degree. As introduced in [20], we utilize the notion of a *geo-dense* graph to characterize such properties, that is, a geometric graph (random or deterministic) with uniform node density across the unit square. It was shown that *random* geometric graphs are *geo-dense* and for radius $r_{\text{reg}} = \Theta(r_{\text{con}})$ all nodes have the same order degree[20]. We formally present the relevant results from [20] in this section, as well as the notion of *bins*, namely the equal size areas that partition the unit square. Such "bins" are the concrete link between lattices and random geometric graphs, essentially forming the lattice backbone of such graphs.

Formally, a **geometric graph** is a graph $G(n,r) = (V,E)$ with $n = |V|$ such that the nodes of V are embedded into the unit square with the property that $e = (u,v) \in E$ if and only if $d(u,v) \leq r$ (where $d(u,v)$ is the Euclidean distance between points u and v). In wireless networks, r naturally corresponds to the broadcast radius of each node. The following formalizes geo-denseness for geometric graphs:

Definition 2.1. [20] Let $G(n,r(n))$ be a geometric graphs[1]. For a constant $\mu \geq 1$ we say that such a class is μ-**geo-dense** if every square bin of size $A \geq r^2/\mu$ (in the unit square) has $\Theta(nA)$ nodes.[2]

The following states the almost *regularity* of geo-dense geometric graphs[20]:

Lemma 2.2. Let $G(n,r)$ be a 2-geo-dense geometric graph with V the set of nodes and E the set of edges. Let $\delta(v)$ denote the degree (i.e number of neighbors) of $v \in V$. Then: (i) $\forall v \in V$ $\delta(v) = \Theta(nr^2)$ and (ii) $m = |E| = \Theta(n^2 r^2)$.

Recall that the critical radius for connectivity r_{con} is s.t $\pi r_{con}^2 = \frac{\log n}{n}$. The following is the relevant lemma that states that *random* geometric graphs with radius at least on the order of that required for connectivity are indeed geo-dense[20]:

Lemma 2.3 (Geo-density of $\mathcal{G}(n,r)$). For constants $c > 1$ and $\mu \geq 2$, if $r^2 = \frac{c\mu \log n}{n}$ then w.h.p. $\mathcal{G}(n,r)$ is μ-geo-dense. That is, w.h.p. (i) any bin area of size r^2/μ in $\mathcal{G}(n,r)$ has $\Theta(\log n)$ nodes, and (ii) $\forall v \in \mathcal{G}(n,r)$, $\delta(v) = \Theta(nr^2)$ and $m = |E| = \Theta(n^2 r^2)$. Further, note that increasing the radius r can only smoothen the distribution further, maintaining regularity.

Given that geo-denseness of connected random geometric graphs is established, we wish to utilize the "binning" directly for its lattice-like properties. As such, we shall introduce the notion of a *lattice skeleton* for geo-dense geometric graphs, including random geometric graphs above connectivity:

Definition 2.4. Let $G(n,r(n))$ be a μ-geo-dense geometric graph for some constant $\mu \geq 1$. Define the μ-lattice skeleton for $G(n,r)$ to be $LS(G(n,r)) = (L,B_{i,j})$ such that $L = (B,E_B)$ is the $\frac{\sqrt{\mu}}{r} \times \frac{\sqrt{\mu}}{r}$ two-dimensional lattice on $\frac{\mu}{r^2}$ bin-points, where each vertex set $B_{i,j}$ represents the set of nodes of $G(n,r)$ that lie in lattice location $\{i,j\}$ of L. Further, for a node $v \in G(n,r)$ with Cartesian coordinates $(x,y) \in [0,1]^2$, denote by $B(v)$ the lattice-bin containing v, namely the bin $B_{i,j}$ such that $i = ceiling(\frac{x\sqrt{\mu}}{r})$ and $j = ceiling(\frac{y\sqrt{\mu}}{r})$.

What is directly clear by geo-denseness is that there is not much variance in the sizes of the bins:

Remark 2.5. Let $LS(G(n,r)) = (L,B_{i,j})$ be the μ-lattice skeleton of a μ-geo-dense geometric graph $G(n,r)$. Then, $\forall i,j, |B_{i,j}| = \Theta(nr^2)$.

---

[1] Either random or deterministic.

[2] Note that if a geometric graph is μ_1-geo-dense then it is also μ_2-geo-dense for any $\mu_2 < \mu_1$. That is, if a distribution is smooth for some granularity x it can only be smoother for a coarser granularity y.

Further, utilizing the choice of $\mu \geq 5$, we may make the stronger statement that the connectivity of the lattice is inherited in the nodes of the overall graph. The justification is simply that r becomes the length of the diagonal connecting the farthest points of adjacent bins, and we formalize combining with Remark 2.5 and Lemma 2.3:

Lemma 2.6. *Let $LS(G(n,r)) = (L, B_{i,j})$ be the 5-lattice skeleton of a 5-geodense geometric graph $G(n,r)$. Then $\forall i,j,k,l$ if $((i,j),(k,l)) \in L$ then $\forall v \in B_{i,j}, u \in B_{k,l}, (v,u) \in G(n,r)$*
In particular, for $c > 1$ if $r^2 \geq \frac{5c}{\pi} r_{con}^2$ then the 5-lattice skeleton $LS(\mathcal{G}(n,r)) = (L, B_{i,j})$ of random geometric graph $\mathcal{G}(n,r)$ satisfies the following:
(i) $\forall i,j |B_{i,j}| = \Theta(nr^2)$ w.h.p.,
(ii) $\forall i,j$ if $v,u \in B_{i,j}$ then $(v,u) \in \mathcal{G}(n,r)$,
(iii) $\forall i,j,k,l$ if link $((i,j),(k,l)) \in L$ then $\forall v \in B_{i,j}, \forall u \in B_{k,l}, (v,u) \in \mathcal{G}(n,r)$, and
(iv) $\forall i,j,k,l$ if $\min\{|i-k|,|j-l|\} \geq 4$ then $\forall v \in B_{i,j}, \forall u \in B_{k,l}, (v,u) \notin \mathcal{G}(n,r)$

From (ii) it is clear that each bin $B_{i,j}$ forms a clique (namely all pairs of nodes within are connected directly by length one paths). From (iii) it follows that a path in the lattice L yields a path in the graph $\mathcal{G}(n,r)$ as well, while (iv) bounds the converse situation in that nodes that lie in bins at least 4 lattice-hops away cannot be directly connected in the graph $\mathcal{G}(n,r)$ either. In particular, (iii) and (iv) yield that pairwise distances between points in the graph $\mathcal{G}(n,r)$ inherit the shortest paths (Manhattan) distances in the corresponding lattice-bins of the lattice-skeleton, up to constant factors. We formalize with the following corollary:

Corollary 2.7. *For $c > 1$ if $r^2 \geq \frac{5c}{\pi} r_{con}^2$ then the 5-lattice skeleton $LS(\mathcal{G}(n,r)) = (L, B_{i,j})$ of random geometric graph $\mathcal{G}(n,r)$ satisfies the following w.h.p.: $\forall u, v \in \mathcal{G}(n,r), dist_{\mathcal{G}(n,r)}(u,v) = \Theta(dist_L(B(u),B(v)) + 1)$ where the function $dist_G$ indicates shortest paths distances in graph G.*

Having established that connectivity and distances for $\mathcal{G}(n,r)$ with radius at least a small constant times r_{con} roughly preserve connectivity and distances in the $\frac{\sqrt{\mu}}{r} \times \frac{\sqrt{\mu}}{r}$ lattice skeleton, let us then consider the number of lattice-nodes $N_{d,L}(v)$ that are at lattice distance exactly d away from v in the lattice L: Clearly, $N_{d,L}(v)$ grows linearly in d by a simple induction on upper and lower bounds. And, the maximum distance to consider is $d = \Theta(\frac{\mu}{r^2})$. Moreover, due to the smooth distribution of random geometric graph nodes in the lattice bins, we must have that the *fraction* $f_{d,L}(B(v))$ of lattice bins at lattice-distance exactly d away from $B(v)$ must be on the same order as the *fraction* of random geometric graph nodes $f_{d,\mathcal{G}(n,r)}(v)$ at hop-distance exactly d away from v. Thus,

$$f_{d,\mathcal{G}(n,r)}(v) = \Theta(f_{d,L}(B(v))) = \Theta(\frac{N_{d,L}(v)}{|L|}) = \Theta(\frac{N_{d,L}(v)r^2}{\mu}) = \Theta(dr^2).$$

Such fractions represent the probability that a node is at distance d away from a given node v. Thus, we may calculate the average path length APL which is the expectation of that probability distribution on the very function d itself:

$$APL(\mathcal{G}(n,r)) \quad = \quad \sum_{d=1}^{D(\mathcal{G}(n,r))} df_{d,\mathcal{G}(n,r)}(v) \quad = \quad \Theta(\sum_{d=1}^{2\frac{\sqrt{\mu}}{r}} df_{d,L}(B(v))) \quad =$$
$$\Theta(\sum_{d=1}^{2\frac{\sqrt{\mu}}{r}} d^2 r^2) = \Theta(r^2 \sum_{d=1}^{2\frac{\sqrt{\mu}}{r}} d^2) = \Theta(r^2 \frac{1}{r^3}) = \Theta(\frac{1}{r})$$

Thus, the average path length for random geometric graphs above the connectivity threshold is the same as the order of the diameter (maximum shortest path lengths) for such graphs, which is $\Theta(\frac{1}{r})$. While the dependence on the radius r in that term may seem optimistic at first, noting that r should be kept as low as possible to reduce energy overhead and interference of the ad-hoc network represented, a realistic constraint on r becomes $r = \Theta(r_{con}) = \Theta(\sqrt{\frac{\log n}{n}})$, namely that achieving degree $\Theta(\log n)$. Thus, APL of reasonable random geometric graphs (of minimal radius guaranteeing connectivity) scales quite badly as $\Theta(\sqrt{\frac{n}{\log n}})$.

2.2 Adding Random Edges

Prior to considering the business of adding random edges to a given initial graph $G_0 = (V_0, E_0)$, note that the set of additional edges E_R and the existing nodes connected by them $V_R \subset V$ forms a graph G_R such that the resulting graph $G = (V, E) = G_0 + G_R$ has $V = V_0 \geq V_R$ and $E = E_0 + E_R$. That is, it is also convenient to view the additional random edges as a new graph G_R superimposed upon the original graph G_0.

Given such a characterization, let us be given a 5-geo-dense geometric graph $G_0 = (V_0, E_0) = G(n,r)$ with 5-lattice-skeleton $(L, B_{i,j})$. In particular, note from Lemma 2.3 that results apply to any $G_0 = \mathcal{G}(n, \Omega(r_{con}))$. Given parameter $k \geq 1$, let vertex set $V_R(k)$ be generated as follows: For any $i, j \leq \frac{\sqrt{5}}{\sqrt{k}r}$ pick a node $v_{i,j}$ uniformly at random from the nodes in the set of bins $\mathcal{B}_{i,j} = \cup_{ki\leq i'\leq(k+1)i, kj\leq j'\leq(k+1)j} B_{i',j'}$[3], and set $v_{i,j} \in V_R(k)$. For the case $k = 0$, let $V_R(0) = V_0$. We now define the various types of random edge sets $E_{R,i}$ for graphs $G_{R,i}(k) = (V_R(k), E_{R,i})$ whose superimpositions upon G_0 we shall consider in this work: Let $E_{R,1}$ be generated as follows: For every node $v \in V_R(k)$ pick 3 neighbors in $V_R(k)$ uniformly at random, discarding situations in which any node has degree greater than 3. Thus, the resulting graph $G_{R,1}(k) = (V_R(k), E_{R,1})$ is the random 3-regular Erdos-Renyi graph defined on vertex set $V_R(k)$. Let $E_{R,2}$ be generated as follows: For every node $v \in V_R(k)$ pick 1 neighbor in V_R uniformly at random. Thus, the resulting graph $G_{R,2}(k) = (V_R(k), E_{R,2})$ is the random 1-out graph defined on vertex set $V_R(k)$.

Similarly to the above, let us define the resulting graphs as follows: Let $G_1 = G_0 + G_{R,1}$. Let $G_2 = G_0 + G_{R,2}$ Essentially, k controls the frequency of special nodes which shall serve as wired link stations. For $k = \Theta(1)$, the frequency is in line exactly with the bins, and thus the occurence of such wired link stations is 1 in every $\Theta(nr^2)$. For $r = \Theta(r_{con})$ that frequency becomes $\Theta(\frac{1}{\log n})$, and for larger broadcast radius it is sparser:

---

[3] This bin union is simply the formalization of a single contiguous bin $k \times k$ as large as the original bins.

Remark 2.8. *Given $k \geq 1$, the frequency of wired link stations $\frac{|V_R(k)|}{|V_0|}$ is $\Theta(\frac{1}{k^2 n r^2})$. Namely, the total number of such stations is $|V_R(k)| = \Theta(\frac{1}{k^2 r^2})$.*[4]

Before proceeding to prove results on average path lengths for $G_1 = G_0 + G_{R,i}$, we note that the manner in which $V_R(k)$ is generated can be simulated approximately by simply choosing a total of logarithmically more wired link stations uniformly at randomly from the original set V_0. This too follows from "coupon collection":

Remark 2.9. *For any k, if every $v \in V_0$ is chosen to be a wired link station with probability $\Theta(\frac{\log(\frac{1}{k^2 r^2})}{n k^2 r^2})$, then, with high probability, for every k^2-bin $\mathcal{B}_{i,j} = \cup_{ki \leq i' \leq (k+1)i, kj \leq j' \leq (k+1)j} \mathcal{B}_{i',j'}$ there exists a vertex $v' \in \mathcal{B}_{i,j}$ such that v' is a wired link station. Moreover, all of the vertices in any k^2 bin are almost-equiprobable and almost-independent whp.*

The last is simply stating that there is not much lost from the uniformly at random choice of a finite set of nodes versus the corresponding Poisson process.

Now, we note that the maximum distance of any node in a k^2 bin to the corresponding wired link station in the k^2 bin is simply bounded by the hop-diameter of the k^2 bin:

Remark 2.10. *Every node is within $\Theta(k)$ hops of the wired link station in its k^2-bin since the k^2-bin simply stretches the Manhattan distances of the original 5-lattice-skeleton by k.*

This remark shall prove relevant in relating inter-node distances in the graph $G = G_0 + G_R$ to inter-node distances in G_R.

2.3   APL and Diameter Bounds for G_1

Thus, we now proceed concerning inter-node distances for the first model $G_{R,1}$:

Remark 2.11. *It is well-known that the random 3-regular graph $G_{R,1}(k)$ is an expander with high probability [10,16]. Therefore, $G_{R,1}(k)$ also exhibits diameter and average path lengths asymptotically at most logarithmic in its vertex set $|V_R(k)|$, with high probability.*

Combining Remarks 2.8, 2.10, and 2.11, we obtain our first bounds on the resulting average and worst-case path lengths:

Theorem 2.12. *The diameter and average path lengths for $G_1(k)$ are as follows: $APL(G_1(k)) = O(Diam(G_1(k))) = O(\min\{2k + \log(\frac{1}{k^2 r^2}), \frac{1}{r}\}) = O(\min\{k + \log(n), \frac{1}{r}\})$.*

Proof. Let $s, t \in V$ be arbitrary. s is within k hops to a node $v \in V_R(k)$, and t is within k hops to a node $w \in V_R(k)$. Moreover, $dist(v, w) = O(\log(|V_R(k)|))$, making the maximum distance at most $2k + \log(|V_R(k)|)$. If r happens to be large, making paths in the original graph (without taking short-cuts in $E_{R,1}$ more convenient, then the maximum distance is the diameter of G_0 which is $\Theta(\frac{1}{r})$.

---

[4] Note that clearly k cannot exceed $\Theta(\frac{1}{r})$.

For *any* broadcast radius r, it is clear that when $k = \Theta(\frac{1}{r})$, which is the maximum allowable k and corresponds to the situation of placing a constant number $\Theta(1)$ wired-stations, there is no asymptotic improvement in diameters and average path lengths. However, for any other intermediate k and the lowest reasonable broadcast radius $r = \Theta(r_{con})$, the following may be noted:

Corollary 2.13. *For $k = O(\log(n))$ and $G_0 = \mathcal{G}(n, \Theta(r_{con}))$, $APL(G_1(k)) = O(Diam(G_1(k))) = O(\log(n))$. On the other hand, for intermediate k such that $k = o(\frac{n}{\log(n)})$ but $k = \omega(\log(n))$, we still obtain asymptotic improvement upon the hop-lengths in G_0 as $APL(G_1(k)) = O(Diam(G_1(k))) = O(k) = o(Diam(G_0))$.*

2.4   APL and Diameter Bounds for G_2

The case for the second model G_2 is not quite as straightforward as that for the first model due to the fact that the random graph $G_{R,2}(k)$ has a positive probability of being disconnected. So, what can we say? It turns out that a lot can be said the moment that $G_{R,2}(k)$ is superimposed directly upon the nodes of *any* connected graph G_{init}:

Theorem 2.14. *[Expansion of $G_{connected} + R_{1-out}$][22] Let G_0 be any connected graph. Then $G = G_0 + R_{1-out}$, which is the graph formed by adding random 1-out edges to every vertex of G_0, is an expander with high probability. In particular, the diameter of G is logarithmic in the vertex set.*

Despite the similarities to the situation for G_1, there is the technical issue that the vertex set for $G_{R,2}(k)$ is not identical to the vertex set for G_0 but rather asymptotically sparser than such. Therefore, we must understand precisely which graph is an expander, and what that yields for the graph G_2. In particular, we need to extract a connected graph $G_{base} = (V_{base}, E_{base})$ that is both a relevant function of the original graph G_0 and such that there is a one-to-one meaningful correspondence between the vertex set V_{base} and $V_R(k)$ to allow application of Theorem 2.14.

Visually, if we could just contract the lattice-sub-skeleton formed from the k^2-bins into just a $|V_R(k)|$ node lattice, and preserve the meaning of such a contraction for paths in the original graph G_0, then we could apply Theorem 2.14 to our contraction, and then reverse the contraction. We may lose the property of constant expansion upon reversing our contraction for some k, but we will still preserve a bound on path lengths. First, let us state the result, then the proof along the idea above:

Theorem 2.15. *The diameter and average path lengths for $G_2(k)$ are as follows: $APL(G_2(k)) = O(Diam(G_2(k))) = O(\min\{k\log(|V_R(k)|), \frac{1}{r}\}) = O(k\log(\frac{1}{r}))$.*

Prior to the theorem proof, let us note the immediate corollary for low broadcast radius, which again notes an exponential improvement in diameter and average path lengths in for poly-logarithmic k:

Corollary 2.16. *For* $k = \Theta(1)$ *and* $G_0 = \mathcal{G}(n, \Theta(r_{con}))$, $APL(G_2(k)) = O(Diam(G_2(k))) = O(\log(n))$. *On the other hand, for intermediate* k *such that* $k = o(\frac{n}{\log^2(n)})$ *but* $k = \omega(1)$, *we still obtain asymptotic improvement upon the hop-lengths in* G_0 *as* $APL(G_2(k)) = O(Diam(G_2(k))) = O(k \log n) = o(\frac{n}{\log(n)}) = o(Diam(G_0))$.

Now to the theorem proof:

Proof (Proof of Theorem 2.15). Consider the graph of node-contractions $G_{contract} = (V_{contract}, E_{contract})$ such that each vertex $v_{i,j} \in V_{contract}$ is precisely the *set* of vertices in k^2 bin $\mathcal{B}_{i,j}$ and the edge set $E_{contract}$ is the two-dimensional lattice appropriately defined on $V_{contract}$. We know from Theorem 2.14 that upon adding random 1-out edges from every $v_{i,j} \in V_{contract}$ we obtain a graph $G' = G_{contract} + R_{1-out}$ with diameter logarithmic in $V_{contract}$. What does this mean?

Consider any two nodes $s, t \in G_0$, say that $s \in \mathcal{B}_{i,j}, t \in \mathcal{B}_{i',j'}$. Let $l = < v_1, v_2, v_3, \ldots, v_{p-1}, v_p >$ be the shortest path between $v_{i,j}$ and $v_{i',j'}$ in $G_{contract}$ (with $v_1 = v_{i,j}, v_p = v_{i',j'}$). First of all, there is an obvious one-to-one correspondence between the vertex sets $V_{contract}$ and $V_R(k)$ which does not change the incidence of the random short-cut edges. The only issue occurs when paths in the contracted graph include a lattice edge, and in that case there is a factor $O(k)$ blow-up in the hop number for the original graph. Thus, for any path l in $G_{contract}$ we may inductively construct the following path l' in $G_{R,2}$ by a sequence of valid sub-path replacements in the place of contracted nodes and the edges between them until no contracted nodes and no contracted edges remain (meaning all nodes are actual vertices of our original graph):

– Base: In accordance with Remark 2.10, replace v_1 in l with the $O(k)$ length shortest-path from s to $v_{i,j}$ in G_0. Similarly, replace v_p in l with the $O(k)$ length shortest-path from $v_{i',j'}$ to t in G_0.

– Inductive case I: For $1 < t < p - 1$, let v_t and v_{t+1} be adjacent nodes in the path l that also remain thus far in our path construction. Uniquely define $v_{x,y} \in v_t, v_{x',y'} \in v_{t+1}$ for $v_{x,y}, v_{x',y'} \in V_R(k)$. Either edge (v_t, v_{t+1}) is a lattice edge or a random edge. If it is a random edge, then replace (v_t, v_{t+1}) of l with the valid wired edge $(v_{x,y}, v_{x',y'})$. Otherwise, if it is a lattice-edge, then there exists a $O(k)$ length shortest-path P between $v_{x,y}$ and $v_{x',y'}$ in G_0 from Lemma 2.6 and Remark 2.10. Thus, in that case, replace (v_t, v_{t+1}) with path P between $v_{x,y}$ and $v_{x',y'}$.

– Inductive case II: Let node $v_{x,y}$ be the node following initial wired-station $v_{i,j}$ in the construction thus far. Due to the construction, it must be that $v_{i,j} \in v_1, v_{x,y} \in v_2$. Therefore, similarly to Inductive case I, either there exists a random wired edge between $v_{i,j}$ and $v_{x,y}$, or there is a length $O(k)$ path between the two in G_0: Replace accordingly. Operate similarly in making the valid replacement between the node $v_{x',y'}$ preceding final wired station $v_{i'j'}$.

Clearly, the above construction is valid, replacing contracted edges with valid sub-paths at each step. Moreover, at every step, the replacement path is at most a $O(k)$ blow up of a contracted edge. Therefore, the constructed valid path in $G_{R,2}$ is also at most a $O(k)$ multiple of the diameter of $G_{contract}$, which we know by Theorem 2.14 to be logarithmic in $|V_R(k)|$, finalizing the proof.

3   Experimental Results

Experiments were conducted for networks of 100 to 1620 nodes. As consistent with the model of the theoretical section, the parameter k was chosen to be 2, and the radius was chosen to be r_{con} exactly. Disconnected $\mathcal{G}(n, r)$ were discarded from consideration. The results can be seen in Figure 1, where the Y-axis is the spectral gap of the normalized Laplacian. Notably, the spectral gap for the random geometric graph approaches zero quickly, whereas the spectra gap for G_1 and G_2 appear to diminish concavely after 500 nodes, possibly even approaching a small constant lower bound. Moreover, note that the number of wired-nodes in comparison to the network size n for n values of 100, 300, 800, 1000, 1300, and 1620 are respectively as follows: 36, 64, 196, 256, 256, 324. The fraction of wired nodes for the network size of 1620 was just $\frac{1}{5}$.

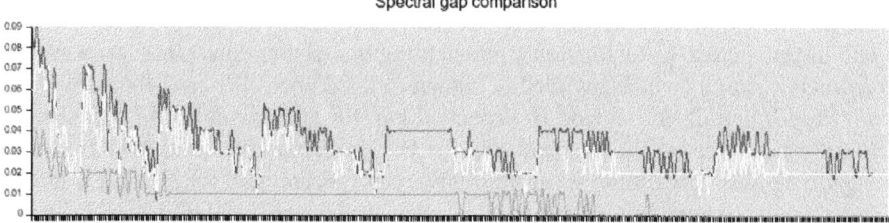

Fig. 1. Spectral Gap comparison for G_1 (Red), G_2 (Yellow), and $\mathcal{G}(n, r)$ (Blue) for network size from 100 to 1620 nodes, radius r_{con}, and k parameter of 2

4   Conclusion and Future Work

We have considered the "small world" effects of adding sparse random edges to an asymptotically diminishing frequency of nodes taken as wired link stations imposed upon a wireless network. In contrast to prior small-worlds work, we care in particular for very sparse link additions and control the sparseness of wired link stations by a parameter k. We establish theoretical proof that under two reasonable models of link additions (G_1 of superimposing a random cubic graph and G_2 of superimposing a random $1 - out$ graph) upon the random geometric graph model of wireless networks an exponential reduction of diameter and average path lengths is obtained even when the wired link stations occur with an inverse poly-logarithmic frequency. We also establish theoretical bounds on the diameter and average path lengths reduction as a general function of k (see Theorems 2.12 and 2.15).

As a related but separate question from the issue of average path lengths, we investigated the improvement in *spectral gap* as well under such models of sparse link additions. The spectral gap, which can refer to either the discrepancy between the first and second eigenvalues of the normalized adjacency matrix representing the graph or to the related measure of the second eigenvalue of the normalized Laplacian representing the graph in the cases of irregular graphs, is known to govern the rate at which the Markov chain corresponding to a natural random walk on the graph converges to the stationary distribution [11,12]: The larger the spectral gap, the faster the convergence to stationarity (namely, the faster the *mixing rate*). It is intuitively clear that the faster the convergence to stationarity for such an oblivious process as a random walk, the greater the intrinsic load-balancing properties of the graph. This quantity of mixing rate is further intricately related to connectivity and resilience properties of the underlying graph measured as the combinatorial quantity of *expansion*. Expansion directly relates to resilience against edge failures, and when the graph is close to regular it also relates to resilience against node failures. In fact one characterization of an *expander graph family* is directly based on spectral gap alone: For a family of graphs parametrized by the size of the vertex set n, the graph family is an *expander graph family* iff the spectral gap is bounded away from zero by a constant, as n approaches ∞. Expander graphs represent extrema for the expansion property. While short average path lengths and short diameter are necessary for graphs to be expanders, these are far from sufficient. Thus, in investigating the improvement in expansion upon adding sparse random edges to a random geometric graph (which by itself is known to have horrible expansion [18,19,20]), we experimentally measured the spectral gap for the normalized Laplacian. We found that whereas the spectral gap for the pure random geometric graph falls rapidly to zero after some n (around 900), the spectral gap for G_1 and G_2 do not fall rapidly to zero but stay bounded away from zero by 0.03 for the duration of the experiment til $n = 1620$. These results are for the sparseness parameter $k = 2$. In either case, a definite discrepancy between the pure wireless spectral gap and the additional link models is exhibited.

Future work concerns establishing theoretical bounds on the spectral gap for G_1 and G_2 as a function of the parameter k. In particular: Is there a k for which G_1 or G_2 is an expander graph family? If so, does the expansion property exhibit a *sharp threshold* based on k? In addition to theoretical bounds, future work includes experimental measures of oblivious routing efficiency and load balancing quality based on random walk simulations of the respective graph families. And, finally, towards answering the question of improvement in resilience for the respective graph families, simulations of non-oblivious routing algorithms in the presence of node and edge failures may also be performed.

References

1. Milgram, S.: The small world problem. Psychology Today 2(1), 60–67 (1967)
2. Newman, M., Barabási, A.L., Watts, D.J.: The Structure and Dynamics of Networks. Princeton University Press, Princeton (2006)

3. Bollobás, B.: Random Graphs. Academic Press, Orlando (1985)
4. Watts, D., Strogatz, S.: Collective dynamics of small-world networks. Nature 393(6684), 440–442 (1998)
5. Penrose, M.D.: Random Geometric Graphs. Oxford Studies in Probability, vol. 5. Oxford University Press, Oxford (2003)
6. Gupta, P., Kumar, P.R.: Critical power for asymptotic connectivity in wireless networks. In: Stochastic Analysis, Control, Optimization and Applications: A Volume in Honor of W. H. Fleming, pp. 547–566 (1998)
7. Helmy, A.: Small worlds in wireless networks. IEEE Communications Letters 7(10), 490–492 (2003)
8. Cavalcanti, D., Agrawal, D., Kelner, J., Sadok, D.F.H.: Exploiting the small-world effect to increase connectivity in wireless ad hoc networks. In: de Souza, J.N., Dini, P., Lorenz, P. (eds.) ICT 2004. LNCS, vol. 3124, pp. 388–393. Springer, Heidelberg (2004)
9. Ye, X., Xu, L., Lin, L.: Small-world model based topology optimization in wireless sensor network. In: Proceedings of the 2008 International Symposium on Information Science and Engieering, vol. 01, pp. 102–106. IEEE Computer Society, Washington, DC (2008)
10. Alon, N.: Eigenvalues and expanders. Combinatorica 6(2), 83–96 (1986)
11. Sinclair, A.: Improved bounds for mixing rates of markov chains and multicommodity flow. Combinatorics, Probability and Computing 1, 351–370 (1992)
12. Jerrum, M., Sinclair, A.: The markov chain monte carlo method: an approach to approximate counting and integration. In: Hochbaum, D. (ed.) Approximations for NP-hard Problems, pp. 482–520. PWS Publishing, Boston (1997)
13. Kempe, D., Dobra, A., Gehrke, J.: Gossip-based computation of aggregate information. In: Proc. of the 44th Annual IEEE Symposium on Foundations of Computer Science, pp. 482–491 (2003)
14. Servetto, S.D., Barrenechea, G.: Constrained random walks on random graphs: routing algorithms for large scale wireless sensor networks. In: Proc. of the 1st Int. Workshop on Wireless Sensor Networks and Applications, pp. 12–21 (2002)
15. Gkantsidis, C., Mihail, M., Saberi, A.: Random walks in peer-to-peer networks. In: Proc. 23 Annual Joint Conference of the IEEE Computer and Communications Societies, INFOCOM (2004)
16. Friedman, J., Kahn, J., Szemerédi, E.: On the second eigenvalue of random regular graphs. In: Proceedings of the Twenty-first Annual ACM Symposium on Theory of Computing, STOC 1989, pp. 587–598. ACM, New York (1989)
17. Füredi, Z., Komlós, J.: The eigenvalues of random symmetric matrices. Combinatorica 1, 233–241 (1981)
18. Boyd, S., Ghosh, A., Prabhakar, B., Shah, D.: Randomized gossip algorithms. IEEE/ACM Trans. Netw. 14(SI), 2508–2530 (2006)
19. Avin, C., Ercal, G.: Bounds on the mixing time and partial cover of ad-hoc and sensor networks. In: Proceedings of the 2nd European Workshop on Wireless Sensor Networks (EWSN 2005), pp. 1–12 (2005)
20. Avin, C., Ercal, G.: On the cover time and mixing time of random geometric graphs. Theor. Comput. Sci. 380(1-2), 2–22 (2007)
21. Chung, F.: Spectral Graph Theory. American Mathematical Society, Providence (1997)
22. Flaxman, A.D.: Expansion and lack thereof in randomly perturbed graphs. Internet Mathematics 4(2), 131–147 (2007)

A Random Walk Approach to Modeling the Dynamics of the Blogosphere

Muhammad Zubair Shafiq and Alex X. Liu

Department of Computer Science and Engineering,
Michigan State University,
East Lansing, MI 48824, U.S.A.
{shafiqmu,alexliu}@cse.msu.edu

Abstract. It is important to develop intuitive and tractable generative models to simulate the topological and temporal dynamics of the blogosphere because these models provide insights about its structural evolution. In such generative models, independent instances of individual bloggers are initiated and these instances interact with each other to simulate the evolution of the blogosphere. Existing generative models of the blogosphere have certain limitations: (1) they do not simultaneously consider the topological and temporal properties, or (2) they utilize the global information about the blogosphere that is typically not available. In this paper, we propose a novel generative model for the blogosphere based on the random walk process that simultaneously considers both the topological and temporal properties and does not utilize the global information about the blogosphere. The results of our experiments show that the proposed random walk based model successfully captures the scale-free nature of both topological and temporal dynamics of the blogosphere.

Keywords: Blogosphere, Random walks, Network Science.

1 Introduction

1.1 Background

The blogosphere constitutes an important niche of online social networks. It consists of blogs and each blog usually contains several posts. As shown in Figure 1, the blogosphere can be envisioned as consisting of two separate networks: (1) blog network, and (2) post network. In these networks, nodes represent blogs and posts respectively; whereas edges are directed hyperlinks from a source node to a destination node.

In recent years, researchers have carried out independent measurement-driven studies to understand various global properties of the blogosphere [6, 7, 9, 12]. These studies have reported several interesting global patterns in the blogosphere, such as distributions of post in-degree, blog in-degree, size of cascades, inter-posting times, and post popularity over time follow power-laws. These discoveries have important implications in terms of evolution of network graphs and information flow, search, and retrieval.

J. Domingo-Pascual et al. (Eds.): NETWORKING 2011, Part I, LNCS 6640, pp. 294–306, 2011.

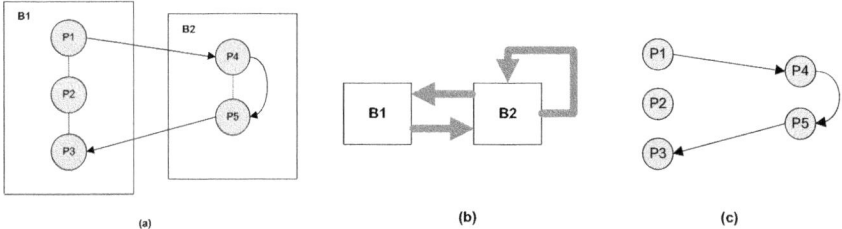

Fig. 1. (a) The blogosphere can be visualized as two networks: (b) blog network, (c) post network

1.2   Problem Statement

A significant amount of research effort has been put in to understand the global properties of the blogosphere. However, little is yet known about the underlying local mechanisms that have led to these topological and temporal patterns at the global scale. One way to understand these local mechanisms is to develop a formal model of individual bloggers in the blogosphere. This model is then replicated for all bloggers and allowed to execute over a given period of time. Finally, global properties of the evolved synthetic network graph are computed and matched to those of real network graphs. A high degree of correlation in compared metrics validate the developed model. In this paper, we aim to develop a realistic, intuitive, and tractable generative model for the blogosphere.

1.3   Proposed Approach

In this paper, we present a generative model based on the random walk process to simulate the evolution of the blogosphere. This model is intuitive and it produces realistic blogosphere networks whose topological and temporal properties match to those discovered in real-world networks. It has the following desirable properties. First, it controls the subtle structural properties of the evolved graph. Second, it does not utilize any global information that may not be locally visible to individual bloggers. Finally, the scale-free distributions evolve only *implicitly* as our model does not explicitly bias link creation probability to nodes' attributes.

1.4   Key Contributions

We summarize our key contributions as follows:

1. We propose a novel multiple random walk based model that can evolve realistic blogosphere networks by mimicking the micro-level interactions of individual bloggers under real-world constraints. To the best of our knowledge, this is the first generative model that *utilizes only locally visible information* to accomplish this task.
2. We propose a novel methodology to quantify the burstiness of blogging activity. Towards this end, we provide a formal analysis of the burstiness of a finitely-bounded, one-dimensional random walk.

3. Our proposed model has the ability to control different structural properties of the generated blogosphere network by varying the length of the random walk.

The rest of the paper is organized as follows. We review the related work in Section 2. In Section 3, we provide background and overview of the proposed random walk based model. In Section 4, we present formal analysis and details of our proposed generative model. In Section 5, we provide experimental results. Finally, we give concluding remarks in Section 6.

2   Related Work

Several research studies have recently focused on the formal modeling of the blogosphere. Some studies have focused on developing formal models of individual bloggers that can evolve graphs with properties similar to those of the real-world blogosphere graphs. Researchers have also tried to model the topological and temporal patterns of the blogosphere. The relevant research studies can be categorized into three types based on the patterns that they analyze or model. We now present a brief overview of each category.

2.1   Topological Properties

In [8], the authors proposed a generative model for the blogosphere called second space. Second space uses several parameters to generate graphs with properties similar to those of real-world graphs. The authors computed multiple topological properties of the simulated blogosphere such as degree distributions, diameter, reciprocity, *etc.* The model developed by second space does not simultaneously consider topological and temporal patterns, which is a major limitation. Moreover, it has a lot of input parameters, which is undesirable. In another study [10, 11], the authors used Kronecker graphs to model the structure of networks. They further proposed KRONFIT algorithm to fit a Kronecker graph generation model to real networks. However, this approach is not specifically designed for modeling the blogosphere.

2.2   Temporal Properties

In [9], the authors performed an analysis of time evolution of the blogosphere using time graphs. They observed that the blogosphere expanded in a bursty manner. Using this observation, the authors created a graph model called the randomized blogspace. This model focuses on temporal patterns observed in the evolution of the blogosphere and does not generalize to other patterns.

2.3   Topological and Temporal Properties

In [5], the authors have proposed the zero-crossing ($\mathcal{ZC}$) model to generate synthetic blogospheres. It is the first work to jointly model temporal and topological properties of the blogosphere. This model also has an intuitive appeal as all procedures closely relate to the mentality of real-world bloggers. The blogosphere generated by $\mathcal{ZC}$ model conforms with several global patterns including but not limited to post in-degree, blog in-degree, and inter-posting times. However, $\mathcal{ZC}$ model has several limitations. First, It cannot control other important structural properties of the evolved graphs such as clustering, centrality, connected components, *etc.* Second, it uses global information (such as total number of in-links to a blog) that is not locally visible to individual bloggers. Finally, the power-laws are explicitly programmed into the model that reduces its intuitive appeal. For example, the probability of creating links to other posts is proportional to their number of in-links (defined as *preferential attachment* process which automatically leads to power laws). In a more intuitive model, a blogger will parse the existing network graph while probabilistically creating new links without any explicit bias.

3 Overview of the Proposed Model

In this section, we first provide the goals of our proposed model and then present its overview. It has been shown earlier that several topological patterns of real-world blogosphere, including blog degree distributions, post degree distributions, inter-posting times, *etc.* follow the power-law distribution [4,15,16,17]. The first goal of our random walk based model is to implicitly produce these power-laws using only local interactions between individual bloggers. We also aim to reproduce the burstiness and self-similarity in the temporal dynamics of the blogosphere using the random walk process [18]. Finally, we want to control various properties of the evolved blogosphere without using multiple parameters.

We now provide an overview of our proposed random walk based model. The *random walk* process constitutes the core of our proposed model [1,14]. To explain the random walk process, an analogy often cited in the literature is of a drunkard walking in city streets with junctions. At each junction, the drunkard chooses one of the possible paths randomly. The rationale behind using the random walk procedure to model the dynamics of the blogosphere is to mimic the local behavior of bloggers in blog reading and writing. When multiple instances of the same random walk based model are invoked, they interact with each other to generate a blogosphere network.

Figure 2 shows an overview of the proposed random walk based model called $\mathrm{m}\mathcal{RW}$. Each instance of $\mathrm{m}\mathcal{RW}$ model corresponds to an individual blogger. $\mathrm{m}\mathcal{RW}$ model consists of five modules: $\mathcal{RW}$-1, $\mathcal{RW}$-2, $\mathcal{RW}$-3, $\mathcal{RW}$-4, and $\mathcal{RW}$-5. $\mathcal{RW}$-1 is a random walk on the discrete one-dimensional number line and aims to mimic the posting behavior of a blogger. A blogger creates a post P every time the zero mark on the number line is crossed.

A blogger has two options after creating a post in $\mathcal{RW}$-1. Either with probability p_l, a blogger decides to include out-links in the created post. Or with probability $1 - p_l$, a blogger publishes a post without any out-links. If a blogger decides to include out-links in the created post, it further has to choose between $\mathcal{RW}$-2 and $\mathcal{RW}$-3, with probabilities $1 - p_e$ and p_e respectively. In $\mathcal{RW}$-2, also called *explore mode*, a blogger aims to create a link to the blog to which it

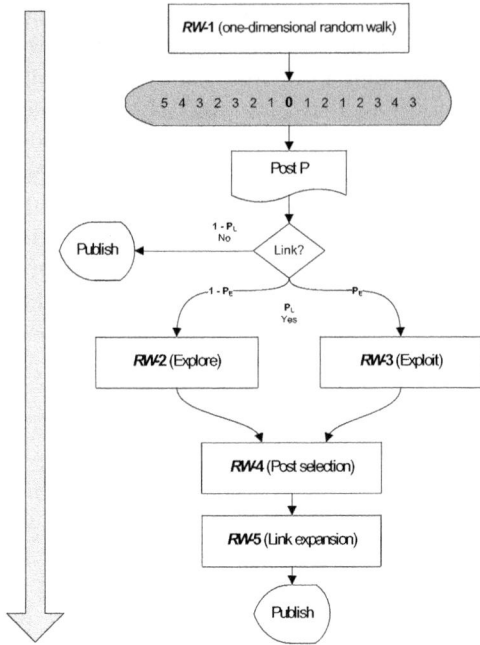

Fig. 2. Overview of the proposed random walk based model

has not linked earlier. In $\mathcal{RW}$-3, also called *exploit mode*, a blogger aims to randomly revisit a previously linked blog. In contrast to $\mathcal{RW}$-1, $\mathcal{RW}$-2 and $\mathcal{RW}$-3 are random walks on the blog graph. At each step of these random walks, a blogger moves to one of the neighbors of the current blog graph node and the blog reached at the end of a finite length random walk is selected for link creation. Furthermore, note that p_l and p_e are randomly sampled from a uniform distribution for every blogger and thus are not input parameters of our proposed model.

After a blog is selected for link creation, a blogger has to choose one of its posts by $\mathcal{RW}$-4. $\mathcal{RW}$-4 is a random walk carried out on the post list of the selected blog starting at the most recent post. The post reached at the end of a finite length random walk is eventually linked in the post P. Finally in $\mathcal{RW}$-5, also called *link expansion*, the blogger conducts a random walk to further link to an arbitrary number of posts referred by the selected post. In contrast to $\mathcal{RW}$-2 and $\mathcal{RW}$-3, $\mathcal{RW}$-4 and $\mathcal{RW}$-5 are carried out on the post graph rather than the blog graph.

In the next section, we provide the details and formal analysis of all five random walks. Here, we iterate that the only input parameter of our $m\mathcal{RW}$ model is the length of the random walk, which controls the properties of the generated blogosphere networks. Moreover, our proposed model simulates local interactions between individual bloggers without utilizing any global information that is not available in the real-world.

4   Formal Analysis of the Proposed Model

4.1   Formal Definitions and Notations

We first formally define the basic concepts and notations that will be utilized in our m$\mathcal{RW}$ model. The blogosphere can be mapped to two separate directed and simple graphs, blog graph $\mathbb{B} = (V_B, E_B)$ and post graph $\mathbb{P} = (V_P, E_P)$. V_B and V_P are sets of *vertices* and E_B and E_P are sets of *edges* for blog and post graphs, respectively. The elements of sets E_B and E_P (edges) are ordered pairs of the form $e = (v_1, v_2)$, where v_1 is source vertex and v_2 is destination vertex, because both $\mathbb{B}$ and $\mathbb{P}$ are directed graphs. A *path* $\mathcal{P}$ of finite length l from vertex u to v on graph is defined as the sequence of edges $e_1, e_2, ..., e_n$. A path does not have any repeated edges. A *walk* $\mathcal{W}$ is also defined as a sequence $e_1, e_2, e_3, ..., e_{n-1}, e_n$, however, it can have repeated edges. Let $E_i = \{e_1, e_2, ..., e_j\}$ denote the set of all possible edges for i^{th} step $(0 < i \leq l)$ of a walk. A walk is called *random walk* $(\mathcal{RW})$ if, at every step i of the random walk, the next edge is uniformly randomly selected from E_i. We now separately provide formal analysis of all five modules of our proposed m$\mathcal{RW}$ model.

4.2 $\mathcal{RW}$-1: One-Dimensional Random Walk Modeling the Temporal Dynamics of Posting Behavior

$\mathcal{RW}$-1 module models the temporal dynamics of a blogger's posting behavior. $\mathcal{RW}$-1 is a random walk on a discrete number line starting at origin. Let $s(t)$ denote the position of a random walker at time tick t. A fair coin is flipped at each time tick $(t \geq 0, s(0) = 0)$. For all head outcomes the position is incremented by a unit $(s(t+1) = s(t) + 1)$, and for tail outcomes the position is decremented by a unit $(s(t+1) = s(t) - 1)$. We also finitely bound the length of random walk. In our initial analysis we assume that the bound is fairly large *i.e.* $n \rightarrow \infty$. For the time tick when current position crosses origin (*i.e.* $s(t) = 0$) a post is published by a blogger. This is also known as *zero-crossing*. Let t_{origin} denote this time tick. Consequently, inter-posting time is denoted by $t_{origin} - 0 = t_{origin}$. In the following text, we provide a mathematical analysis of two important properties of $\mathcal{RW}$-1.

Probability Distribution of Inter-posting Times: We want to find the probability distribution function of zero-crossing, *i.e.* returning to origin. Note

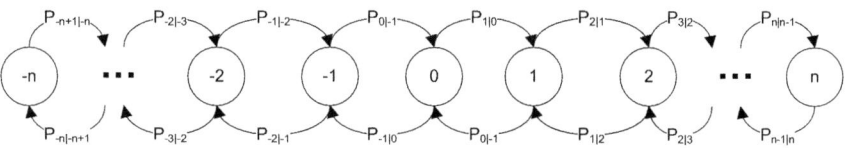

Fig. 3. $\mathcal{RW}$-1: One-dimensional random walk

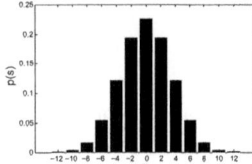 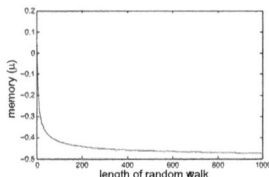 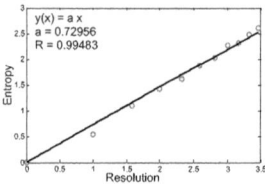

(a) Probability distribution (b) Variation in memory (c) Entropy plot to char-
function of final position af- with respect to walk length acterize self-similarity in
ter $l(=12)$ steps   posting times

Fig. 4. Properties of $\mathcal{RW}$-1 module

that the probability of returning to origin is zero for odd number of time ticks.
Also the probability of zero-crossing is $2^{-t}\binom{t}{t/2}$ for even number of time ticks.
It can be shown after applying Stirling's formula to inter-posting time: $p \propto t^{-a}$,
where a is a positive constant. This shows that the distribution of inter-posting
time should follow a power law in the graphs generated by our model.

Burstiness of Inter-posting Times: To mathematically model burstiness,
we model the probability distribution function of zero-crossing after l steps. Towards this end, we first model the random walk process using a Markov chain
shown in Figure 3. Note that $n \gg l$ and conditional probabilities $P_{0|1}, P_{1|0}, \ldots =$
$\frac{1}{2}$. It can be empirically shown using the Pascal's triangle that the probability
distribution of landing at state s after l steps decays quickly for the increasing
values of s (see Figure 4(a)) [14]. Equivalently, we can show using the Stir-
ling's formula that: $P(s,l) = \frac{2}{\sqrt{2\pi l}} e^{\frac{-s^2}{2l}}$. Since we are interested in finding the
probability distribution of zero-crossing, $i.e.$ $s = 0$. Hence, we can simplify this
relationship to: $P(0,l) = \frac{2}{\sqrt{2\pi l}}$.

In order to model the burstiness of zero-crossing process, we collapse all non-
zero states to a single state for simplification. Moreover, we exploit the symmetry
of the problem to combine positive and negative states. As a result, we get a
1st order, 2 state, discrete time Markov chain. It is also known as the Gilbert-
Elliot model [13]. The states here represent the position after l steps. Using the
above-mentioned model, state 0 refers to origin on the number line and state 1
jointly models the rest of non-origin positions. A suitable measure to model the
burstiness of the Gilbert-Elliot model was proposed in [13]. It is called memory
μ and is defined as: $\mu = 1 - P_{0|1} - P_{1|0}$, where $-1 \le \mu \le 1$. $\mu = 0$ corresponds to
zero memory, $\mu > 0$ corresponds to $persistent$ memory, and $\mu < 0$ corresponds to
$oscillatory$ memory. Note that $P_{1|0} = P_{0|0} = 0.5$ and $P_{0|1} = 1 - P_{1|1}$; therefore,
the relationship for μ can be reduced to: $\mu = P_{1|1} - \frac{1}{2}$. After plugging the
probability of staying in state 1 after l steps from our previous analysis (while
assuming homogeneity), we get: $\mu = \frac{2}{\sqrt{2\pi l}} - \frac{1}{2}$. Figure 4(b) shows the variation
in memory of zero-crossing with respect to walk length. Note that $\mu \to -0.5$ as
$l \to \infty$.

Given the posting times in a real-world blog network, a well-known method
to empirically quantify the burstiness of inter-posting times is using the entropy

plot. The entropy plot of a self-similar process is linear and the slope of the entropy plot quantifies the burstiness. The slope of the entropy plot varies in the range $[0, 1]$, where 0 refers to a highly bursty process and 1 indicates a periodic (non-bursty) process. In [5], the authors showed that the slope of the entropy plot was approximately 0.88 for a sample of the blogosphere, which is in the middle of the two extremes. This observation could not be explained by their proposed model.

On the other hand, the burstiness of inter-posting times predicted by our proposed model varies as the function of the length of random walk. Therefore, we can tune this parameter to generate realistic blogosphere networks. To perform a one-to-one comparison between the burstiness predicted using our proposed model and that empirically measured using the entropy plot, we generate a sample blogosphere with a total of $100,000$ blogs, $500,000$ posts, and the length of random walk was set to 10. Figure 4(c) shows the entropy plot of inter-posting timings in a blogosphere network generated using $\mathfrak{m}\mathcal{RW}$ model. It is evident from the regression fit that the plot is almost linear while the slope of the fit is about $0.7 < 1$, which is close to what was observed for the real-world blogosphere in [5]. Furthermore, our model can also establish the type of burstiness. The negative sign shows that the burstiness of zero-crossing process is in fact oscillatory. This shows that the probability of remaining in a given state is lower than the steady-state probability of being in that state [13].

4.3 $\mathcal{RW}$-2/3: Random Walk on Blog Graph (Explore/Exploit)

$\mathcal{RW}$-2 and $\mathcal{RW}$-3 modules model the linking behavior of a blogger. Some previous studies have proposed to choose blogs proportional to their in-degree [5]. However, in real-world scenarios, a blogger does not know all blogs linking to a given blog. Rather, only a small fraction of in-linking blogs may be known. In fact, many well-known blogging platforms, such as Blogger and Wordpress, explicitly provide some of the in-linking blogs. We conclude that the total number of in-links to a blog is a global information and is not locally visible to an individual blogger. Therefore, a realistic and intuitive generative model of the blogosphere should only use information locally visible to individual bloggers to mimic their linking behavior. In this paper, we propose to use random walk on the directed blog graph to model the linking behavior of bloggers while using only locally visible information. The properties of random walks on graphs have been well-studied in previous research [16, 17].

$\mathcal{RW}$-2 module mimics the explore mode of a blogger and initiates at a randomly chosen vertex v in the blog graph. The blog reached at the end of a random walk is selected. The length of random walk l for every blogger is selected at the start of blogging activity. For the ith step of random walk $(i \leq l)$, let $\mathbb{O}_v(i)$ and $\mathbb{I}_v(i)$ denote the set of all outgoing and incoming links respectively for vertex $v(i)$. Also, let $\mathcal{I}(i) \in \mathbb{I}_v(i)$ denote the singleton set containing the incoming traversal link for i^{th} step of the random walk. The outgoing traversal link at step i is uniform-randomly chosen from the set of links $\mathbb{S}_i$, which is defined as: $\mathbb{S} = \mathbb{O}_v(i) \cup \mathcal{I}(i)$. $\mathcal{I}(i)$ provides 'immediate backtrack' functionality via incoming traversal link.

$\mathcal{RW}$-3 module captures the "exploit" mode of a blogger. It is the same as $\mathcal{RW}$-2 module except for the fact that the walk always starts at the corresponding node of blogger conducting the walk. We show later in Section 4.6 that the probability of link creation in both $\mathcal{RW}$-2 and $\mathcal{RW}$-3 is proportional to in-degree while using only locally visible information.

4.4 $\mathcal{RW}$-4: Random Walk on Reverse Chronological Post List

In $\mathcal{RW}$-4 module, a blogger chooses a post from the blog selected in modules $\mathcal{RW}$-2 or $\mathcal{RW}$-3. The post list is typically ordered in the reverse-chronological order and the links between posts are undirected. The length of random walk l in $\mathcal{RW}$-4 is the same as used in $\mathcal{RW}$-1/2/3. The random walk is always initiated at the post $P1$ which is denoted by $v(1)$. Note that $deg(v(i)) = 2$, $\forall i \in \{2, 3, ..., l-1\}$.

The formal analysis of $\mathcal{RW}$-4 is similar to that of $\mathcal{RW}$-1. $\mathcal{RW}$-4 is essentially one-sided subset of $\mathcal{RW}$-1 if $l \gg k$, where k is the size of post list. The distribution of final position s is similar to the one plotted in Figure 4(a). It is evident that the more recent posts have higher probability of being selected for link creation and this procedure is reflective of real-world behavior of bloggers.

4.5 $\mathcal{RW}$-5: Random Walk on Post Graph (Link Expansion)

$\mathcal{RW}$-5 module models the link expansion behavior, which is found in real-world blogs. In link expansion, a blogger recursively refers to some out-links of the selected post. In $\mathcal{RW}$-5 module, a blogger probabilistically creates links to posts linked by the post that is selected by $\mathcal{RW}$-4. This random walk is initiated at the selected post on the out-directed post graph. The length of random walk l is the same as used in all other random walks. For the ith step of random walk ($i \leq l$), let $\mathbb{O}_v(i)$ denote the set of all outgoing links for vertex $v(i)$. Also, let $\mathcal{I}(i) \in \mathbb{I}_v(i)$ denote the singleton set containing the incoming traversal link for i^{th} step of random walk. The outgoing traversal link is uniform-randomly chosen from the set of links $\mathbb{S}_i$ which is defined as: $\mathbb{S}_i = \mathbb{O}_v(i) \cup \mathcal{I}(i)$. We provide the formal analysis for $\mathcal{RW}$-5 in Section 4.6.

4.6   Formal Analysis of $\mathcal{RW}$-2/3/5

We now provide a formal analysis of random walks on blog and post graphs in modules $\mathcal{RW}$-2, $\mathcal{RW}$-3, and $\mathcal{RW}$-5. Note that random walk procedure uses only locally visible information. The goal of our formal analysis is to show that the random linking via random walks is equivalent to the well-known preferential attachment principle found in real-world blogs [15].

Towards this end, let $P(v)$ denote the probability that vertex v is chosen for linking at the end of random walk which starts from vertex w out of total N vertices. Also let $P(w)$ denote the probability that vertex w is chosen for start of random walk, so $P(w) = 1/N$. Using Bayes rule, we can show that: $P(v) = \frac{P(v|w)P(w)}{P(w|v)}$. We note that the destination vertex at every step of random walk is chosen randomly from the set of existing links, $i.e.$, $P(v|w) = 1/deg(w)$.

Similarly, it can be shown that $P(w|v) = 1/deg(v)$. We combine these observations to get the following result: $P(v) = \frac{deg(v)}{Ndeg(w)}$. Note that if w is chosen randomly then $deg(w) = \mu_{deg}$, where μ_{deg} is the average degree of the graph which is constant. We conclude that: $P(v) \propto deg(v)$. This result is essentially the mathematical formulation of the preferential attachment principle.

5   Experimental Results

In this section, we gauge the properties of the blogosphere networks generated using our proposed m$\mathcal{RW}$ model. We analyze various properties of static snapshots of the generated blog networks.

5.1   Inter-Posting Time

In Section 4.2, our formal analysis showed that inter-posting times of simulated bloggers are self-similar and that the distribution of inter-posting times follows a power-law. Figure 5(a) shows the distribution of inter-posting times for a randomly selected blogger in our proposed model with length of random walk equal to 10. Note that we have sorted the post-index with respect to its inter-posting time. The data clearly follows a straight line on a log-log scale. The high value of goodness-of-fit parameter $R \approx 0.98$ for the regression line confirms that the distribution of inter-posting times follow a power law distribution. This indicates the fact that blogging activity can be characterized by long periods of inactivity separated by sudden and short periods of activity.

5.2   Blog In-Degree

In Figure 5(b), we show the blog in-degree distribution of a randomly chosen generated blogosphere. Here we have sorted the blog-index with respect to its degree. A power-law curve is fitted on the data at log-log scale. The high value of goodness-of-fit parameter $R \approx 0.95$ shows that we have reasonable confidence in this observation. This highlights the fact that only a few blogs receive large number of in-links and a majority of blogs remain unnoticed.

5.3   Post In-Degree

Figure 5(c) shows the distribution of post in-degree for one of the blogospheres generated using m$\mathcal{RW}$ model. A reasonable portion of the observed data follows the straight line of the log-log scale. The high value of goodness-of-fit parameter $R \approx 0.98$ highlights the scale-free nature of the graph evolved using our proposed model. Similar to our observation for blog in-degree, this also highlights that only a few posts get linked by a large number of other posts.

5.4   PageRank

PageRank is a well-known measure which forms the core of Google web search engine developed by Brin et $al.$ [2,3]. PageRank value is essentially an importance

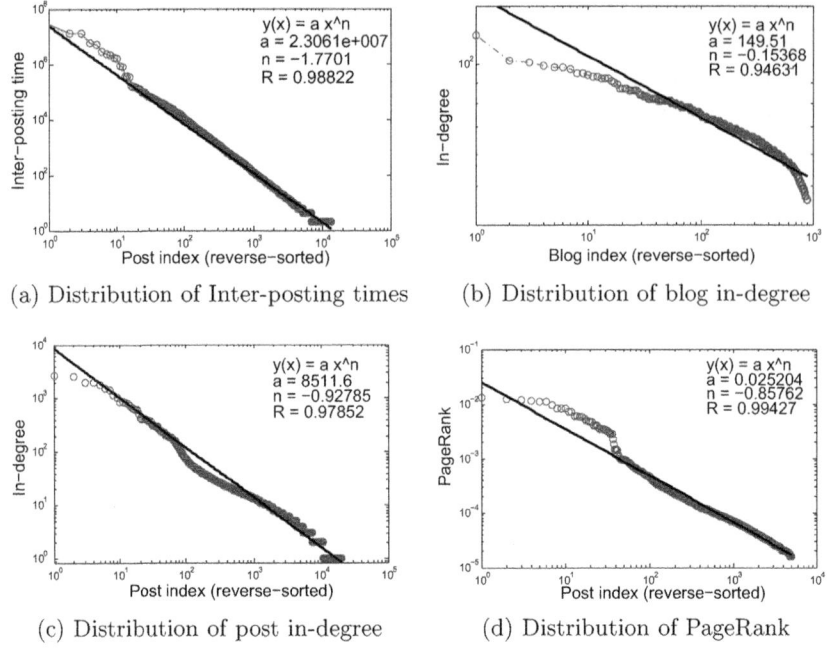

(a) Distribution of Inter-posting times   (b) Distribution of blog in-degree

(c) Distribution of post in-degree   (d) Distribution of PageRank

Fig. 5. Properties of Static Blog Snapshots

weight assigned to every node in a network. The links from blogs with larger in-degrees are considered relatively more significant in PageRank computation. PageRank value is established from the principal eigenvector of the adjacency matrix of a graph. In our analysis, we have found that the distribution of PageRank values of blogs follows a power-law. Figure 5(d) shows the distribution of PageRank of blogs for one of the blogospheres generated using our proposed $m\mathcal{RW}$ model. It is evident that the PageRank distribution follows a straight line on the log-log scale and has a high value of goodness-of-fit parameter $R \approx 0.99$.

6   Conclusion

In this paper, we have presented a novel and intuitive model using multiple random walks for generating the blogosphere. This model overcomes the limitations of previously proposed models. Furthermore, the results of our experiments show that the blogosphere generated using our proposed $m\mathcal{RW}$ model possess well-studied properties of the real-world blog graphs. In future, we plan to explore the control provided by our proposed model over other structural properties of the generated blogosphere using a single input parameter *i.e.* the length of random walk.

The successful modeling of topological and temporal dynamics of the blogosphere using random walk has several several interesting applications: For example, graph extrapolation can be done by predicting the trajectory of individual random walkers (bloggers) constituting the blogosphere. The only parameter to be estimated in this regard is the length of random walk. (2) The developed model can also be utilized to study and predict the effect of *active probing* (*e.g.* applying certain constraints on link creation) on topological and temporal properties of networks.

Acknowledgements

The authors would like to thank Dr. Habib Salehi for valuable comments and suggestions on an initial draft of the paper.

This material is based in part upon work supported by the National Science Foundation under Grant Number IIS-0968495. Any opinions, findings, and conclusions or recommendations expressed in this material are those of the authors and do not necessarily reflect the views of the National Science Foundation.

References

1. Aldous, D., Fill, J.: Reversible Markov Chains and Random Walks on Graphs. Book Draft (2001)
2. Brin, S., Pag, L.: The anatomy of a large-scale hypertextual web search engine. Computer Networks and ISDN Systems 33, 107–117 (1998)
3. Brin, S., Page, L., Motwami, R., Winograd, T.: The pagerank citation ranking: bringing order to the web. Tech. rep., Computer Science Department, Stanford University (1998)
4. Evans, T.S., Saramaki, J.P.: Scale-free networks from self-organization. Physical Review E 72(026138) (2005)
5. Gotz, M., Leskovec, J., McGlohon, M., Faloutsos, C.: Modeling blog dynamics. In: AAAI ICWSM (2009)
6. Gruhl, D., Guha, R., Liben-Nowell, D., Tomkins, A.: Information diffusion through blogspace. In: WWW, pp. 491–501 (2004)
7. Gurzick, D., Lutters, W.G.: From the personal to the profound: understanding the blog life cycle. In: CHI, pp. 827–832 (2006)
8. Karandikar, A., Java, A., Joshi, A., Finin, T., Yesha, Y., Yesha, Y.: Second space: Generative model for the blogosphere. In: AAAI ICWSM, pp. 198–199 (2008)
9. Kumar, R., Novak, J., Raghavan, P., Tomkins, A.: On the bursty evolution of blogspace. In: WWW, pp. 568–576 (2003)
10. Leskovec, J., Chakrabarti, D., Kleinberg, J., Faloutsos, C., Ghahramani, Z.: Kronecker graphs: An approach to modeling networks. arXiv:0812.4905v2 (2009)
11. Leskovec, J., Faloutsos, C.: Scalable modeling of real graphs using kronecker multiplication. In: ICML (2007)
12. Leskovec, J., McGlohon, M., Faloutsos, C., Glance, N., Hurst, M.: Cascading behavior in large blog graphs. In: Jonker, W., Petković, M. (eds.) SDM 2007. LNCS, vol. 4721. Springer, Heidelberg (2007)
13. Mushkin, M., Bar-David, I.: Capacity and coding for the Gilbert-Elliot channels. IEEE Transaction on Information Theory 35(6), 1277–1290 (1989)

14. Rudnick, J., Gaspari, G.: Elements of the random walk: an introduction for advanced students and researchers. Cambridge University Press, Cambridge (1944)
15. Saramaki, J., Kaski, K.: Scale-free networks generated by random walkers. Physica A: Statistical Mechanics and its Applications 341, 80–86 (2004)
16. Vazquez, A.: Knowing a network by walking on it: emergence of scaling. Statistical Mechanics (2000)
17. Vazquez, A.: Growing network with local rules: Preferential attachment, clustering hierarchy, and degree correlations. Physical Review E 67(056104) (2003)
18. Vazquez, A., Oliveira, J.G., Dezso, Z., Goh, K.I., Kondor, I., Barabasi, A.L.: Modeling bursts and heavy tails in human dynamics. Physical Review E 73(036127) (2006)

A Nash Bargaining Solution for Cooperative Network Formation Games

Konstantin Avrachenkov[1], Jocelyne Elias[2], Fabio Martignon[3],
Giovanni Neglia[1], and Leon Petrosyan[4]

[1] INRIA Sophia Antipolis
{Konstantin.Avrachenkov,Giovanni.Neglia}@sophia.inria.fr
[2] Paris Descartes University
jocelyne.elias@parisdescartes.fr
[3] University of Bergamo
fabio.martignon@unibg.it
[4] St. Petersburg State University
lapetr@apmath.spbu.ru

Abstract. The Network Formation problem has received increasing attention in recent years. Previous works have addressed this problem considering almost exclusively networks designed by selfish users, which can be consistently suboptimal.

This paper addresses the network formation issue using cooperative game theory, which permits to study ways to enforce and sustain cooperation among agents. Both the Nash bargaining solution and the Shapley value are widely applicable concepts for solving these games. However, we show that the Shapley value presents three main drawbacks in this context: (1) it is non-trivial to define meaningful characteristic functions for the cooperative network formation game, (2) it can determine for some players cost allocations that are even higher than those at the Nash Equilibrium (i.e., if players refuse to cooperate), and (3) it is computationally very cumbersome.

For this reason, we solve the cooperative network formation game using the Nash bargaining solution (NBS) concept. More specifically, we extend the NBS approach to the case of multiple players and give an explicit expression for users' cost allocations. Furthermore, we compare the NBS to the Shapley value and the Nash equilibrium solution, showing its advantages and appealing properties in terms of cost allocation to users and computation time to get the solution.

Numerical results demonstrate that the proposed Nash bargaining solution approach permits to allocate costs fairly to users in a reasonable computation time, thus representing a very effective framework for the design of efficient and stable networks.

Index Terms: Network Formation, Cooperative Game Theory, Coalition, Nash bargaining solution, Shapley value.

1 Introduction

The Network Formation problem has become increasingly important given the continued growth of computer networks such as the Internet. The design of such

J. Domingo-Pascual et al. (Eds.): NETWORKING 2011, Part I, LNCS 6640, pp. 307–318, 2011.

networks is generally carried out by a large number of self-interested actors (users, Internet Service Providers . . .), all of whom seek to optimize the quality and cost of their own operation.

Over the past years, the network formation problem has been tackled almost exclusively from a non-cooperative point of view. Recent works [1,2,3,4,5] have modeled how independent selfish agents can build or maintain a large network by paying for possible edges. Nash equilibria in such games, however, can be much more expensive than the optimal, centralized solution. This is mainly due to the lack of cooperation among network users, which leads to design costly networks.

The underlying assumption in all the above works is that agents are completely non-cooperative, isolated entities. However, this assumption could be not entirely realistic, for example when network design involves long-term decisions (e.g., in the case of Autonomous Systems peering relations). It is more natural that agents will discuss possible strategies and, as in other economic markets, form coalitions taking strategic actions that are beneficial to all members of the group. Moreover, incentives could be introduced by some external authority (e.g., the network administrator, government authority) in order to increase the users' cooperation level.

Preliminary works, like [6,7], tried to overcome this limitation by incorporating a socially-aware component in the users' utility functions. This solution, though, can be insufficient to obtain cost-efficient networks in all scenarios. In fact, it has been demonstrated in [6] that, quite surprisingly, highly socially-aware users can form stable networks that are much more expensive than the networks designed by purely selfish users.

To address the above issues, in this paper we formulate the network formation problem as a *cooperative* game, where groups of players (named *coalitions*) coordinate their actions and pool their winnings; consequently, one of the problems is how to divide the cost savings among the members of the formed coalition.

The *Shapley value* and the *Nash bargaining solution* are widely applicable solution concepts for cooperative games. The former has appealing properties, since it provides a unique and fair solution [8]. The Nash bargaining approach, on the other hand, studies situations where two or more agents need to select one of the many possible outcomes of a joint collaboration [9,10]. Examples include wage negotiation between an employer and a potential employee, or trade negotiation between two countries. Each party in the negotiation has the option of leaving the table, in which case the bargaining will result in a disagreement outcome. The Nash bargaining solution (NBS) is a very effective tool to model interactions among negotiators, and is unique for bargaining games satisfying Pareto optimality, symmetry, scale independence, and independence of irrelevant alternatives [9,10].

However, we will show that the Shapley value presents several drawbacks in this context: (1) it is non-trivial to define meaningful characteristic functions for the cooperative network formation game, (2) the cost allocation determined by the Shapley value can be, in some cases, even costlier than that obtained at some Nash equilibrium, and (3) for our network formation game, it cannot be determined in a reasonable computation time.

For these reasons, we propose a *Nash bargaining approach* to solve the cooperative network formation problem. More specifically, as a key contribution, we extend the Nash bargaining solution for the cooperative network formation problem to the case of multiple players with linear constraint, and give explicit expressions for users' cost allocations. To the best of our knowledge, the derived explicit expressions are new.

Furthermore, we perform a thorough comparison of the proposed Nash bargaining solution with other classic approaches like the Shapley value and the Nash equilibrium solutions, using different network scenarios.

Numerical results demonstrate that the proposed Nash bargaining solution can compute efficient cost allocations in a short computing time, thus representing a very effective tool to plan efficient and stable networks.

The main contributions of this work can therefore be summarized as follows:

- the formulation of the network formation problem as a cooperative game, where players cooperate to reduce their costs.
- The proposition of a novel Nash bargaining solution for the n-person cooperative network formation problem, which has appealing properties in terms of planning efficient networks and cost allocations in a reasonable computation time.
- A comparison of the proposed approach with classic solutions, like the Shapley value and the Nash equilibrium concepts, in large-size network topologies.

The paper is organized as follows: Section 2 discusses related work. Section 3 introduces the cooperative network formation game, the proposed Nash bargaining solution. Section 4 presents numerical results that demonstrate the effectiveness of the NBS approach in different realistic network scenarios. Finally, Section 5 concludes this paper.

2   Related Work

The network formation problem has been addressed in several recent works, mainly in the context of non-cooperative games [1,2,6]. The works in [3,4,11,12] have further considered coordination issues among players.

The so-called *Shapley network design game* is proposed in [1]. In this non-cooperative network formation game, each player chooses a path from its source to its destination, and the overall network cost is shared among the players in the following way: each player pays for each edge a proportional share $\frac{c_e}{x_e}$ of the edge cost c_e, where x_e is the number of players that choose such edge. In [6], the Shapley network design game is extended, adding a socially-aware component to users' utility functions.

The survey article in [11] presents the most notable works on network formation in *cooperative games*; furthermore, the existence of networks that are stable against changes in link choices by any coalition is studied in [13]. In [14],

Andelman et al. analyze strong equilibria with respect to players' scheduling as well as a different class of network creation games in which links may be formed between any pair of agents. For these latter games, strong Nash equilibria (i.e., equilibria where no coalition can improve the cost of each of its members) achieve a constant Price of Anarchy, which is defined as the ratio between the cost of the worst Nash equilibrium and the social optimum. Strong Nash equilibria ensure stability against deviations by every conceivable coalition of agents. A similar problem is considered in [12], where nodes can collaborate and share the cost of creating any edge in the host graph.

The works in [3,4] study the existence of strong Nash equilibria in network design games under different cost sharing mechanisms. More specifically, the authors in [3] show that there are graphs that do not admit strong Nash equilibria, and then give sufficient conditions for the existence of approximate strong Nash equilibria.

The idea of using the Nash bargaining solution in the context of telecommunication networks has been considered in different networking scenarios. Such approach was first presented for packet-switched data networks by Mazumdar et al. [15]. The concept of Nash bargaining solution is used by Yaiche et al. [16] to derive a price-based resource allocation scheme that can be applied to the available bit rate service in ATM networks. In [17] the authors propose a scheme to allocate subcarrier, rate, and power for multiuser orthogonal frequency-division multiple-access systems. The approach considers a fairness criterion, which is a generalized proportional fairness based on Nash bargaining solutions and coalitions.

The reader is referred to the next section, to the book by Muthoo [9] and the paper by Nash [10] for a general introduction to the Nash bargaining solution concept.

3   Cooperative Network Formation Game: Formulation and Solutions

This section illustrates the cooperative network formation game considered in this work, and describes the proposed Nash bargaining solution (NBS). A review of the Shapley value approach is preliminarily proposed for comparison reasons.

3.1   Network Model

We are given a directed graph $G = (V, E)$, where each edge e has a nonnegative cost c_e; each player $i \in \mathcal{I} = \{1, 2, \ldots n\}$ is identified with a source-destination pair (s_i, t_i), and wants to connect his source to the destination node with the minimum possible cost. Note that c_e represents the *total* edge cost, which is shared among the players according to the allocation algorithms we will describe in the following.

We consider a cooperative game in strategic form $G = \langle \mathcal{I}, A, \{J^i\} \rangle$, where $\mathcal{I}$ is the set of players, A_i is the set of actions for player i, $A = A_1 \times \ldots \times A_n$, and J^i is the objective (cost) function, which player i wishes to optimize (minimize).

In a cooperative game, players bargain with each other before the game is played. If an agreement is reached, players act according to such agreement, otherwise players act in a non-cooperative or antagonistic way. Note that the agreements must be binding, so players are not allowed to deviate from what is agreed upon.

3.2   The Shapley Value Solution

We now review the Shapley value solution approach, and discuss meaningful definitions for the characteristic function.

The Shapley value is a widely applied concept for solving cooperative games. It is a possible way to allocate the total costs among the members of a coalition, taking into account their different importance for the coalition. The main advantage of the Shapley value is that it provides a solution that is both unique and fair: it is unique in the class of subadditive cooperative games (see definition below); it is fair in a sense that it satisfies a series of axioms intuitively associated with fairness (see [8]). However, while these are both desirable properties, the Shapley value has one major drawback: for many coalition games, including our network formation game, it cannot be determined in a reasonable time. We shall discuss computation aspects in more detail below.

A Shapley function ϕ is a function that assigns to each possible characteristic function v a vector of real numbers, i.e.,

$$\phi(v) = [\phi_1(v), \ldots, \phi_i(v), \ldots, \phi_n(v)], \tag{1}$$

where $\phi_i(v)$ represents the cost of player i in the game.

The characteristic function, v, is a real-valued function that associates with every non-empty subset $\mathcal{S}$ of $\mathcal{I}$ (i.e., a coalition) a real number $v(\mathcal{S})$, the cost of $\mathcal{S}$; $v(\mathcal{S})$ must satisfy the following properties[1]:

1. $v(\emptyset) = 0$.
2. (Subadditivity) if $\mathcal{S}$ and $\mathcal{T}$ are disjoint coalitions ($\mathcal{S} \cap \mathcal{T} = \emptyset$), then $v(\mathcal{S}) + v(\mathcal{T}) \geq v(\mathcal{S} \cup \mathcal{T})$.

This latter property means that cooperation can only help but never hurt.

Note that defining the characteristic function is not straightforward for the cooperative network formation game considered in this work, since a "natural" definition can violate the subadditivity property, as we will discuss in the following.

The three definitions reported hereafter "naturally" arise in our networking problem as candidate characteristic functions:

1. Players in $\mathcal{S}$ and players in $\mathcal{I} - \mathcal{S}$ form two separate coalitions. Each coalition tries to minimize the total cost for its members, taking into account the selfish behavior of the other coalition. A Nash equilibrium is reached, and $v(\mathcal{S})$ is defined as the total cost for members in $\mathcal{S}$ at this equilibrium.

---

[1]   The second one is required to guarantee the uniqueness of the Shapley value solution.

2. The value of the coalition S is defined as its *security level*, i.e. as the minimum total cost that S can guarantee to itself when members in $\mathcal{I} - S$ act collectively in order to maximize the cost for the coalition S.
3. The value of coalition S is equal to the minimum cost that its members would incur if players in $\mathcal{I} - S$ would be absent.

We note that, in our specific game, these three definitions give increasing value to a coalition S. In fact, when players in $\mathcal{I} - S$ minimize their own cost (first definition), their path choices cannot be as bad for S as when they try to maximize the cost for S (second definition). Still, when players in $\mathcal{I} - S$ are present, they are obliged to select paths to connect their source-destination pairs, and some of these links may be used also by players in S, so that $v(S)$ is smaller in the second definition than in the third.

To better illustrate the differences underlying these definitions, let us consider the hexagon network scenario of Figure 1, with 6 links and 3 players having the following source-destination pairs: (s_1, t_1), (s_2, t_2) and (s_3, t_3). All link costs are equal to 1, except for link $t_3 \rightarrow t_2$, which has a cost equal to $1 - \epsilon$, ϵ being a very small constant.

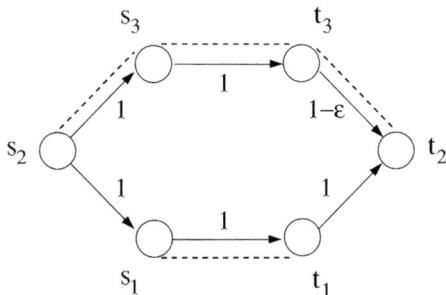

Fig. 1. Hexagon network topology: the 3 players must connect their source-destination nodes (s_i, t_i). The optimal solution, which in this case coincides with both the Nash equilibrium point and the Nash bargaining solution, is illustrated with dashed lines.

Table 1 reports, for each of the three above definitions, the corresponding characteristic function values. It can be easily checked that definition (1) does *not* lead to a characteristic function, since the subadditivity property is not satisfied (for example, $v(12) + v(3) < v(123)$), and therefore it cannot be used to compute Shapley values. Indeed, with such definition, cooperation among players can lead to costlier solutions. On the other hand, definitions (2) and (3) lead to characteristic functions.

Theorem 1. *In the Cooperative Network Formation Game, the security level (definition 2) and the minimum cost of the coalition (definition 3) satisfy the axioms of characteristic function.*

Proof: See Research Report [18].

Table 1. Hexagon network scenario: characteristic function values, $v(\mathcal{S})$, for definitions (1), (2) and (3)

Coalition ($\mathcal{S}$)	Characteristic Function value ($v(\mathcal{S})$)		
	Definition (1)	Definition (2)	Definition (3)
$\emptyset$	0	0	0
1	1	1	1
2	2.5-ϵ	2.5-ϵ	3-ϵ
3	0.5	1	1
12	3	3	3
13	1.5	1.5	2
23	3-ϵ	3-ϵ	3-ϵ
123	4-ϵ	4-ϵ	4-ϵ

To calculate the Shapley function, suppose we form the grand coalition (the coalition containing all n players) by entering the players into this coalition one at a time. As each player enters the coalition, he is charged the cost by which his entry increases the cost of the coalition he has entered. The cost a player pays by this scheme depends on the order in which the players enter. The Shapley value is just the average cost charged to the players if they enter in a completely random order, i.e.

$$\phi_i = \sum_{\mathcal{S} \subset \mathcal{I}, i \in \mathcal{S}} \frac{(|\mathcal{S}| - 1)!(n - |\mathcal{S}|)!}{n!}(v(\mathcal{S}) - v(\mathcal{S} - \{i\})). \tag{2}$$

It can be proved that the problem of computing the Shapley value is an NP-complete problem. Polynomial methods, based on sampling theory, have been proposed in [19] for approximating the Shapley value; these estimations, though, are efficient only if the worth of any coalition $\mathcal{S}$ can be calculated in polynomial time, which is not the case for our problem.

In fact, even using the approximation methods proposed for example in [19], it is necessary to compute the worth of an extremely large number of coalitions, which is computationally very cumbersome, while as we see next, our proposed Nash bargaining solution needs only computing the worth of the grand coalition.

3.3   The Nash Bargaining Solution (NBS)

Since the computation cost of the Shapley value can be extremely high in network scenarios with many players, in this paper we consider another approach to cooperative game: Nash bargaining. We will show that the computation of the Nash bargaining solution is very light.

Let u_i denote the maximal acceptable cost that user i is willing to pay. In the present work we suggest the following three options:

1. the cost for user i to connect its source-destination nodes in a purely non-cooperative game (i.e., the Nash equilibrium solution);

2. the cost for user i to connect its source-destination nodes in a zero-sum game where all the other players are trying to maximize the cost of user i;

3. the cost for user i to connect its source-destination nodes when there is no other player.

The vector $u = \{u_1, u_2, \ldots u_n\}$ is also denoted as the *disagreement point* of the cooperative network formation game (i.e., what will happen if players cannot come to an agreement). Clearly, the cost achieved by every player at any agreement point (every possible outcome of the bargaining game) has to be at most equal to the cost achieved at the disagreement point.

We now derive a Nash bargaining solution for allocating the total network cost to users. To this aim, we extend the well-known two-player NBS concept to the n-player network formation game, considering transferable network costs, providing explicit expressions. This assumption means that the players or the system administrator can redistribute the total cost among the players.

Let u_{soc} denote the total network cost resulting from social optimization. This can be computed, for example, formulating the generalized Steiner Tree problem [20] with an Integer Linear Program, using a mathematical programming model (like AMPL), and solving it with a commercial solver (like CPLEX). Solving such problem provides the least-cost network topology that connects all source-destination pairs.

Then, the Nash bargaining solution can be given in explicit form.

Theorem 2. *The Nash bargaining solution for player i, α_i is given by the following expression:*

$$\alpha_i = u_i - \frac{\sum_k u_k - u_{soc}}{m}, \tag{3}$$

where m coincides with the number of players n (i.e., $m \equiv n$) if we allow for negative costs (i.e., some α_i values are negative, which means that some players are actually paid to ensure their participation). Otherwise, if only non-negative costs are allowed (or equivalently, if no positive transfers are permitted), m is defined as the largest integer for which the following inequality is satisfied:

$$\frac{1}{m-1}\left(\sum_{i=1}^{m-1} u_i - u_{soc}\right) < u_m \tag{4}$$

having assumed, without loss of generality, that players are ordered such that $u_1 \geq u_2 \geq \ldots \geq u_n$.

Proof: See Research Report [18].

We would like to emphasize that in the first case α values can be positive or negative, while in the second case α values are non-negative. In particular, m gives the number of non-zero α values, i.e., $\alpha_1, \alpha_2, \ldots \alpha_m$ are positive and given by expression (3), while $\alpha_{m+1}, \ldots \alpha_n$ are equal to zero.

4   Numerical Results

This section reports the numerical results obtained applying our proposed Nash bargaining solution (NBS) to cooperative network design games played in various network scenarios, including simple network instances and more complex random topologies. The NBS, computed as illustrated in the previous section, is compared both to the cost allocation provided by the Shapley value, as well as to a Nash equilibrium solution. This latter is determined in the non-cooperative network formation framework proposed in [1], revised in Section 2, starting from the empty network and using a best response algorithm where each user greedily minimizes its path cost until an equilibrium is reached.

We assume that positive transfers are allowed. To compute the Shapley value, we further assume that the worth of a coalition S is the minimum cost that its members would incur if players in $\mathcal{I} - S$ would be absent (definition 3). This allows us to consider the "worst case", i.e. the costlier definition for a coalition, as discussed before. As for the disagreement point u_i in the NBS, we reasonably assume that it is the cost for user i to connect its source-destination nodes in a purely non-cooperative game (i.e., the Nash equilibrium solution). However, we underline that our proposed NBS approach is general and can be applied to any problem setting. The investigation of the impact of other characteristic function and disagreement point choices is left as future research issue.

Let us first consider the simple network scenario already illustrated in Figure 1, with 6 links and 3 players. The optimal network cost is here $u_{soc} = 4 - \epsilon$, and coincides with the cost of the network formed at the Nash Equilibrium Point (NEP). The Nash equilibrium and the Shapley value solutions for this scenario are reported in Table 2, together with the Nash bargaining solution, which in this case coincides with the NEP.

Numerical results show that the solution given by the Shapley value for player 3 ($\frac{5-2\epsilon}{6} \approx 0.83$) is costlier than that of the Nash equilibrium, 0.5. We further observe that even defining the value of a coalition as its security level (definition 2, Section 3.2) leads to the same Shapley values reported in Table 2. As a consequence, the Shapley value solution is somehow *unstable* for all the considered definitions of the characteristic function, since some players (i.e., player 3 in this scenario) can deviate to reduce their cost. This is surprising, because the Shapley value satisfies the *individual rationality* property, so that the Shapley value allocation is always preferable for each player than playing alone. The

Table 2. Hexagon network scenario with 3 players. The table reports the cost paid by each player at the Nash Equilibrium Point, the Shapley value and the Nash bargaining solution. The total network cost is equal to $4 - \epsilon$ for all allocation algorithms.

Algorithm	(s_1, t_1)	(s_2, t_2)	(s_3, t_3)
NEP	1	2.5 - ϵ	0.5
Shapley value	$\frac{5+\epsilon}{6} \approx 0.83$	$\frac{14-5\epsilon}{6} \approx 2.33$	$\frac{5-2\epsilon}{6} \approx 0.83$
NBS	1	2.5 - ϵ	0.5

apparent paradox originates from the fact that the value of the single player coalition has been defined either as the cost incurred if all other players are absent (definition 3), or as its security value, considering that all the other players are trying to maximize its cost (definition 2). In reality, at the Nash equilibrium, the cost of player i is smaller than such values and, as we have shown, it can be even smaller than the Shapley value imputation.

The same behavior can be observed also in more general topologies. To show this, we considered random network scenarios generated as follows: we randomly extract the position of N nodes, uniformly distributed on a square area with edge equal to 1000. As for the network links, which can be bought by players to connect their endpoints, we consider random geometric graphs, where links exist between any two nodes located within a range R. The link cost is set to its length.

Table 3 and Figure 2 illustrate the results obtained in a random geometric graph scenario with 50 nodes, range $R = 500$ (which means approximately more than 1200 links) and, respectively, 10 and 15 source-destination pairs (players). The table and figure report the costs for the players reached at the Nash equilibrium, the Shapley value as well as our proposed Nash bargaining solution. The total network cost is reported in the last column; note that such value corresponds, for the Shapley value and the Nash bargaining allocation algorithms, to the socially optimal solution (u_{soc} parameter), which can be obtained as explained in Section 3.)

It can be observed that, in all scenarios, at least 2 players (marked in bold in the table, with arrows in Figure 2) have a Shapley value that is higher than the Nash equilibrium cost. However, the cost saving between the NEP and the optimal cost (which is approximately 700 and 1250 for the $n = 10$ and $n = 15$ scenarios, respectively) could be re-distributed, which is what the Nash bargaining solution does, increasing the appeal of the cost sharing solution.

Obviously, since both the Shapley value and the NBS distribute the social cost (u_{soc}) among the players, there will be players whose allocation is costlier under the NBS than with the Shapley value allocation. This happens, in the numerical examples we considered, for players that have a large cost at the Nash equilibrium. However, every player is always better off under the NBS allocation than at the Nash equilibrium, since cost savings are redistributed.

Furthermore, we observe that computing the Shapley value for $n = 15$ players took several weeks of computation on the workstation used to obtain the numerical results reported in this paper, i.e., an Intel Pentium 4 (TM) processor with

Table 3. Random geometric network scenario with 10 players. The table reports the cost paid by each player at the Nash Equilibrium Point, the Shapley value and the Nash bargaining solution. The total network cost is also reported.

Algorithm	P1	P2	P3	P4	P5	P6	P7	P8	P9	P10	Total cost
NEP	283.0	149.4	235.3	824.4	714.8	450.5	674.0	195.6	186.0	266.9	3979.9
Shapley value	260.5	**170.6**	**253.9**	717.1	472.8	387.5	508.0	142.5	183.8	175.6	3272.2
NBS	212.3	78.6	164.6	753.6	644.0	379.7	603.2	124.8	115.3	196.2	3272.2

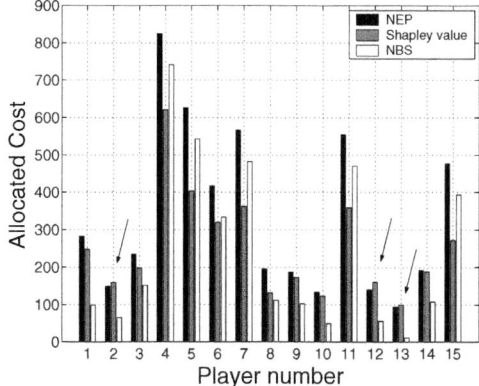

Fig. 2. Random geometric network scenario with 15 players. The figure reports the cost paid by each player at the NEP, the Shapley value and the NBS. The total network cost is equal to 5076.0 at the NEP and to 3802.7 for the Shapley value and NBS allocations.

CPUs operating at 3 GHz and with 1024 Mbyte of RAM. Therefore, computing the Shapley value for a larger number of players is practically infeasible in such network scenario. On the other hand, our proposed n-person Nash bargaining solution is very simple to calculate, and could be computed within a few minutes in all considered network scenarios, thus representing a practical and efficient solution to the network formation problem.

5   Conclusion

In this paper we proposed a novel and efficient Nash bargaining solution for the cooperative network formation problem with n players. Our solution has very appealing properties in terms of planning efficient networks and determining cost allocations in a very short computation time.

We compared our proposed solution to classic approaches, like the Shapley value and the Nash equilibrium concepts, in simple and large-size network topologies, with an increasing number of players.

Numerical results demonstrate that our approach permits to achieve very effective cost allocations, thus representing an efficient an promising framework for the planning of stable networks.

References

1. Anshelevich, E., Dasgupta, A., Kleinberg, J., Tardos, E., Wexler, T., Roughgarden, T.: The price of stability for network design with fair cost allocation. In: Proc. of the 45th Annual Symposium on Foundations of Computer Science (FOCS), Rome, Italy, October 17-19, pp. 295–304 (2004)
2. Chen, H.L., Roughgarden, T.: Network design with weighted players. In: Proc. of the 18th ACM Symposium on Parallelism in Algorithms and Architectures (SPAA 2006), Cambridge, MA, USA, July 30-August 2 (2006)

3. Albers, S.: On the value of coordination in network design. In: Proc. of the 19th Annual ACM-SIAM Symposium on Discrete Algorithms (SODA), San Francisco, CA, USA, pp. 294–303 (2008)
4. Epstein, A., Feldman, M., Mansour, Y.: Strong equilibrium in cost sharing connection games. In: Proc. of the 8th ACM Conference on Electronic Commerce, San Diego, CA, USA, pp. 84–92 (June 2007)
5. Chen, H.L., Roughgarden, T., Valiant, G.: Designing networks with good equilibria. In: Proc. of the 19th Annual ACM-SIAM Symposium on Discrete Algorithms, pp. 854–863 (2008)
6. Elias, J., Martignon, F., Avrachenkov, K., Neglia, G.: Socially-Aware Network Design Games. In: Proc. of INFOCOM 2010, San Diego, CA, USA (2010)
7. Azad, A.P., Altman, E., ElAzouzi, R.: From Altruism to Non-Cooperation in Routing Games. In: Proc. of Networking and Electronic Commerce Research Conference, Lake Garda, Italy (October 2009)
8. Aumann, R.J., Myerson, R.B., Roth, A.: The Shapley Value. Game-Theoretic Methods in General Equilibrium Analysis, 61–66 (1994)
9. Muthoo, A.: Bargaining theory with applications. Cambridge Univ. Press, Cambridge (1999)
10. Nash Jr., J.F.: The bargaining problem. Econometrica 18(2), 155–162 (1950)
11. van den Nouweland, A.: Models of network formation in cooperative games. Cambridge Univ. Press, Cambridge (2005)
12. Demaine, E.D., Hajiaghayi, M., Mahini, H., Zadimoghaddam, M.: The Price of Anarchy in Cooperative Network Creation Games. In: Proc. of STACS 2009, Freiburg, Germany, February 26-28, pp. 301–312 (2009)
13. Jackson, M., van den Nouweland, A.: Strongly stable networks. Games and Economic Behavior 51, 420–444 (2005)
14. Andelman, N., Feldman, M., Mansour, Y.: Strong price of anarchy. In: Proceedings of the 18th Annual ACM-SIAM Symposium on Discrete Algorithms (SODA 2007), New Orleans, Louisiana, January 7-9 (2007)
15. Mazumdar, R., Mason, L.G., Douligeris, C.: Fairness in network optimal flow control: Optimality of product forms. IEEE Trans. Comm. 39(5), 775–782 (1991)
16. Yaiche, H., Mazumdar, R.R., Rosenberg, C.: A game theoretic framework for bandwidth allocation and pricing in broadband networks. IEEE/ACM Transactions on Networking 8(5), 667–678 (2000)
17. Han, Z., Ji, Z., Ray Liu, K.J.: Fair Multiuser Channel Allocation for OFDMA Networks Using Nash Bargaining Solutions and Coalitions. IEEE Trans. Comm. 53(8), 1366–1376 (2005)
18. Avrachenkov, K., Elias, J., Martignon, F., Neglia, G., Petrosyan, L.: A Nash bargaining solution for Cooperative Network Formation Games. In: INRIA Research Report no. 7480 (December 2010),
http://hal.archives-ouvertes.fr/inria-00544527/en/
19. Castro, J., Gomez, D., Tejada, J.: Polynomial calculation of the Shapley value based on sampling. Computers & Operations Research 36(5) (2009)
20. Khan, M., Kuhn, F., Malkhi, D., Pandurangan, G., Talwar, K.: Efficient distributed approximation algorithms via probabilistic tree embeddings. In: Proc. of the 27th Symposium on Principles of Distributed Computing, pp. 263–272 (2008)

Optimal Node Placement in Distributed Wireless Security Architectures

Fabio Martignon[1], Stefano Paris[2], and Antonio Capone[2]

[1] Department of Information Technology and Mathematical Methods,
University of Bergamo
fabio.martignon@unibg.it
[2] Department of Electronics and Information,
Politecnico di Milano
{paris,capone}@elet.polimi.it

Abstract. Wireless mesh networks (WMNs) are currently accepted as a new communication paradigm for next-generation wireless networking. They consist of mesh routers and clients, where mesh routers are almost static and form the backbone of WMNs.

Several architectures have been proposed to distribute the authentication and authorization functions in the WMN backbone. In such distributed architectures, new mesh routers authenticate to a key management service (consisting of several servers, named *core nodes*), which can be implemented using threshold cryptography, and obtain a temporary key that is used both to prove their credentials to neighbor nodes and to encrypt all the traffic transmitted on wireless backbone links.

This paper analyzes the optimal placement of the core nodes that collaboratively implement the key management service in a distributed wireless security architecture. The core node placement is formulated as an optimization problem, which models closely the behavior of real wireless channels; the performance improvement achieved solving our model is then evaluated in terms of key distribution/authentication delay in several realistic network scenarios.

Numerical results show that our proposed model increases the responsiveness of distributed security architectures with a short computing time, thus representing a very effective tool to plan efficient and secure wireless networks.

Index Terms: Wireless Mesh Networks, Security, Distributed Architecture, Optimal Node Placement.

1 Introduction

Wireless Mesh Networks (WMNs) have been accepted as a new communication paradigm able to provide a cost-effective means to deploy all-wireless network infrastructures [1]. Network nodes of WMNs, named mesh routers, provide access to mobile users like access points in wireless local area networks, and they relay information hop by hop like routers, using the wireless medium.

J. Domingo-Pascual et al. (Eds.): NETWORKING 2011, Part I, LNCS 6640, pp. 319–330, 2011.
© IFIP International Federation for Information Processing 2011

As the mesh networking technology has become popular, the research community has proposed innovative security solutions to meet the requirements of WMNs. However, only few works consider the communication and computational overhead caused by the proposed protocols in their design.

In WMNs, two different security areas can be identified: one related to the *access* of user terminals which can be provided using standard techniques [2], and the other related to network devices in the *backbone* of the WMN.

Backbone security is a crucial issue. Mesh networks typically employ low-cost devices that cannot be protected against removal, tampering or replication. As a consequence, the research community has proposed several architectures to authenticate and authorize the mesh devices.

Centralized security solutions, like those proposed in [3,4], can exhibit lower costs than distributed approaches; however, they are characterized by a single point of failure (the key management server), which can be exploited by an adversary to attack and subvert the whole network.

To overcome this problem, some preliminary distributed solutions have been proposed in the wireless ad hoc and mesh network research fields [5,6,7]. The essence of all distributed security architectures lies in the necessity for all mesh routers to transmit periodically to a subset of nodes (which we refer to as *core nodes*) the authentication request to be authorized to join the network. The authorization usually consists of some cryptographic information necessary to create the temporary key that all mesh routers use to encrypt the traffic transmitted over the wireless backbone.

The deployment of the core nodes is therefore a key element for the performance of all distributed wireless security architectures. In fact, each generic mesh router should be sufficiently close to a subset of core nodes, in order to collect in the shortest possible time their partial authorization responses necessary to obtain the entire cryptographic information.

In this paper, we provide a general framework to increase the responsiveness of all distributed wireless security architectures, by focusing on the optimal placement of core nodes.

In an effort to understand how this issue impacts the performance of WMNs, our work makes the following unique contributions. (1) First, we propose a novel and efficient Integer Linear Programming model that optimizes the core node positions, reducing the overall authentication delay in a WMN where a distributed security architecture is implemented. (2) Second, we use an effective link metric that, differently from existing works, models the features of commercial wireless cards. The metric consider both the available channel rate and the total delay that packet transmissions experience over wireless links. (3) Third, we evaluate the proposed model in several realistic network scenarios, comparing the responsiveness of the two architectures proposed in [4] and [7]. Numerical results demonstrate that our proposed model permits to improve consistently the WMN responsiveness. Furthermore, our model can be solved to the optimum in a short computing time, even for large-scale network topologies, thus representing an important tool to design secure and efficient distributed wireless mesh networks.

The paper is structured as follows: Section 2 discusses related work. Section 3 briefly presents two typical security architectures, one fully distributed, the other centralized, which will be used in the Numerical Results section to gauge the performance of our proposed node placement model. Section 4 describes the Integer Linear Programming model that finds the optimal placement of core nodes. Section 5 discusses numerical results that show the effectiveness of our approach in a set of realistic network scenarios. Finally, conclusions and future research issues are presented in Section 6.

2   Related Work

Several works investigate the use of cryptographic techniques to achieve high fault tolerance against network partitioning. In [5] and [8], two different approaches are presented to allow specific coalitions of devices to act together as a single certification authority, whereas in [9] a hierarchical key management architecture is proposed to obtain an efficient establishment of distributed trust. Capkun et al. [10] propose a fully self-organized public-key management scheme that does not rely on any trusted authority to perform the authentication of other peer nodes. The public key management schemes proposed in [6] and [11] further enhance the security of distributed approaches by using proactive secret sharing and fast verifiable share redistribution techniques which permit to update periodically the secret shares.

Even though these distributed systems improve the network fault tolerance by removing the single point of failure introduced by centralized schemes, they are not very efficient in terms of computational or communication overhead. On the other hand, the centralized architecture proposed in [4] (MobiSEC), provides both access control for mesh users and routers with a negligible impact on the network performance. However, this latter solution is characterized by a single point of failure (the key management server), which can be exploited by an adversary to attack and subvert the whole network.

In summary, we underline that none of the above solutions considers the performance optimization issue in its design, while in this paper we provide a general framework to increase the responsiveness of all distributed security architectures.

The problem of finding the optimal places of network nodes that perform specific tasks in WMNs has been extensively studied in literature. In particular, this problem can be considered as a variation of the Capacitated Facility Location Problem (CFLP), which has been studied in the field of Operations Research [12]. Several works have extended the CFLP in order to model the constraints of the wireless environment and maximize the network throughput [13], or optimize the gateway placement while enabling an efficient reuse of the available resources [14]. Since CFLP is NP-hard, several heuristic algorithms have been proposed to efficiently solve such problems [15,16].

To the best of our knowledge, this paper is the first that addresses the core node placement problem in WMNs protected by distributed security architectures. This is performed using an optimization tool, while simulation results confirm the performance improvement derived from applying our proposed node placement strategy.

3   Centralized versus Distributed Security Architectures

This section briefly introduces two security architectures, one fully centralized (named MobiSEC [4]), the other completely distributed (named DSA-Mesh [7]). Their performance will be compared in the Numerical Results section under the proposed node placement model. The goal is simply to illustrate the main differences between centralized and distributed security approaches in WMNs. We underline that our proposed model is general and can be applied to optimize the node placement of any distributed wireless security architecture.

MobiSEC and DSA-Mesh adopt a similar approach to protect the backbone of a WMN, that is, all mesh routers obtain the same temporary secret which is used both to prove their credentials to neighbor nodes and to encrypt all the traffic transmitted on the wireless backbone links.

In MobiSEC, backbone security is provided as follows: each new router that needs to connect to the mesh network first authenticates to the nearest mesh router exactly like a client node, gaining access to the WMN. Then it performs a second authentication, connecting to a Key Server able to provide the credentials to join the mesh backbone. Finally, the Key Server distributes the information needed to create the temporary key that all mesh routers use to encrypt the traffic transmitted over the wireless backbone.

On the other hand, DSA-Mesh is a completely distributed architecture, since it distributes the Key Server functionalities among a group of core nodes using threshold cryptography [7]. In the DSA-Mesh architecture, the Key Server consists of n special mesh routers (the *core* routers), which collaboratively generate the new session secret and provide it to the other backbone nodes (the *generic* mesh routers). The employment of threshold cryptography permits to reduce the overhead of the authentication and key management protocols, since it enables t out of n core mesh nodes to perform this operation jointly, whereas it is infeasible for at most $t - 1$ nodes to do so, even by collusion. Throughout the paper, we will use the notation (n, t) to indicate such a threshold cryptographic system. Since we assume that a WMN contains t tamper-resistant nodes, n can be at most equal to $2t - 1$ in order to make it impossible to compromise the t nodes necessary to recover the session secret. For this reason, and for maximizing at the same time the network reliability to node failures, the most reasonable setting, adopted in the following, is $n = 2t - 1$.

A generic mesh router, after entering in the radio range of a mesh router already connected to the wireless backbone, broadcasts its first request to the entire network to obtain the secret used in the current session by the other routers that form the backbone, and the time when it was generated.

Each core node that receives the request from a generic node, after verifying the authenticity of the request, sends back the session secret and the timestamp encrypted with the public key of the generic node, and signs the message with its partial secret of the *key service* private key.

The generic node, after receiving at least t different responses, combines them and verifies the digital signature of the message using the *key service* public key. If the digital signature of the message is correct, then the generic node decrypts the message and obtains the secret used by all mesh routers to create the key sequence of the current session. Finally, the generic node, based on the instant at which it joins the backbone, computes the key currently used to protect the wireless backbone and its remaining validity time.

Therefore, in the DSA-Mesh architecture the core nodes distribute the information needed to create the temporary key used by all mesh routers.

Figure 1 shows an example WMN with the message exchanges performed between generic and core nodes. A (5,3) threshold scheme is adopted, that is, there are $t = 3$ tamper resistant nodes out of $n = 5$ core nodes; black and white circles represent core and generic nodes, respectively.

For the sake of clarity, we draw only the messages necessary to compute the signature, which are represented by solid arrows for requests (from generic to core nodes), and by dashed arrows for core node responses.

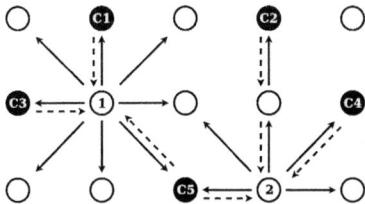

Fig. 1. Distributed security architecture: a (5,3) threshold scheme is adopted in this example WMN, where black and white circles represent, respectively, core and generic nodes. Solid lines represent requests, while dashed lines represent core node responses.

4   Optimal Core Node Placement

This section formulates the core node placement in distributed wireless security architectures as an optimization problem. In particular, an Integer Linear Programming (ILP) model is provided to solve the placement problem.

The input consists of a set of network node locations, N, the number t of messages needed to reconstruct the session secret (which is equal to the number of tamper-resistant nodes in the WMN), the total number $n = 2t - 1$ of core nodes, and finally a *distance* (or *cost*) d_{ij} between each pair of network nodes $i, j \in N$.

The problem consists in placing the n core nodes in order to minimize the maximum distance between each generic node and the farthest of the t closest core nodes. The rationale behind minimizing the *maximum* distance between generic and core nodes lies in the fact that only when t core node messages are collected, each generic node can discover the key used to encrypt data on the WMN backbone.

The setting of the distance function is a key element for an effective modeling of the behavior of a real WMN. For this reason, we propose to define it as the expected transmission delay, which depends both on the data rate (transmission time) and channel length (propagation time). More precisely, we compute the path loss of each wireless link, and then we evaluate the achievable transmission rate according to that provided by the data sheet of the Wistron CM9 commercial wireless cards (based on Atheros chipset). Note that almost all wireless cards based on the same Atheros chipset are characterized by the same sensitivity.

In the next section we will compare our proposed metric with a simple scenario in which the wireless link costs are all set to 1, regardless of their data rate and length.

The received power (P_r) and the corresponding path loss (PL_{dB}) of a wireless link having length d can be computed according to the following equations:

$$P_r = P_t \cdot \left(\frac{\lambda}{4\pi}\right)^2 \cdot \left(\frac{1}{d}\right)^\gamma = P_t \cdot \left(\frac{c}{4\pi f}\right)^2 \cdot \left(\frac{1}{d}\right)^\gamma$$

$$PL_{dB} = 10 \cdot log_{10}\left(\frac{P_t}{P_r}\right) = 10\gamma \cdot log_{10}(d) + 20 \cdot log_{10}(f) - 32.45$$

(1)

where the transmission power P_t is equal to 0.1 W, while the frequency f is set to 5.18 MHz; γ is the path loss exponent.

The achievable transmission rate of the wireless link e, r_e, is evaluated comparing its path loss with those listed in Table 1.

Table 1. Achievable transmission rate as a function of the CM9 wireless card sensitivity

PL (dB)	88	87	85	83	80	75	73	71
Rate (Mbit/s)	6	9	12	18	24	36	48	54

The transmission time of the link, t_e, is then approximated simply dividing the message length, L, by the transmission rate: $t_e = \frac{L}{r_e}$. Note that in real WMN implementations, the rate of each link can be easily provided by the underlying MAC protocol.

The overall transmission time of the path P_{ij} connecting any two mesh routers i and j can therefore be evaluated as the sum of the transmission times of all the links that belong to P_{ij}:

$$d_{ij} = \sum_{e \in P_{ij}} t_e = L \cdot \sum_{e \in P_{ij}} 1/r_e.$$

(2)

We observe that, since in distributed security architectures the authentication messages are actually transmitted at the highest priority level, the queueing delays they experience are almost negligible. Hence, expression (2) represents a good approximation of the overall delay experienced by these messages.

Having defined link and path costs, we now introduce the decision variables used in our ILP model: y_i indicates which network nodes are selected as core nodes, whereas x_{ij} provides the assignment of generic nodes to core nodes. More precisely:

$$y_i = \begin{cases} 1 & \text{if generic mesh router } i \text{ is selected as core node} \\ 0 & \text{otherwise} \end{cases}$$

$$x_{ij} = \begin{cases} 1 & \text{if generic mesh router } j \text{ is assigned to core node } i \\ 0 & \text{otherwise} \end{cases}$$

Given the above definitions, the optimal core node placement problem can be stated as the follows:

$$\min{(u)} \tag{3}$$

$$s.t. \sum_{i \in N} x_{ij} = t \qquad \forall j \in N \tag{4}$$

$$\sum_{i \in N} y_i = n \tag{5}$$

$$x_{ij} \le y_i \qquad \forall i, j \in N \tag{6}$$

$$u \ge x_{ij} d_{ij} \qquad \forall i, j \in N \tag{7}$$

$$x_{ij} \in \{0, 1\} \qquad \forall i, j \in N \tag{8}$$

$$y_i \in \{0, 1\} \qquad \forall i \in N \tag{9}$$

The objective function (3) minimizes the maximum distance (referred to as u) between each generic node and the farthest of the t closest core nodes. Constraints (4) ensure that each node $j \in N$ is assigned exactly to t core nodes, while constraints (5) ensure that exactly $n = 2t - 1$ mesh routers are selected as core nodes. Constraints (6) restrict generic node assignments only to selected core nodes (i.e., they ensure that whenever a node j is assigned to a core node i, this latter must necessarily have been selected as core node). Constraints (7) define the lower bound on the maximum distance between any node i and core node j, which is the quantity being minimized as objective function. Finally, constraints (8) and (9) ensure the integrality of the binary decision variables.

We observe that our problem is NP-hard, since it generalizes the t-neighbor n-center problem [17]. However, we were able to solve it to the optimum for realistic, large WMN scenarios using the CPLEX commercial solver; moreover, several polynomial time approximation algorithms that achieve a constant approximation factor have been proposed to solve different versions of this problem [17].

Finally, we note that our problem can be easily extended to take into account an incremental deployment of the WMN. In this regard, when topology changes occur (i.e., mesh routers join/leave the WMN) the network can be reconfigured considering the placement of already installed core nodes as fixed, while the position of eventual additional core routers is optimized adaptively. The investigation of the reconfiguration problem is left as future research issue.

5   Numerical Results

In this section we demonstrate the effectiveness of our proposed node placement model, measuring the performance improvement that derives from the optimal placement of core nodes in a fully distributed WMN security architecture, namely DSA-Mesh (reviewed in Section 3). A comparison with a centralized approach (MobiSEC) is also provided.

We simulated various WMN scenarios using Network Simulator (ns v.2). Packets are routed over shortest-paths, which are statically computed for all node pairs; this is meant to reduce the overhead due to routing protocols, allowing us to evaluate more precisely the effect of our node placement model on the network performance.

We consider as performance figure the *delay* necessary to complete the periodic authentication protocol performed by the two security architectures, which provides an indication of the protocol responsiveness. In particular, we analyze the average and maximum delays experienced by all generic nodes to receive the response from the Key Server (in MobiSEC) or the last response from the core nodes (in DSA-Mesh). The optimal placement of the core nodes for each network scenario is obtained solving the ILP model described in Section 4 with the CPLEX solver.

The computational time we measured using a Pentium 4 with 3.0 Ghz and 2 GByte of RAM was always inferior to 20 seconds for network topologies comprising up to 40 nodes. We further tested our model with larger WMN scenarios, including up to 100 nodes, and the computational time never exceeded 10 minutes. Hence, the proposed model can be used to effectively design secure WMNs, and can be further envisaged to perform autonomic reconfiguration on highly reconfigurable network scenarios.

The aforementioned average and maximum delays were measured considering both Grid and Random network scenarios. More specifically, each Grid topology is composed of N nodes placed over a $2000\ m \times 2000\ m$ area, with $N \in \{35, 40\}$. Random topologies are obtained by randomly scattering $N \in \{35, 40\}$ nodes over a $800\ m \times 800\ m$ area.

To evaluate the behavior of the two architectures in realistic traffic conditions, we set up background UDP data transfers (with packet size equal to 1000 bytes) that involve disjoint network links.

The maximum transmission range of each wireless node is equal to $250\ m$; the carrier sensing range is $550\ m$ when using the highest power level, equal to 0.1 W (these are ns v.2 default values). The maximum channel capacity is set to 54 Mbit/s, in accordance with commercial wireless card specifications (see Table 1). The reception threshold, the carrier sense and the capture thresholds are set to -64 dBm, -82 dBm and 10 dB, respectively. The path loss exponent, γ, is set to 2, and we leave as future issue the investigation of the model sensitivity to such parameter. All nodes use the same wireless channel since ns v.2 does not support natively multi-channel or multi-interface wireless nodes.

For each scenario we performed 10 independent measurements, achieving very narrow 0.95 confidence intervals, which we do not show for the sake of clarity. The simulation time on which we evaluated the performance was equal to 3000 seconds.

Figure 2 shows the *average delay* (for grid and random networks, respectively) measured by all generic nodes as a function of the network load imposed by the background UDP traffic. The lines identified by *"center"* and *"corner"* labels report the results obtained with the centralized security architecture (MobiSEC): the first line corresponds to a network configuration where the Key Server is installed at the center of the analyzed topology; on the other hand, the "corner" line reports the results obtained installing the Key Server at the topology border. The other curves in Figure 2 show the results obtained by the distributed

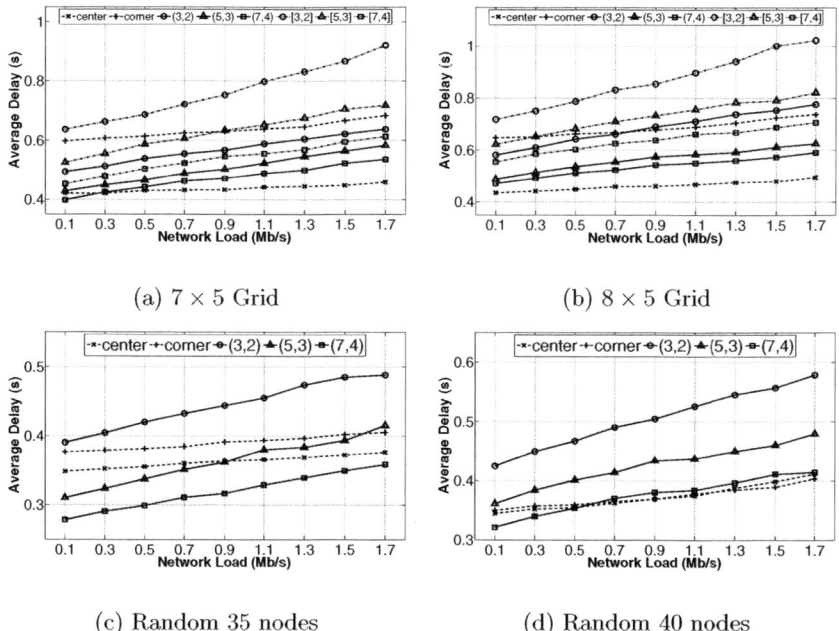

(a) 7 × 5 Grid

(b) 8 × 5 Grid

(c) Random 35 nodes

(d) Random 40 nodes

Fig. 2. Average Delay (seconds) measured in Grid and Random topologies with 35 and 40 nodes

security architecture (DSA-Mesh) using different (n, t) threshold schemes, viz. $(3, 2)$, $(5, 3)$ and $(7, 4)$. Solid lines correspond to the authentication delay measured when the core nodes are placed optimally, solving our ILP model (these curves are represented with the notation (n, t) in the legend), whereas dotted-dashed lines show this performance figure when the core nodes are simply installed at the topology corners (these curves are represented with the notation $[n, t]$). In this latter case, core nodes are placed in the corners of the grid network and, if $n > 4$, the remaining core nodes are placed randomly on the grid.

It can be observed that the average delay increases with increasing network load. At high traffic loads, the centralized architecture is more responsive than DSA-Mesh, when the Key Server is placed at the center of the network, since waiting for one authentication reply is less time-consuming than collecting t responses from the core nodes. However, when the Key Server is installed at the topology border (the "corner" curve) the centralized architecture exhibits higher delays, which are larger than those experienced by DSA-Mesh when the optimal placement is employed, regardless of the (n, t) scheme deployed. We further note that the optimal node placement permits to obtain consistent performance improvements, for any security scheme considered.

Figures 2(a) and 2(b) further show that the distributed architecture with seven core nodes placed in the topology corners (i.e., the $[7, 4]$ threshold scheme) exhibits in average a lower authentication delay than the centralized architecture with a single server (the "corner" curve). Hence, when a WMN operator cannot

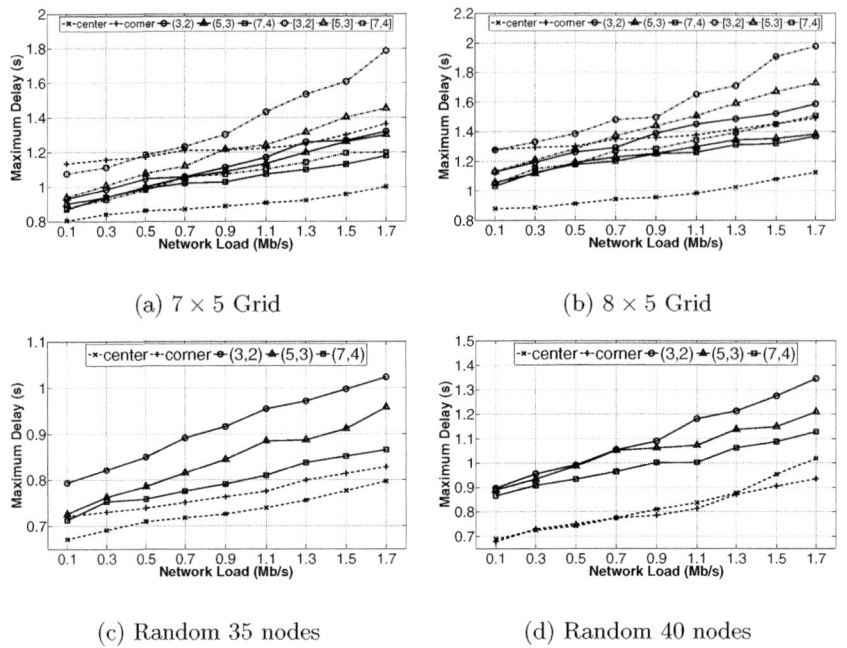

(a) 7 × 5 Grid

(b) 8 × 5 Grid

(c) Random 35 nodes

(d) Random 40 nodes

Fig. 3. Maximum Delay measured in Grid and Random topologies with 35 and 40 nodes

optimize the network design prior to its deployment, or when the WMN grows in an autonomous fashion, a distributed architecture with a high number of core nodes should be privileged with respect to a centralized system, since it provides better performance during the network operation.

Note that placing the Key Server in remote places, such as the corner node of a grid topology, can be the unique option in many network scenarios, where all the management services run on high-end machines.

In the Random topologies, the performance improvement of DSA-Mesh with respect to the centralized architecture is less evident than in the Grid scenarios, since the random network diameters are quite limited. In fact, the square area (800 m × 800 m) over which the network nodes are distributed, bounds the distance of the worst path, and therefore the maximum value assumed by the authentication delay. As a consequence, the centralized architecture exhibits in all network scenarios similar performance. For sake of clarity, in these figures we did not show the results obtained placing core nodes at the topology edges, since they follow the same trends illustrated in the grid topologies of Figures 2(a) and 2(b).

To obtain a more complete comparison, we also measured the *maximum delay*, which provides an indication of the worst-case performance of the DSA-Mesh architecture. Figures 3 shows such performance metric for grid and random networks composed of 35 and 40 nodes. The curves in these scenarios follow a trend similar to that obtained in the average case discussed above. More specifically,

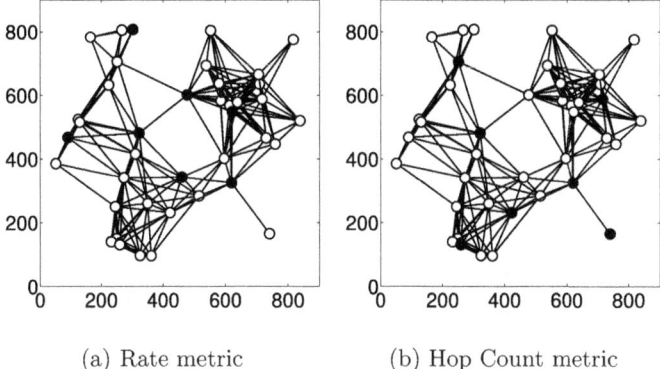

(a) Rate metric   (b) Hop Count metric

Fig. 4. Optimal placement of core nodes (the black circles) in a random topology with $N = 40$ nodes. Two different cost metrics are compared, namely the hop count and the overall transmission time (rate metric).

the optimal placement of the core nodes reduces considerably the maximum delay experienced by generic nodes, for any threshold scheme considered. On the other hand, the optimal node placement obtained using the model proposed in this paper permits to outperform a centralized architecture where the authentication and key distribution server is placed at the border of the network topology.

Finally, Figure 4 illustrates the optimal placement of the core nodes (the black circles) in a random topology scenario with 40 nodes and a (7,4) threshold scheme, comparing the metric introduced in Section 4 (the overall transmission time, referred to as Rate Metric) to a simple cost metric (the Hop Count, where the cost of each link is equal to 1). It can be observed that the Hop Count metric privileges paths with a low number of long wireless links (which, consequently, have low transmission rates), whereas the transmission time metric chooses routes with a high number of relatively short wireless links. As a consequence, the Hop Count metric tends to spread the core nodes over the network topology due to the limited distance among all node pairs, whereas the transmission time metric places the core nodes in areas highly crowded of network nodes.

Note that the different placements illustrated in Figure 4 lead to an average delay that is up to 30% higher for the Hop Count than for the Rate metric we propose in this paper.

6   Conclusion

This paper tackled the problem of determining the optimal placement of the core nodes that collaboratively implement the key management service in a distributed wireless security architecture. The core node placement has been formulated as an optimization problem, modeling closely the behavior of real wireless channels; the performance improvement achieved solving our model has been evaluated in terms of key distribution/authentication delay (average and maximum) in several realistic network scenarios.

Numerical results show that our proposed modeling framework permits to increase both the average and the worst-case responsiveness in distributed architectures designed for WMNs with a short computing time, thus representing a very effective tool to design efficient and secure wireless networks.

References

1. Akyildiz, I.F., Wang, X., Wang, W.: Wireless mesh networks: a survey. Elsevier Computer Networks 47(4), 445–487 (2005)
2. IEEE Standard 802.11i. Medium Access Control (MAC) security enhancements, amendment 6. IEEE Computer Society (2004)
3. Zhang, Y., Fang, Y.: Arsa: An attack-resilient security architecture for multihop wireless mesh networks. IEEE Journal on Selected Areas in Communications 24(10), 1916–1928 (2006)
4. Martignon, F., Paris, S., Capone, A.: Design and Implementation of MobiSEC: a Complete Security Architecture for Wireless Mesh Networks. Elsevier Computer Networks 53(12), 2192–2207 (2009)
5. Yi, S., Kravets, R.: Moca: Mobile certificate authority for wireless ad hoc networks. In: Annual PKI Research Workshop, PKI 2003 (2003)
6. Kim, J., Bahk, S.: Meca: Distributed certification authority in wireless mesh networks. In: IEEE CCNC, pp. 267–271 (2008)
7. Martignon, F., Paris, S., Capone, A.: DSA-Mesh: a Distributed Security Architecture for Wireless Mesh Networks. Wiley Security and Communication Networks 4(3), 242–256 (2011)
8. Luo, H., Zerfos, P., Kong, J., Lu, S., Zhang, L.: Self-securing ad hoc wireless networks. In: IEEE ISCC, pp. 567–574 (2002)
9. Xu, G., Iftode, L.: Locality driven key management architecture for mobile ad-hoc networks. In: IEEE MASS, pp. 436–446 (2004)
10. Capkun, S., Buttyan, L., Hubaux, J.-P.: Self-organized public-key management for mobile ad hoc networks. IEEE Trans. on Mobile Computing, 52–64 (2003)
11. Wua, B., Wua, J., Fernandeza, E.B., Ilyasa, M., Magliveras, S.: Secure and efficient key management in mobile ad hoc networks. In: IEEE IPDPS (2005)
12. Nauss, R.M.: An improved algorithm for the capacitated facility location problem. Journal of the Operational Research Society 29(12), 1195–1201 (1978)
13. Aoun, B., Boutaba, R., Iraqi, Y., Kenward, G.: Gateway placement optimization in wireless mesh networks with QoS constraints. IEEE Journal on Selected Areas in Communications 24(11), 2127–2136 (2006)
14. Targon, V., Sansò, B., Capone, A.: The joint Gateway Placement and Spatial Reuse Problem in Wireless Mesh Networks. Computer Networks (2009)
15. He, B., Xie, B., Agrawal, D.P.: Optimizing the Internet gateway deployment in a wireless mesh network. In: IEEE MASS (2007)
16. Amaldi, E., Capone, A., Cesana, M., Filippini, I., Malucelli, F.: Optimization models and methods for planning wireless mesh networks. Computer Networks 52(11), 2159–2171 (2008)
17. Khuller, S., Pless, R., Sussmann, Y.J.: Fault tolerant k-center problems. Theoretical Computer Science 242(1), 237–246 (2000)

Geographical Location and Load Based Gateway Selection for Optimal Traffic Offload in Mobile Networks

Tarik Taleb[1], Yassine Hadjadj-Aoul[2], and Stefan Schmid[1]

[1] NEC Europe Ltd, Heidelberg, Germany
{tarik.taleb,stefan.schmid}@neclab.eu
[2] University of Rennes-1, Rennes, France
yhadjadj@irisa.fr

Abstract. To cope with the rapid increase of data traffic in mobile networks, operators are looking for efficient solutions that ensure the scalability of their systems. Decentralization of the network is one of the key solutions. With such solution, small-scale core network nodes (gateways) are locally deployed to serve the local community of users, in a decentralized fashion. In this paper, we devise methods that enable User Equipments (UEs), both in idle and active mode and while being on the move, to always have optimal Packet Data Network (PDN) connections in such decentralized networks. The proposed methods are compared based on their impacts on current 3GPP standards and the benefits of the overall approach are evaluated through simulations. Encouraging results are obtained.

Keywords: SIPTO, 3GPP Network, traffic offload, mobile network.

1 Introduction

Along with the ever-growing community of mobile users and the tremendous increase in the traffic associated with a wide plethora of emerging bandwidth-intensive mobile applications, mobile operators are facing a challenging task to accommodate such huge traffic volumes, beyond the original network capacities[1][2]. The challenge becomes more significant considering the fact that the Average Revenues per Users (ARPU) are getting lower given the trend towards flat rate business models. Operators are thus investigating cost-effective methods for accommodating such traffic with minimal investment to the existing infrastructure.

Decentralizing the mobile network is one of the key solutions. With such solution, small-scale core network nodes, e.g., Packet Data Network Gateways (PDN-GWs), and Serving GWs (S-GWs), are locally deployed to serve the local community of users, in a decentralized fashion[1]. The benefits of such decentralized mobile network are manifold. Indeed, with such decentralized networks, operators will be able to optimize usage of their resources by selectively breaking out IP traffic whenever and

---

[1] In this paper, the focus is on the Evolved Packet System (EPS) [4][5] but the general description can be equally applied to the General Packet Radio Service (GPRS). In this case, Serving GPRS Support Node (SGSN) would map on to S-GW and MME, and Gateway GPRS Support Node (GGSN) would map on to P-GW.

J. Domingo-Pascual et al. (Eds.): NETWORKING 2011, Part I, LNCS 6640, pp. 331–342, 2011.
© IFIP International Federation for Information Processing 2011

as near to the edge of operator's network as possible. This concept is in line with Selected IP Traffic Offload (SIPTO) [3], a topic currently under discussion within 3GPP. It also fits the changing paradigm for traffic routing, inspection and charging by operators (e.g., the trend towards flat rates and differentiation into "dumb", bit-pipe traffic and value-added services). It is also in line with the quest for a network architecture flatter than what has been achieved with the Evolved Packet System (EPS) [4][5]. Effectively, by breaking out selected traffic at entities close to the moving terminal and thus also at the edge of the network (e.g., Radio Network Controller (RNC) in 3G, eNodeB in the EPS), operators will be able to avoid overloading their scarce core network resources (i.e., GGSNs, SGSNs, and P/S-GWs).

The discussions and analysis in the 3GPP standardization group currently focus on the definition of the architecture, i.e., on where the point of local breakout/traffic offload should be placed. Issues regarding security, charging, mobility, traffic control/handling, and optimal gateway selection are yet to be investigated. In this paper, we focus on finding adequate solutions that address the latter issue.

The optimal gateway selection problem is illustrated in Fig. 1. As will be discussed later, the issue can be solved relatively easily in case of UEs being in ECM-Idle (EPS Connection Management) state (i.e. idle mode). However, in this paper, we will also present a solution for optimal gateway reselection for UEs being in ECM-connected state (i.e., active mode). Here, we assume UEs to have the capability to support multiple PDN connections to multiple Access Point Names (APNs). Fig. 1(a) depicts the case of a UE initiating a PDN connection to a particular P-GW via a Long Term Evolution (LTE) radio cell (i.e., eNB1). At a later instant, and after performing handoff to a distant eNB (i.e., eNB2 in Fig. 1(b)), the UE initiates a new IP session to the same APN while the old IP session is still active. Following the current standards [6], the UE cannot have multiple IP sessions via different PDN connections to the same APN over the same access. Therefore, the new IP session (i.e., session 2) will be established via the old P-GW (i.e., P-GW1) as depicted in Fig. 1(b). This is clearly not an optimal decision given the fact that another more optimal P-GW (P-GW 2) is available in the visited area. The optimality of the new PDN gateway can be assessed in terms of both geographical proximity to the UE as well as expected load. A radical solution to this issue could be to set up the new IP sessions via the optimal gateway and to migrate the existing sessions to the new gateway. However, this solution would have significant impact on the user experience due to service disruption. As an alternative, we suggest that the UE would keep the old IP sessions via the old P-GWs but sets up new IP sessions (to the same APN) via the currently most optimal gateway. We propose and compare three methods whereby the setup of new IP sessions via a newly selected gateway is initiated by UE, MME, or P-GW.

Indeed, when a UE, accessing a particular APN using a given PDN connection, performs handoff to a new radio cell, and/or a more optimized P-GW (e.g., in terms of load, geographical proximity relative to UE, etc.) becomes available, the UE shall be able to set up a new PDN connection to the more optimal P-GW when it initiates a new IP session. This paper is about devising a mechanism that enables UEs to establish another optimized PDN connection (e.g., upon a trigger from the network or when judged appropriate by the UE) to the same APN, and a mechanism with which a UE would bind IP flows/connections/sessions (when they are established) to their corresponding PDN connections. In this way, the UE always remembers which IP flow/connection/session uses which PDN connection. In should be noted that in this paper a session is defined based on the IP address of the peer node, application type, and underlying protocol types, and that a session can be associated with more than one protocol type (e.g., SIP, RTP, and RTCP).

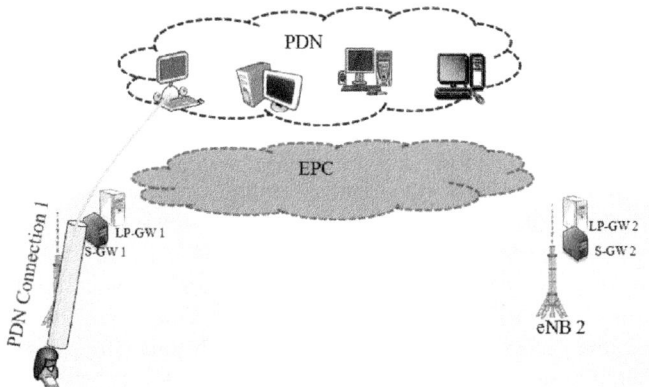

(a)   PDN Connection 1 initiated at time T when UE is connected to eNB1

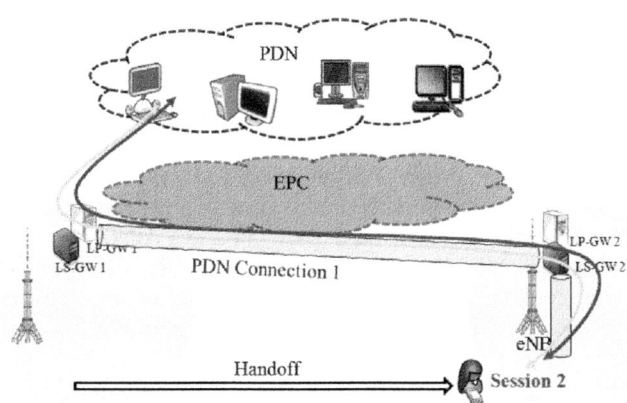

(b)   After performing handoff to eNB2, UE initiates a new session via the old P-GW

Fig. 1. Limitation of current standards

The remainder of this paper is organized as follows. Section 2 presents the state of the art. Section 3 presents our proposed gateway selection mechanisms and that is for UEs in both ECM-idle mode and ECM-connected mode, respectively. The implementation issues of the proposed solutions are also discussed. Section 4 evaluates the overall approach and showcases its technical benefits. The paper concludes in Section 5.

2   State of the Art

The Evolved Packet Core (EPC) is designed to encompass different 3GPP accesses (i.e., 2G, 3G and LTE) as well as non-3GPP access (e.g., WiMAX, CDMA2000 ©,

1xRTT, etc.). The richness of EPC accesses gave birth to a new 3GPP entity called ANDSF (Access Network Discovery Selection Function) that assists UEs to find the best or most suitable access out of the many available ones [7]. It has also led to different interesting 3GPP study items whereby UEs are allowed to have simultaneous accesses to different networks using different access technologies. In [8], the 3GPP SA2 group started a study item to investigate different possibilities for dynamic IP flow mobility between 3GPP access and one, and only one, non-3GPP access. The study of solutions to support routing of different PDN connections through different access systems is also in scope. Some of the solutions are being standardized in the technical specifications [9]. In [8], a work item is proposed to allow a UE, equipped with multiple network interfaces, to establish multiple PDN connections to different APNs via different access systems and to selectively transfer PDN connections between the accesses with the restriction that multiple PDN connections to the same APN shall be kept in one access. Whilst there is also another work, impacting different technical specifications, that aims for enabling UEs to establish and disconnect multiple PDN connections to the same APN uniformly across the EPS, a mechanism that indicates to the network or to the UE when it is beneficial to set up a new IP session over a new PDN connection via the same access is overlooked. Indeed, in current solutions, a UE, supporting multiple APNs, may have different PDN connections, each associated with a different APN (e.g., Internet, IMS, WLAN, etc). When a UE is using a PDN connection to access a particular APN, that PDN connection (to the APN in question) does not change until the UE becomes in idle mobility. In other words, P-GW relocation is recommended only during idle mobility, to avoid service disruption, because when the P-GW changes, the old PDN connection is simply torn down. Additionally, as long as the UE has a PDN connection (to access a particular APN) and is in active mode, the UE will always use the same P-GW to set up any new IP sessions to the same APN.

As stated earlier, mobile operators are aiming for the decentralization of their networks. In this context, a mechanism that indicates to the network when it is beneficial for a UE to set up a new IP session via a new PDN connection will be highly required. Its importance becomes further vital knowing the interest of operators in offloading "dump" traffic as locally as possible to achieve the goals of SIPTO [3]. In this paper, we propose a set of mechanisms that enable a UE to know how and when to establish a new optimized PDN connection for launching new IP sessions to a particular APN. This is done without impacting/compromising the on-going (old) PDN connections to the same APN.

3   Local GW Selection

In this paper, two states of UEs are separately discussed: ECM-idle mode and ECM-connected mode. In the latter, we specifically consider the case of UEs supporting multiple PDN connections. The objective is to trigger UEs to re-establish PDN connections (e.g., those subject to SIPTO) when it is beneficial for both the network and the UEs to reselect another nearby local P-GW, e.g., in case the UE traveled a significant distance from the original P-GW.

3.1 Triggering Local GW Selection for UEs in Idle Mode

In the EPS the concept of "always-on" was adopted. This means that once a UE has established a PDN connection, it remains configured in the network even in case the UE goes idle. This feature has the great advantage that the UE can communicate right-away when it becomes active and does not require re-establishment of a PDN connection (incl. allocation of a new IP address) at the time when the user wants to communicate. As stated earlier, the problem with this in the context of SIPTO or decentralized networks in general is the following:

1. Once a UE has established a PDN connection to a well-positioned Local P-GW (in the core or close to the radio station), it maintains this connection – and therefore also the associated "local" GW – until it explicitly disconnects from this PDN.

2. This implies that if the user moves a significant distance away from the initial (H)eNB, the chosen local P-GW providing the access towards the service network (i.e. PDN) may not be optimal anymore.

Fig. 2 illustrates this deficiency for a user that first connects to a service via a macro cell (eNB1) and a "near-by" P-GW in the core network (step 1), and then moves a significant distance away while in idle mode. Without the optimization, the UE remains connected to the P-GW originally chosen, although a local P-GW (LP-GW1) could offer a more optimal access to the service.

To resolve this problem, this paper suggests that the UE simply re-establishes the PDN Connections (e.g., those subject to SIPTO traffic) during idle mode mobility when it has moved away a significant distance from the originally selected "local" P-GW. By simply re-establishing the PDN Connection, the default P-GW selection mechanism would ensure that the UE will again be connected to a local P-GW that is geographically/topologically close to the user's location. As a result, the traffic subject to SIPTO would be again "broken out" or "offloaded" at the most suitable location from the user and operator point of view.

To re-establish a PDN connection, a UE can be either triggered by the network via, for example, a flag during the Tracking Area Update (TAU) procedure [5] or the network could simply disconnect the PDN connection. However, in the latter case, the

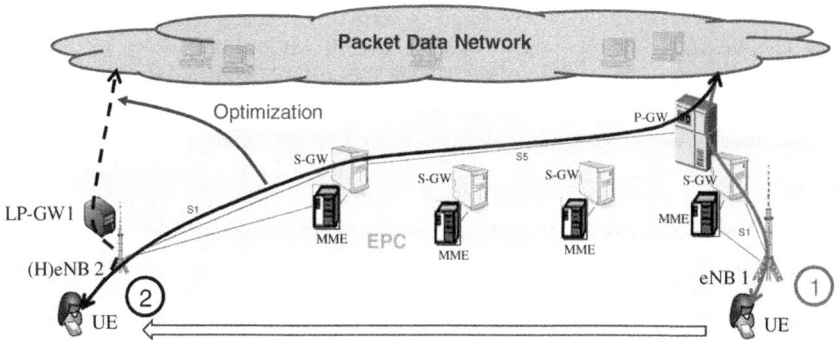

Fig. 2. Expected optimization for UEs in idle mode

UE would first have to establish a new PDN connection (incl. IP address allocation, tunnel configuration, etc.) before it can become active. This would break the "always on" concept and also negatively impact the user experience as extra delay is incurred. Nevertheless, since the proposed PDN connection re-establishment procedure is proposed to take place only during idle-mode mobility, no degradation of the service quality or service disruption would be introduced.

The open issue is to identify when a UE should re-establish a PDN connection. The following solutions are feasible:

- **Option 1** - Periodically – after a configurable time period: The issue with this solution is that stationary UE will introduce a lot of extra signaling overhead without any gain for the operator.

- **Option 2** - Upon Tracking Area Update – whenever the UE changes tracking area: The change of tracking area ensures that the UE has actually moved away from the original location. It is, however, not clear whether the re-establishment would actually lead to a different Local GW.

- **Option 3** - Upon indication from the network – whenever the network considers it beneficial: The network indicates to the UE (e.g., as part of idle-mode mobility procedures or with a special cause for the PDN disconnection) when a particular PDN connection should be re-established. Since the operator has the knowledge of the network topology and the available local P-GWs, it can indicate to the UE when the re-establishment of a PDN connection is worthwhile – i.e., when this leads to a better P-GW selection and thus to a more optimal path.

3.2   Triggering Local GW Selection to Accommodate New IP Sessions of UEs in Active Mode

As discussed earlier, in current standards, as long as a UE has a PDN connection to access a particular APN through a particular access type, the UE will maintain the same P-GW to set up any new IP sessions to the same APN. Motivated by the example of Fig. 1, discussed in the introduction, we argue that when a UE, accessing a particular APN using a given PDN connection, performs handoff to a new base station, and a more optimal P-GW (e.g., in terms of load, geographical/topological proximity relative to UE) becomes available, the UE should be able to set up a new PDN connection to the more optimized P-GW when the UE initiates new IP sessions to the same APN.

Fig. 3 depicts the envisioned solution. Indeed, after handoff, and when a new, optimal P-GW becomes available, a UE establishes a PDN connection to the new P-GW to establish any new IP session. The on-going sessions keep using the old PDN connection. In the figure, we apply our solution to a scenario whereby the UE performs handoff to an area where another optimal P-GW becomes available. However, the solution can be also applied even if the UE remains in the same area (i.e., cell) and the P-GW it is connected to becomes non-optimal (e.g., because it is currently highly loaded) and another P-GW (e.g., a less loaded one) becomes optimal.

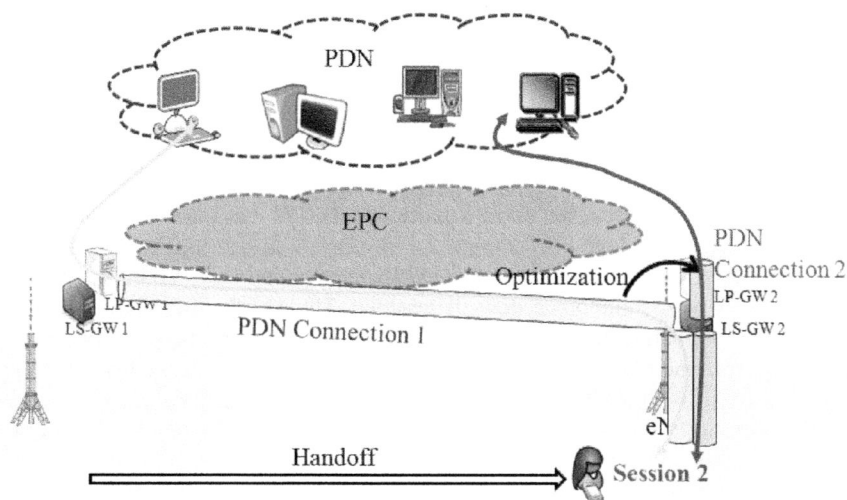

Fig. 3. Setting up new IP sessions via optimal PDN connections while old IP sessions using the old PDN connection are not compromised

The key question is how to trigger the UEs to establish a new PDN connection for new IP sessions while keeping the old ones on an already available PDN connection. To cope with this issue, we propose the following three methods and qualitatively compare them.

MME-initiated: The MME may apply different mechanisms to check if there are any more optimal P-GWs available (e.g., take into account network and GW load information or UE mobility prediction). When a more optimal P-GW is available, the MME could use existing signaling message during handover to indicate to the UE that for new IP sessions to the APN in question, it should consider establishment of a new PDN connection. This indication can be in the form of a flag, based on which the UE establishes a new optimized PDN connection when it wants to initiate a new IP session, or it can be in the form of an explicit indication of the IP address of the optimal P-GW (e.g., as part of the TAU procedure).

UE-initiated: In this solution, when a UE wants to set up a new IP session to an APN with which it has an ongoing PDN connection, it queries MME (e.g., using NAS signaling) if it should use the existing PDN connection or consider a new one. Querying MME can be also done based on other events, such as when a UE performs a number of handoffs, after a particular period of time, after the UE moves for a certain distance based on location information, after the UE enters into a new tracking area and/or a specific area during a specific time, etc. Compared to the MME-initiated approach, the UE-initiated solution may generate unnecessary queries to the MME. However, both these solutions require, apart from the modifications for allowing multiple PDN connections to the same APN through the same access type, only very minimal extensions of the standard signaling interfaces, i.e. a new indicator (flag) between UE and MME.

P-GW initiated: For this solution, two alternatives can be envisioned. In the first one, when the current P-GW realizes that the UE is to be better serviced by another P-GW, it simply rejects any requests for any new IP sessions. The rejection can be done via a new error message. This operation intuitively requires that P-GWs have the ability to filter traffic per IP flow/session. It also requires that P-GWs have knowledge on the optimality of a set of other P-GWs (e.g., only neighboring ones).

In the second alternative, when a particular P-GW starts running under specific conditions (e.g., at a load exceeding a certain threshold), it notifies a selected set of UEs (with ongoing connections to the P-GW) to establish new PDN connections with other P-GWs to accommodate new IP sessions. This can be done by designing new and specific signaling messages using S5/8 and S11 interfaces, in addition to NAS signaling from the MME to the UE, by introducing a flag in data packets, or including a flag in existing signaling messages between PGW and UE (e.g., PCO – Protocol Configuration Options [10]). Whilst achieving these requirements is not impossible, they admittedly add some level of complexity to P-GWs and some minor ones to UEs.

Finally, to support the establishment of a new (more optimal) PDN connection for new IP sessions, UEs need the ability to bind new IP sessions (when they are established) to a specific PDN connection or P-GW. A straightforward solution for such binding would be a mapping based on the destination IP address of the peer, application type, and protocol types. In this way, a UE always knows which IP flow/session uses which PDN connection. When all IP flows/sessions associated with a particular PDN connection are finished, the UE can trigger the release of the corresponding PDN connection. For this purpose, a timer based solution can be adopted, which simply tears down a PDN connection when no packets are sent for some time.

4   Performance Evaluation

In this section, we evaluate the proposed solutions and highlight their technical benefits. In the performance evaluation, the focus is on the case of active UEs. The conducted simulations were run for 100s; a duration long enough to ensure that the system has reached its stable state. They are based on the network simulator ns-3 [11] using a network topology as depicted in Fig. 4. We simulate 20 UEs, distributed uniformly around eNB1 over a surface of 2000 x 2000 m2. All simulated nodes are moving over the coverage areas of eNB1 and eNB2, changing their point of attachment to the network once the signal of the target eNB becomes stronger than that of the source eNB.

At the beginning of the simulation, each UE initiates an ON-OFF application, with ON time set to one second, sending data packets at a rate randomly selected from within the interval [75:150] Kbps, simulating applications ranging from VoIP to video streaming (e.g., YouTube). The packet payload length is set to 256 bytes. Upon moving to a new cell, UEs initiate new IP sessions with the same characteristics as described above.

The first metric used to evaluate the efficiency of the proposed solution consists in the transmission buffer length of P-GW1. Fig. 5 demonstrates that the conventional approach, which uses the old P-GW for the new IP sessions, experiences increased buffer lengths, resulting in several buffer overflows and also increasing the flows' latencies as depicted in Fig.7. Indeed, with the conventional approach, UEs do not consider nearby P-GWs when establishing new IP sessions. Instead, they establish their new IP sessions via the old gateway, which increases its load, resulting in buffer overflows and longer latencies.

As a direct consequence of buffer overflows, the aggregated packets loss, shown in Fig. 6, increases significantly which may degrade the quality of the different services. In contrast, by exploiting local P-GWs to accommodate new IP sessions, the proposed

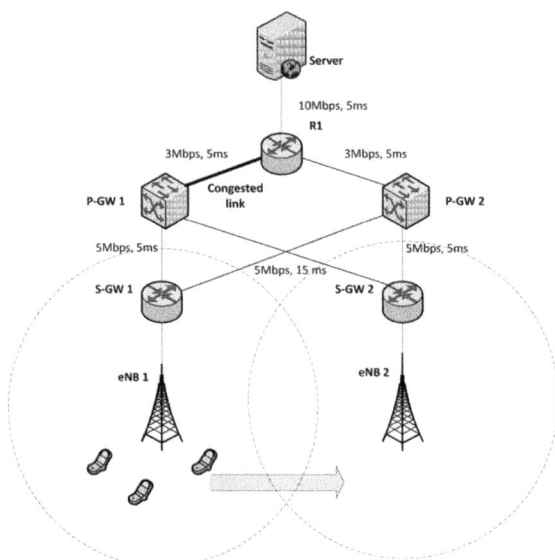

Fig. 4. The considered network topology

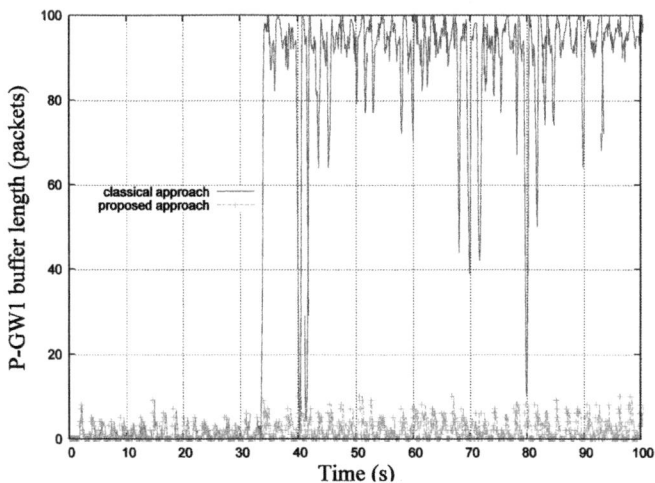

Fig. 5. The transmission buffer length at P-GW1

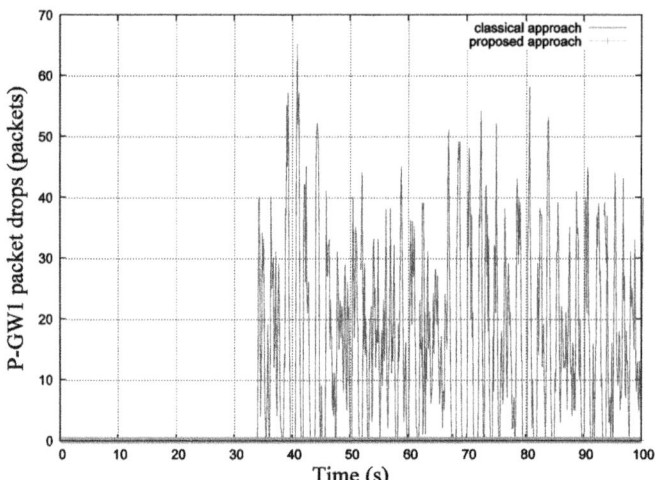

Fig. 6. The packet drops at P-GW1

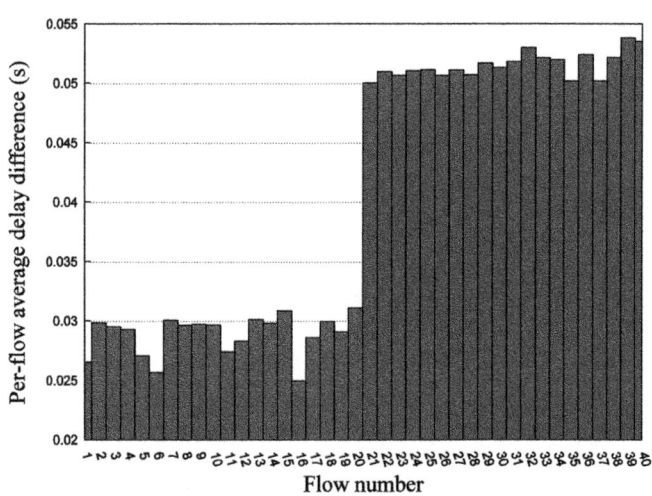

Fig. 7. The difference between the latencies of the classical and the proposed approaches

approach distributes better the traffic among the available P-GWs. This feature helps in avoiding congestion at the old P-GW, as demonstrated in Figs. 5 and 6. Indeed, the proposed approach exhibits no buffer overflows and almost null packet losses.

Fig. 7 depicts the difference between the average latencies experienced by each application in case of the conventional and the proposed approaches. The delay differences exhibited by the first 20 flows are approximately 30ms and correspond mainly to the buffering delay as the application packets traverse the same path to the server. Indeed, in both approaches, the first 20 flows are not impacted by gateway selection: they are created before the UEs' movement to the other eNB. Flows ranging from 21 to 40 are newly created after the handoff of UEs. They are therefore impacted by gateway selection. The flows 21-40 exhibit higher delay differences, in the order of approximately 50ms. In fact, when using the proposed approach, the system always selects optimal gateways, which significantly decreases the average delay for each flow. It should be recalled that already-established applications continue using the old P-GW.

Fig. 8 depicts the packet loss experienced by the different flows in case of the conventional approach (i.e., as almost no loss was experienced in case of the proposed approach). The loss is mainly a result of buffer overflows experienced at congested P-GWs. The loss may become more significant along with the increase in the number of mobile terminals, terminals' mobility, and/or the application's data transmission rates.

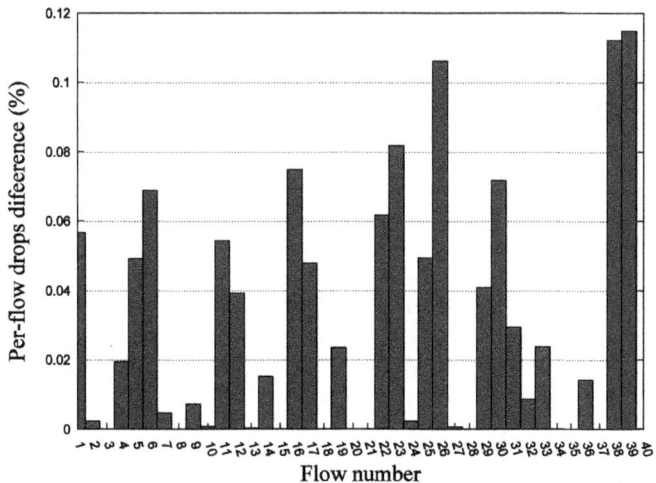

Fig. 8. The difference between the drops of the classical and the proposed approaches

5 Concluding Remarks

In this paper, we highlighted the need and benefits for always optimizing the path of mobile terminals (UEs) to their corresponding anchor gateways, namely P-GWs in EPS, in case of decentralized mobile operator networks and/or networks that adopt traffic offload strategies. Adequate methods were devised for UEs in ECM-idle mode and those in ECM-connected mode. In the latter case, we compared three methods that trigger a UE to first establish a new and more optimal PDN connection before creating new IP sessions, without compromising the old IP sessions. Admittedly, the devised methods involve some additional complexity at different core network nodes (depending on the solution). However, the benefits for operators, verified through simulations, clearly justify the required enhancements.

Whilst the main motivation behind this work consists in supporting the decentralization of future mobile operator networks and the envisioned traffic offload strategies, the devised solutions can also assist in energy saving and efficient load balancing.

Tailoring our proposed methods to such objectives forms the future directions of our research work in this area.

Acknowledgment

The work described in this paper is partially supported by the national French project ANR VERSO ViPeer.

References

1. Cisco Visual Networking Index: Global Mobile Data Traffic Forecast Update, 2009-2014, White Paper (February 2010)
2. MOBILE TRAFFIC GROWTH + COST PRESSURES = NEW SOLUTIONS?, New Mobile (January 2010)
3. TS Group Services ad System Aspects; Local IP Access and Selected IP Traffic offload (Rel. 10), 3GPP TR 23.829 V1.1.0 (May 2010)
4. 3rd Generation Partnership Project, Architecture enhancements for non-3GPP accesses Rel 10, 3GPP TS 23.402 V10.0.0 (June 2010)
5. 3rd Generation Partnership Project, General Packet Radio Service (GPRS) enhancements for Evolved Universal Terrestrial Radio Access Network (E-UTRAN) access, 3GPP TS 23.401 V10.0.0 (June 2010)
6. 3rd Generation Partnership Project, Non-Access-Stratum (NAS) protocol for Evolved Packet System (EPS); Stage 3, 3GPP TS 24.301 V9.3.0 (June 2010)
7. 3rd Generation Partnership Project, Access Network Discovery and Selection Function (ANDSF) Management Object (MO), 3GPP TS 24.312 V9.1.0 (March 2010)
8. 3rd Generation Partnership Project, Multi Access PDN connectivity and IP flow mobility, 3GPP TR 23.861 V1.3.0 (February 2010)
9. 3rd Generation Partnership Project, IP flow mobility and seamless Wireless Local Area Network (WLAN) offload; Stage 2, 3GPP TS 23.261 (June 2010)
10. 3rd Generation Partnership Project, Mobile radio interface Layer 3 specification; Core network protocols; Stage 3, 3GPP TS 24.008 V9.3.0 (June 2010)
11. ns3, The ns-3 Network Simulator, http://www.nsnam.org/

Femtocell Coverage Optimisation Using Statistical Verification

Tiejun Ma and Peter Pietzuch

Department of Computing,
Imperial College London,
London SW7 2AZ, United Kingdom
{tma,prp}@doc.ic.ac.uk

Abstract. Femtocells are small base stations that provide radio coverage for mobile devices in homes or office areas. In this paper, we consider the optimisation of a number of femtocells that provide joint coverage in enterprise environments. In such an environment, femtocells should minimise coverage overlap and coverage holes and ensure a balanced traffic workload among them. We use statistical verification techniques to monitor the probabilistic correctness of a given femtocell configuration at runtime. If there is any violation of the desired level of service, a self-optimisation procedure is triggered to improve the current configuration. Our evaluation results show that, compared with fixed time, interval-based optimisation, our approach achieves better coverage and can detect goal violations quickly with a given level of confidence when they occur frequently. It can also avoid unnecessary self-optimisation cycles, reducing the cost of self-optimisation.

Keywords: Femtocell, Self-Optimisation, Statistical Verification.

1 Introduction

Femtocells [5] are cellular base stations that cover small areas of tens of meters. They are low cost and low power devices that are normally installed by consumers in homes for better indoor mobile voice and data reception [3, 5]. A femtocell can be considered as a wireless cellular access point, which transfers data traffic through the home broadband connection to the operator's core network. By 2014, the deployment of femtocells is expected to reach around fifty million, providing service to more than a hundred million people [1].

An important challenge for femtocells is to optimise their radio coverage area dynamically. The goal is to achieve a desired level of performance for mobile transmission, avoid undesired interference and reduce power consumption. Providing optimal femtocell signal coverage is important to improve users' mobile usage experience as well as reduce service cost expenditure. Since femtocells are deployed by users themselves, they must also be able to self-configure all required parameters during operation with minimal user intervention [6].

J. Domingo-Pascual et al. (Eds.): NETWORKING 2011, Part I, LNCS 6640, pp. 343–354, 2011.

In enterprise environments, a number of femtocells may be deployed together to achieve joint coverage. This is also done to cover a large area while balancing user load, minimising coverage gaps without signal and wasteful coverage overlaps between multiple femtocells. Self-optimisation adapts the configuration of radio parameters of femtocells at runtime, for example, by setting the power level of their radio signals. Achieving good configuration is challenging due to the diversity and dynamic nature of the deployment environment. Since femtocells are deployed in a decentralised fashion, self-optimising femtocells might interference with each other. In addition, users of femtocells might move around at random, changing the workload and requiring dynamic hand-over between femtocells.

Current femtocell optimisation methods derive the average value of overlap, gap and workload within a fixed time interval and then compare this value to a predefined threshold, known as the optimisation goal [7,9]. This approach achieves reasonable performance and adaptability using an evolving algorithm as shown in [9]. However, its convergence rate is slow because self-optimisation is only triggered at fixed time intervals. When violations of optimisation goals are rare, the algorithm cannot prevent unnecessary re-computation of parameters, which wastes energy.

In contrast to observing the average value of a desired property, *statistical verification* [11, 12, 15] is a technique that checks the probabilistic correctness and satisfaction of a system against its desired behaviour. It has two major benefits: first, it adopts a formal mathematical specification to describe the desired system behaviour without ambiguity; second, it measures a property more accurately than the average value because it considers the probabilistic likelihood that a property is violated. The possibility of false positives and negatives can be bounded below a required level [15]. When property violations occur frequently, a fast response to trigger re-optimisation is needed rather than waiting for a given timeout (i.e., false negatives). When violations are rare, statistical verification can also avoid unnecessary self-optimisation (i.e., false positives). Thus, statistical verification reduces energy consumption as well as durations of instability caused by self-optimisation.

In this paper, we show how to use a statistical verification technique to achieve self-optimisation goals for coverage optimisation in a joint femtocell scenario. In particular, the contributions of the paper are:

– We model a femtocell network as a stochastic discrete event-based system and formalise its desired behaviour using stochastic temporal logic. This model describes properties rigorously and can be verified using statistical verification.
– We propose a statistical verification technique based on hypothesis testing for verifying properties in femtocell coverage optimisation. When a property violation is detected, self-optimisation is triggered.
– We evaluate our approach in simulation and show that, compared with a fixed time, interval-based approach, it is effective and can provide a 25%–38% improvement in terms of reduced coverage overlaps and gaps.

The remainder of the paper is organised as follows. In the next section, we describe related work on femtocell coverage optimisation and statistical verification. Section 3 introduces the goals of femtocell coverage optimisation in more detail. Section 4 provides the formal model and definitions for femtocell coverage optimisation. Section 5 presents evaluation results. We give conclusions and possible future work in Section 6.

2   State of the Art

2.1   Femtocell Coverage Optimisation

Due to the predicted adoption of femtocells, researchers have begun to consider the problem of coverage optimisation [4,6,7,8]. However, all of this work focuses on single femtocell coverage optimisation for small-area residential users rather than on multiple femtocells that achieve joint coverage in large enterprise environments. The goals are to provide good indoor coverage, preventing signals from leaking outdoors [6], and to increase the flexibility in deployment locations [7,8].

For multiple femtocells, the main optimisation goal is to reduce coverage overlaps and gaps, as well as to balance the workload among femtocells. Ho et al. [9] propose to use of genetic programming for coverage optimisation of a group of femtocells. Their approach finds a suitable signal power to control a femtocell's coverage area. The authors show that genetic programming with adjustment of signal power at a fixed time intervals can balance the average workload of femtocells, with only small coverage gaps and overlaps. Their work is closest to ours in terms of the scenario considered for joint femtocells coverage optimisation. However, our focus is on selecting an optimal trigger time for the signal power adjustment algorithm using statistical verification. We want to carry out fast yet accurate violation detection with probabilistic guarantees of optimisation goals. This avoids the drawbacks of approaches that use fixed time intervals and average values, which may take a long time to achieve optimisation goals and cannot avoid unnecessary self-optimisations.

2.2   Statistical Verification

Sen et al. [12] first considered verification of systems based on statistical hypothesis testing. The authors assume that the system has already been deployed so that execution traces cannot be controlled (black-box system). Verification is based on the observation of system behaviour, and they show the effectiveness of verification models based on continuous time Markov Chains. The idea of verifying a black-box system would be suitable for femtocell coverage optimisation when the self-optimisation algorithm is unknown, such as the evolving algorithm proposed in [9].

Younes and Simmons [15] extend Sen et al.'s approach in order to verify probabilistic and time-based properties of black-box systems, as generalised by semi-Markov decision processes. The authors adopt the *sequential probability ratio test* (SPRT) and verify system traces as stochastic discrete event based models. Such a model is suitable for modelling violations of femtocell optimisation goals. Thus we use a stochastic discrete event based model in this paper. Lee et al. [11,13] propose a monitoring and checking framework based on SPRT, which verifies the probabilistic correctness of a system based on its implementation at runtime. The authors show that such an approach can be adopted to verify probabilistic properties of wireless sensor networks. The idea of monitoring and checking implementations at runtime is suitable for the coverage optimisation of femtocells. Therefore, we adopt the same SPRT-based verification approach.

Next we give a short overview on the SPRT statistical verification approach. SPRT was first developed by Wald [14] as a likelihood-based test for observed data. It is designed for continuous monitoring. In practice, it is often evaluated at frequent but discrete time intervals. SPRT defines a *Null* (H_0) hypothesis that expresses the accepted value for the parameter under test and the *Alternative* (H_1) hypothesis as the unacceptable value for such parameter. Let $\mathbb{N}^+$ be the set of positive integers. SPRT takes a sample of pre-determined size $m \in \mathbb{N}^+$ and, based on the result obtained from drawing this fixed-size sample, it makes a decision regarding these two hypotheses as follows.

Let X_t be a random variable representing the number of observations at time t. With the classical SPRT, tests are performed continuously at every time t as additional data are collected. The test statistic is the likelihood or log-likelihood ratio defined as

$$\frac{p_{1m}}{p_{0m}} = \prod_{t=1}^{m} \frac{P(X_t|H_1)}{P(X_t|H_0)} \text{ or } \log(\frac{p_{1m}}{p_{0m}}) = \sum_{t=1}^{m} \frac{P(X_t|H_1)}{P(X_t|H_0)}. \tag{1}$$

This test statistic is sequentially monitored for all values of $t > 0$, until either H_0 is rejected by $\frac{p_{1m}}{p_{0m}} \geq (1-\beta)/\alpha$ or until H_1 is rejected by $\frac{p_{1m}}{p_{0m}} \leq \beta/(1-\alpha)$, where α, β are type I and type II errors. With this stopping rule, the Null hypothesis is falsely rejected with probability α when it is true, while the Alternative hypothesis is falsely rejected with probability β when it is true.

SPRT only needs a minimum number of external event observations to reach decisions with bounded probability of false positives and negatives [14]. This allows SPRT to be applied to black-box systems and make fast yet accurate decisions. Moreover, SPRT is computationally inexpensive. All of these features make it particularly suitable for the femtocell coverage optimisation problem.

3   Coverage Optimisation in Femtocells

Fig. 1 shows a simple femtocell network with four femtocells jointly providing mobile signal coverage in an office area. Each femtocell uses home broadband access as a backhaul to the mobile operator's core network [5]. For a femtocell, the coverage radius $\mathcal{R}$ indicates its radio coverage area and the change of signal power δ is used to adjust the radius. When a user equipment (UE), such as a mobile device, enters the femtocell network, it performs a handover to a femtocell, which is then used to transmit data. This results in improved mobile service and traffic offloading from the core network. However, there may be coverage gaps and coverage overlap between femtocells. When a UE is not covered by any femtocell (coverage gap), it has to handover to the core network again for its data

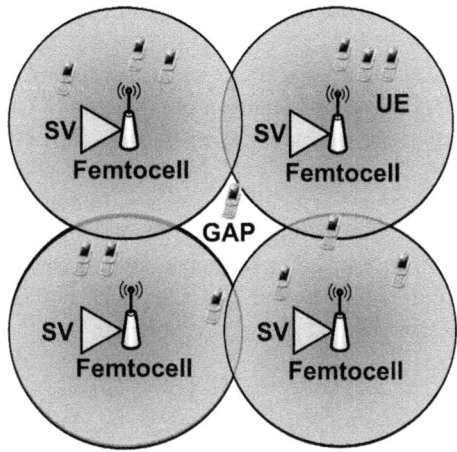

Fig. 1. Illustration of the Femtocell Coverage Optimisation Problem

transmission. When a UE is covered by more than one femtocell (coverage overlap), interference increases and this results in energy being wasted. Thus both coverage overlaps and gaps are undesired behaviour and should be avoided.

Each femtocell should serve a similar number of UEs to achieve load balancing. Therefore, the aim of femtocell coverage optimisation is to minimise coverage overlap and coverage gap, and also to balance the workload among femtocells. Solving this problem involves adjusting the signal power in order to control the coverage area of femtocells and deciding when such a signal power adjustment should be triggered. In this paper, we mainly address the later problem by adopting statistical verification. The following is a list of desired optimisation goals that we consider:

- Minimising the femtocell coverage gaps D. Coverage gaps can be inferred from the number of handovers between femtocells and the core network, which represents the number of times a UE enters a coverage gap;
- Balancing the load L among femtocells in a group to prevent overload or under-utilisation;
- Minimising the coverage overlaps O by reducing the signal power.

Here we give a simple mathematical formulation of this problem. Let D_{thr}, O_{thr}, L_{thr} be the required upper bound for femtocell coverage gap, overlap and load, respectively. Let $\mathbb{F}(D;\ O;\ L;\ D_{thr};\ O_{thr};\ L_{thr}) \xrightarrow{\delta} \mathcal{R}$ be a function (in short as $\mathbb{F} \xrightarrow{\delta} \mathcal{R}$), of which the input parameters are the current coverage performance observations (D, O, L) and the expected coverage performance $(D_{thr}, O_{thr}, L_{thr})$. Then the output parameter is a suitable power adjustment δ, which enables a femtocell to achieve the expected $\mathcal{R}$ that satisfies $(D_{thr}, O_{thr}, L_{thr})$. Such a signal power re-computation decision is made independently by each individual femtocell.

4   Statistical Verification for Femtocell Networks

To achieve the optimisation objectives given in Section 3, verification goals are used to check the probabilistic satisfaction of coverage optimisation. These verification goals can be represented as *eventually* $\forall n \in \textbf{Femtocells} \models (D < D_{thr})$ $\wedge (L < L_{thr}) \wedge (O < O_{thr})$ where $n \in \mathbb{N}^+$ is the number of femtocells in the network. These properties are typically expressed in temporal logics. We adopt the continuous stochastic logic (CSL) [2] as a formalism for expressing quantitative properties. This is because CSL is semantically capable of specifying property such as "with probability of at most 0.1" ($\mathbb{P}_{\leq 0.1}(\Phi)$). For the femtocell coverage optimisation problem, the desired properties are formulated as Equations (2)–(4):

$$\phi := \mathbb{P}_{\geq \theta_D}(true \ \mathbb{U}^{\leq t}(D < D_{thr})), \tag{2}$$

$$\varphi := \mathbb{P}_{\geq \theta_L}(true \ \mathbb{U}^{\leq t}(L < L_{thr})), \tag{3}$$

$$\psi := \mathbb{P}_{\geq \theta_O}(true \ \mathbb{U}^{\leq t}(O < O_{thr})), \tag{4}$$

where $\mathbb{U}^t$ is the bounded until path formula and t is a specified time. We regard t as t_{now} to indicate that a property should be satisfied when it is evaluated. θ_D, θ_L and θ_O represent the desired level of confidence of each property. For example, Equation (2) represents the probability of D being less than D_{thr} with the confidence of no less than the required level θ_D until now. Similar meaning applies to Equations (3)–(4) for O and L. If any of these properties are violated, then $\mathbb{F} \xrightarrow{\delta} \mathcal{R}$ should be triggered.

4.1   Modelling Overlap, Load and Gap Violations as Stochastic Discrete Event Processes

For checking probabilistic satisfaction of given properties, such as Equations (2)–(4), statistical verifiers need to be evaluated against the trace of occurred coverage overlap, coverage gap and overload events. Here we give a formal description of how the trace of these events is modelled.

Let Ω be a sample space. $\mathcal{F}$ is a collection of subsets of Ω. Let $\mathbb{P} : \mathcal{F} \to [0, 1]$ be a *probability measure* defined on $\mathcal{F}$. Let $(\Omega, \mathcal{F}, \mathbb{P})$ be a probability space. A stochastic process is a collection of random variables $X = \{X_t : t \in \mathbf{T}\}$ with index set $\mathbf{T}$.

For the property of a femtocell, there are two types of states, in which a property is either violated (*true*) or not (*false*). Thus it can be represented as a Bernoulli process, where each random variable X is a Bernoulli trial, mapping $\Omega \to \{true, false\}$, such that $X = true$ with probability p and $X = false$ with probability $1 - p$ ($p \in \mathbb{P}$). The probability distribution of a sum of n Bernoulli trials with parameter p can be represented as a binomial distribution with parameter n and p as follows:

$$B(n, p) = \sum_{j=0}^{x} \binom{n}{j} p^j (1 - p)^{n-j}. \tag{5}$$

For femtocell coverage overlap O, load L and gap D, let X_O, X_L and X_D be three random variables. X_O represents the occurrence of an overlap event, where $X_O = false$ when no overlap event occurred, $X_O = true$ when an overlap event occurred. The maximum allowed probability $\mathbb{P}(X_O = true)$ of such a Bernoulli process is O_{thr}. Similarly, X_L and X_D represent the occurrence of an overload or gap event, respectively.

In this a way, we model system states of a femtocell as a stochastic discrete event system (denoted as $\mathcal{M}$), which is similar to the work introduced in [15]. The state space of such a system can be represented as sequential observations at times $t_1, t_2, \cdots$, for example, a sequence of overlap events. The state change is discrete rather than continuous and is triggered when these events are detected. The verification procedure is to compute whether the occurrence probability of these events is above the predefined threshold. Thus the verification of each desired property shown in Equations (2)–(4) can be achieved by using the verification method shown in Equation (1).

It is worth mentioning that, in the real world, X_O, X_L and X_D are not completely independent. When reducing a femtocell's coverage area, overlap and load are reduced, but the possibility of gaps happening is increased and vice-versa. However, when triggering self-optimisation using statistical verification, the decision is made independently according to the observation of each property. Thus we consider X_O, X_L and X_D as independent observations.

4.2   Femtocell Coverage Verification Procedure

Based on the stochastic discrete event models from the previous section, we describe in detail how to detect violations using statistical verifiers.

For a femtocell, let trigger $\mathbb{F} \xrightarrow{\delta} \mathcal{R}$ at an arbitrary time t be a binary decision $\mathcal{D}_t \in \{true, false\}, t \in \{1, 2, \cdots\}$, where $true$ represents triggering this function and $false$ represents the opposite decision. Let V_1, V_2, $\cdots$, V_N ($N \in \mathbb{N}^+$) be verifiers. Let P_1, P_2, $\cdots$, P_N be properties, which are used to describe desired behaviours of a femtocell network. Each verifier V_i monitors a given property P_i ($i \in N$). At time t, a verifier V_i observes a random variable X_t according to P_i. Then, let H_0^i be the Null hypothesis of the observed property P_i that is violated and H_1^i be the Alternative hypothesis.

During the verification procedure, if H_0^i holds, then $D_t = true$, otherwise if H_1^i holds, $D_t = false$. Based on the verification rule (see Equation (1)), we have the following forms:

$$\text{accept } H_0^i \text{ if } p_t^i \leq \log(\frac{\beta}{1 - \alpha});$$

$$\text{accept } H_1^i \text{ if } p_t^i \geq \log(\frac{1 - \beta}{\alpha}); \tag{6}$$

$$\text{continue if } log(\frac{\beta}{1 - \alpha}) \leq p_t^i \leq \log(\frac{1 - \beta}{\alpha}),$$

where p_t^i is the $\log(\frac{p_{1m}}{p_{0m}})$ in Equation (1), α and β are the type I and type II errors respectively (α, $\beta = 1 - \theta$), as introduced in Section 2.2.

Algorithm 1. Statistical verification based coverage optimisation algorithm

---

1. **List of verification goals in Equations (2)–(4):** $\mathcal{M} \models \phi, \varphi, \psi$;
2. Input threshold θ for coverage gap, overlap and load $D_{thr}; O_{thr}; L_{thr}$;
3. **Start statistical verifiers V_i for observing each given property**
4. For each property $\in \{\phi, \varphi, \psi\}$, when an observation X_t is received
5. **Verify each property against Equation (6)**
6. **if** a given property is violated ($\mathcal{M} \not\models \{\phi, \varphi, \psi\}$) **then**
7.   Trigger the self-optimisaton algorithm ($\mathbb{F} \xrightarrow{\delta} \mathcal{R}$).
8.   Output optimised parameters δ
9.   Clean observations X_ts for the verified property
10. **end if**
11. Continue the statistical property monitoring

With the verification procedure from Equation (6), each femtocell in the network has one or more verifiers and makes the decision of triggering $\mathbb{F} \xrightarrow{\delta} \mathcal{R}$. For example, $H_0^D : \mathbb{P}(X_D = true) \geq D_{thr}$ represents the hypothesis that the occurrence probability of coverage gap exceeds the given threshold ($\mathcal{M} \not\models \phi$ or $\mathcal{M} \models \neg\phi$) and $H_1^D : \mathbb{P}(X_D = true) \leq D_{thr}$ represents the alternative hypothesis ($\mathcal{M} \models \phi$). When the H_0^D is accepted, a violation is detected, so the decision $\mathcal{D}_t$ should be set to be *true* in order to trigger $\mathbb{F} \xrightarrow{\delta} \mathcal{R}$.

Algorithm 1 gives the pseudo code for the statistical verification-based femtocell coverage optimisation procedure. Lines 1–2 specify properties as verification goals with a given confidence level θ. In lines 3–4, verifiers are initiated $V_1, \cdots, V_N$ for each given property. Lines 5–10 specify that if a violation is detected, $\mathbb{F} \xrightarrow{\delta} \mathcal{R}$ is triggered. Here we keep $\mathbb{F} \xrightarrow{\delta} \mathcal{R}$ as a high-level abstraction, which can be substituted by any algorithm designed for this type of problem. After the optimisation procedure finishes, the previous observation history is cleared to refresh the verifiers. Eventually the femtocell network should become stable and achieve $\{\phi, \varphi, \psi\}$.

5   Evaluation

The goal of our evaluation is to examine the effectiveness of our statistical verification approach in terms of reducing the probability of coverage overlaps and gaps, and achieving optimisation quickly as well as avoiding unnecessary re-optimisation. We investigate the impact of the proposed *statistical verification-based approach* (noted as *SVA*). The results obtained are compared with the performance of the *fixed-time interval approach* (noted as *FIA*). The underlying power adjustment algorithms are considered to be the same.

We implement a simulator for a four-femtocell network deployed within a plane office area as introduced shown in Fig. 1. The statistical verification module is implemented as part of Matlab 2009b. The occurred coverage overlap or gap events are collected by each femtocell and passed to its statistical verifier as input. For each statistical verifier, Equation (6) is implemented as the main

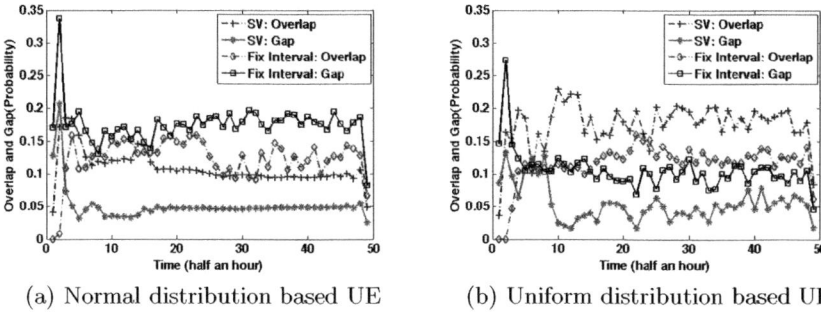

(a) Normal distribution based UE   (b) Uniform distribution based UE

Fig. 2. Overlap and Gap Probabilities for the Femtocell Network

function where its likelihood function follows Equation (1) and the probability distribution of the input data follows Equation (5). The violation of desired coverage overlap or gap threshold is used as the Null hypothesis in Equation (6).

For the initial deployment, we assume that the femtocells are already placed through automatic planning [10]. We ignore the required property of workload L_{thr} for simplicity. We implement UEs, which randomly appear in the femtocell network, to simulate user behaviour. We set the number of UEs to be 24, which has the same ratio of UEs and femtocells as in [9], where there are 50 femtocells and 300 UEs. We assume that the average time of a UE at a given location within the femtocell network is 90 seconds, chosen from an exponential distribution.

It is necessary to make assumptions about the spatial location of UEs. A normal distribution is a good choice due to the Central Limit Theorem—sufficiently many location samples from UEs can be approximately normal distributed. A uniform distribution is also a good model for the behaviour in an office environment because UEs are often equally distributed around the office area. We assume that UEs move within a square area around the centre of femtocells. Such a setting results in a higher overlap and gap possibilities than that of UEs moving across all possible locations in a femtocell network. Here we apply the same settings as in [9], setting $D_{thr} = 0.1$ and $O_{thr} = 0.3$. We adopt $\theta_D = \theta_O = 0.95$ as the verification confidence level.

5.1 Evaluation Results

The evaluation experiments were run on a MacOS 10.5.8 machine with an Intel 2.53 GHz core 2 CPU with 4 GB of RAM. We simulate 24 hours in the lifetime of a four-femtocell network.

Fig. 2 shows the impact of the occurrence probability of coverage overlap P_O and gap P_D under the SVA and FIA schemes. The results show that the SVA achieves a better satisfaction by lowering P_O and P_D. The SVA drastically reduces P_O by 25% and P_D by 38% under UEs following a normal distribution (Fig. 2(a)) compared to FIA. A similar 26% improvement in reducing the P_D for UEs with uniformly distributed movements is shown Fig. 2(b). However, SVA

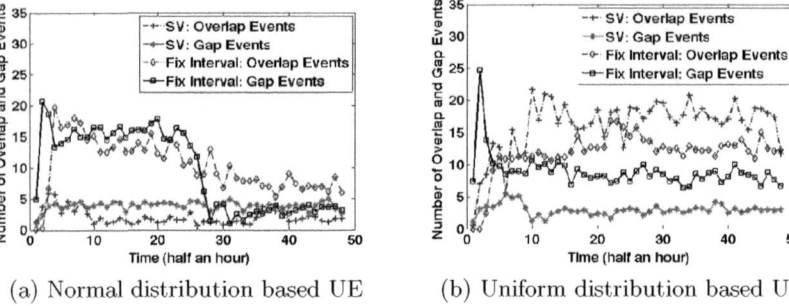

(a) Normal distribution based UE   (b) Uniform distribution based UE

Fig. 3. Average Number of Overlap and Gap Events per Femtocell

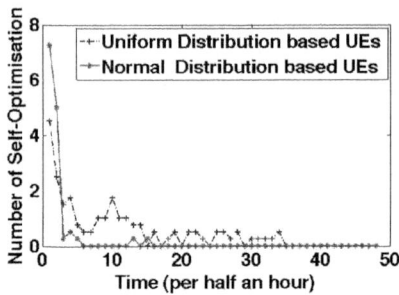

Fig. 4. Number of Self-Optimisations per 0.5 hour

does not improve P_O and the overlap is more likely to be increased by about 24% compared to FIA. Although P_O tends to increase, the overlap is still lower than a threshold $O_{thr} \leq 0.3$. Furthermore P_D is lower than P_D of FIA.

Analogous to the occurrence probability, the number of occurred overlap events X_O and gap events X_D can be also observed under two UE movement schemes. Fig. 3 shows that SVA successfully reduces the number of occurred overlap events and gap events in comparison to FIA. For UEs with normally distributed movements, Fig. 3(a) shows that SVA lowers the number of generated overlap and gap events. The number of gap events is also reduced under uniformly distributed movements (Fig. 3(b)).This in agreement with the result from Fig. 2. Both results demonstrate that SVA achieves a better optimisation than FIA.

We also record the number of triggered power adjustments ($\mathbb{F} \xrightarrow{\delta} \mathcal{R}$). For FIA, within each fixed time interval, there is always one power adjustment if the desired level of overlap or gap probability has not been achieved. Fig. 4 shows the number of power adjustments per half hour using SVA. Note that for both normal and uniform distributions of UE movements, SVA detects violation initially rapidly and avoids unnecessary self-optimisations as whole network approaches a stable optimal state.

5.2 Discussion

A limitation of our statistical verification approach is that, if a desired property is infeasible to achieve and violations occur very often, power adjustments are triggered frequently without being able to converge to the optimisation goal. To avoid such a problem, a limit on the number of self-optimisations during a given time interval can be added to SVA. This might slow down the convergence speed of the optimisation but it prevents the system from becoming unstable.

Moreover, different office environments might require different coverage overlap, gap and verification confidence values. A smaller overlap or gap threshold makes it harder for the femtocell network to achieve its self-optimisation goal. Higher verification confidence means that a verifier requires more evidence (i.e., overlap or gap events) to detect violations. It would be interesting to explore the impact of this parameter and we plan to do this as part of future work.

6 Conclusions and Future Work

In this paper, we have shown how to use statistical verification to manage violations of optimisation goals in order to provide joint coverage in femtocell networks. Our proposed approach is based on a sequential likelihood ratio hypothesis test. It improves selection time to trigger the signal power adjustment procedure of femtocells, and in turn improves the satisfaction of desired coverage performance. Our evaluation results show that our approach performs better than a fixed time interval-based technique, resulting in a better satisfaction of the desired upper bound of overlap and gap occurrence probabilities. Our approach can help optimise a femtocell network faster and avoid unnecessary signal power adjustments when violations are rare. All these features are necessary for guaranteeing fast, accurate and autonomous femtocell coverage self-optimisation.

For future work, we will consider scenarios with larger enterprise femtocell networks and changes in femtocell locations in order to evaluate the scalability, convergence speed and stability of statistical verification. In addition, we want to exploit peak and off-peak usage patterns in femtocell networks, leading to more realistic workloads of daily and weekly patterns of mobile devices.

Acknowledgements

The work reported in this paper has formed part of the Flexible Networks area of the Core 5 Research Programme of the Virtual Centre of Excellence in Mobile & Personal Communications, Mobile VCE (www.mobilevce.com), and has been jointly funded by Mobile VCE's industrial member companies and the UK Government, via the Engineering and Physical Sciences Research Council.

References

1. Femtocell Market Status (2010), http://www.femtoforum.org/femto/Files/File/InformaFemtocellMarketStatusQ12010F.pdf

2. Aziz, A., Sanwal, K., Singhal, V., Brayton, R.: Verifying Continuous-time Markov Chains. In: Alur, R., Henzinger, T.A. (eds.) CAV 1996. LNCS, vol. 1102, pp. 269–276. Springer, Heidelberg (1996)
3. Chandrasekhar, V., Andrews, J., Gatherer, A.: Femtocell Networks: A Survey. IEEE Communications Magazine 46(9), 59–67 (2008)
4. Choi, S., Lee, T., Chung, M., Choo, H.: Adaptive Coverage Adjustment for Femtocell Management in a Residential Scenario. Management Enabling the Future Internet for Changing Business and New Computing Services, 221–230 (2009)
5. Claussen, H., Ho, L., Samuel, L.: An Overview of the Femtocell Concept. Bell Labs Technical Journal 13(1), 221 (2008)
6. Claussen, H., Ho, L., Samuel, L.: Self-Optimization of Coverage for Femtocell Deployments. In: Proc. Wireless Telecom. Symp., pp. 278–285
7. Claussen, H., Pivit, F.: Femtocell Coverage Optimization Using Switched Multi-Element Antennas. In: IEEE Int. Conf. on Comm., pp. 1–6 (2009)
8. Duran, A., Carrasco, G.: UMTS Femtocell Performance in Massive Deployments: Capacity and GoS Implications. Bell Labs Technical Journal 14(2), 185–202 (2009)
9. Ho, L., Ashraf, I., Claussen, H.: Evolving Femtocell Coverage Optimization Algorithms Using Genetic Programming. In: 20th IEEE In. Sym. on Personal, Indoor and Mobile Radio Comm., pp. 2132–2136. IEEE, Los Alamitos (2010)
10. Molina, A., Athanasiadou, G., Nix, A.: The Automatic Location of Base-stations for Optimised Cellular Coverage: A New Combinatorial Approach. In: IEEE 49th Vehicular Technology Conference, vol. 1, pp. 606–610. IEEE, Los Alamitos (2002)
11. Sammapun, U., Lee, I., Sokolsky, O.: RT-MaC: Runtime Monitoring and Checking of Quantitative and Probabilistic Properties. In: 11th IEEE Int. Conf. on Embedded and Real-Time Comp. Sys. and Apps., pp. 147–153 (2005)
12. Sen, K., Viswanathan, M., Agha, G.: On statistical model checking of stochastic systems. In: Etessami, K., Rajamani, S.K. (eds.) CAV 2005. LNCS, vol. 3576, pp. 266–280. Springer, Heidelberg (2005)
13. Sokolsky, O., Sammapun, U., Regehr, J., Lee, I.: Runtime Verification for Wireless Sensor Network Applications. In: Runtime Verification (2008)
14. Wald, A.: Sequential Tests of Statistical Hypotheses. The Annals of Mathematical Statistics 16(2), 117–186 (1945)
15. Younes, H., Simmons, R.: Statistical Probabilistic Model Checking with a Focus on Time-Bounded Properties. Information and Computation 204(9), 1368–1409 (2006)

Points of Interest Coverage with Connectivity Constraints Using Wireless Mobile Sensors

Milan Erdelj[1], Tahiry Razafindralambo[1], and David Simplot-Ryl[2]

[1] INRIA Lille - Nord Europe
firstname.lastname@inria.fr
[2] University of Lille 1
david.simplot@univ-lille1.fr

Abstract. The coverage of Points of Interest (PoI) is a classical requirement in mobile wireless sensor applications. Optimizing the sensors self-
deployment over a PoI while maintaining the connectivity between the
sensors and the sink is thus a fundamental issue. This article addresses
the problem of autonomous deployment of mobile sensors that need to
cover a predefined PoI with a connectivity constraints and provides the
solution to it using Relative Neighborhood Graphs (RNG). Our deployment scheme minimizes the number of sensors used for connectivity thus
increasing the number of monitoring sensors. Analytical results, simulation results and real implementation are provided to show the efficiency
of our algorithm.

1 Introduction

Wireless sensor networks have received a lot of attention in recent years due
to their potential applications in various areas such as environment monitoring
[12,4]. Covering and monitoring events from the environment in a given area are
difficult tasks. Indeed, sensors have to be correctly placed to monitor the events
and a connection between the monitoring sensors and a base station (sink) have
to be kept to report data.

In this context, sensor placement can be divided into off-line and online
schemes. Although off-line deployments can provide optimal placement of sensors, they require precise knowledge of the events' locations. Online deployments
can cope with this drawback but are only feasible when sensors have motion capabilities. However, the main advantage of online deployments is the possibility
to obtain particular topologies which can provide properties such as connectivity.

Sensor placement related to coverage issues is intensively studied in the literature, and can be divided into three categories. *The full coverage* problem aims
at covering the whole area. Sensors are deployed to maximize the covered area
[6]. *The barrier coverage* problem aims at detecting intrusion on a given area.
Sensors have to form a dense barrier in order to detect each event that crosses
the barrier [8]. *Point of Interest coverage* aims at monitoring specific points in
the field of interest [9]. Different examples and results related to the deployment
of sensors can be found in [18].

J. Domingo-Pascual et al. (Eds.): NETWORKING 2011, Part I, LNCS 6640, pp. 355–366, 2011.
© IFIP International Federation for Information Processing 2011

Previous works on Points of Interest (PoI) coverage using mobile sensors, such as [9], do not consider the use of a base station where sensors have to report data and to which a sensor have to be permanently connected either directly or in a multi-hop fashion. The use of a base station in PoI coverage increases the deployment complexity since a connectivity constraint is added. In this article, we report a solution that solves the PoI coverage problem. We consider a network composed by mobile sensors and a base station (data sink). We also assume that at the beginning of the deployment the sensors are connected to the base station. In our deployment solution, connectivity is the main constraint and, therefore, is maintained all along the deployment procedure by a local control of the topology.

In our proposed solution, each sensor moves toward a PoI but also maintains the connectivity with a subset of its neighboring sensors. The use of the Relative Neighborhood Graph (RNG) reduction for choosing the subset of neighbors and local connection preservation provides a global connectivity of the network. Once the global connectivity can be provided locally, we want the sensors to deploy in such a way that the number of sensors used for connectivity is minimized and the number of sensors that cover the PoI is maximized.

The main contribution of this paper is a deployment algorithm that has the following properties:

- Our algorithm achieves *PoI coverage*. Examples of static, moving and multiple PoI coverage are provided.
- *Connectivity* between each sensor and the base station is kept all along the deployment procedure.
- Our algorithm is *local* i.e., every decision taken is based on local neighborhood information only and does not require synchronization.
- It is *efficient*, it minimizes the number of connectivity sensors and maximizes the number of covering sensors.

The rest of this paper is organized as follows: Section 2 provides background which includes state of the art, assumptions, definitions and a problem statement. Section 3 describes our deployment algorithm with its properties. Simulation results are given in Section 4 where we consider static, moving and multiple PoI coverage. Real implementation of our algorithm using Wifibots[2] is presented in Section 5 and conclusions are drawn in Section 6.

2   Background and Assumptions

2.1   State of the Art

In this section, papers about deployment and self-deployment of wireless sensor networks are reviewed and we shortly extend this state of the art to mobile robots deployment. As our main focus is PoI coverage with connectivity constraint, we only cite papers that consider these two properties. Moreover, we consider the deployment of mobile sensors but more interested readers can refer to [6,13] for static random deployment strategies, to [10,3] for off-line computation of sensor placement and to [18] or [15] for complete surveys. There are mainly three ways to optimize the deployment or the placement of mobile sensors: the coverage pattern [16], grid quorum [7] and virtual force based movements [5].

This paper belongs to the virtual force based movement category, where sensors are repelled or attracted each other by using virtual forces like electromagnetic particles. The sensors move step by step and the virtual forces are computed based on the set or a subset of neighboring sensors. With this deployment strategy connectivity and PoI coverage can be provided.

In this article, we consider an environmental monitoring application. In many cases, monitoring the whole area might be unnecessary. Therefore, monitoring some points of interest increases the sensing performance and reduces the deployment cost. Surprisingly, very few works consider the actual problem of PoI coverage. To the best of our knowledge, the only work that consider PoI coverage is [9]. In [9], authors propose an algorithm to monitor some specific points periodically. Unlike the work presented in this paper, results from [9] do not consider connectivity issue. In [11], authors developed an algorithm to deploy the sensors around a PoI following a triangle tesselation. In this work, the PoI is not covered by all the sensors and is only used as a focus point.

Our work considers single and multiple PoI coverage where connectivity has to be kept between the sensors that cover the PoI and a base station (or sink). Moreover, we increase the connectivity constraint and provide an algorithm in which connectivity is kept all along the deployment procedure.

2.2   Motivation and Preliminaries

By considering the environmental monitoring, we assume that an event is detected by an external entity and not by the mobile wireless sensor network itself. Moreover, we assume that this external entity can precisely define the event's location. When the event is detected, it's position is sent to the mobile sensors through a fixed base station. The mobile sensors, then, self-deploy to monitor this event and report data (such as temperature, humidity, video, etc.) to the sink in a multi-hop fashion. Furthermore, it is possible to have more that one event and these events could be mobile. To dynamically adapt to the changing requirements, the deployment algorithm must provide properties such as connectivity all along the deployment procedure. We use the following definitions and notations for the network model:

Definition 1. *Let $G(V, E)$ be the graph representing the sensor network. V is the set of vertices each one representing a sensor. $E \subseteq V^2$ is the set of edges; $E = \{(u, v) \in V^2 \mid u \neq v \wedge d(u, v) \leq R\}$, where $d(u, v)$ is the euclidean distance between sensors u and v and R is the communication range. $G(V, E)$ is our model of the sensor network.*

Assumption 1. *We assume that each sensor has its position denoted by $(x(u), y(u))$ for sensor u. This position can be provided by any internal mechanisms or external entities such as GPS.*

Assumption 2. *We assume that at the beginning of the deployment the sensors are randomly spread out around the base station at a maximum distance of $d < R/4$ from the base station. This condition ensures that the network is connected.*

2.3 Relative Neighborhood Graph

The Relative Neighborhood Graph (RNG) [14] is a graph reduction method. Given an initial graph G, the RNG graph extracted from G is a graph with a reduced number of edges but the same number of vertices. Let the sensors be the vertices of the initial graph and that there exists an edge between two vertices if the two sensors can communicate directly. We assume here that the communication between two sensors is possible only if the distance between them is less than a given communication range. To build an RNG from an initial graph G, an edge that connects two sensors is removed if there exists another sensor that is at a lower distance from both sensors.

Using the RNG reduction has two main advantages. First, the RNG reduction can be computed locally by each sensor since sensors only need the distances with its neighbors [14]. Second, given that the initial graph is connected, the RNG reduction is also connected. These two properties are important for scalability and connectivity preservation. Indeed, to preserve the connectivity of the whole network, each sensor has to preserve the connectivity with its neighbors that are part of the RNG graph. In our algorithm, we use these properties to preserve connectivity and to ease the movement computation.

3   Deployment Algorithm for PoI Coverage

3.1 Basic Idea

At the beginning of the deployment, all the sensors are in the communication range of the base station and all the sensors are also within communication range of each other. Each sensor moves independently from the other sensors. The sensors are not synchronized, motion decisions are taken individually and all the sensors run the same algorithm. It is important to notice here that the base station can compute an optimal placement and can provide this location to each sensor which can move toward this optimal position. However by doing so, it is hard to ensure that the network is connected all along the deployment procedure. Therefore, when tracking a moving PoI, some sensors may not have an up-to-date position and placement.

In order to cover the PoI, the sensors move toward one predefined point that may be chosen randomly within the set of PoIs. The direction of a sensor is given by the following unit vector : $\overrightarrow{\Delta} = \overrightarrow{d_p}/||\overrightarrow{d_p}||$, where $\overrightarrow{d_p}$ is the vector toward the PoI. The movement vector of a sensor is thus $\overrightarrow{m} = d \cdot \overrightarrow{\Delta}$, where d is the maximum distance that the sensor covers while maintaining connectivity with its RNG neighbors. If $d^{+}(u)$ is the distance between sensor u and its farthest RNG neighbor, $d \leq (R - d^{+}(u))/2$, where R is the communication range. This condition ensures that, when considering worst case movements, all the RNG neighbors stay connected. After the computation is finished, sensor u moves according to vector $\overrightarrow{m}$.

In order to avoid an infinite small movements of sensors, we add a condition on d. If $d < \epsilon_1$, with $\epsilon_1 > 0$ then we set $d = 0$. Note also that for termination purpose, if the distance between a sensor and the PoI is lower than a given threshold ϵ_2, with $\epsilon_2 > 0$ (related to sensing range), the sensor stops moving.

3.2 Deployment Algorithm

Our deployment process is formally described in Algorithm 1. The algorithm is divided into three parts which are related to three important aspects of deploying a fleet of mobile sensors. The first part is related to the coverage requirements. The second part considers connectivity preservation and the sensor's movement is performed in the third part. Since these three parts are independent, it is simple to modify each part independently from the others. It is thus easy to modify the direction computation while using the same path planning algorithm and maintaining connectivity.

Algorithm 1. Deployment process

Part 1 – Direction computation on sensor u:
1: $\overrightarrow{\Delta} = \overrightarrow{d_p}/\|\overrightarrow{d_p}\|$
Part 2 – Distance/speed computation on sensor u:
1: $d = (R - d^+(u))/2$
2: $\nu = \frac{d}{\delta}$, δ is the periodicity of movement decision, ν the speed
Part 3 – Motion of sensor u:
1: Move to new position using: speed ν, direction $\overrightarrow{\Delta}$ and distance d.
2: Take obstacle into account.

3.3 Algorithm Properties

Theorem 1. *Connectivity. If at time $t = T$ the graph is connected, $\forall t = i, i > T$ the resulting graph at time $t = i$ is connected.*

Proof. In an asynchronous environment, sensors can run algorithm 1 at any time. Let u and v be two sensors and u and v are connected a time $t = T$. Let $u \in RNG(v)$, $v \in RNG(u)$ and $d(u,v) = d^+(u)$. Let us assume that two sensors run Algorithm 1 at the same time and that they are moving in the opposite direction of each other. The maximum distance covered by sensor v depends on $d(u,v)$. Since $d(u,v) \leq d^+(v)$ the maximum distance covered by sensor v is $d_v = (R - d^+(v))/2 \leq (R - d^+(u))/2$. Therefore, the maximum distance between sensor u and v after their respective movement is $d(u,v) + (R - d^+(u))/2 + (R - d^+(v))/2 \leq R$. Thus, after their respective movement, sensors u and v are still connected. If the connection to the farthest RNG neighbor is maintained, the connection to closer RNG neighbors is also maintained and if the connectivity with RNG neighbors is kept, network connectivity is thus also kept [14].

Theorem 2. *Straight line deployment. Let b be the base station, p be the PoI and let us assume that sensor u is not on the segment $[b, p]$. The distance h between a sensor and the segment $[b, p]$ is strictly decreasing.*

Proof. At each step of the deployment, sensor u moves toward the PoI. Since the direction of the sensor is $\overrightarrow{up}$, where u is the sensor's position and the covered distance is $d \geq 0$, the distance between a sensor and the PoI is strictly decreasing. As a consequence, the distance between the sensor and the segment $[b, p]$ is also decreasing. It is worth noting that when the sensor $u \in [b, p]$, it remains in the segment during movement and $h = 0$.

Theorems above show that our algorithm preserves connectivity all along the deployment procedure. Furthermore, we bring proof that, at the end of deployment, sensors used for connectivity are more likely to form a straight line toward the PoI and to be at the distance of $R - \epsilon_1$ from their neighbors.

4   Deployment Simulations

This section shows the performance evaluation results of our algorithm regarding the static, moving and multiple PoI coverage. Simulations were performed using WSNet[1]. In the simulations, δ is set to $5s$ and we set the communication range to be equal to the sensing range, but this assumption can be easily modified without affecting the behavior of the deployment. In this paper, we mainly focus on connectivity for PoI coverage. Therefore, comparisons with other works are hard to provide since literature lacks similar algorithms.

4.1   Static PoI

Deployment example. Figure 1 shows an example of the deployment's evolution where the PoI is located at position $[70, 100]$. After $180s$, the deployment is finished. In the simulation setup, the sensors move during five seconds and compute a new direction after their movements. This figure shows that the sensors form a straight line between the base station and the PoI which reduces the number of sensors used for connectivity preservation and therefore increases the number of sensors involved in coverage.

Coverage quality. The Figure 2(a) presents the number of covering sensors w.r.t. the distance between the PoI and the base station. In the simulation, the base station is considered as a sensor which is not mobile. That is, we consider 20 sensors including the base station. This figure shows that the number of sensors used for connectivity is minimized and that the number of covering sensors is maximized. For example, when the PoI is at distance 40, we need 3 sensors for

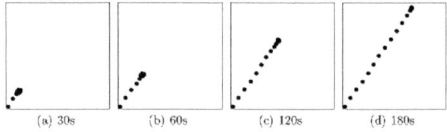

(a) 30s   (b) 60s   (c) 120s   (d) 180s

Fig. 1. Evolution of sensors' positions depending on time. In this simulation there are 20 sensors with a range of 10 on a square of 100×100. The PoI is located at $[70, 100]$.

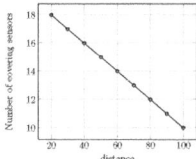
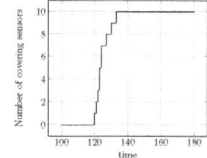
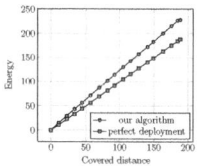

(a) Number of covering sensors w.r.t distance.

(b) Number of covering sensors w.r.t time.

(c) Energy consumed w.r.t distance.

Fig. 2. Coverage quality, deployment speed and energy consumption

connectivity at distances 10, 20, 30 and the base station at distance 0, which means that 4 sensors are used for connectivity and 16 sensors cover the PoI.

Deployment speed. The Figure 2(b) plots the number of covering sensors depending on time. In this simulation, PoI is at distance 100 and 20 mobile sensors are considered. A movement decision is taken every $\delta = 5s$. This figure shows that the first PoI is covered by at least one sensor after $120s$. Note here that we check the coverage every $1s$. This means that the first covering sensor has a mean speed of $0.75m/s$ ($90m$ covered distance after $120s$).

Energy consumption. To evaluate the energy consumed by each sensor during the deployment, we consider a simple energy model where the energy consumed by a sensor u is: $E(u) = d\alpha + \beta$, where d is the covered distance and α and β are constants (here, $\alpha = 1$, $\beta = 1$). This simple energy model considers the distance covered by a sensor but also penalizes multiple small movements. Figure 2(c) shows the energy consumption of each sensor for a deployment of 20 sensors and a PoI at $[100, 100]$. This figure shows that the energy consumption is linear depending on the covered distance. Moreover, our scheme consumes small amount of energy since (for example) for a covered distance of $105m$, 130 energy units are needed. We can notice that a sensor can cover $R/2 = 5m$ in every movement decision period since it has to maintain connectivity with its neighbors. Therefore, the sensor needs at least $105/5 = 21$ iterations to cover $105m$. The energy consumed by the sensor is at least $E(u) = 105 \times 1 + 1 \times 21 = 126$ which is very close to 130.

4.2 Moving PoI

Deployment strategies and examples. We present three different strategies for covering a new PoI when the sensors are already deployed. In the first strategy, STR1, the sensors first move back to the base station before deploying toward the new location of the PoI. This strategy provides a high coverage quality but increases the deployment duration and the amount of energy consumed. In the second strategy, STR2, the sensors try to move directly toward the location of the PoI without going back to the base station. This strategy reduces the time needed to cover the new PoI but also reduces the coverage quality since an

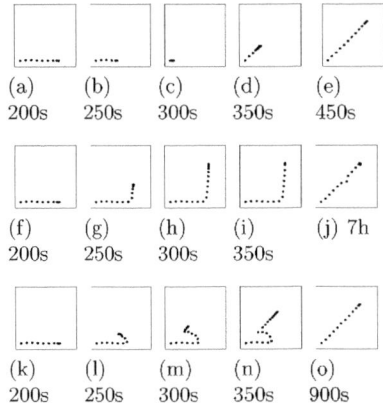

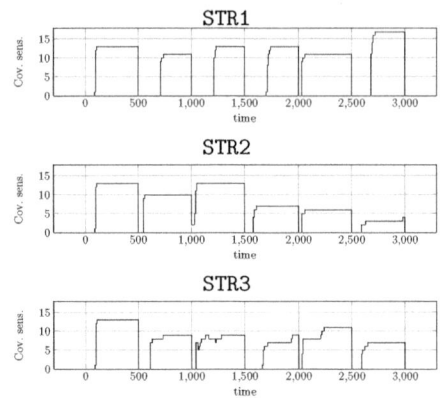

Fig. 3. Sensors' positions depending on time. In this simulation there is 20 sensors with a range of 10 on a square of 100×100. The PoI is first located at $[70, 0]$ and then at $[70, 70]$ after $200s$.

Fig. 4. Number of covering sensors w.r.t time. Simulation parameters: $R = 10$, 20 sensors including the base station. The simulation lasts $3000s$. A new location of the PoI is chosen every $500s$.

increasing number of sensor is needed to preserve connectivity. The third strategy, STR3, is a mix of STR1 and STR2, in which sensors first move toward the segment $[b, p]$ and after, when the distance between the particular sensor and the segment is lower than $R/4$, toward the PoI. This strategy combines the advantages of STR1 and STR2.

The Figure 3 shows the example of deployment for different tracking strategies. Figures 3(a) to 3(e) show the deployment using STR1. We can see from this set of figures that after $450s$ the deployment reaches it's end and that the first covering sensor reaches the PoI between $[350 - 450]s$. Figures 3(f) to 3(j) show the deployment using STR2. This set of figures shows that the deployment terminates after 7 hours but that the PoI is reached after $300s$. The long termination time is mainly due to the fact that sensors can only make small movements since they are at a distance close to R of each other. Figures 3(k) to 3(o) show that the deployment using STR3 terminates after $900s$ and that the PoI is first reached between $[350 - 900]s$, which is a consequence of this deployment strategy being a trade-off between STR1 and STR2.

Deployment performance. We run a simulation of $3000s$ with 20 sensors and move the PoI at a random location every $500s$. The Figure 4 plots the number of covering sensors depending on time, coverage quality and (re)deployment speed for these three strategies. We can see from Figure 4 that each new PoI location is covered by at least one sensor for each strategy and that, from the coverage quality point of view, STR1 performs very good compared to other strategies. From the redeployment speed point of view, STR2 shows very good performances. We can see that between $[1000 - 1500]s$ the PoI is covered at most after $10s$ (we sample the number of covering sensors every $10s$). For STR1, $200s$ are needed and for STR3, 30s are needed. We can notice here that at time between $[500 - 1000]$ the PoI is located at $[93, 27]$ and between $[1000 - 1500]s$ it is at $[75, 1]$.

4.3   Multiple PoIs

Deployment strategies and examples. We have developed two strategies for the coverage of multiple PoIs. In our first strategy, F1, we consider each PoI independently. In this case, the base station is responsible of dividing and assigning the set of sensors to each particular PoI. The assignment of a subset of sensors to a given PoI can be done regarding the distance between the PoI and the base station or based on some other criteria. Second strategy, F2, considers the set of PoIs as a whole. The base station defines some intermediate points that have to be reached by the sensors before effectively covering a PoI. For instance, when two PoIs have to be covered, an intermediate point could be the gravity center of the two PoIs and the base station. Note that we can also consider the Steiner tree to choose intermediate points [17].

The Figure 5 shows the deployment example for F1 and F2 respectively. In these simulations, we use 30 sensors and two PoIs at $[90, 50]$ and $[50, 90]$. For F1, we consider that the set of sensors is divided into two subsets and each subset is assigned to one PoI. The Figure 5 shows that for F1 the deployment terminates after 180s and that the PoIs are considered independently. For F2, we choose the gravity center of the PoIs and the base station as an intermediate point. In F2 (as in F1), each sensor is also assigned to a given PoI by the base station. However, before moving toward its PoI, the sensor needs to reach the intermediate point.

Deployment performance. Figure 6 plots the number of covering sensors depending on time for two strategies. This figure shows the trade-off performance between deployment speed and coverage quality. Indeed, F1 outperforms F2 regarding deployment speed since the two PoIs are covered by at least one sensor at 140s for F1 and this value is 160s for F2. However, the coverage quality provided by F2 is better than the coverage provided by F1. Note that for F1, the number of covering sensors is not equal for two PoIs since we consider 30 sensors in our simulation, including the base station. Therefore, 14 sensors are dedicated

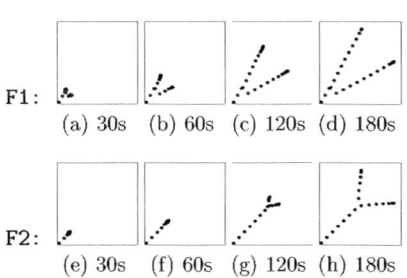

Fig. 5. Sensors' positions with multiple PoIs depending on time.

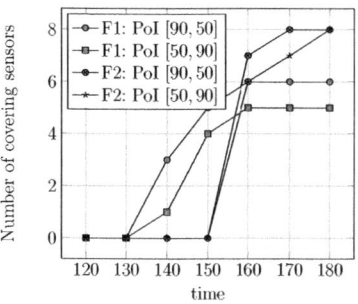

Fig. 6. Number of covering sensors w.r.t time for F1 and F2.

to one PoI and 15 sensors to the other. This is not the case for F2 since a subset of sensors is used in common for connectivity.

5   Implementation

This section shows real example of deployment and implementation of our algorithm in order to prove the proposed concept. We also show in this section that even with real radio condition, obstacle condition and without accurate position information our algorithm performs well. Mobile sensors used in this work are Wifibot mobile robotic platforms (visit Wifibot website [2] for more details).

Our experiments were ran indoor and we used dead reckoning technique using motor encoders to get robots' positions during the deployment. The Wifibot has an 802.11a/b/g interface which is used in our experiments to send periodic messages containing position data to surrounding robots. It is important to notice that due to indoor conditions and radio instability it was hard to evaluate the real communication range of the robots. Therefore, we fixed it arbitrarily and discarded messages received from robots that are out of communication range.

We have also implemented a simple obstacle avoidance since Wifibots are equipped with two IR proximity sensors. When a robot encounters an obstacle it stops moving and considers the obstacle in it's next direction computation. The right (left) hand rule is used to bypass the obstacle depending on the value from each IR sensor. The integration of an efficient obstacle avoidance scheme in our algorithm is left for future work. It is also important to notice that in order to alleviate the effect of messages losses, we increase the frequency of Hello messages. Due to space limitations we will not describe the Wifibot implementation of our algorithm in details.

Figure 7 shows the example of deployment for a single PoI as presented in Section 4.1 with an obstacle. In this experimentation, we have 3 Wifibots and a base station at $[0,0]$. The PoI is at $[0,11]$ and the communication range is set to $4.5m$ while all other parameters are the same as in simulations. Figure 8 shows results regarding multiple PoI coverage as presented in Section 4.3. In this experimentation we used 8 Wifibots with a range of $15m$ and two PoIs at $[25,45]$ and $[45,25]$.

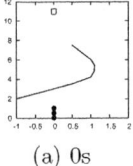
(a) 0s

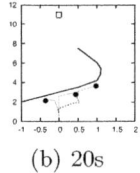
(b) 20s

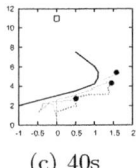
(c) 40s

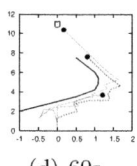
(d) 60s

Fig. 7. Wifibot deployment with an obstacle

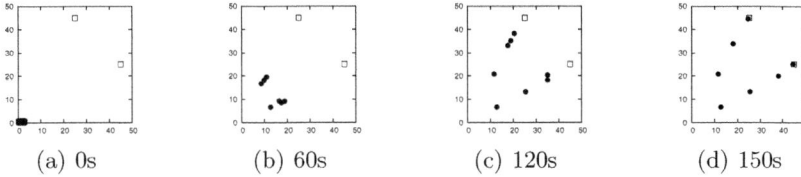

(a) 0s   (b) 60s   (c) 120s   (d) 150s

Fig. 8. Multiple targets at $[20, 45]$ and $[45, 20]$ with comm. range of $15m$ (8 robots).

6   Conclusion

We present an algorithm for Point of Interest (PoI) coverage with mobile wireless sensors. In our algorithm, the sensors must cover the PoI while maintaining the connectivity with a fixed base station. The algorithm is distributed, needs only local information at each sensor, does not require synchronization and is divided into three parts. In the first part, the sensor computes its direction. In the second part, the distance that has to be covered by the sensor and its speed is computed. The third part is devoted to sensor's motion. Unlike other algorithms described in the literature, our algorithm maintains the connectivity all along the deployment procedure and therefore allows the tracking of mobile PoI. The connectivity maintenance of our algorithm is done by using the Relative Neighborhood Graph (RNG). Indeed, if a graph G is connected, the RNG extracted from G is also connected. Hence, during their movements, the sensors only have to keep the connection with their RNG neighbors to keep the whole graph connected. Moreover, the RNG computation uses only local information.

We evaluate the performances of our algorithm regarding the number of sensors that covers the PoI, the deployment speed, and the energy consumption. We also provide some proofs about the connectivity preservation, the algorithm's termination and the shape of the resulting graph (straight line). We provide some results regarding the coverage of moving PoI and multiple PoIs. Moreover, we implement our algorithm on Wifibots and show that our algorithm can be easily implemented and can work in real conditions by using a simple collision avoidance scheme rule and by alleviating message losses. The next step of this work is to consider the coverage of multiple moving PoIs and to consider the effect of having more than one base station.

Acknowledgments

This work is partially funded by the French National Research Agency (ANR) under the VERSO RESCUE (ANR-10-VERS-003) project and the French Institut National de Recherche en Informatique et Automatique (INRIA) under the research action MISSION.

References

1. wsnet.gforge.inria.fr
2. www.wifibot.com
3. Bai, X., Kumar, S., Xuan, D., Yun, Z., Lai, T.: Deploying wireless sensors to achieve both coverage and connectivity. In: International Symposium on Mobile Ad hoc Networking and Computing (ACM Mobihoc), New York, USA, pp. 131–142 (2006)
4. Barrenetxea, G., Ingelrest, F., Schaefer, G., Vetterli, M., Couach, O., Parlange, M.: Sensorscope: Out-of-the-box environmental monitoring. In: International Conference on Information Processing in Sensor Networks (IPSN), Los Alamitos, USA, pp. 332–343 (2008)
5. Batalin, M., Sukhatme, G.: Spreading out: A local approach to multi-robot coverage. In: International Symposium on Distributed Autonomous Robotic Systems (DARS), Fukuoka, Japan, pp. 373–382 (2002)
6. Carle, J., Simplot-Ryl, D.: Energy-efficient area monitoring for sensor networks. Computer 37(2), 40–46 (2004)
7. Chellappan, S., Gu, W., Bai, X., Xuan, D., Ma, B., Zhang, K.: Deploying wireless sensor networks under limited mobility constraints. IEEE Transactions on Mobile Computing 6(10), 1142–1157 (2007)
8. Chen, A., Kumar, S., Lai, T.: Designing localized algorithms for barrier coverage. In: ACM International Conference on Mobile Computing and Networking (ACM Mobicom), New York, USA, pp. 63–74 (2007)
9. Cheng, W., Li, M., Liu, K., Liu, Y., Li, X., Liao, X.: Sweep coverage with mobile sensors. In: IEEE International Parallel & Distributed Processing Symposium (IEEE IPDPS), Miami, USA, pp. 1–9 (2008)
10. Krause, A., Guestrin, C., Gupta, A., Kleinberg, J.: Near-optimal sensor placements: maximizing information while minimizing communication cost. In: International Conference on Information Processing in Sensor Networks (IPSN), Nashville, USA, pp. 2–10 (2006)
11. Li, X., Frey, H., Santoro, N., Stojmenovic, I.: Focused-coverage by mobile sensor networks. In: IEEE International Conference on Mobile Adhoc and Sensor Systems (IEEE MASS), Macau, China, pp. 466–475 (2009)
12. Martinez, K., Ong, R., Hart, J.: Glacsweb: a sensor network for hostile environments. In: IEEE Conference on Sensor and Ad Hoc Communications and Networks (IEEE SECON), Santa Clara, USA, pp. 81–87 (2004)
13. Simplot-Ryl, D., Stojmenovic, I., Wu, J.: Energy-efficient backbone construction, broadcasting, and area coverage in sensor networks. In: Handbook of Sensor Networks, pp. 343–380 (2005)
14. Toussaint, G.T.: The relative neighbourhood graph of a finite planar set. Pattern Recognition 12(4), 261–268 (1980)
15. Wang, B., Lim, H.B., Ma, D.: A survey of movement strategies for improving network coverage in wireless sensor networks. Computer Communications 32(13-14), 1427–1436 (2009)
16. Wang, Y.-C., Hu, C.-C.: Efficient placement and dispatch of sensors in a wireless sensor network. IEEE Transactions on Mobile Computing 7(2), 262–274 (2008)
17. Winter, P.: Steiner problem in networks: a survey. Networks 17(2), 129–167 (1987)
18. Younis, M., Akkaya, K.: Strategies and techniques for node placement in wireless sensor networks: A survey. Ad Hoc Networks 6(4), 621–655 (2008)

A Deep Dive into the LISP Cache and What ISPs Should Know about It

Juhoon Kim*, Luigi Iannone, and Anja Feldmann

Deutsche Telekom Laboratories – Technische Universität Berlin,
Ernst-Reuter-Platz 7, 10587 Berlin, Germany
jkim@net.t-labs.tu-berlin.de

Abstract. Due to scalability issues that the current Internet is facing, the research community has re-discovered the Locator/ID Split paradigm. As the name suggests, this paradigm is based on the idea of separating the identity from the location of end-systems, in order to increase the scalability of the Internet architecture. One of the most successful proposals, currently under discussion at the IETF, is LISP (Locator/ID Separation Protocol). A critical component of LISP, from a performance and resources consumption perspective, as well as from a security point of view, is the LISP Cache. The LISP Cache is meant to temporarily store mappings, i.e., the bindings between identifiers and locations, in order to provide routers with the knowledge of *where* to forward packets. This paper presents a thorough analysis of such a component, based on real packet-level traces. Furthermore, the implications of policies to increase the level of security of LISP are also analyzed. Our results prove that even a timeout as short as 60 seconds provides high hit ratio and that the impact of using security policies is small.

Keywords: Locator/ID Separation, Next Generation Internet, Addressing and Routing Architectures, Measurements.

1 Introduction

In the last years, there has been an increasing interest in the Locator/ID Split paradigm, which is recognized to be a strong candidate to become a fundamental building block of the Next Generation Internet. Differently from the current architecture, where one single namespace, namely the IP address space, is used for both indentifying and locating end-systems, in the Locator/ID Split paradigm two different namespaces are used: the *ID space* and the *Locator space*, to respectively identify and locate end-systems. Such a separation aims at solving the scalability issues that the current Internet is facing [19], mainly concerning the continuously increasing BGP routing table [1], but also concerning addressing, mobility, multi-homing [22], and inter-domain traffic engineering [25].

The main effort to tackle these issues has started three years ago, when the Routing Research Group (RRG) has been rechartered to explore the possibility

---

* Corresponding author.

J. Domingo-Pascual et al. (Eds.): NETWORKING 2011, Part I, LNCS 6640, pp. 367–378, 2011.
© IFIP International Federation for Information Processing 2011

to enhance the Internet architecture by introducing some form of Locator/ID Split. Among the numerous proposals introduced since then [16], the most successful so far are ILNP (Identifier/Locator Network Protocol [5]), which has been adopted by the RRG, and LISP (Locator/ID Separation Protocol [7]), which has been chartered as Working Group in the IETF and has been already deployed in an international test-bed [2]. On the one hand, all the proposals improve the Internet architecture, in a way or another, by some form of Locator/ID Split. On the other hand, they introduce the necessity to distribute and store bindings between IDs and Locators (i.e., mappings), which is the key critical component that can make the difference. Hence, the importance of evaluating what is the *cost* (i.e., memory consumption, overhead, ...) of storing and distributing these mappings. In addition, the way they are managed does also have an impact on the robustness of the proposed architecture with respect to security threats.

Focusing on LISP as reference protocol, in this paper we present a thorough analysis of the mapping cache. In a previous work, Iannone et al. analyzed a Netflow traffic trace from a small/medium sized university campus [11]. We build our work on the same methodology, but develop a new emulator in order to go beyond previous results and provide a deeper analysis. For this purpose, we use two 24-hour packet-level traces from a large European ISP, taken at different periods of the year. Further, Saucez et al. [24], in their security analysis of the LISP protocol, highlighted that a number of attacks can be avoided by slightly modifying the policy that LISP uses to encapsulate and decapsulate packets. Hence, we evaluate what are the implications, from a scalability point of view, of running LISP in such a more secure manner, which we call *symmetric LISP*.

The contributions of this paper are many-fold. We thoroughly analyze the behavior of a LISP Cache for traffic of a large number of DSL customers. Our results show that it is possible to maintain a high hit ratio with relatively small cache sizes, showing that there is no use in large timeouts. Compared to [11], we show how the LISP Cache has good scalability properties. We analyze what and how much would change if instead of running vanilla LISP, i.e., as defined in the specifications, symmetric LISP is used in order to increase security. Our results are important for ISPs in order to identify and quantify the resources needed to deploy LISP, with respect to the level of resiliency that they want to guarantee.

The remainder of this paper is structured as follows. In Section 2 we present the related work. In Section 3 we provide an overview of LISP in order to highlight how the LISP Cache works and the motivation for symmetric LISP. Section 4 describes how we collected and analyzed the traces used to obtain the results presented in Section 5. Finally, we draw our conclusions in Section 6.

2   Related Work

The concept of Locator/ID Split was already discussed, but never widely explored, in the late 90s ([23], [10]); rather used to solve specific issues, like cryptographic security with HIP (Host Identity Protocol [20]) or multi-homing for IPv6 with Shim6 [21]. The situation has changed after the rechartering of the RRG, with quite a number of proposals being discussed, either based on tunneling (e.g., LISP [7], HAIR [8]) or on address rewriting (e.g., ILNP [5], Six/One [26]). Despite the plethora of proposals, very few works have tackled the evaluation of such a critical component as the cache.

Kim et al. [27] studied route caching based on Netflow data, showing how caching is feasible also for large ISPs. However, they used a generic approach, not focused on LISP, and a limited cache size. Iannone et al. [11] have performed an initial study, without considering security issues, based on a single Netflow trace of a small/medium sized university campus. We adopt a similar methodology to provide comparable results, which we discuss in Section 6. Nevertheless, because we base our results on two 24-hour packet level traces captured in different periods we are able to provide a deeper analysis, including security aspects.

In the work of Zhang et al. [9], the authors propose a simple LISP Cache model with bounded size, using a Least Recently Used (LRU) policy to replace stale entries.[1] Their evaluation is based on a 24-hour trace of outbound traffic of one link of the CERNET backbone, the China Education and Research Network. In our analysis we take a different approach, assuming that the cache is not limited in size. This is a reasonable assumption since, as we will show later, even for large ISPs there is no need to use large caches and by tuning the expiration timer, caches do not grow so large as to hit memory limits of current routers.

Jakab et al. [14] propose a LISP simulator able to emulate the behavior of different mapping systems. However, they focused on the evaluation of the latency of their own mapping system solution, namely LISP-TREE, and compared it to other approaches but neglecting the analysis of the LISP Cache.

3 LISP Overview

The main idea of the Locator/ID Separation Protocol (LISP [7]) is to split the IP addressing space in two orthogonal spaces, one used to identify the end-hosts, and one used to locate them in the Internet topology. By re-using IP addresses, LISP is incrementally deployable and meant to be used on the border routers of stub domains, like in the scenario presented in Figure 1. Stub domains use internally an IP prefix, called EID-Prefix as for End-system IDentifier, which is part of the ID space. This prefix does not need to be globally routable, since it is used only for routing in the local domain. On the contrary, the core of the Internet, known as Default Free Zone (DFZ), will use globally routable IP addresses that are part of the locator space. In particular, the IP addresses used by stub domains' border routers on their upstream interfaces (toward the provider) are part of such a space and represent the Routing LOCators (RLOCs), since they allow to *locate* EID in the Internet. The binding between an EID-Prefix and the set of RLOCs that locate it is a *mapping*. EID-Prefixes are no longer announced in the DFZ, allowing a size reduction of BGP's routing table. Evaluating such kind of benefits is beyond the scope of this paper. Further information can be found in the work of Quoitin et al. [22] and Iannone et al. [12].

---

[1] In the rest of the paper we will use the term *LISP Cache* and *cache* interchangeably.

In the context of LISP, end-hosts generate normal IP packets using EIDs as source and destination addresses, where the destination EID is obtained, for instance, through a DNS query. LISP then tunnels those packets from the border router of the source domain to the border router of the destination domain, using the RLOCs as source and destination addresses in the outer header. For tunneling, LISP uses an IP-over-UDP approach, with the addition of a LISP-specific header between the outer UDP header and the inner (original) IP header.[2] In the LISP terminology, the border router in the source domain performing the encapsulation is called Ingress Tunnel Router (ITR), while the one in the destination domain, performing the decapsulation, is called Egress Tunnel Router (ETR). In general, they are just named xTRs. To perform tunneling operations routers use two data stores, namely the LISP Database and the LISP Cache.

The LISP Database stores mappings that bind the local EID-Prefixes (i.e., inside the local domain) to a set of RLOCs belonging to the xTRs deployed in the domain. The purpose of the LISP Database is two-fold. For outgoing packets, if a mapping exists for the source EID it means that the packet has to be LISP encapsulated and the source RLOC is selected from the set of RLOCs associated to the source EID-Prefix. For incoming packets, if a mapping exists for the destination EID, then the packets are decapsulated. The LISP Database is statically configured on each xTR. Its size is directly proportional to the number of the EID-Prefixes and xTRs that are part of the local domain. Due to its static nature and limited size, the LISP Database does not present any scalability issue and is not further analyzed in this paper.

The LISP Cache temporarily stores the mappings for EID-Prefixes that are not part of the local domain. This is necessary to correctly encapsulate outgoing packets, in particular to select the RLOC to be used as destination address in the outer header. Mappings are stored only for the time that are used to encapsulate packets, otherwise, after a timeout, they are purged from the cache. While critical for the dimensioning of the system, the LISP specification does not provide any value for this timeout, leaving the choice (or responsibility) to the implementers and system administrators. It should be clear that since the entries in the cache can expire, they are also entered in an on-demand fashion. This makes the cache so critical, since its content, size, and efficiency is totally traffic driven. In particular, the first outgoing packet, destined to an EID for which there is no mapping in the cache, triggers a cache-miss. Such an event, in turn, triggers a query message that the ITR sends to the Mapping Distribution System.[3] The latter is a lookup infrastructure designed to retrieve the mapping for the destination EID of the packet that triggered the query. This is the same principle of DNS, which allows to retrieve the IP address(es) of a server from its

---

[2] The LISP header contains information about RLOCs' reachability and traffic engineering. Details on this header can be found in the protocol specification ([7], [13]).

[3] Note that, in case of cache-miss, the packet that triggered it cannot be encasulated, since there is no mapping available. The LISP specifications do not explicitly describe what to do with the packet, however, it is out of the scope of the present work to evaluate this particular aspect.

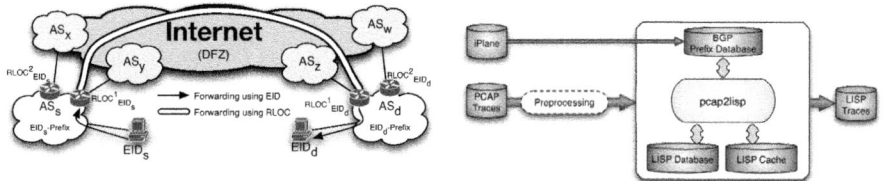

Fig. 1. Example of LISP deployment and communication

Fig. 2. Structure of the *pcap2lisp* emulator

FQDN (Fully Qualified Domain Name). Insofar, several Mapping Distribution Systems have been proposed, but only the BGP-based ALT [6] is an official working item of the LISP Working Group. It is out of the scope of this paper to compare different mapping systems. Further information can be found in the work of Jakab et al. [14] and Mathy et al. [18].

The attentive reader should have noticed that there is a fundamental asymmetry in LISP when performing encapsulation and decapsulation operations. Indeed, while doing encapsulation the ITR uses the cache to locate what to put in the outer header, on the contrary, the ETR, while decapsulating the packet, does not use the cache. Rather, it just performs the checks described previously for the LISP Database and eventually decapsulates the packet.

This asymmetry leaves the LISP protocol vulnerable against specific attacks on the ETR, as pointed out by the work of Saucez et al. [24]. The authors describe attacks exploiting data packets as well as attacks that leverage on fields of the LISP header. Attacks are not limited to classic DoS and spoofing, but regard also cache poisoning, i.e., injecting wrong information into the cache, or cache overflow, i.e., saturating the cache and increasing the number of cache-misses. The analysis they performed, led the authors to conclude that to increase the level of security of the whole LISP architecture, the best solution is to use a symmetric model with a drop policy. Put differently, the authors suggest to only allow encapsulation *as well as* decapsulation if mappings are present in both the LISP Cache *and* the LISP Database. Otherwise, when the mapping is not in the cache a miss is generated and the packet is dropped. The rationale behind this suggestion is the assumption that the mapping distribution system is secured and trusted, hence, sanity checks can be performed at both ends, when encapsulating and when decapsulating. In particular, for the latter operation it means that if the packet contains information that is not coherent with the content of the cache, it is dropped. The implications of introducing the symmetric model to increase security are two-fold: on the one hand, performing additional checks when performing decapsulation may reduce the performances, however this is a common price for security mechanisms. On the other hand, this means that the size of the cache, its dynamics, and the control traffic overhead is increased.

In the second half of this paper, we will refer to the symmetric model as *symmetric LISP* and we will assess its impact as compared to vanilla LISP to evaluate what are the trade-offs to increase security in the LISP protocol.

4   LISP Emulation

To evaluate LISP we implemented *pcap2lisp*, which is meant to emulate the behavior of LISP's xTRs. The emulator is essentially designed to be fed with pcap-formatted traffic data and mimics as much as possible the LISP architecture; hence we based its implementation on two main modules (cf., Figure 2). The LISP Database, which is a manually configured list of internal network prefixes (EIDs). The LISP Cache, which stores EID-Prefixes and related statistic. Besides these two modules there is a central logic that creates the correct statistics, periodical reports that are written in logs, and that overviews the correct management of the cache timeouts. In addition, we use a local BGP prefixes database, fed with the list of BGP prefixes published by the iPlane Project [17]. The database is used to group EID-to-RLOCs mappings with the granularity of existing BGP prefixes, because, as for today, there is no sufficient information to predict what will be the granularity of mappings in a LISP-enabled Internet. This BGP granularity follows the methodology proposed in [11], thus making the results comparable. Furthermore, such an approach allows using the BGP database as a black box returning the mappings needed to feed the cache.

We fed *pcap2lisp*with two sets of anonymized traffic data collected within a large European ISP for 24 hours in April 2009 (APR09) and in August 2009 (AUG09) from a vantage point covering more than 20,000 DSL lines and already validated in other studies ([4,15]).

5   Results

Similarly to the work in [11], we used three different cache timeout values, respectively 60 seconds, 180 seconds (i.e., three minutes), and 1,800 seconds (i.e., 30 minutes). The reason why we choose 60 seconds, instead of 300 minutes like in [11], is because that work already proved that the 300 minutes timeout value is inefficient considering the hit ratio vs. the cache size.

It is useful starting by identifying the working set, in this case represented by the number of observed BGP prefixes in our measurement environment. Figure 3 depicts the total number of contacted prefixes per minute, as well as the breakdown of incoming, outgoing, and bi-directional traffic. What we can identify in Figure 3 is that the majority of the observed prefixes (i.e., 70.1% in average in both traces) are bi-directional. One thing to keep in mind is that in the case of vanilla LISP, incoming packets do not play any role in the LISP cache, as explained in Section 3. On the contrary, when using symmetric LISP, incoming prefixes have an impact on the cache since the ETR does need a mapping. Hence, the 11.4% (average value in both traces) of incoming prefixes are a key differentiation between the two versions of LISP.

5.1   Vanilla LISP

A key benefit of LISP is the reduction of the routing table size in the DFZ [22], however, it exists a trade-off between the reduction of the routing table size and

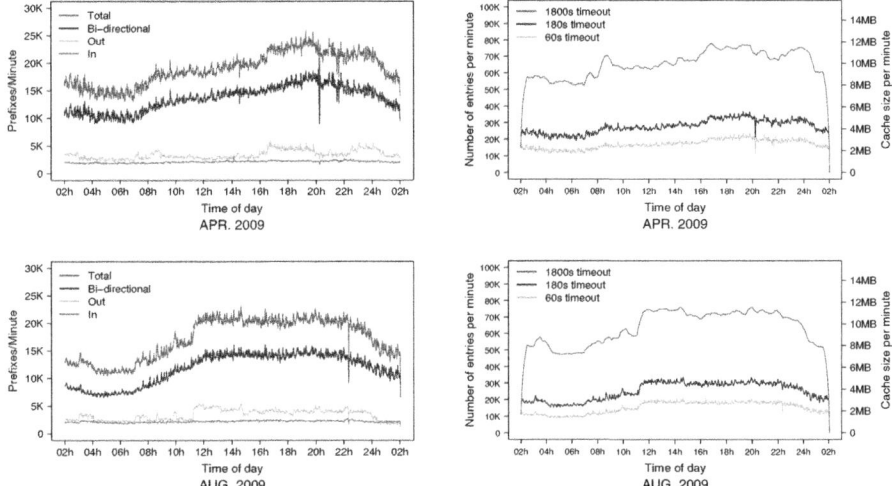

Fig. 3. Report of the number of correspon-   **Fig. 4.** Number of entries and cache size
dent prefixes per minute   (assuming two RLOCs) per minute

the LISP Cache size. The analysis of the caching *cost* is important for estimating
the actual benefits of LISP. Figure 4 illustrates the number of entries and the
size of the cache measured per one-minute time bin. Note that there are two
measurement scales for the y-axis in the plot. While the left scale indicates the
number of cache entries, the right scale indicates the size of the cache expressed
in MBytes. The drop of the cache size at the end of the plot is due to the fact
that we run the emulation until all cache entries are expired due to the timeout.

We can observe that the number of entries and the size of the cache are pretty
low for the 60 and 180 seconds timeout cases (respectively slightly more than
10,000 and 20,000) and their curves show a spiky behavior. The 1,800 seconds
timeout is much higher but with smoother changes, however, it takes about 30
minutes to reach the stable working set (almost 60,000 entries).

In the plot, and differently from [11], we calculate the size of the cache using
the size of the data structure of a real LISP implementation, namely Open-
LISP [3]. In OpenLISP each entry of the cache is a radix node containing all the
information of the mapping and pointing to a chained list of RLOCs nodes. Thus,
the size C_{size} of the cache is given by: $C_{size} = N_E \times (Radix_s + N_R \times RLOC_s)$.
Where N_E and N_R represent respectively the number of entries in the cache
and the number of RLOCs per entry. $Radix_s$ is the size of a radix node (56
bytes in OpenLISP), while $RLOC_s$ is the size of the RLOC node (48 bytes in
OpenLISP). For the results in Figure 4 we assume two RLOCs per mapping.

Figure 5 depicts traffic volume and the traffic overhead of LISP expressed in
MBytes/sec. This plot is based on 60 seconds timeout, because, as we will show
next, the traffic overhead is in inverse proportion to the timeout value. Therefore,
the estimation of the traffic overhead reported in this analysis shows the maximum
values. The line plotted in positive values indicates outgoing traffic, whereas the

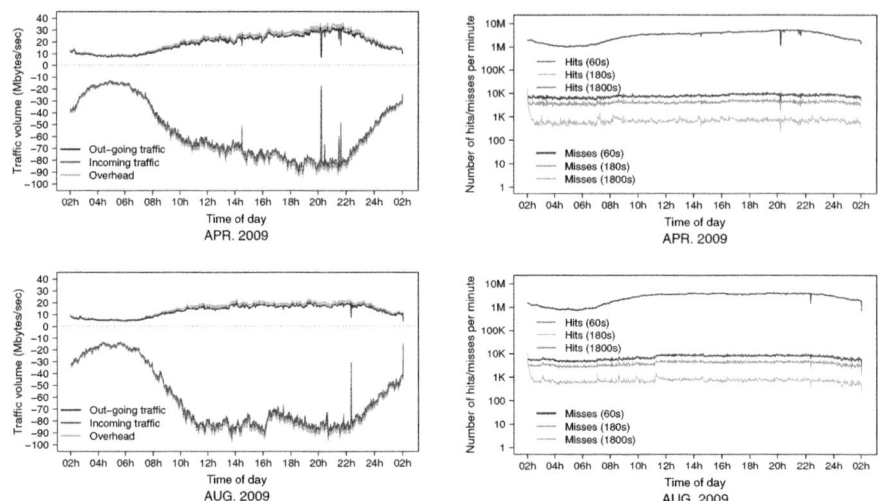

Fig. 5. Traffic volume and overhead due to LISP (60 seconds timeout value)

Fig. 6. Number of hits and misses per minute

line plotted in negative values indicates incoming traffic. The shading over and under the traffic line indicates the traffic overhead. The overall overhead O can be evaluated by adding the overhead due to the LISP encapsulation, i.e., the number of packets N_P multiplied the size of the prepended header E_H and the volume of traffic generated to request and receive the missing mappings. Assuming the one single request/reply exchange is performed for each miss, the latter is given by the number of misses N_M multiplied by the size of the Map-Request message M_{REQ}, for outgoing traffic, or the size of the Map-Reply, for incoming traffic, which is given by a fixed size M_{REP} plus the size of each RLOCs record R times the number of RLOCs N_R. Note that the Map-Request and Map-Reply are the standard messages defined by LISP to request and receive a mapping from the mapping distribution system, independently from the specific instance of this latter. Putting everything together, for outgoing traffic we have: $O_{out} = N_{P_{out}} \times E_H + (N_M \times M_{REQ})$; while for the incoming traffic we obtain: $O_{in} = N_{Pin} \times E_H + (N_M \times (M_{REP} + N_R \times R))$. From the LISP specifications [7] the size of E_H is 36 bytes, the size of M_{REQ} is (without any trailing data) 24 bytes, the size of M_{REP} (without any trailing data) is 28 bytes, to which we need to add 12 bytes for each RLOC record R. With these numbers is possible to plot the overhead in Figure 5, which is between 3.6% and 5.2% for APR09 and 3.6% and 4.6% for AUG09.

As the previous analysis shows, the selection of an appropriate timeout value is of prime importance for the efficiency of the cache. Thus, we further analyze the implications of different timeout values. We first evaluate the efficiency of cache by investigating the ratio between cache misses and cache hits. Figure 6 depicts the number of hits and misses in different timeout values. Even with the y-axis in logarithmic scale it is not possible to distinguish the lines of the cache hits, since they are overlapping each other. Hence, we need to look at the difference in the

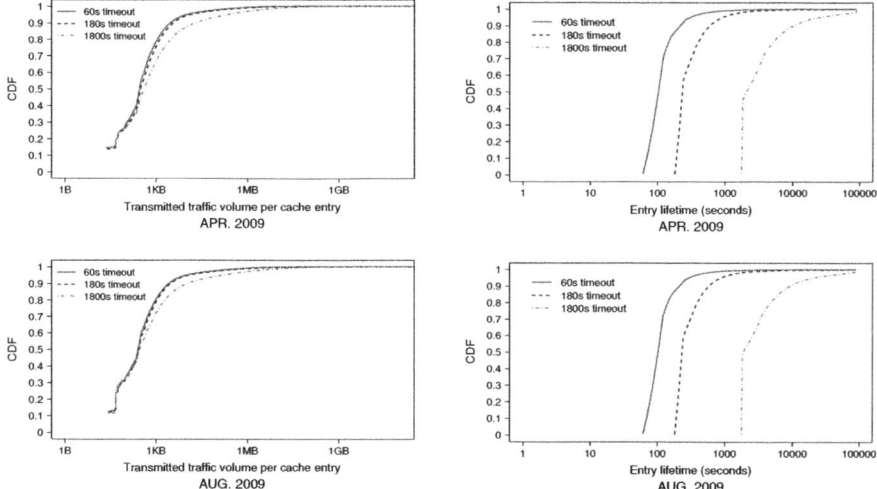

Fig. 7. Cumulative Distribution Function of the traffic volume per cache entry

Fig. 8. Cumulative Distribution Function of entries' lifetime

number of cache misses. Despite the identifiable difference, it is hard to avoid the claim that the benefit from the longer timeout value is insignificant compared to its cost in terms of size. Indeed, the hit ratio is always higher than 99% for every timeout value, while the size can be more than 5 times bigger when we compare the 60 seconds case with the 1,800 seconds case.

To confirm that a large cache is not useful, in Figure 7 we plot the CDF of the traffic volume forwarded by each entry, showing that the vast majority of cache entries carry less than 1 MBytes of traffic. Furthermore, Figure 8 shows how at least 50% of the entries have a lifetime only slightly longer than the timeout value. Both CDFs are the consequence of the well-known Internet phenomena that a small number of prefixes are responsible for the majority of Internet traffic. Coupled with the evidence mentioned above, we draw the inference that the minimum timeout value (60 seconds) is the most cost-beneficial.

5.2 Symmetric LISP

With respect to the security threats discussed in Section 3, we evaluate the extra cache size and the extra traffic overhead when the symmetric model is used. Since the extension of the mapping mechanism to incoming traffic is the main idea of the symmetric model, an increase in the number of entries and the size of the cache is expected and confirmed by the plot in Figure 9. For a clear comparison, the cache size shown in Figure 4 is presented again in gray color (lighter gray in black and white print). From Table 1, we see that there is a 6.8% to 9.9% increase in the cache size, when the symmetric model is used. Interestingly, the higher the timeout value, the lower the increase; we argue that it is due to the fact that longer timeouts increase entries' reuse probability.

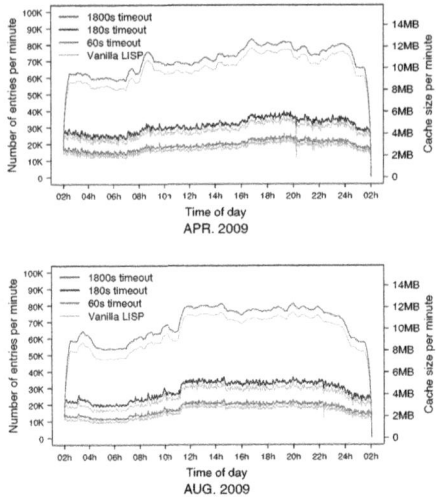
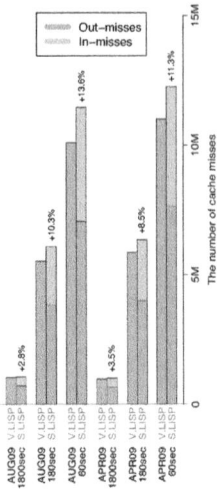

Fig. 9. Extra overhead caused by the security enhancement

Fig. 10. Cache-miss for vanilla LISP and symmetric LISP

Table 1. Cache size (in Mbytes) increase due to security enhancement (two RLOCs)

Timeout	APR09			AUG09		
	Vanilla	Symmetric	Difference	Vanilla	Symmetric	Difference
60 sec	3.47	3.76	+8.36%	3.04	3.34	+9.87%
180 sec	5.32	5.74	+7.89%	4.85	5.29	+9.07%
1800 sec	11.29	12.06	+6.82%	10.99	11.76	+7.01%

Now, we expand the analysis to include the extra traffic overhead. Recall that each cache-miss, even for incoming traffic, triggers at least two messages (Map-Request and Map-Reply). Figure 10 compares the number of cache-miss under vanilla LISP and symmetric LISP. The most interesting result shown in the plot is the fact that the number of miss caused by incoming packets does not correspond to the increase in the number of cache-miss. Indeed, we observe that 2.8% to 13.6% of cache-miss are additionally produced under symmetric LISP, but the percentage of cache-miss due to incoming packets is much higher, while the number of cache-miss due to outgoing packets shrinks.

When translated into traffic overhead, the increase of miss generates a small increase in bandwidth consumption. Nevertheless, after applying the same computation like for vanilla LISP, the measured increase in bandwidth is always lower than 0.5%, hence not critical if not negligible.

6   Conclusion

The scalability issues of the current Internet have triggered an important amount of research on the Locator/ID Split paradigm. LISP, in particular, has gained momentum and seems to be a strong candidate to be widely adopted. The critical component in the LISP architecture, from a scalability and security point of view, is the LISP Cache. By using two 24-hour traces of a large European ISP, we thoroughly analyzed this component. In the present work we adopted the same methodology as in [11], in this way, the two works nicely complement and validate each other. An interesting comparison with the previous work is about the size of the cache vs. the number of users. In our case the size of the cache is almost doubled, however, in our traces there is more than three times the number of users than in the traces analyzed in [11]. This suggests that the average size of the LISP cache does not grow linearly with the number of end-systems. This is an important result, since it confirms that the LISP cache has good scalability properties. Future work will focus on further analyzing this aspect.

We went further in the analysis, mainly, but not only, in two points: the analysis of very short cache timeout and the analysis of the symmetric LISP model, which improves security. When the timeout value is as small as 60 seconds, the measurements prove that efficiency of the cache is still very high, with more than 99% hit-ratio, while the size reduces almost by half compared to the 180 seconds timeout case, and it is almost five times smaller compared to than the 1,800 seconds timeout case.

Another main contribution of the present work concerns the symmetric LISP model. The presented analysis shows that the increase in the size of the cache and the generated overhead is the order of 13% (for the 60 second timeout case). This looks like a very low cost compared to the security benefits, which have a very high value for ISPs and vendors.

Acknowledgments

This work is partially funded by the Trilogy Project (http://www.trilogy-project. org), a research initiative (ICT-216372) of the European Community under its Seventh Framework Program.

References

1. BGP Routing Table Analysis Report, http://bgp.potaroo.net
2. International LISP Infrastructure, http://www.lisp4.net
3. The OpenLISP Project, http://www.openlisp.org
4. Ager, B., Schneider, F., Kim, J., Feldmann, A.: Revisiting cacheability in times of user generated content. In: 13th IEEE Global Internet Symposium (March 2010)
5. Atkinson, R.: ILNP Concept of Operations. IETF - Internet Engineering Task Force, draft-rja-ilnp-intro-03.txt (February 2010)
6. Farinacci, D., Fuller, V., Meyer, D., Lewis, D.: LISP Alternative Topology (LISP+ALT). IETF - Internet Engineering Task Force, draft-ietf-lisp-alt-04.txt (April 2010)
7. Farinacci, D., Fuller, V., Meyer, D., Lewis, D.: Locator/ID Separation Protocol (LISP). IETF - Internet Engineering Task Force, draft-ietf-lisp-07.txt (April 2010)

378   J. Kim, L. Iannone, and A. Feldmann

8. Feldmann, A., Cittadini, L., Mühlbauer, W., Bush, R., Maennel, O.: HAIR: Hierarchical Architecture for Internet Routing. In: The Workshop on Re-Architecting the Internet (ReArch 2009) (December 2009)
9. Zhang, H., Chen, M., Zhu, Y.: Evaluating the Performance on ID/Loc Mapping. In: The Global Communications Conference (Globecom 2008) (November 2008)
10. Hiden, R.: New Scheme for Internet Routing and Addressing (ENCAPS) for IPNG. IETF - Internet Engineering Task Force, RFC 1955 (June 1996)
11. Iannone, L., Bonaventure, O.: On the Cost of Caching Locator/ID Mappings. In: 3rd International Conference on Emerging networking EXperiments and Technologies (CoNEXT 2007) (December 2007)
12. Iannone, L., Levä, T.: Modeling the Economics of Loc/ID Split for the Future Internet. Future Internet Assembly (FIA) Book (April 2010)
13. Iannone, L., Saucez, D., Bonaventure, O.: LISP Map Versioning. IETF - Internet Engineering Task Force, draft-ietf-lisp-map-versioning-00.txt (September 2010)
14. Jakab, L., Cabellos-Aparicio, A., Coras, F., Saucez, D., Bonaventure, O.: LISP-TREE: A DNS Hierarchy to Support the LISP Mapping System. IEEE Journal on Selected Areas in Communications (September 2010)
15. Kim, J., Schneider, F., Ager, B., Feldmann, A.: Today's usenet usage: Characterizing NNTP traffic. In: 13th IEEE Global Internet Symposium (March 2010)
16. Li, T.: Recommendation for a Routing Architecture. IRTF - Internet Research Task Force, draft-irtf-rrg-recommendation-08.txt (May 2010)
17. Madhyastha, H.V., Isdal, T., Piatek, M., Dixon, C., Anderson, T., Krishnamurthy, A., Venkataramani, A.: iPlane: An Information Plane for Distributed Services. In: 7th Symposium on Operating Systems Design and Implementation (OSDI 2006). USENIX Association (November 2006)
18. Mathy, L., Iannone, L.: LISP-DHT: Towards a DHT to map Identifiers onto Locators. In: The Workshop on Re-Architecting the Internet (ReArch 2008) (December 2008)
19. Meyer, D., Zhang, L., Fall, K.: Report from the IAB Workshop on Routing and Addressing. IETF - Internet Engineering Task Force, RFC 4984 (September 2007)
20. Moskowitz, R., Nikander, P.: Host Identity Protocol (HIP) Architecture. IETF - Internet Engineering Task Force, RFC 4423 (May 2006)
21. Nordmark, E., Bagnulo, M.: Level 3 Multihoming Shim Protocol for IPv6. IETF - Internet Engineering Task Force, RFC 5533 (June 2009)
22. Quoitin, B., Iannone, L., de Launois, C., Bonaventure, O.: Evaluating the Benefits of the Locator/Identifier Separation. In: 2nd ACM/IEEE Workshop on Mobility in the Evolving Internet Architecture (MobiArch 2007) (August 2007)
23. Saltzer, J.: On the Naming and Binding of Network Destinations. IETF - Internet Engineering Task Force, RFC 1498 (August 1993)
24. Saucez, D., Iannone, L., Bonaventure, O.: LISP Security Threats. IETF - Internet Engineering Task Force, draft-saucez-lisp-security-01.txt (July 2010)
25. Saucez, D., Donnet, B., Iannone, L., Bonaventure, O.: Interdomain Traffic Engineering in a Locator/Identifier Separation context. In: The Internet Network Management Workshop (INM 2008) (October 2008)
26. Vogt, C.: Six/One: A Solution for Routing and Addressing in IPv6. IETF - Internet Engineering Task Force, draft-vogt-rrg-six-one-01.txt (November 2007)
27. Kim, C., Caesar, M., Gerber, A., Rexford, J.: Revisiting route caching: The world should be flat. In: Moon, S.B., Teixeira, R., Uhlig, S. (eds.) PAM 2009. LNCS, vol. 5448, pp. 3–12. Springer, Heidelberg (2009)

Data Plane Optimization in Open Virtual Routers

Muhammad Siraj Rathore, Markus Hidell, and Peter Sjödin

Telecommunication Systems (TS) Lab, School of ICT, KTH-Royal Institute of Technology,
16440 Kista, Sweden
{siraj,mahidell,psj}@kth.se

Abstract. A major challenge in network virtualization is to virtualize the components constituting the network, in particular the routers. In the work presented here, we focus on how to use open source Linux software in combination with commodity hardware to build open virtual routers. A general approach in open router virtualization is to run multiple virtual instances in parallel on the same PC hardware. This means that virtual components are combined in the router's data plane, which can result in performance penalty. In this paper, we investigate the impact of the design of virtual network devices on router performance in Linux namespace environment. We identify performance bottlenecks along the packet data path. We suggest design changes to improve performance. In particular, we investigate modifications of the "macvlan" device, and analyze the performance improvements in terms of packet forwarding. We also investigate how the number of virtual routers and virtual devices within a physical machine influence performance.

Keywords: network virtualization, virtual router, SoftIRQ, NAPI, Softnet API.

1   Introduction

With the continuous growth of the Internet to new areas, services and applications, the demands increase on the ways in which we organize and manage networks. Future networks need to be flexible, support a diversity of services and applications, and should be easy to manage and maintain. One way of addressing these requirements is to *virtualize* routers. Router virtualization involves running several router instances on the same physical hardware, in a way that allows each instance to appear as a separate, independent router. This makes it possible to support a multitude of services, management disciplines, and protocols in parallel on different virtual routers.

A general approach to router virtualization is to use computing virtualization techniques to run multiple operating systems as guests in parallel on the same hardware, and let each guest run one instance of the router software. In *open source virtual routers*, these operating systems are based on open source software that can be combined with commodity PC hardware.

Broadly virtualization techniques can be categorized into hypervisor-based and container-based virtualization. With hypervisor-based virtualization (such as KVM and Xen), the hardware is virtualized so that each guest runs its own operating system. In contained-based virtualization (sometimes also called jails or operating

J. Domingo-Pascual et al. (Eds.): NETWORKING 2011, Part I, LNCS 6640, pp. 379–392, 2011.

system virtualization), the operating system is partitioned into multiple domains, where each guest runs in its own domain. Examples of container-based virtualization include FreeBSD Jail, Linux-VServer, OpenVZ, and Linux Namespaces. Container-based virtualization is more "light weight" in the sense that it is based on partitioning of the operating system resources, something that could potentially be achieved with no or little extra processing overhead. The drawback is that all guests need to run the same operating system, as they are sharing the same operating system kernel. Hypervisor-based virtualization is more flexible in this respect but incurs a larger overhead, which limits the number of concurrent guests.

The purpose of our work is to study how container-based virtualization can be used for router virtualization. We focus on contained-based virtualization because of its potential for low processing overhead and support for many simultaneous guests. The starting point for our work is the observation that there is a significant performance penalty for performing packet processing in guests. The performance penalty comes from the level of indirection between network interfaces and virtual routers: When a packet arrives on a network interface, the corresponding virtual router should be identified, and the packet should be redirected to the virtual router. The way in which this redirection is performed has fundamental impact on overall performance.

A common solution is to use existing kernel components such as software bridges and virtual interfaces. In previous work we have shown that this is a costly solution in terms of performance, as it introduces considerable processing overhead [9]. A more promising approach is to use the *macvlan* device – a kernel object specifically designed for support of multiplexing and demultiplexing of packets between physical and virtual interfaces based on MAC addresses. The purpose of this paper is to investigate how a virtual router framework can be designed around the macvlan device and Linux Namespaces. We find that macvlan exhibits undesirable behavior at overload – when traffic load increases above a certain point, the effective throughput goes down. This is a considerable disadvantage of any routing platform, and we therefore propose a revised version, called NAPI-macvlan. The performance of NAPI-macvlan is studied in terms of throughput, scalability, and virtual router isolation properties. We demonstrate that NAPI-macvlan has superior performance, compared to macvlan, and does not exhibit the negative behavior at overload.

The rest of this paper is organized as follows: Section 2 surveys related work on virtual router platforms. Thereafter, Section 3 describes a Linux virtual router framework based on macvlan devices and Namespaces. This section also introduces the NAPI-macvlan and gives the rationale behind its design. Section 4 presents and analyses performance measurements of virtual router configurations using macvlan and NAPI-macvlan. Finally, Section 5 concludes the paper.

2   Related Work

There are many examples of work, where different virtualization technologies are evaluated as virtual router platforms. For instance, Xen virtualization has been investigated in detail [1] and it has been shown that Xen can achieve a considerable forwarding rate in the privileged domain, but that guest domain forwarding results in poor performance. Studies on data plane virtualization using Xen can also be found

[2], where packet forwarding through guest domain is suggested in order to virtualize the data plane. Another work demonstrates how to make efficient use of multicore commodity hardware for virtual routers [3]. It also identifies performance bottlenecks associated with the currently available commodity hardware.

VINI [4] presents a virtual network infrastructure for network experimentation in a realistic and controlled environment. For router virtualization, VINI uses User-Mode Linux (UML) and Click to define custom data planes. The data plane runs in user space, something which gives great flexibility, but the implementation is also subject to a significant performance penalty.

Another platform for virtual networks is Trellis [5]. Router virtualization in Trellis is based on customized components that have been introduced to improve forwarding rates. The resulting throughput is compared with virtual routers based on Xen and Openvz, it is shown that higher throughput can be achieved in Trellis.

An alternative virtual router enabling technology is to use a source code merging scheme as a mechanism to define custom data planes for virtual routers [6]. Source code is a language used to define the packet path for each virtual router. The packet path is specified in terms of networking functions that are connected together. Click and Linux VServer is used to provide virtual routers platform.

PdP (Parallel data Plane) presents a virtual network platform to achieve high speed packet processing [7]. It runs control and data planes in guest machines for better isolation and flexibility. In order to boost the forwarding rate, PdP uses an architecture where multiple guest machines perform packet forwarding in parallel.

The Crossbow architecture [8] is yet another example of a virtual network platform. The focus is network resource virtualization to achieve fair bandwidth sharing among various virtual instances. Virtualization of physical network interfaces is proposed using different virtual devices (such as VNIC and virtual switch).

3   Linux Based Virtual Routers

A virtual router sends and receives packets on virtual interfaces. Besides this, a virtual router is just like a physical router: is has a routing table, routing protocols, packet filtering rules, management interface, and so on. A virtual router runs in a host environment, which is responsible for allocating resources to the virtual router, and for managing these resources. There can be multiple virtual routers in the same host environment, sharing the available resources, as illustrated in Fig. 1. Even though a virtual router communicates on virtual interfaces, its purpose is often to process packets that appear on the physical interfaces in the host environment. This means that the host environment needs to redirect packets between physical and virtual interfaces. When a packet is received on a physical interface, the host environment checks the packet in order to identify the virtual router to which the packet should be redirected, and makes the packet available to the virtual router on one of its virtual interfaces, When the virtual router has processed the packet and determined the next hop, the packet is placed on an outgoing virtual interface from where it is finally transmitted on a physical interface.

The redirection of packets between physical and virtual interfaces introduces a layer of indirection that is not present in a physical router. This functionality can be implemented in several ways. One common configuration is to use the regular software bridge module in the Linux kernel to interconnect the interfaces, as shown in [9]. The software bridge provides a general-purpose switching function that allows packets to be switched between interfaces (virtual or physical) based on MAC addresses and MAC address learning. This solution is general in the sense that it allows packet to be switched between any pair of interfaces, and it is an attractive solution being based on a well-known software component already available in the kernel. However, for the purpose of redirecting packets between virtual and physical interfaces, it introduces a considerable amount of overhead. This, in turn, incurs performance penalties, something that was investigated in previous work [9]. Similar solutions that have been used are the virtual switch [14] and the short bridge [5]. It is also possible to use, for example, IP routing and Network Address Translation (NAT) for traffic to and from virtual machines, but those are not suitable for virtual routers.

A promising solution from a performance point of view is to replace the software bridge with a multiplexing/demultiplexing module. This is a more restricted solution, but potentially more efficient, since it can move packets from physical interface to virtual with less overhead. In the following we will investigate this solution closer. We start by examining the packet processing path in a Linux-based virtual router in more detail.

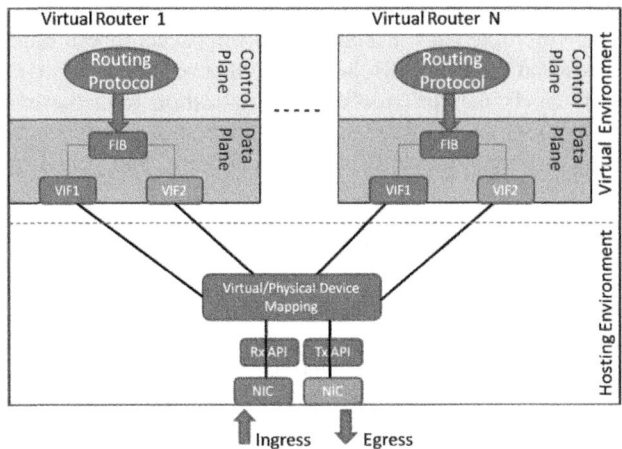

Fig. 1. General design of Linux-based virtual routers

3.1 Packet Processing Path

When a packet is received on a **Network Interface Card (NIC)**, the packet is transferred to a receive buffer in main memory, and an RX (receive) interrupt is generated. Interrupt processing is costly and can have large impact on forwarding performance [10]. Modern NICs provide features like interrupt coalescing and interrupt throttling for mitigating the negative effects of interrupt handling [11]. Current Linux use an RX interrupt processing scheme called NAPI [12], adopted from kernel 2.4.20 [13]. Instead of using one hardware interrupt per packet, NAPI combines interrupts with polling so that multiple packets can be processed within a single hardware interrupt. The NAPI interrupt handler accumulates packets in receive buffers, and schedules a software interrupt (SoftIRQ) to trigger processing of a batch of incoming packets.

After a SoftIRQ, the packet is delivered towards a virtual router via the corresponding virtual interface (VIF). Virtual interfaces are exactly like physical interfaces except that they are completely implemented in software. Therefore, no hardware interrupts are involved. A common example of a virtual interface is *veth*—the virtual Ethernet device. It has its own MAC address and an administrator has full access to its configuration in terms of MAC address, IP address assignment etc. Another example of virtual interfaces is the *macvlan* device, which will be discussed in more detail below. Both veth and macvlan are the part of the Linux kernel (2.6.x).

When the virtual router has processed a packet from its incoming virtual interface, the packet will be scheduled for transmission on an outgoing NIC. The outgoing NIC has a transmit (TX) queue where outgoing packets are placed. The NIC is then informed that outgoing packets are ready for transmission. The NIC's DMA engine fetches the packet from host memory and transmits it onto the physical media.

3.2   The macvlan Virtual Interface

In previous work we investigated how a careful selection of virtual interface and approach for redirecting packets between virtual and physical interfaces can improve the overall system performance [9]. We compared a macvlan-based virtual router with veth/bridge-based virtual router using both OpenVZ and Linux namespaces virtualization environments. We concluded that, in comparison with veth/bridge, a macvlan based virtual router is far less CPU demanding and can achieve higher throughput. In addition, it shows better behavior in overload situations.

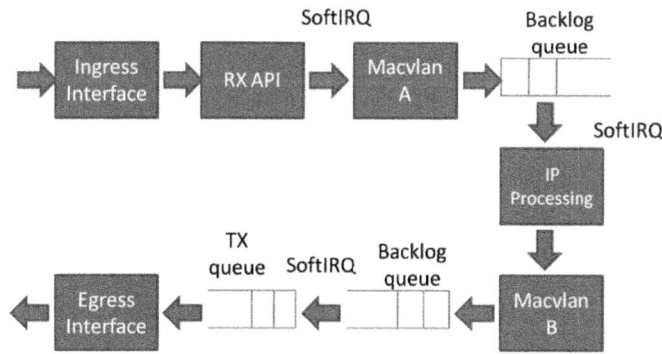

Fig. 2. Data plane for a macvlan based virtual router

The macvlan device provides a mechanism to define multiple virtual interfaces on top of a single physical interface. Each virtual interface is bound to a physical interface and has its own MAC address. A MAC address table is maintained for the virtual/physical address mapping. Fig. 2 depicts the packet data path for a macvlan-based virtual router. A packet received in a physical ingress interface enters into the virtual router through a virtual interface, macvlan A. After MAC level processing, the packet is queued in a *backlog queue* and a SoftIRQ is scheduled. When the SoftIRQ occurs, the packet is fetched from the queue for processing in the virtual router. As a result of the processing, an outgoing virtual interface is determined, and the packet is handed over to this virtual interface, macvlan B. There, the packet is buffered in another backlog queue and a new SoftIRQ is scheduled. Finally, upon the new SoftIRQ, the packet is moved from the backlog queue to the transmission queue of the physical egress interface.

Like a physical interface, a virtual interface always delivers a packet for receive interrupt processing after interface level processing. It can be observed in Fig. 2 that there is a backlog queue following each macvlan device. This design stems from the use of the Softnet API (a predecessor of NAPI). Upon completion of interface level processing, the macvlan device calls Softnet API, which buffers the packet in a backlog queue and schedules a SoftIRQ for further packet processing [12]. The Softnet API performance limitations are well known for physical interfaces [10], and can create livelock situations at high traffic loads. We believe that the use of the Softnet API for virtual interfaces has the following design weaknesses:

- Backlog queue congestion can cause serious throughput degradation for bridge/veth based virtual routers [9].
- The backlog queue is maintained on a per CPU basis, which means that virtual routers running on the same CPU will share the same queue. This may result in resource contention, something that can corrupt isolation properties between virtual routers. It may also limit overall system scalability.
- There are multiple queuing points along the data path. This may cause unnecessary delays in packet processing, and lead to inefficient usage of CPU resources.

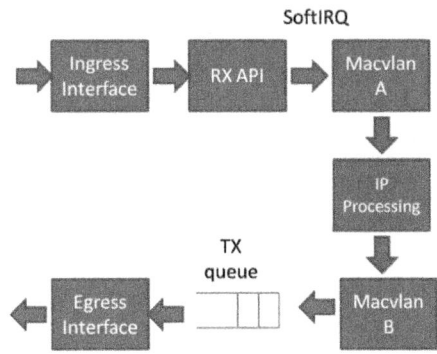

Fig. 3. Data plane for a NAPI-macvlan based virtual router

In order to address these issues, we introduce two changes in the data path:

- The backlog queues are eliminated from the data path.
- Packet receive processing is modified to that a packet is carried from the ingress interface, through virtual router, to the TX queue of outgoing physical interface in a single SoftIRQ.

The modified data path is illustrated in Fig. 3. We have implemented this data path as a modified macvlan device, called *NAPI-macvlan*. It is different from a macvlan device in the sense that it uses NAPI (netif_receive_skb) instead of Softnet API so that packets are delivered directly to the next processing module without being stored in an intermediate queue.

4   Performance Evaluation

This section presents a performance evaluation of macvlan and NAPI-macvlan based virtual router platforms from different performance perspectives, considering throughput, scalability, isolation and latency. In all test cases, we use Linux namespaces as the foundation for virtualization. We relate the performance of a virtual router to regular IP forwarding in a non-virtualized Linux based router. Throughout this section, the latter is denoted "IP Forwarder", and we use it as a reference to study the effects of applying virtualization.

4.1   Experimental Setup

We adopt a standard method to examine router performance in conformance with RFC 2544 [16]. A source machine generates network traffic that passes through a device under test (DUT), which forwards towards the destination machine. The nodes are connected using 1 Gb/s links.

We use an AMD Phenom quad core 3.2 GHz machine as traffic generator (GEN). The machine is equipped with 4GB of memory and two Gigabit Ethernet Intel® PRO/1000 PT server adapters (controller chip 82571GB). Another machine with the same specifications is used as destination (SINK). We use an Intel dual core 2.6 GHz machine (E8200) as the DUT. The machine has 4GB memory and a quad port Gigabit Ethernet Intel PRO/1000 PT server adapter. The DUT is running Linux kernel version 2.6.33-Netnext. All network interfaces are running NAPI-aware network drivers. Interrupt throttling is turned off for all interfaces. We use pktgen [17], an open source tool for traffic generation and throughput computation at sink. For all tests, 64 byte packets are generated. The DUT is running a single CPU, unless something else is specified.

4.2   Throughput

Case I
We start with a simple scenario: a DUT with two physical interfaces, forwards packets from one interface to another (unidirectional traffic flow). The results are show in Fig. 4. With this scenario, non-virtualized IP forwarding reaches a maximum throughput of 785 kpps. This provides the baseline forwarding performance as a reference for the other measurements. Fig. 4 also shows the throughput for virtual routers with two virtual interfaces, using macvlan devices as well as NAPI-macvlan. One virtual interface is connected to the ingress physical interface while the other virtual interface is connected to the egress physical interface. Fig. 4 shows that the macvlan-based virtual router attains up to 690 kpps while the the NAPI-macvlan-based virtual router achieves around 700 kpps.

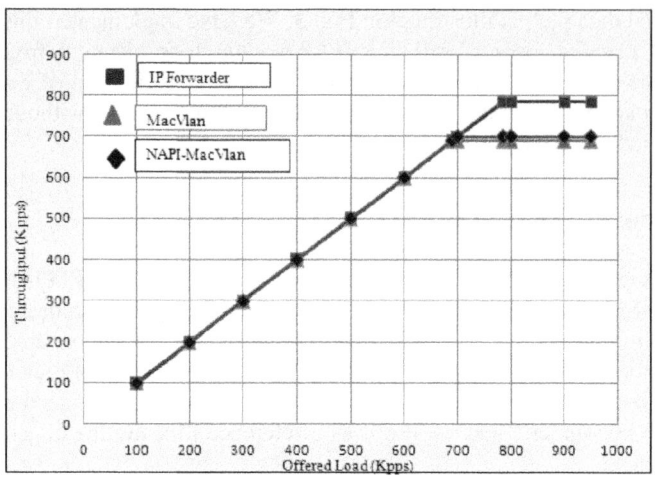

Fig. 4. Load vs throughput (Case I)

Case II

For the next step we add a physical interface to the setup. We offer load on two interfaces and use the third as egress interface. A single CPU core is thus processing traffic belonging to three physical interfaces. For this scenario, we consider aggregate offered load on the system (on both physical interfaces). The results are shown in Fig. 5 (three interfaces). The native IP Forwarder reaches a packet rate of 785 kpps. The macvlan-based virtual router achieves a peak rate of 600 kpps at an offered load of around 600 kpps. Then, as the load increases, the packet rate will drop down to 480 kpps. The NAPI-macvlan-based virtual router peaks at 700 kpps, which is sustained as the offered load increases—a significant performance difference under overload.

To understand the throughput difference between macvlan and NAPI-macvlan, we study packet drop locations along the data path. In case of macvlan (Fig. 2), we find that packets are dropped at two locations: in the backlog queue after the incoming macvlan device (after "Macvlan A" in Fig. 2) and on the ingress physical interface. When the load is increased beyond 600 kpps, packet drop starts in the backlog queue. Increasing offered load results in more packet drop and throughput degradation, but packets are still accepted by the ingress interfaces. This situation remains until the aggregated load on the ingress interfaces reaches 1000 kpps. Above this load level, the ingress physical interface starts dropping packets, and from that point a throughput of 480 kpps is maintained for increasing load, as shown in Fig. 5. This type of behavior has been explained in detail for veth/bridge-based virtual routers [9]. In case of NAPI-macvlan, the only packet drop location is on the ingress physical interface. When the load is increased above 700 kpps (Fig. 5), the ingress interfaces start dropping packets, and a throughput of 700 kpps is sustained for higher loads. Clearly, this more graceful overload behavior is much more preferable.

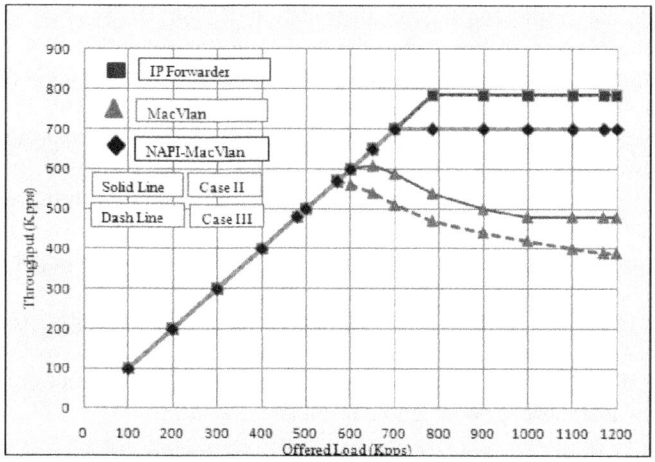

Fig. 5. Load vs throughput (II, III)

An interesting conclusion that can be made from this test case is that the cost of dropping packets in software is significant compared to dropping packets directly on the ingress interfaces. In Case II, we explain the gradual throughput degradation of macvlan virtualization for loads between 600 and 1000 kpps by the fact that packets are still accepted by the ingress interfaces and accordingly the CPU has to spend cycles both on packets that get dropped and on packets that get forwarded. Spending CPU cycles on packets that eventually will be dropped is clearly a waste of resources, adding overhead and reducing performance. This behavior cannot be seen for NAPI-macvlan virtualization since the only drop location is on the physical interface. Once a packet has been accepted by the macvlan device from the physical interface, the packet will be processed throughout the entire data path and delivered to the egress interface.

Case III

In this test case, we add yet another physical interface to the DUT and spread the offered load over three ingress interfaces while having a single egress interface. In this case, throughput for macvlan degrades even further. The backlog queue is now shared by even more ingress interfaces, something which increases the drop rate at the CPU level. The performance of NAPI-macvlan, on the other hand, is not affected by adding yet another physical ingress interface.

4.3 Scalability

Our first scalability study is to analyze the impact of running multiple virtual routers on the same physical platform. Therefore, we extend the experimental setup of case III with an increasing number of virtual routers. We create up to 32 virtual routers, each with four virtual interfaces (three ingress and one egress). Each ingress physical interface is now shared by up to 32 ingress virtual interfaces. A pktgen script is used to send 10 million packets in a row through one virtual router at a time. We do not observe any noticeable throughput degradation in the DUT when adding virtual routers. It can be seen in Fig. 6 that the maximum throughput for the DUT remains almost constant irrespectively of the number of existing virtual routers (1 to 32). This behavior is the same both for macvlan and NAPI-macvlan. The tests indicate a promising scalability property of Linux namespaces. Another important scalability concern is to study the impact of IP route lookup on forwarding performance. Until now we consider a unidirectional virtual router with a single routing table entry—a valid test setup but not a practical scenario. We move towards a more realistic scenario by considering bidirectional traffic flows together with a larger routing table. We extend the setup of case I and update the virtual router with 512 routing table entries. We offer load on both physical interfaces, which is forwarded towards each other through the virtual router (i.e. bidirectional). As a first step, we offer load with the same IP destination in all packets. In this case, routing information is available in routing cache and there is no need to consult routing table.

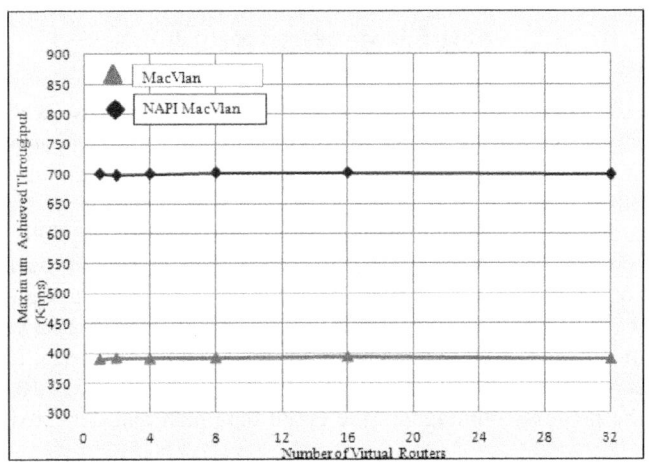

Fig. 6. Throughput vs. no. of virtual routers

The results are shown in Fig. 7 (single destination, 2 virtual interfaces). The NAPI-macvlan router achieves 700 kkpps and macvlan router reaches 660 kpps. A minor drop rate in the backlog queue can be observed for macvlan. As the next step, we offer load with 512 different IP destinations to see the impact of having the CPU make a route lookup with misses in the route cache. Fig. 7 (multiple destinations, 2 virtual interfaces) shows a minor throughput degradation for NAPI-macvlan (680 kpps). However, for macvlan packet drops in the backlog queue are now becoming substantial. As a result, throughput drops to 500 kpps. In the above scenario, the virtual router has only two virtual interfaces so there are only two entries for virtual/physical device mapping. All packets arrive on one interface and are transmitted on the other. Such a setup will not allow studies of the impact of virtual/physical device mapping, something which is important from scalability perspective. Furthermore, a virtual router with two virtual interfaces may not always be useful. We increase the number of virtual interfaces in the setup. For each physical interface, 256 virtual interfaces are created (512 in total). The bidirectional load is

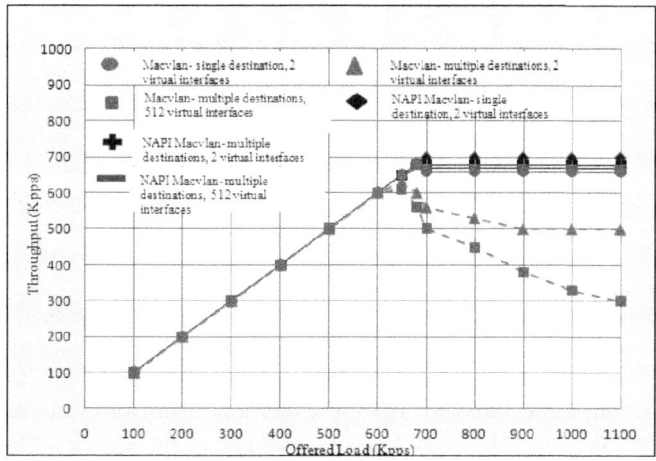

Fig. 7. Throughput for 2, 512 VIFs

offered with 512 different IP destinations so that different egress virtual interfaces will be used for different destinations. We investigate the impact of device mapping.

The results are shown in Fig. 7 (512 virtual interfaces). It shows a small throughput degradation for NAPI-macvlan (670 kpps), compared to the earlier peak rate of 700 kpps. For macvlan, throughput degrades to 300 kpps as a result of more severe backlog congestion. We observe that when the computational burden increases on the CPU, the backlog congestion becomes more adverse.

Fig. 8, gives a more detailed picture of throughput versus the number of virtual interfaces. The number of IP destinations in the traffic flow is increased along with the number of virtual interfaces. The impact is minor for the NAPI-macvlan based router. However, throughput drops sharply for the macvlan case.

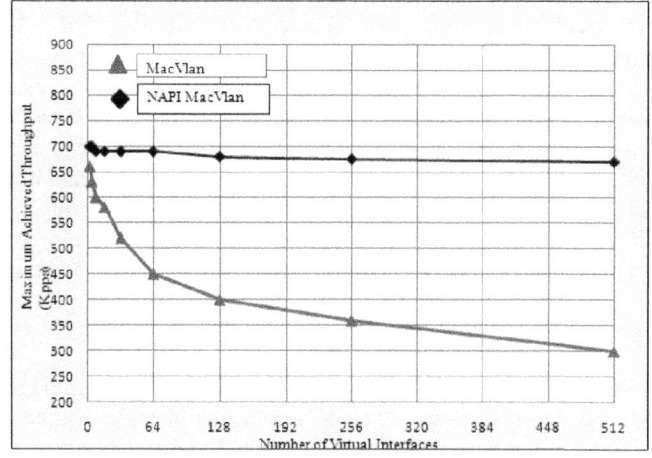

Fig. 8. Throughput vs VIFs

4.4 Isolation

In the previous subsection we verified that an increasing number of passive virtual routers had almost no effect on the overall throughput performance for neither macvlan nor NAPI-macvlan (Fig. 6). However, it is also important to investigate how multiple active virtual routers, running on the same CPU, might affect each other's operation. We refer to this as isolation properties. In this subsection, we study these isolation properties through an experiment where we analyze how a high offered load on one virtual router might influence the operation of another virtual router sharing the same CPU. The setup has two virtual routers, each one with a dedicated pair of physical interfaces and unidirectional traffic flow.

As a first step, we offer a load of 300 kpps on the ingress physical interfaces of both virtual routers at the same. We observe an aggregate throughput of 600 kpps. It shows that both virtual routers are working independently without affecting each other. The result is the same for both macvlan and NAPI-macvlan virtual routers. Thereafter, we increase the offered load to 1000 kpps on the ingress physical interface of VR1 while still offering 300 kpps on the ingress interface of VR2. For this scenario, the throughput results are presented in Table 1.

Table 1. Isolation between virtual routers

Setup	Packet rate (kpps)					
	Offered load			Throughput		
	VR1	VR2	Total	VR1	VR2	Total
Macvlan	1000	300	1300	480	55	535
NAPI-Macvlan	1000	300	1300	350	300	650

We can note that the overall throughput for NAPI-macvlan is higher than for macvlan virtualization (650 kpps vs 535 kpps). Moreover, the overload on VR1 in the macvlan case results in serious performance degradation in VR2. In the NAPI-macvlan, on the other hand, no such effects can be seen.

The explanation to this difference in isolation can again be described by the backlog queue that is present in macvlan virtualization. Since the backlog queue is on a per CPU basis, it is shared between the two virtual machines. So, even though we have isolation at the physical interface level, this isolation cannot be preserved between VR1 and VR2 because of the shared backlog queue. For NAPI-macvlan, VR1 and VR2 do not have any drop location in common. Therefore, packet drops occur only on the physical interfaces and the isolation properties can be preserved.

4.5 Latency

Latency is an important parameter for many network applications. Pktgen provides a utility to compute packet latency. It records packet transmission time at the traffic generator and then packet reception time at the sink. The difference provides the packet latency. We use the case I setup for latency measurements. The test is conducted for a fixed amount of time (120 sec). A load of 600 kpps is offered and we make sure that all packets are received at the sink (i.e., no packet drop occurs here). Fig. 9 displays the average latency for each configuration. It can be seen that the IP forwarder and NAPI-macvlan virtualization have the same latency. However, the average latency is doubled for macvlan virtualization.

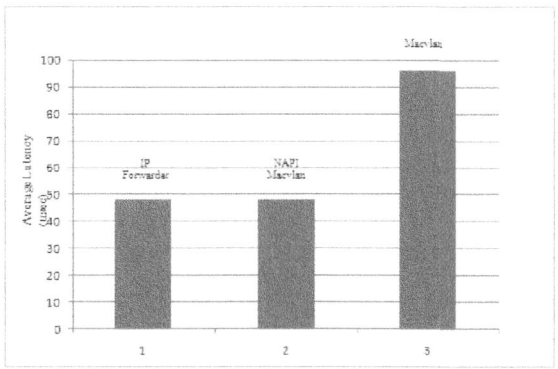

Fig. 9. Latency measurements for virtual setups

The reason for the higher average delay for the macvlan setup is that the macvlan uses the earlier mentioned Softnet API, incurring an overall processing delay. In case of both original IP forwarding and NAPI-macvlan, a packet is forwarded within a single SoftIRQ. This reduces the packet processing delay and results in a lower average latency.

5   Conclusion and Future Work

We have evaluated performance of virtual router platforms based on Linux Namespaces. The virtual routers were using the macvlan device (virtual interfaces) which is the part of the main stream Linux kernel 2.6. We analyzed the router's data plane and pointed out that backlog queuing can form a severe performance bottleneck. We proposed an alternative data plane by eliminating the backlog queue. To achieve this, we modified the macvlan device and introduced a variant denoted "NAPI-macvlan". We compared the performance of macvlan and NAPI-macvlan based virtual routers. We achieved better forwarding rates using NAPI-macvlan, particularly in different kinds of overload situations. Furthermore, in contrast to macvlan, NAPI-macvlan based routers proved superior when it comes to preserving isolation properties. It was also demonstrated that NAPI-macvlan based virtual routers improve the scaling properties. Finally, a considerable improvement in latency was also observed while using NAPI-macvlan. In our future work, we plan to evaluate NAPI-macvlan based routers using multiple CPU cores and multi-queue NICs.

Acknowledgments. We would like to thank Robert Olsson from TSLab KTH and Jens Laas from Uppsala University, for their valuable input during the work.

References

1. Egi, N., Greenhalgh, A., Handley, M., Hoerdt, M., Mathy, L., Schooley, T.: Evaluating Xen for virtual routers. In: PMECT 2007 (August 2007)
2. Anhalt, F., Primet, P.: Analysis and experimental evaluation of data plane virtualization with Xen. In: IEEE 5th ICNS (November 2009)
3. Egi, N., Greenhalgh, A., Handley, M., Hoerdt, M., Mathy, L.: Towards High Performance virtual routers on commodity hardware. In: ACM CoNext (December 2008)
4. Bavier, Feamster, N., Huang, M., Patterson, L., Rexford, J.: In VINI Veritas: Realistic and Controlled Network Experimentation. In: SIGCOMM 2006: Proceedings of ACM SIGCOMM 2006 Conference, Pisa, Italy, September 11-15 (2006)
5. Bhatia, S., Motiwala, M., Muhlbauer, W., Valancius, V., Bavier, A., Feamster, N., Peterson, L., Rexford, J.: Trellis: A Platform for Building Flexible, Fast Virtual Networks on Commodity Hardware. In: ACM ROADS 2008, Madrid, Spain, December 9 (2008)
6. Keller, E., Green, E.: Virtualizing the data plane through source code merging. In: PRESTO 2008: Proceedings of the ACM Workshop on Programmable Routers for Extensible Services of Tomorrow (2008)
7. Liao, Y., Yin, D., Gao, L.: PdP: Parallelizing Data Plane in Virtual Network Substrate. In: ACM VISA 2009, Barcelona, Spain, August 17 (2009)
8. Liao, Y., Yin, D., Gao, L.: Crossbow: From Hardware Virtualized NICs to Virtualized Networks. In: ACM VISA 2009, Barcelona, Spain, August 17 (2009)
9. Rathore, S., Hidell, M., Sjödin, P.: Performance Evaluation of Open Virtual Routers. In: IEEE GlobeCom Workshop on Future Internet, Miami, USA, December 10 (2010)
10. Bianco, A., Finochietto, J.M., Galante, G., Mellia, M., Neri, F.: Open-Source PC-Based Software Routers: A viable Approach to High-Performance Packet Switching. In: Ajmone Marsan, M., Bianchi, G., Listanti, M., Meo, M. (eds.) QoS-IP 2004. LNCS, vol. 3375, pp. 353–366. Springer, Heidelberg (2005)
11. Intel: Interrupt moderation using Intel Gigabit Ethernet controllers (Application Note 450), http://download.intel.com/design/network/applnots/ap450.pdf (last accessed April, 2010)
12. Salim, J.H., Olsson, R., Kuznetsov, A.: Beyond softnet. In: Proceedings of the 5th Annual Linux Showcase & Conference (ALS 2001), Oakland, CA, USA (2001)
13. Rio, M., et al.: A map of the networking code in Linux kernel 2.4.20. Technical Report DataTAG-2004-1, FP5/IST DataTAG Project (March 2004)
14. Pfaff, B., Petit, J., Koponen, T., Amidon, K., Casado, M., Shenker, S.: Extending Networking into the virtualization layer. In: ACM Sigcomm HotNets (September 2009)
15. Soltesz, S., Poltz, H., Fiuczynski, M., Bavier, A., Patersson, L.: Container-based Operating System Virtualization: A Scalable, High-performance Alternative to Hypervisors. In: EuroSys 2007: Proceedings of the 2nd ACM EuroSys Conference, March 21-23 (2007)
16. RFC 2544 Benchmarking methodology for interconnecting devices, http://tools.ietf.org/html/rfc2544 (last accessed April, 2010)
17. Olsson, R.: pktgen the Linux packet Generator. In: Proceedings of the Linux Symposium, Ottawa, July 20-23, vol. 2, pp. 11–24 (2005)

Performance Comparison of Hardware Virtualization Platforms

Daniel Schlosser, Michael Duelli, and Sebastian Goll*

University of Würzburg, Institute of Computer Science,
Chair of Communication Networks, Würzburg, Germany
{schlosser,duelli,goll}@informatik.uni-wuerzburg.de

Abstract. Hosting virtual servers on a shared physical hardware by means of hardware virtualization is common use at data centers, web hosters, and research facilities. All platforms include isolation techniques that restrict resource consumption of the virtual guest machines. However, these isolation techniques have an impact on the performance of the guest systems. In this paper, we study how popular hardware virtualization approaches (OpenVZ, KVM, Xen v4, VirtualBox, VMware ESXi) affect the network throughput of a virtualized system. We compare their impact in a dedicated and a shared host scenario as well as to the bare host system. Our results provide an overview on the performance of popular hardware virtualization platforms on commodity hardware in terms of network throughput.

Keywords: hardware virtualization, analysis, commodity hardware, network throughput, isolation.

1   Introduction

Hardware virtualization means abstracting functionality from physical components and originated already in the late 1960s. Recently, the importance of virtualization has drastically increased due to its availability on commodity hardware, which allows multiple *virtual guests* to share a physical machine's resources. The resource access is scheduled and controlled by the *virtual machine monitor* (VMM), also called *hypervisor*. Different virtualization platforms have been implemented and are widely used in professional environments, e.g. data centers and research facilities like G-Lab [1], to increase efficiency and reliability of the offered resources. The resources that may be used by a virtual guest system as well as lower performance bounds regarding these resources are determined in *service level agreements* (SLA) between providers and customers. The implementation of the VMM influences how well the resource allocation complies with these SLAs and to which extent different guests interfere with each other. Hence,

---

* This work was funded by the Federal Ministry of Education and Research (BMBF) of the Federal Republic of Germany (Förderkennzeichen 01BK0917, GLab). The authors alone are responsible for the content of the paper.

J. Domingo-Pascual et al. (Eds.): NETWORKING 2011, Part I, LNCS 6640, pp. 393–405, 2011.
© IFIP International Federation for Information Processing 2011

a provider has to know the performance limitations, the impact factors, and the key performance indicators of the used VMM to ensure isolation and SLAs.

Hardware virtualization is used in two basic scenarios. In the *dedicated* scenario, only a single virtual guest is run on a physical host. By means of hardware virtualization, the guest system is made independent of the underlying physical hardware, e.g., to support migration in case of maintenance or hardware failures. Hardware virtualization is also used to consolidate the resources of the physical host among multiple virtual guest systems in a *shared* scenario. While the performance of a virtualized system can be directly compared to the bare host system in the dedicated scenario, the performance of a virtual guest may be also affected by interference with other virtual guests in a shared scenario.

The performance cost of virtualization and *isolation*, i.e. the prevention of virtual guest interference, has been widely studied for CPU, memory, and hard disk usage. Nowadays, most commodity servers have built-in hardware support for virtualization enabling fast context switching in the CPU and resource restriction for the memory. However, the *input/output* (I/O) system, especially network throughput, is still a crucial factor.

Therefore, we focus on the network throughput of virtualized systems as the performance metric in this paper. We apply this metric on a non-virtualized Linux and on a virtualized version of this system using several popular hardware virtualization platforms, i.e. OpenVZ, KVM, Xen v4, VirtualBox, VMware ESXi. Furthermore, we install a second virtual guest and analyze the effects when the second virtual guest is idle, running some CPU/memory intensive tasks, or is transmitting packets.

The remainder of this paper is structured as follows. In Section 2, we present the different virtualization concepts used by the considered VMMs and related work. We explain our measurement setup and methodology in Section 3. Results for the dedicated scenario considering only a single guest are described in Section 4. An analysis of two virtual guests in shared scenarios is discussed in Section 5. In Section 6, we draw conclusions and give an outlook on future work.

2   Background and Related Work

In this section, we describe the techniques used by the considered VMMs to virtualize the underlying hardware and give an overview on related work.

2.1   Virtualization Concepts

The recent popularity of virtualization on commodity hardware created a plethora of virtualization platforms with different complexity and environment-dependent applicability. An overview of virtualization platforms is given in [20].

Several ways to realize hardware virtualization have evolved, which differ in the required adaptations of the guest system. For instance, a VMM may run on bare hardware (called *Type 1*, *native*, or *bare metal*) or on top of an underlying/existing *operating system* (OS) (called *Type 2* or *hosted*). Also *hybrids* between Type 1 and 2 are possible. Finally, OS-level virtualization is a special

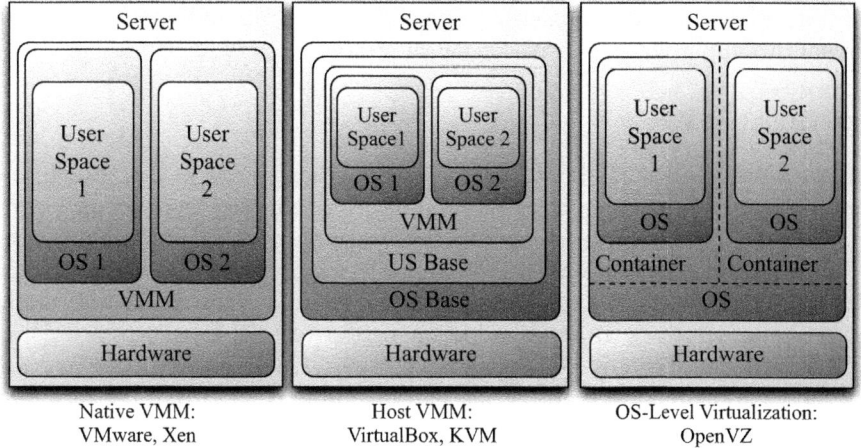

Fig. 1. Virtualization concepts of the considered VMMs

form, where the guests are run in a container of the OS operating directly on the hardware.

Native VMMs and hosted VMMs provide a *full virtualization*, i.e. the guest system does not have to be modified. This means that the virtualized guest operating system is working as if it would run directly on the hardware. But whenever the guest wants to access a hardware resource, which is only a virtualized resource, the VMM has to interrupt the guest and 'emulate' this hardware access. This affects the performance of the guest system. To alleviate these problems, the guest system can be modified, e.g. with special drivers, to redirect resource accesses to function calls of the VMM. This method is called *para-virtualization* and may improve the performance of the guest. OS level virtualization solves this resource access problem by using the same modified kernel in the guest containers as in the hosting OS. Hence, this technique has the advantage that the overhead for virtualization is quite low compared to the other solutions. However, the OS and the kernel of the guest is predetermined.

In this work, we consider five popular hardware virtualization platforms which can be categorized as

- *native virtualization*, represented by VMware ESXi [19] and Xen v4 [21],
- *host virtualization*, represented by VirtualBox [16] and KVM [12], and
- *operating system-level virtualization*, represented by OpenVZ [15].

Figure 1 depicts how the considered virtualization platforms encapsulate virtual guests and indirect access to resources for applications, which run in user space.

Fair access to the physical hardware is important and known to work for resources that can be split into quotas, e.g. memory, disk space, or processor cycles per second. However, access to any of these separately restricted resources involves the *input and output* (I/O) of data which typically shares a single bottleneck, e.g. an internal bus. Especially, network I/O fairness is a crucial issue which involves another shared physical resource, the *network interface card* (NIC). In this paper, we consider the packet forwarding throughput of the aforementioned virtualization platforms as a performance metric.

2.2   Related Work

An overview of virtualization techniques and mechanisms as well as their history and motivation is given in [8]. Several publications consider *resource access fairness* when multiple guest systems run on a single host. In [14], virtualization was done using Xen. The authors show that while the processor resources are fairly shared among guests, the scheduling of I/O resources is treated as a secondary concern and decreases with each additional guest system. The authors of [4] use a NetFPGA-based prototype to achieve network I/O fairness in virtual machines by applying flexible rate limiting mechanisms directly to virtual network interfaces. Fairness issues in software virtual routers using *Click* are considered in [9]. The impact of several guests with and without additional memory intensive tasks on packet forwarding is measured. More details and performance tuning for the Click router is provided in [10].

As mentioned before, the performance and fairness heavily depend on the used VMM. Hence, several comparisons of mostly two VMMs have been conducted. We focus on the references considering network I/O performance which altogether consider packet forwarding performance with different packet lengths from 64 to 1500 bytes. In [18], a comparison of VMware ESX Server 3.0.1 and open-source Xen 3.0.3 with up to two guests was conducted by VMware using the Netperf packet generator. They show that Xen heavily lags behind VMware ESX. This study was repeated with VMware ESX Server 3.0.1 and open-source Xen Enterprise 3.2.3 in [22] by Xen Inc. which showed that a specialized Xen Enterprise lags by a few percent. The authors of [7] consider multiplexing of two data flows with different scheduling mechanisms on a real Cisco router, a virtual machine host, in two VM guests using Xen, and a NetFPGA packet generator. They show the dependency of the software routers on the packet length. Measurements of OpenVZ and User Mode Linux (UML) were conducted in [5] on wireless nodes of the Orbit test bed. UDP throughput and FTP performance is measured, but it is also stated that the hardware is not the best choice for virtualization. In [6], the network virtualization platform *Trellis* is presented and packet-forwarding rates relative to other virtualization technologies and native kernel forwarding performance are compared. The considered virtualization technologies comprise Xen, Click, OpenVZ, NetNS, and the Trellis platform. The Linux kernel module *pktgen* [13] is used for packet generation. In their results, Xen has the lowest performance followed by OpenVZ.

To the best of our knowledge, this is the first paper to compare the packet forwarding performance of today's most popular hardware virtualization platforms on a common commodity server system without special hardware requirements, e.g. NetFPGA cards, or fine-tuned virtualization parameters, such that comparisons can be easily repeated.

3   Hardware Setup and Measurement Scenarios

In the following section, we give an overview on the hardware setup of our measurements and provide technical details on the considered scenarios.

3.1   Hardware Setup

For our measurements, we use seven Sun Fire X4150 x64 servers from the G-Lab testbed [1], each with 2×Intel Xeon L5420 Quad Core CPU (2×6 MB L2 Cache, 2.5 GHz, 1333 MT/s FSB, and 8×2 GB RAM. The *unit under test* (UUT) is equipped with eight 1 Gigabit ports, while all other servers are equipped with only four 1 Gigabit ports. The UUT is connected on each NIC to three other servers each with a single connection using Cat.5e TP cable or better. The setup and interconnection of the servers is illustrated in Figure 2.

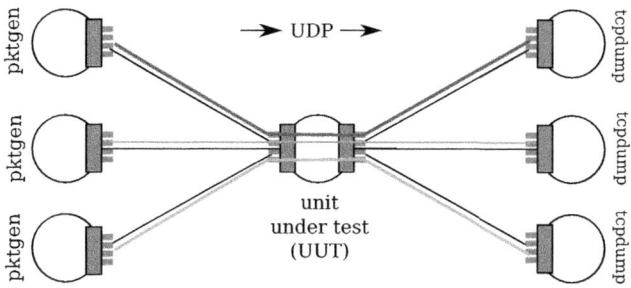

Fig. 2. The hardware setup

Each machine is running Debian 5.0.4 ("lenny") with Linux kernel 2.6.26 compiled for the amd64 (x86_64) architecture. Only for the test of Xen we needed to adjust the base system with Debian 5.0.5 ("lenny") with Linux kernel 2.6.31 for amd64. Since VMware ESXi installs directly on the host machine, the choice of a custom operating system is not applicable in this case. We investigate the following versions of the different VMMs with a Debian "lenny" as a guest system and the given Kernel installed:

- KVM 0.9.1 (kvm-72) [12]: Guest running Linux 2.6.26 for amd64
- OpenVZ 3.0.22 [15]: Guest running same OS kernel as host
- VirtualBox 3.2.8 [16]: Guest running Linux 2.6.26 for amd64
- VMware ESXi 4.1.0 [19]: Guest running Linux 2.6.26 for amd64
- Xen development version of July 16th, 2010 [21]: Guest running Linux 2.6.26 for i386 (i686; x86) with *physical addressing extensions* (PAE) support.

For the KVM and VirtualBox machines, virtual NICs are using the "virtio" para-virtualized network driver. For OpenVZ and Xen, para-virtualization is achieved implicitly through the OS-level or para-virtualization approach taken by OpenVZ and Xen, respectively. For VMware ESXi, both the "E1000" full-virtualized network driver and the "VMXNET 3" para-virtualized network driver (using the VMware Tools on the guest) are considered.

3.2   Measurement Metrics and Methodology

In all scenarios, we investigate the packet forwarding throughput performance
of the different virtualization solutions on the UUT. In each experiment, network traffic is generated using the packet generator integrated into the Linux
kernel (pktgen [13]). It sends out UDP packets which are forwarded by the UUT
to another destination server where the traffic is recorded by tcpdump [3] and
discarded afterwards. After stopping packet generation, the packet generator
reports how many packets it was able to send. We calculate the "offered bandwidth" from this value and the time in which the traffic has been sent as well as
the "measured throughput" from the data of the tcpdump file.

For our performance evaluation, we set up two different scenarios, a dedicated
host scenario and a shared host scenarios, which we describe in the next sections.

3.3   Dedicated Host Scenario – Impact of Virtualization

In this scenario, we measure the packet forwarding performance of the Linux
system without any virtualization (*raw*) and compare it with a setup where
the same system is installed inside the tested VMM. We consider up to three
traffic sources. It has to be noted that each traffic stream through the UUT does
not interfere with any other stream outside the server, i.e. each traffic stream
enters the server over a different network interface and also leaves the system via
another distinct network interface. Each network interface is directly bridged to
a corresponding virtual interface of the guest system.We consider packets with
an Ethernet frame size of 64, 500, 1000, and 1500 byte for each data stream.

For each packet size, a target bandwidth of 10, 100, 200, 400, 600, 800, 900,
and 1000 Mbit/s was run nine times for statistical evidence. Since pktgen does
not allow for setting a target bandwidth, but instead manages different loads
of traffic by waiting a certain amount of time between sending two consecutive
packets, this delay had to be calculated. For target bandwidth b [Mbit/s] and
target packet size s [byte], the resulting delay [ns] would be $8000 \times (s+24)/b$. This
formula works well for large packet sizes, but the kernel is not able to generate
small packets fast enough. Hence, we adopted the inter packet delay to 1 and
0 ns to maximize the generated traffic and send as many packets as possible.

3.4   Shared Host Scenario – Impact of Other Guest Systems

The consolidation of multiple virtual systems on a single physical host may have
an impact on the network performance of the virtualized guests. Thus, we add
a second virtual guest to the VMM, which runs another Debian "lenny" and
repeat all measurements. Initially, the second guest does not run any processes
generating additional load (*no load*). In order to investigate if CPU intensive
processes in another guest systems has an influence of the performance of our
UUT, we generate variable amounts of CPU and virtual memory load with the
stress tool (version 0.18.9) [2] and considered two more cases. In the first case,
called *light load*, stress was running with parameter --cpu 1, corresponding

to a single CPU worker thread. In the second case, called *heavy load*, stress was running with parameters --cpu 8 --vm 4, corresponding to 8 CPU worker threads, and 4 virtual memory threads which allocate, write to, and release memory. In case of full virtualization (KVM, VirtualBox, VMware, and Xen), each virtual guest was assigned a single (virtual) CPU out of the 8 physical CPU cores installed on the server.

Besides CPU and memory load, we targeted the influence of a second guest systems network load, even if the guests do not share physical network interfaces. We focus in this scenario on Ethernet frame sizes of 64 byte, as the previous scenarios revealed this packet size to be most critical. Hence, we reconfigured the UUT to have only two virtual interfaces, which are bridged to different physical interfaces. We adjust the second guest system to have also two virtual interfaces, which are bridged to two distinct physical interfaces. We run the same tests as before, but this time we send traffic through the second guest. We adjust the traffic in such a way that it is 50% and 95% of the maximum rate, which we measured in the test with the second guest being idle. For this test, we also measure the throughput of the second guest to investigate if the second guest system can still forward the offered data rate.

4   The Performance Cost of Virtualization

Depending on the used VMM, the network throughput performance of the virtual guests differs significantly. In Figure 3, we plot the bandwidth the traffic generator produced ("offered bandwidth") against the throughput the central test system was able to forward ("measured throughput") averaged over all nine repetitions of a test. The error bars, which in many cases can hardly be seen, present the 95% confidence intervals for the mean values. The graphs for the 1500 byte packets reveal no difference between the system without virtualization and most of the VMMs. Only the VirtualBox system is not able to forward all packets. For smaller packet sizes, i.e. 1000 bytes and 500 bytes, the difference between the different VMMs is clearly visible. The fact that the offered bandwidth for all packet sizes above 500 bytes increases up to the maximum of the link, i.e. 1 Gbit/s, shows that the VMMs are able to fetch the packets from the network interface. However, in all cases in which the measured bandwidth is lower than the offered bandwidth, a significant number of packets get lost in the *unit under test* (UUT). Looking at Figure 3d, two additional effects can be noticed for the smallest used packet sizes, i.e. 64 bytes. Not all graphs extend to the same length on the y-axis. This means that the traffic generator was not able to offload the same amount of traffic to the UUT in all cases. This behavior is caused by *layer-2 flow control*. Whenever the frame buffer of the network card is fully occupied, the network card sends a notification to the sending interfaces to throttle down the packet rate. The maximum throughput of the UUT is never affected by this behavior. We kept it activated, as it provides another option to the test system to react on traffic rates it cannot handle besides dropping packets. Please note that in all cases in which the layer-2 flow control is actively reducing the sent packets, the test system already drops packets. This behavior,

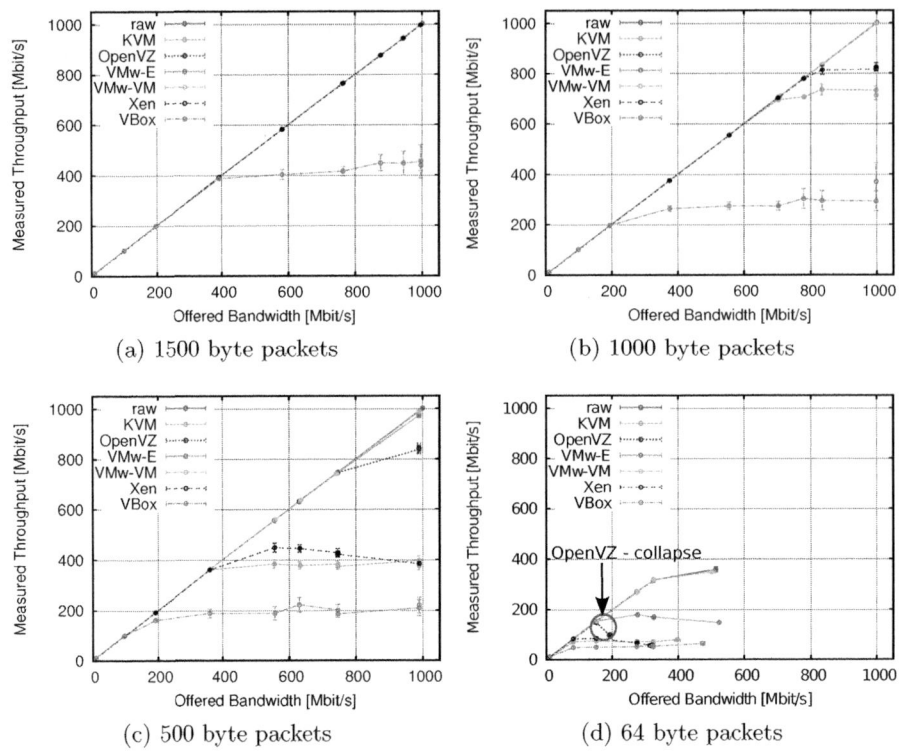

Fig. 3. Offered bandwidth vs throughput of VMMs for different packet sizes

therefore, never influences the maximum throughput of the system. The second observable effect is the behavior of OpenVZ when forwarding 64 byte packets. The OpenVZ system achieves a maximum throughput at a given input rate and collapses afterwards. We also notice this effect in other scenarios with 64 byte and 500 byte packet sizes. But in these cases, the throughput rate does not only collapse but is heavily varying for some offered traffic rates.

Next, we consider how the throughput of the UUT scales if we increase the number of traffic generators. Figure 4 depicts the maximum achieved throughput with one, two, and three traffic sources. Note that as we have seen, e.g. in Figure 3c, the maximum achieved throughput is not necessarily recorded at the highest offered bandwidth. The bar plot for the UUT without any VMM reveals that in all cases, except the one sending the 64 byte packets, the performance of the raw system scales linearly, i.e. we measure a throughput of 1 Gbit/s, 2 Gbit/s, and 3 Gbit/s. In the case of 64 byte packets, Figure 4d also reveals a limitation of the raw system, as the accumulated throughput for two and three traffic generators is the same. It has to be noted that in this case the raw system uses layer-2 flow control to throttle down the traffic generators. But the raw system is able to forward all received packets, in contrast to all VMMs which accept up to 600 Mbit/s input rates but drop packets down to the depicted maximum accumulated throughput. Another behavior which cannot be seen in the bar

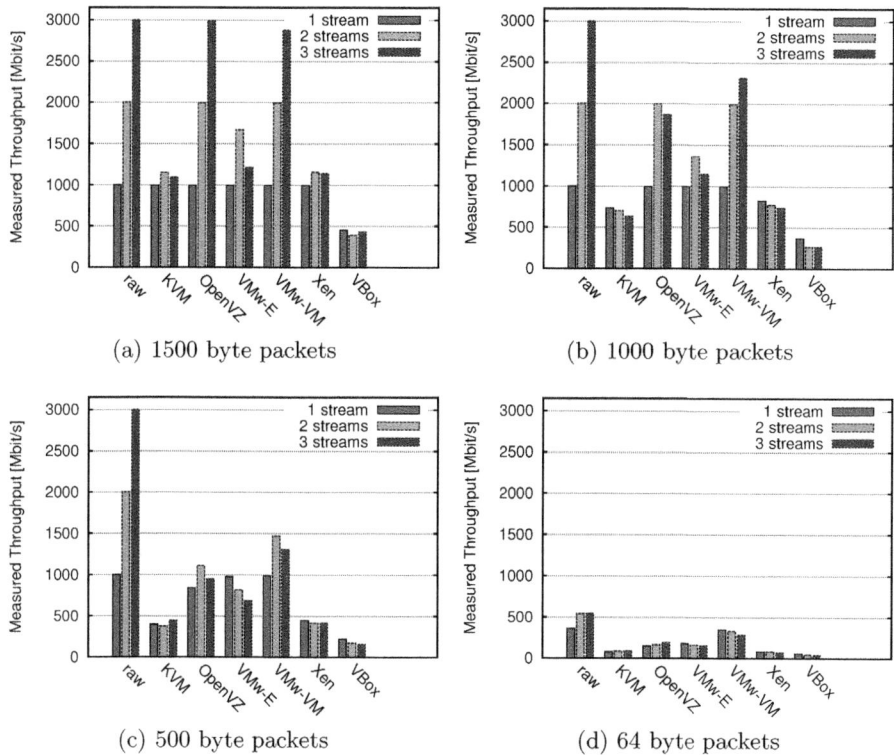

(a) 1500 byte packets

(b) 1000 byte packets

(c) 500 byte packets

(d) 64 byte packets

Fig. 4. Maximum accumulated throughput of VMMs

plots is that we measured a significant difference in the mean throughput of the three data streams for the VMware system without the para-virtualized driver.

The results in Figure 4 show that most VMMs have a throughput limitation, which varies with the packet size. All VMMs except OpenVZ reveal this limitation even for 1500 byte packet sizes. Only OpenVZ is able to forward packets of this size with the same performance as the raw system. However, if we look at the results for the other packet sizes, we clearly see the impact of virtualization even for OpenVZ. For smaller packet sizes, i.e. 500 byte and below, the VMware with the para-virtualized VMXNET3 driver (abbreviated by VMw-VM in the figures) outperforms all other solutions.

In conclusion, we see that the performance costs of virtualization gets higher, the more small packets are sent and the higher the bandwidth on the UUT is. Also OpenVZ, which only provides OS-level virtualization, has a noticeable influence on the systems' performance in these cases. The biggest problem identified is that the VMMs try to take as many packets of the network card as possible and therefore prevent layer-2 flow control to limit incoming data to a rate, which can be handled by the system. Hence, the majority of packets is lost before entering the guest system.

5   Isolation of Guest Systems

Initially, we conducted tests with a second virtual guest Linux that was only idle. These tests showed that the difference to the scenario with only a single guest are mostly negligible. Only OpenVZ showed a difference in the shape of its collapse for high offered bandwidth rates with small packets. Hence, we focus in this paper on the more interesting effects. More details are provided in [17].

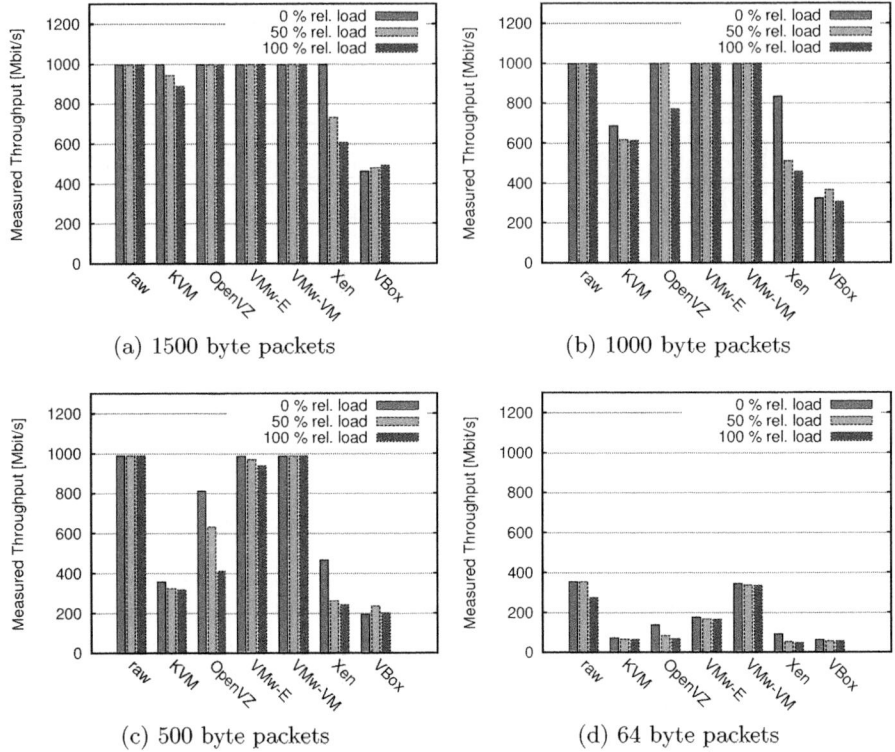

Fig. 5. Maximum throughput considering a second guest transmitting small packets relative to the maximum throughput in the dedicated scenario, c.f. Figure 4d

The impact of a second virtual guest, which is performing CPU and memory intensive tasks, is also small. We, therefore, provide only the difference in maximum achieved throughput in Table 1. It has to be noted that positive values in Table 1 correspond to an increase of throughput under these conditions. For example, Xen forwards in any considered scenario more packets, if a second virtual guest is running CPU or CPU and memory intensive tasks. Only OpenVZ is negatively affected by more than 10% relative forwarding performance considering small packets. This effect might be caused by the fact that OpenVZ, in the original implementation, does not foresee the option to restrict the execution of processes to predefined CPUs. Thus, the stress processes are run on all CPUs

Table 1. Averaged relative throughput changes considering a single traffic source and a second guest running CPU/Memory intensive processes,cf. Section 3.4

packet	KVM		OpenVZ		VM-E1000		VM-VMXNET3		Xen		VirtualBox	
size	light	heavy	light	heavy	light	heavy	light	heavy	light	heavy	light	heavy
64	4.2%	-0.4%	1.0%	-18.2%	-2.7%	-4.5%	-0.5%	-0.6%	6.8%	7.9%	2.1%	2.0%
500	5.5%	4.0%	1.5%	-11.4%	-4.4%	-5.3%	0.0%	-0.1%	3.7%	3.4%	-2.0%	-4.0%
1000	5.1%	5.6%	0.0%	0.0%	0.0%	0.0%	0.0%	0.0%	0.8%	1.7%	3.6%	-4.8%
1500	0.0%	0.0%	0.0%	0.0%	0.0%	0.0%	0.0%	0.0%	0.0%	0.0%	3.4%	2%

Table 2. Averaged offered bandwidth (in) and throughput measured(out) for 2nd guest in background at 50% relative load, c.f. Figure 4d

packet	raw		KVM		OpenVZ		VM-E1000		VM-VMXNET3		Xen		VirtualBox	
size	in	out	in	out	in	out	in	out	in	out	in	out	in	out
1500	153.1	153.1	37.1	37.1	72.1	72.1	71.8	71.8	152.2	151.9	37.1	36.8	37.1	36.7
1000	153.0	153.0	37.1	37.1	72.1	72.1	71.8	71.8	152.0	151.8	37.1	36.6	37.1	36.0
500	153.0	153.0	37.1	37.1	72.1	68.1	71.8	71.8	152.0	151.7	37.1	36.1	37.1	36.2
64	153.0	153.0	37.1	37.0	72.1	62.5	71.8	71.8	152.1	151.8	37.1	35.4	37.1	36.8

Table 3. Averaged offered bandwidth (in) and throughput measured(out) for 2nd guest in background at 95% relative load, c.f. Figure 4d

packet	raw		KVM		OpenVZ		VM-E1000		VM-VMXNET3		Xen		VirtualBox	
size	in	out	in	out	in	out	in	out	in	out	in	out	in	out
1500	275.8	270.4	72.1	63.2	127.0	127.0	126.6	126.6	274.4	268.3	72.0	56.2	72.1	42.1
1000	275.6	270.4	72.1	61.3	127.0	102.7	126.7	126.6	274.3	267.6	72.1	50.9	72.1	41.5
500	275.9	271.4	72.1	60.8	125.9	80.1	126.6	126.6	274.3	267.5	72.1	48.9	72.1	40.2
64	274.9	269.3	72.1	64.6	124.4	62.7	126.6	126.6	274.4	268.4	72.0	47.2	72.1	42.4

compared to the other scenarios, where the `stress` processes are strictly bound to the CPU of the second process. It has to be noted that both OS containers in OpenVZ are started with the "CPUUNIT" parameter, which means equal sharing of CPU resources. This result is surprising and we intend to study it closer in future work.

However, the impact of a second guest is clearly recognizable, whenever it is handling small packets. Figure 5 depicts our measurement results for the measurement scenario in which the second guest forwards small packets, cf. Section 3.4. Even if the UUT is only handling 1500 byte packets KVM and Xen achieve a lower maximum throughput. But if we consider the results for streams of 1000 byte and 500 byte packets, also OpenVZ and VMware without a para-virtualized network driver are affected, whereas the raw system and the VMware guest with VMXNET3 driver still handle all packets.

For 64 byte packets it seems as if the VMware system with VMXNET3 driver outperforms the unvirtualized system. But this is not the case. Due to the fact that we send packets through the second guest relative to the performance we measured before, the raw system handles 270 Mbit/s of 64 byte packets in the background and forwards all packets. The VMware system is only forwarding about 200 Mbit/s in the second guest and is buying the higher throughput of the test system by a loss rate of about 20%.

The offered bandwidth and the measured throughput for the second guest in this experiment are provided for both load scenarios in Table 2 and Table 3, respectively. These measurements reveal that the second guest is also negatively affected. If both guests are forwarding 64 byte packets, loss rates up to 50% can be experienced. The most severe impact can again be seen for OpenVZ. It seems as if the well known problem of the Linux kernel to handle small packets, c.f. [11], has an even larger impact when using OS-level virtualization. This effect has to be considered when planning experiments on experimentation facilities using this kind of virtualization, as it might cause a high variance in the results of experiments, depending on the processes in other guest systems.

6   Conclusion

In this paper, we analyzed the impact of virtualization for a single system in terms of packet forwarding throughput. We compared KVM, OpenVZ, VMware, Xen, and VirtualBox to a raw host system and the impact of another virtual guest system performing CPU and memory intensive tasks as well as forwarding of small packets. Our results revealed that the impact of virtualization is noticeable for most virtualization platforms even when only considering the throughput on a single network interface. When increasing the number of interfaces and traffic sources, each VMM revealed a throughput bottleneck, which depends on the size of the packets being transmitted but is significantly lower than the forwarding capacity of the raw system. The impact of a second guest, which is idle or running only CPU and memory intensive tasks, is rather small. However, a second guest that forwards small packets even at comparably low rates is able to influence the performance of a virtualized system severely. Our findings show that virtualized systems and the traffic they handle need to be monitored. In professional system, network monitoring and exact measurements of the used systems enable the diagnosis of performance bottlenecks and definition of relocation strategies. But also for research facilities, in which many scientists share the resources of a test bed, it is crucial to record all influences which might be introduced by another test running on the same test bed, in order to provide credible scientific results. In future work, we will integrate the results from this paper into performance models for virtualized systems and virtual software routers to analyze and optimize throughput and isolation.

Acknowledgments. The authors would like to thank Prof. Tran-Gia for the stimulating environment which was a prerequisite for this work.

References

1. German-Lab (G-Lab), http://www.german-lab.de
2. Stress tool, http://weather.ou.edu/~apw/projects/stress/
3. tcpdump, http://www.tcpdump.org

4. Anwer, M.B., Nayak, A., Feamster, N., Liu, L.: Network I/O fairness in virtual machines. In: Proceedings of the 2nd ACM SIGCOMM Workshop on Virtualized Infrastructure Systems and Architectures (VISA), New York, USA, pp. 73–80 (2010)

5. Bhanage, G., Seskar, I., Zhang, Y., Raychaudhuri, D., Jain, S.: Analyzing router performance using network calculus with external measurements. In: TridentCom 2010 (May 2010)

6. Bhatia, S., Motiwala, M., Muhlbauer, W., Mundada, Y., Valancius, V., Bavier, A., Feamster, N., Peterson, L., Rexford, J.: Trellis: a platform for building flexible, fast virtual networks on commodity hardware. In: Proceedings of the 2008 ACM CoNEXT Conference, New York, USA, pp. 72:1–72:6 (2008)

7. Bredel, M., Bozakov, Z., Jiang, Y.: Analyzing router performance using network calculus with external measurements. In: 18th International Workshop on Quality of Service (IWQoS), pp. 1–9 (June 2010)

8. van Doorn, L.: Hardware virtualization trends. In: Proceedings of the 2nd International Conference on Virtual Execution Environments (VEE), New York, USA, pp. 45–45 (2006)

9. Egi, N., Greenhalgh, A., Handley, M., Hoerdt, M., Huici, F., Mathy, L.: Fairness issues in software virtual routers. In: Proceedings of the ACM Workshop on Programmable Routers for Extensible Services of Tomorrow (PRESTO), New York, USA, pp. 33–38 (2008)

10. Egi, N., Greenhalgh, A., Handley, M., Hoerdt, M., Huici, F., Mathy, L.: Towards high performance virtual routers on commodity hardware. In: Proceedings of the 2008 ACM CoNEXT Conference, New York, USA, pp. 20:1–20:12 (2008)

11. Han, S., Jang, K., Park, K., Moon, S.: PacketShader: a GPU-accelerated software router, vol. 40, pp. 195–206. ACM, New York (2010)

12. KVM: Kernel-based virtual machine, http://www.linux-kvm.org/

13. Olsson, R.: pktgen the Linux packet generator. In: Proceedings of Linux Symposium (2005)

14. Ongaro, D., Cox, A.L., Rixner, S.: Scheduling I/O in virtual machine monitors. In: Proceedings of the fourth ACM SIGPLAN/SIGOPS International Conference on Virtual Execution Environments (VEE), New York, USA, pp. 1–10 (2008)

15. OpenVZ: Open virtualization, http://openvz.org/

16. Oracle: Virtualbox, http://www.virtualbox.org/

17. Schlosser, D., Duelli, M., Goll, S.: Performance Comparison of Common Server Hardware Virtualization Solutions Regarding the Network Throughput of Virtualized Systems. Tech. rep., University of Würzburg (2011)

18. VMware: A Performance Comparison of Hypervisors (2007), http://www.vmware.com/pdf/hypervisor_performance.pdf

19. VMware ESXi, http://www.vmware.com/products/vi/esx/

20. Wikipedia: Comparison of platform virtual machines — Wikipedia, The Free Encyclopedia (2010), http://en.wikipedia.org/w/index.php?title=Comparison_of_platform_virtual_machines

21. Xen, http://www.xen.org/

22. XenSource Inc.: A Performance Comparison of Commercial Hypervisors (2007), http://www.vmware.com/pdf/hypervisor_performance.pdf

A Novel Scalable IPv6 Lookup Scheme Using Compressed Pipelined Tries

Michel Hanna, Sangyeun Cho, and Rami Melhem

Computer Science Department,
University of Pittsburgh,
Pittsburgh, PA, 15260, USA
{mhanna,cho,melhem}@cs.pitt.edu

Abstract. An IP router has to match each incoming packet's IP destination address against all stored prefixes in its forwarding table. This task is increasingly more challenging as the routers have to: not only keep up with the ultra-high link speeds, but also be ready to switch to the 128-bit IPv6 address space while the number of prefixes grows quickly. Commercially, many routers employ Ternary Content Addressable Memory (TCAM) to facilitate fast IP lookup. However, TCAMs are power-eager, expensive, and not scalable. We advocate in this paper to keep the forwarding table in trie data structures that are accessed in a pipeline manner. Especially, we propose a new scalable IPv6 forwarding engine based on a multibit trie architecture that can achieve a throughput of 3.1 Tera bits per second.

Keywords: IPv6, Tries Compression, Next Generation Internet.

1 Introduction

In the IP lookup (or forwarding) problem, the router has to match the destination address of every incoming packet against its forwarding table to determine the packet's next hop on its way to the final destination [22]. An entry in the forwarding table, or an IP prefix, is a binary string of a certain length followed by wild card (don't care) bits and an output port. The actual matching requires finding the LPM or the Longest Prefix Matching [18]. Recently the problem is getting more significance given the anticipated switch from the 32-bit IPv4 to the 128-bit IPv6 [1].

The main research streams that deal with the packet forwarding problem are algorithm-based and hardware-based. The most well-known algorithm-based solutions use the binary trie data structures [6, 16, 18, 20]. A trie is: "a tree-like data structure allowing the organization of prefixes on a digital basis by using the bits of prefixes to direct the branching" [18]. Figure 1(a) shows and example of an 8-bit address IP forwarding table, where $a, b, \cdots, m$ are symbols given to the prefixes for identification. In Figure 1(b) we show the equivalent binary trie of the IP forwarding table given in Figure 1(a). The main advantages of a trie-based solution are that they provide simple time and space bounds. However,

J. Domingo-Pascual et al. (Eds.): NETWORKING 2011, Part I, LNCS 6640, pp. 406–419, 2011.

	Prefix	Port
a	000*****	0
b	000101**	1
c	0001111*	2
d	0010****	0
e	00111***	2
f	0110****	0
g	01111***	2
h	1*******	0
i	1001****	0
j	11011***	2
k	110101**	1
l	111101**	1
m	1111111*	2

(a)

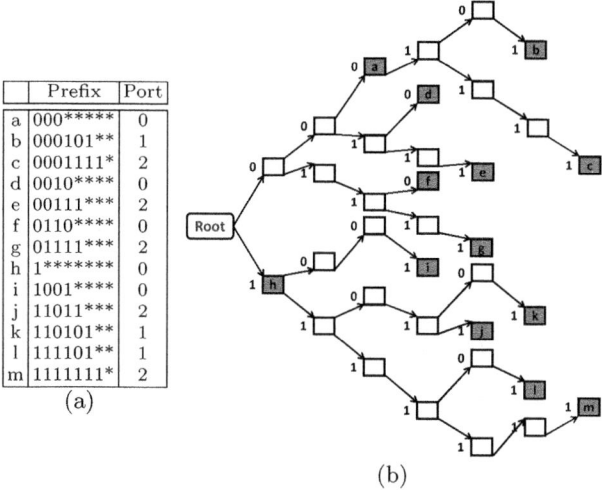

(b)

Fig. 1. (a) An example of an 8-bit address space forwarding table. (b) Its binary trie representation.

with the 128-bit IPv6 prefixes, both trie height and enumeration become an issue when the prefixes are stored inside the nodes.

Hardware-based packet forwarding engines are divided into many classes. The first class uses the Ternary Content Addressable Memory (TCAM), which has been the *de facto* standard for the packet forwarding application [18, 19]. A TCAM is a fully-associative memory that can store 0, 1 and don't care bits. In a single clock cycle, a TCAM chip finds the longest prefix that matches the address of the incoming packet by searching all stored prefixes in parallel. Nevertheless, TCAM has serious deficiencies: high power consumption, poor scalability to long IPv6 prefixes and lower operating frequency compared to other memory technologies [2].

The class of hash-based hardware packet forwarding engines has become popular recently [9,10,11]. The hash-based engines are promising because they offer constant search time. However, inefficiency rises when two or more keys are mapped to the same bucket, which is called "collision". One way to handle collisions is by *chaining*, which makes each bucket of the the hash table a linked list. The most obvious downside of chaining is the unbounded memory access time [5].

The last class of hardware solutions is based on the multibit trie representation of the forwarding table. In a multibit trie, one or more bits are scanned in a fixed or variable *strides* to direct the branching of the children [18, 22]. Figure 2(a) shows a three-level fixed stride equivalent of the IP Figure 1(a). The trie root in Figure 2(a) has 8 children or rows, since we use the first 3 bits at level one for branching. Each node in the multibit trie is either empty or stores a prefix. Multibit trie-based solutions have the following features: (1) they are easily mapped into a hardware pipeline [2], (2) they reduce the lookup time greatly compared to binary trie [13], and (3) they have low power consumption [13].

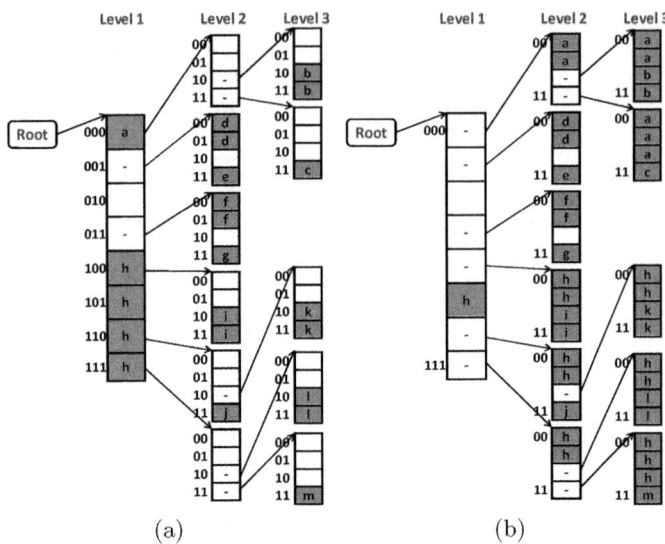

(a)   (b)

Fig. 2. (a) Multibit trie for Figure 1(a) with strides $\{3, 2, 2\}$. (b) Its leaf-pushed trie.

A multibit trie reduces the memory lookup time of a unibit trie by decreasing its height [18, 22]. The prefixes in a multibit trie have to be expanded into a set of allowed lengths, through a process called "Controlled Prefix Expansion" (CPE) [20]. Gupta et al. [8] propose a two-level hardware multibit trie, of 24 and 8 bits strides, for IPv4 packet forwarding. The scheme's lookup time is 2 memory cycles at the memory cost of at least 33MB. In general, the strides of a k-level multibit trie will be denoted by $s = \{s_1, \cdots, s_k\}$, where s_l is the number of bits used at level l.

Prefixes can be represented as *address ranges* that in some cases overlap [18]. The prefixes that do not overlap are "disjoint" prefixes [20, 23]. If all prefixes in the forwarding table are disjoint, then we call them "independent" [23]. The independent prefix sets are important because there is only one prefix that matches any incoming packet, thus avoiding the LPM calculation. Any prefix set is transformed into an independent set using a technique called *leaf pushing* [20, 23]. Figure 2(b) shows the leaf-pushed multibit trie for Figure 1(a), where we copy (or push) the prefixes from the intermediate nodes to the leaves. For example, prefix 'a' in Figure 2(a) is copied two times at level 2 and four times at level 3 in Figure 2(b).

In this paper we propose a novel IP forwarding scheme based on the compressed multibit trie framework. The main goal is to avoid any prefix matching during the IP lookup process while achieving scalability to the 128-bit IPv6 and relaxing the memory requirement. We reduce the memory footprint by introducing a new two-phase inter-node compression algorithm. Unlike existing compression algorithms [6, 16], we do not use any encoding or bitmaps that have adverse effects on the run time and usually complicate the incremental update process. By using an SRAM pipelined architecture, we estimate that our scheme

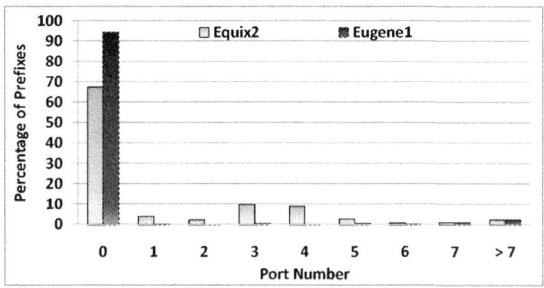

Fig. 3. Prefixes distribution of Eugene1 and Eqix2 vs. output ports numbers

can process 4.9 Giga packets per second and at the same time use less than 1.0 MB of memory for real IPv6 tables with 10.4K prefixes on average.

The rest of this paper is organized as follows: In Section 2 we describe our new IP forwarding scheme. The experimental results and evaluation are given in Section 3. Before we conclude and discuss the future work in Section 5, we talk about prior art in Section 4.

2 Our New Forwarding Scheme

An interesting observation is that the outcome of any IP lookup operation is an interface (or output port) number [4, 22]. Real-life backbone routers contain a relatively small number of output ports (usually a few tens). Note that there is a difference between the next hop information, which is an IP address of a router, and the output port which is just a physical port. The routers assign many next hop addresses to one output port [4, 22]. This means that the number of unique next hop information is greater than the number of output ports. Throughout this work we use both terms "next hop information" and "output port" interchangeably.

Furthermore, we find that the distribution of the prefixes among the output ports exhibits a certain skewness. Figure 3 plots the distribution of the prefixes against the output ports of tables "Eugene1" and "Equix2", chosen as representatives from the IPv6 tables listed in Table 1 (Section 3). We find that 94% of the prefixes share port 0 as for Eugene1, while 68% prefixes share port 0 for Equix2. Hence, if we replace the prefixes in a trie representation of the forwarding table with their associated port numbers, we have a great opportunity to reduce the total number of nodes in this trie using inter-node compression. However, before we describe our inter-node compression algorithm, we build a leaf-pushed, port-based multibit trie which is trimmed to remove redundant leaves.

2.1 Constructing the Uncompressed Multibit Trie

Figure 4 depicts our trie before applying our inter-node compression technique for the forwarding table in Figure 1(a). We call this trie "Trimmed, leaf-pushed,

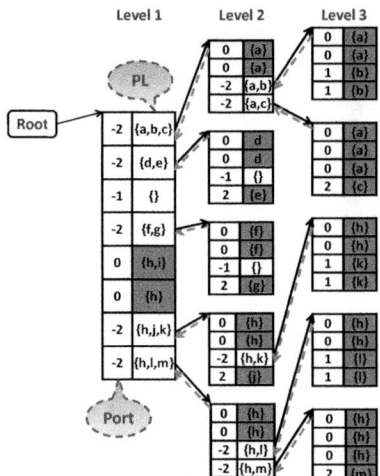

Fig. 4. Uncompressed multibit trie for Figure 1(a)

port-based multibit trie", since it is a leaf-pushed multibit trie in which we replace the prefixes with their output port numbers. If there is a subtrie that has the same port number for all of its leaves, then we trim it keeping in its place only one leaf with that port number. Herein, we will refer to this trie as "uncompressed" trie since our final trie is compressed (Section 2.2).

Each node at level l of a k-level multibit trie with strides $s = \{s_1, \cdots, s_k\}$ is a data structure of type "**UTN**", uncompressed trie node, which consists of a single dimension array, $T[\ ]$ of size 2^{s_l}, plus a back pointer, bp, to the node that points to the current node. For example, the trie in Figure 4 has $k = 3$ levels and a set of strides $s = \{3, 2, 2\}$. Each row in $T[\ ]$ contains three variables: PL, a list of prefixes to keep the prefixes that are mapped to this row, $port$, to store the port of this row if all of its PL prefixes have the same port, and ptr to store a forward pointer to a next level UTN node (if any). The $port$ field is set to -1 if the row is empty and to -2 if the row has a pointer to the next level. Note that the first column in each trie node in Figure 4 is almost identical to the leaf-pushed trie of Figure 2(b) after replacing the prefixes with their ports.

The idea is simple: construct a leaf-pushed multibit trie, then replace each prefix with a port number. During this process, sometimes the prefixes that are mapped to a certain row have identical port numbers. As an example, consider row number 4 at level 1 of Figure 4 where both prefixes h and i map to port '0'. Thus, we do not need to expand the trie to the second level as shown in Figure 2(b); instead we just stop at level 1 for this row. Thus, we remove the second level leaf that originally stemmed from row 4 of level 1 and replace it with one port, 0, at this row. Hence, the number of nodes in the leaf-pushed trie in Figure 2(b) is reduced. The dotted arrows in Figure 4 represent back pointers, while the solid ones represent regular next level, *forward*, pointers. The back pointers are essential for our inter-node compression Algorithm 2 (Section 2.2).

Algorithm 1. Uncompressed Multibit Trie Construction Algorithm

---

UTN * **BuildNode**(Prefixes List SP, int l, UTN $*cp$)
{
   Define N = new UTN node with $N.T[\ ]$ = new array of size 2^{s_l}
   and $N.bp = cp$ /*record back pointer*/
   for (each $p \in SP$) **do** {
    if($|p| < S_l$), **then** {
     Expand p to length S_l, add the expanded prefixes
     to SP and remove any redundancy } }
   for (each $p \in SP$) **do** {
    r = row index specified by bits S_{l-1} to S_l of p
    $N.T[r].PL = N.T[r].PL \cup \{p\}$ }
   for($r = 1$; $r \leq 2^{s_l}$; r++){ /*each row*/
    if($N.T[r].PL$ is empty), **then** {
     $N.T[r].port = -1$ } /*empty row*/
    else {
     if($\forall(p,q) \in N.T[r].PL$, $p.port == q.port$), **then** {
      $N.T[r].port = p.port$ }
     else {
      Let p_L be the longest prefix in $N.T[r].PL$
      if($|p_L| \leq S_l$), **then**{
       $N.T[r].port = p_L.port$ }
      else { $N.T[r].port = -2$ /*mark as pointer*/
       $N.T[r].ptr = $ **BuildNode**($N.T[r].PL$, $l+1$, $N.T[r]$) }
    } } }
   return N
}

Algorithm 1 has one function, **BuildNode()**, which takes three arguments: a list of prefixes SP, an integer l = level number, and a UTN current pointer, cp. We start by passing a prefix list SP = all the forwarding table prefixes, $l = 1$ and a $NULL$ pointer from the $main()$ function to $BuildNode()$, which returns a pointer, $Root$, to a Uncompressed trie. $BuildNode()$ begins by allocating a new UTN pointer, N, as a new node, records its back pointer and allocates its $T[\ ]$ as an array of size 2^{s_l}. It expands any prefix from SP that has a length less than S_l, a cumulative stride[1], and adds only the unique prefixes to the original prefix set SP. In other words, if any of the expanded prefixes already exist in SP, we delete the expanded prefix since it inherits its original prefix's length. Each prefix is mapped to a row r, where $r = 1, \cdots, 2^{s_l}$, in the $N.T[\ ]$ array and is stored in its associated prefix list, $N.T[r].PL$.

If no prefix is mapped to a certain row, r, we set $N.T[r].port$ to -1 (indicating an empty row). When a packet is matched at this row we apply the default route [4]. If all the prefixes that are stored in r have the same output port number, we simply set $N.T[r].port$ to that port. In case all the prefixes that are mapped to r have lengths less than or equal to S_l, we choose the longest prefix,

---

[1] A cumulative stride, $S_l = \sum_{i=1}^{l}(s_i)$, is the number of bits that are used till trie level l.

say p_L, and set $N.T[r].port$ to $p_L.port$. Note that we define $|p|$ to be the length of prefix p and $|SP|$ to be the size of the set of prefixes SP. The recursion in Algorithm 1 occurs when there is at least one prefix that is mapped to r and has a length longer than S_l. In this case, we set $N.T[r].port$ to -2 (indicating that this row has a pointer to the next level) then we set $N.T[r].ptr$ to a new instance of the $BuildNode()$ by passing the set of prefixes that are mapped to this row, $N.T[r].PL$, the next level number, $l + 1$, and the current table row pointer, $N.T[r]$.

2.2   Inter-Node Compression

In Algorithm 2 we construct a compressed trie from our uncompressed trie generated by Algorithm 1. Our algorithm aims at removing redundancy in trie nodes by checking if there exist two or more nodes at the same level that have identical contents. Since our compression algorithm combines different nodes at the same level, we call it "inter-node" compression and the resulting compressed trie is called *Inter-Node Compressed Trie* (**INCT**).

We start with a copy of our uncompressed trie at level $l = k$ and scan each pair of tables at this level to see if there exists identical tables. If we find two identical nodes, we delete one of them and use its back pointer to update the forward pointer of its parent node. We keep doing this at each level in a backward fashion till we stop at the root. Figure 5(a) represents the three-level trie of Figure 1(a), while Figures 5(b) and (c) depict how we perform compression on this trie level by level. We do not show the back pointers in Figures 5(b) and (c) as they are not needed after the compression. We define the "compression ratio" as the percentage of the total number of trie rows after the compression divided by the total number of trie rows before the compression. The compression ratio in this example is $\frac{28}{48} = 58.3\%$.

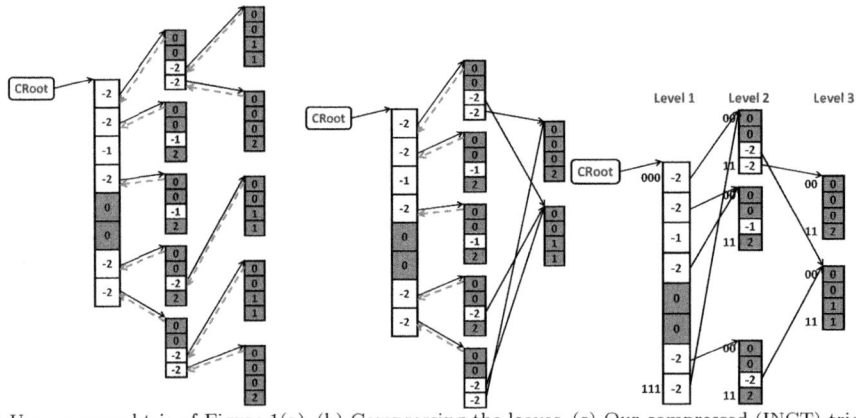

(a) Uncompressed trie of Figure 1(a) (b) Compressing the leaves (c) Our compressed (INCT) trie

Fig. 5. Compressed trie construction from its corresponding uncompressed trie

---

Algorithm 2. The Inter-Node Compression Algorithm

CompressTrie(*UTN* *CRoot)
{
   Construct the arrays *noPtrs*[ ], *P*[ ][ ]
   for($l = k; l > 1; l − −$) {
     for($i = 1; i < (noPtrs[l] − 1); i + +$) {
       for($j = i + 1; j < noPtrs[l]; j + +$) {
        if(**IdenticalNodes**($P[l][i], P[l][j]$)) **then**{
         /*use P[l][j]'s back pointer to point to $P[l][i]$*/
         *($P[l][j].bp$).$ptr = P[l][i]$
         Delete node $P[l][j]$ }
   } } }
}

The main idea of Algorithm 2 is to make a copy, *CRoot*, from the uncompressed trie, *Root*, then compress it and move the compressed trie without the back pointers and the *PL* fields to the forwarding plane. For this purpose we need to define *noPtrs*[ ] as a global array of counters of size k to keep track of how many nodes we have per trie level. In addition, we define an array of *UTN* pointers, *P*[ ][ ], that stores the pointers to all the nodes for each trie level, which is used only during the compression and is deleted afterwards.

Function **CompressTrie**() of Algorithm 2 does the actual compression. It starts from the uncompressed trie copy, *CRoot*, then builds the *P*[ ][ ] arrays. The function then sequentially scans the trie levels starting from the leaves to the root to see if there are two identical nodes, which is done by calling **IdenticalNodes**(). *IdenticalNodes*() is a function that takes two *UTN* node pointers as arguments and returns *true* if and only if the two nodes have exactly the same contents, otherwise, it returns *false*. If we find two identical nodes $P[l][i]$ and $P[l][j]$, we use the back pointer of $P[l][j]$ to adjust its forward pointer to point to $P[l][i]$ and delete $P[l][j]$.

2.3 INCT Packet Lookup

In this section we show how to search for a port number in our INCT trie-based IP forwarding scheme. The great advantage is that we do not have to match the incoming packet address against any prefix. During a packet lookup, we split the destination address into k chunks and use each chunk as a row index, r_l, where $l = 1, \cdots, k$, to a certain table. We start from the *CRoot*, and for each row $r_l \in CRoot.T[\ ]$, if $Root.T[r_l].port \geq 0$, we direct the packet to this port and we stop scanning the trie. However, if $CRoot.T[r_l].port$ equals -1, then we apply the default route. Finally, if $CRoot.T[r_l]$ .port equals -2, then we use $CRoot.T[r_l].ptr$ to move to the next next level and recursively repeat the process until either find a port or an empty row.

Table 1. Statistics of IPv6 tables on June 2010. **H** is the number of output ports.

Table	Size	H	Table	Size	H
Equix1	3,180	9	Linx2	37,282	13
Equix2	3,215	9	Quagga1	3,464	7
Eugene1	3,211	16	Quagga2	3,299	4
Eugene2	3,233	15	Wide1	5,412	2
Linx1	36,366	13	Wide2	5,470	2

3   Experimental Evaluation

To evaluate our proposed ideas we built a simulation environment written in C++ and employed 10 real IPv6 files dated June 2010 from the routing information service (RIS) project [17]. Table 1 lists the tables, along with their sizes and their number of unique next hops.

The memory requirement at each row of either uncompressed trie or INCT trie is encoded in 3 bytes, where we use 2 bytes for a pointer and one byte for a port number. An empty row is encrypted by resetting all 3 bytes to zeros. Two bytes per pointer gives us the ability to address $2^{16} = 65.5$K nodes per level.

3.1   INCT Evaluation

Selecting the Strides. A common practice in pipelined trie forwarding engines is to use a large initial stride [12]. This is due to the fact that most prefixes in Internet IPv6 forwarding tables are of length 24 bits or longer [4,14,22]. In this work we follow this common practice. We aim to have a small number of trie levels to limit the delay incurred by any incoming packet.

First, we select the number of strides for two representative IPv6 forwarding tables to find the best number of strides. Figure 6 shows the result of varying the number of strides from 5 to 9 for the two files Linx1 and Eugene2. We find that the same trend that having a larger number of strides leads to lower memory utilization. However, since we want to reduce the packet's delay, we choose $k = 7$ as a tradeoff between memory utilization and delay. To select the actual strides' values, we use a simple brute-force heuristic to reduce the total memory requirements of our INCT trie as we mentioned before. In the following subsections we will use the notation "$Xyz(m)$" to distinguish different trie schemes, where Xyz is the trie scheme name and m is the number of trie levels (i.e., height). For example, Uncompressed(7), means our uncompressed trie with a height of 7.

INCT Performance. Figure 7 shows the memory requirements of both INCT(7) and Uncompressed(7) trie for the IPv6 tables given in Table 1. The average total memory for INCT(7) is 466KB, while Uncompressed(7) trie has an average of 1.0MB. This means that we achieve on average a compression ratio of 44.7% or an average space saving of 55.3%. The maximum memory requirement belongs

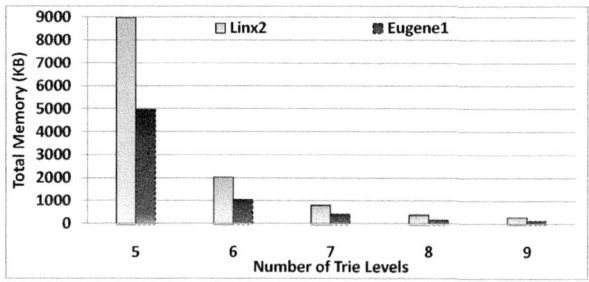

Fig. 6. Number of INCT trie levels vs. total memory (KB) for Linx2 and Eugene1

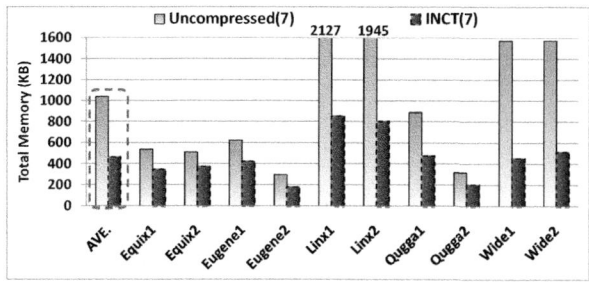

Fig. 7. Memory (KB) of INCT(7) vs. Uncompressed(7) trie for tables given in Table 1

to table Linx1, 857KB for INCT(7) and 2.0MB for Uncompressed(7) trie. The smallest space savings is 27% for table Equix2, while the largest space savings is 71% for table Wide1. The actual compression ratio varies from file to file due to the distribution of the output ports among the prefixes.

INCT vs. Other Compression Schemes for IPv6. Figure 8 shows the memory utilization of MIPS(57), binary INCT(57), Lulea(6) (which is almost like the Tree Bitmap scheme [7]) and INCT(6). Since IPv6 is currently using only 64 bits out of its 128 bits for actual address prefixes while the other 64 bits are to identify MAC address [14,15], we assumed that Lulea IPv6 version would have 6 strides: $\{16, 16, 8, 8, 8, 8\}$. In addition, we assumed that MIPS's stride set would start by an initial stride of 8 bits then scans the rest of the 64 bits, 56 bits, one bit at a time. Note that INCT(6) is using the same strides as Lulea(6), while binary INCT(57) is using the same strides of MIPS(57).

MIPS [23] is the first technique that utilized the limited number of ports in Internet routers to lower the number of prefixes in a forwarding table. The main goal of MIPS is to store the prefix set inside a TCAM chip in any order. MIPS builds a leaf-pushed binary trie to obtain an independent set of prefixes, replacing prefixes by their next hops which are then replaced with their corresponding port numbers. As a secondary effect MIPS reduces the number of prefixes needed to be stored. Our trie construction algorithm, Algorithm 1, can be considered as a generalization of the MIPS trie construction algorithm for the multibit trie

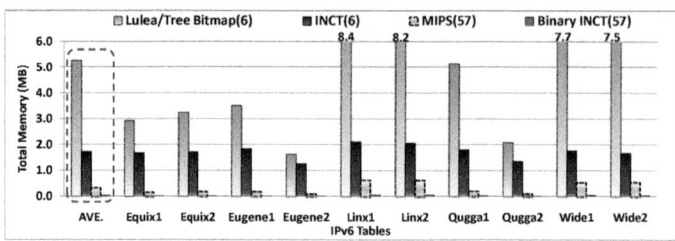

Fig. 8. Memory (MB) of Lulea(6)/Tree Bitmap, INCT(6), MIPS(57) and Binary INCT(57) for tables in Table 1

case. MIPS also keeps its trie in the control plane for update handling, while it retains a copy of the trie prefixes in the TCAM lookup engine. Our scheme does not store the prefixes in the fast forwarding data plane.

The performance is measured as the total memory used by each scheme. In general, MIPS(57)'s average total memory is 339KB, while Lulea(6)'s is 5.3MB. Binary INCT(57) is smaller by 88% than MIPS(57) on average and INCT(6) is smaller by 67% than Lulea(6).

3.2   INCT Forwarding Engine Performance Estimation

In this section we estimate the actual processing rate of our INCT packet forwarding engine using the standard CACTI memory version 5.3 simulator [21]. We use the high performance 32nm SRAM technology with one read port and one write port.

The CACTI results of the pipelined architectures are shown in Figure 9. For each IPv6 table we choose the maximum size of each trie level to be simulated. This is why the total size of each trie is larger than its previously reported average size. The IPv6 INCT(7) saves 60.6% of area and 41.5% of total read dynamic energy over the Uncompressed(7). The stage frequency equals the minimum of the maximum operating frequency among the trie levels, which is 4.9GHz. In other words, the pipelined architecture's estimated throughput is 4.9Giga packets per second, which means a loss of 8% comparing to Uncompressed(7).

	Uncompressed(7)	INCT(7)	Savings and Losses %
Total RAM Size (MB)	2.11	0.85	59.8
Total Access Time (ns)	3.74	2.85	23.8
Pipeline Stage Frequency (GHz)	5.29	4.90	-7.4
Total Read Dynamic Energy (nJ)	0.09	0.05	3.0
Total Read Dynamic Power at Max Freq. (W)	0.54	0.29	45.9
Total Area (mm²)	5.42	2.14	60.6

Fig. 9. The CACTI simulations for INCT(7) vs. Uncompressed(7)

For minimum packet size of 80 bytes, the INCT(7) throughput is $4.9 \times 80 \times 8 = 3.1$Tbps (Tera bit per second). The total delay incurred by a packet, which is the summation of the access time of each pipeline stage, is 2.9ns for INCT(7).

4   Prior Art and Related Work

Many optimizations are suggested to reduce the binary trie lookup time like the path-compressed trie which benefits from the fact that a binary trie may have long sequences of single-child nodes, then skip scanning some bits to reduce average lookup time [18]. Crescenzi et al. [18] use full prefixes expansion to 32 bits, then construct two-level multibit tries of 16 bits each. Their compression technique is based on repetitions elimination of the LPM of each subtrie at the second level only. They use 1.2MB to store a lookup table of 40K prefixes.

The Lulea algorithm [6] uses *bitmap* compression to eliminate data redundancy in three levels, leaf-pushed tries with fixed strides of $s = \{16, 8, 8\}$ [22]. The scheme has two data structures per node: bitmap and an array that contains the compressed data. For each row of a trie, Lulea stores '1' in the bitmap if the current and previous row values are not equal while pushing the current value to the compressed data vector, otherwise it stores '0'. Though the Lulea scheme is reported to store 32K prefixes in 160KB memory, it has a lookup time between 4 and 12 memory cycles and suffers from very high update time [6,22]. Eatherton et al. [7] modify the Lulea scheme to facilitate incremental updates while keeping the compression ratio as low as that of Lulea's. The authors propose two bitmaps per trie node, one for all internally stored prefixes and one for external pointers, which allow them to reduce the update time through avoiding leaf pushing [7,22]. They store a table of 41K prefixes in 450KB.

Note that all these schemes are "intra-node" compression schemes, while our scheme is an "inter-node" compression scheme. Though these preceding algorithms are devised mainly to handle the 32-bit IPv4, they could be utilized also for the 128-bit IPv6 address [14]. In fact, very few trie-based solutions are explicitly proposed to handle IPv6 [3,15]. In [15], the authors propose a hybrid data structure that combines binary tree (trie), a segment table and a route bucket (hash table) to store an average of 645 real IPv6 prefixes in 390KB with a worst case access latency of six memory cycles. The authors in [3] propose an architecture similar to [15] in the sense that they use a hybrid data structure comprised of: direct lookup table, hash table and multibit trie that uses a new Tree bitmap compression. They achieve a throughput of 160Gbps utilizing on-chip SRAM and off-chip DRAM memory of 4.9MB for synthetic IPv6 tables.

Recently, many researchers are proposing trie-based advanced pipeline architectures [2,12,13]. Their architecture optimizations are based on the minimization of the stage maximum memory, the minimization of the total memory, or the maximization of the throughput. Most of these solutions use dynamic programming algorithms to map trie nodes from to pipeline stages. The work in [13] constructs a pipelined architecture from fixed stride tries since variable stride tries are hard to implement and maintain. The authors provide dynamic programming algorithms to find the optimal multibit trie for a given table. The optimization in [13] is based on minimizing the maximum memory required for a single stage. The pipeline depth is between 2 and 8 stages and each stage uses between 60KB and 80KB for an IPv4 forwarding table of 100K prefixes.

Jiang and Prasanna [12] addressed the same problem with a different strategy. Their architecture consists of multiple pipelines that have between 20 and 25 stages, each with 18KB of RAM. The main idea is to have more stages than the actual number of the trie levels for flexibility. They report an overall a throughput of 3.2Tbps, which is boosted by using an intelligent caching technique that allows processing 8 packets in parallel. The aforementioned schemes are orthogonal to our proposed scheme and can be used to balance memory distribution among stages before applying our compression.

5   Conclusions and Future Work

In this paper we introduced a novel packet forwarding scheme that uses multi-bit trie inter-node compression to reduce the total memory. We showed that our scheme has a better compression ratio than MIPS, Lulea and Tree bitmap compression schemes. In addition, most previously proposed schemes store their prefixes (or a pointer to them) in the forwarding trie, which we do not. On average, our pipelined INCT architecture, INCT(7), consumes 466KB and can achieve a throughput of 3.1Tbps.

We believe that this work could be applied to other packet processing domains such as packet classification and deep packet inspection. Also, we need to consider systematic methods to select the trie strides rather than just reducing the memory. Since our INCT utilizes an inter-node compression technique, we can apply other intra-node compression techniques (e.g., Lulea [6]) in addition to achieve more compression ratio. Furthermore, we can increase the throughput by using smart caching techniques that take into account the traffic characteristics as well as the trie shape.

Acknowledgment

This work was supported in part by a NSF grant (CCF-0952273).

References

1. Arano, T.: IPv4 Address Report. Potaroo Projection (2010),
   http://www.potaroo.net/tools/ipv4/index.html
2. Baboescu, F., Tullsen, D., Rosu, G., Singh, S.: A Tree Based Router Search Engine Architecture with Single Port Memories. ACM Sigarch Com. Arch. 33(2) (2005)
3. Bando, M., Chao, J.: Flashtrie: Hash-based Prefix-Compressed Trie for IP Route Lookup Beyond 100Gbps. In: IEEE Infocom (2010)
4. Chao, H.J., Liu, B.: High Performance Switches and Routers. Wiley, Chichester (2007)
5. Cormen, T., Leiserson, C., Rivest, R., Stien, C.: Introduction to Algorithms. McGraw-Hill, New York (2003)

6. Degermark, M., Brodnik, A., Carlsson, S., Pink, S.: Small forwarding tables for fast routing lookups. ACM Sigcomm (1997)
7. Eatherton, W., Varghese, G., Dittia, Z.: Tree Bitmap: Hardware/Software IP Lookups with Incremental Updates. ACM Sigcomm Comp. Rev. 34(2) (2004)
8. Gupta, P., Lin, S., Mckeown, N.: Routing Lookups in Hardware at Memory Access Speeds. In: IEEE Infocom (1998)
9. Hanna, M., Demetriades, S., Cho, S., Melhem, R.: CHAP: Enabling Efficient Hardware-based Multiple Hash Schemes for IP Lookup. In: IFIP Networking (2009)
10. Hanna, M., Demetriades, S., Cho, S., Melhem, R.: Progressive Hashing for Packet Processing Using Set Associative Memory. In: IEEE/ACM ANCS (2009)
11. Hanna, M., Demetriades, S., Cho, S., Melhem, R.: Advanced Hashing Schemes for Packet Forwarding Using Set-Associative Memory Architectures. Journal of Distributed and Parallel Computing (JPDC) 71, 1–15 (2011)
12. Jiang, W., Prasanna, V.: Multi-Terabit IP Lookup Using Parallel Bidirectional Pipelines. In: ACM Computing Frontiers (2008)
13. Kim, K.S., Sahni, S.: Efficient Construction of Pipelined Multibit-Trie Router-Tables. IEEE Trans. on Comp. 56(1) (2007)
14. Li, Y.K., Pao, D.: Comparative Studies of Address Lookup Algorithms for IPv6. In: IEEE ICACT (2006)
15. Li, Z., Zheng, D., Ma, Y.: Tree, Segment Table, and Route Bucket: A Multistage Algorithm for IPv6 Routing Table Lookup. In: IEEE Infocom (2007)
16. Nilsson, S., Karlsson, G.: IP-Address Lookup Using LC-Tries. IEEE J. on Sel. Areas in Comm. 17(6) (1999)
17. RIS. Routing Information Service, http://www.ripe.net/ris/
18. Ruiz-snchez, M., Biersack, E., Dabbous, W.: Survey and Taxonomy of IP Address Lookup Algorithms. IEEE Network 15(2) (2001)
19. Shah, D., Gupta, P.: Fast Updating Algorithms for TCAMs. IEEE Micro Mag. 21(1) (2001)
20. Srinivasan, V., Varghese, G.: Fast Address Lookups Using Controlled Prefix Expansion. ACM Trans. Comp. Sys. 17(1) (1999)
21. Thoziyoor, S., Muralimanohar, N., Ahn, J.H., Jouppi, N.P.: CACTI 5.1: An Integrated Cache Timing, Power, and Area Model. Technical report, HP Labs
22. Varghese, G.: Network Algorithmics: An Interdisciplinary Approach to Designing Fast Networked Devices. Morgan Kaufmann, San Francisco (2005)
23. Wang, G., Tzeng, N.-F.: TCAM-Based Forwarding Engine with Minimum Independent Prefix Set (MIPS) for Fast Updating. In: IEEE ICC (2006)

oBGP: An Overlay for
a Scalable iBGP Control Plane

Iuniana Oprescu[1,5,6], Mickaël Meulle[1], Steve Uhlig[2],
Cristel Pelsser[3], Olaf Maennel[4], and Philippe Owezarski[5,6]

[1] Orange Labs, 38-40, rue du Général Leclerc
92794 Issy-les-Moulineaux Cedex 9, France
{mihaela.oprescu,michael.meulle}@orange-ftgroup.com
[2] Deutsche Telekom Laboratories & Technische Universität Berlin,
Ernst-Reuter-Platz 7, 10587 Berlin, Germany
steve@net.t-labs.tu-berlin.de
[3] Internet Initiative Japan, Jinbo-cho Mitsui Bldg.,
1-105 Kanda Jinbo-cho, Chiyoda-ku, Tokyo 101-0051, Japan
cristel@iij.ad.jp
[4] University of Loughborough, Department of Computer Science,
Haslegrave Bldg., Loughborough, LE3TU, United Kingdom
olaf@maennel.net
[5] CNRS, LAAS, 7 avenue du colonel Roche, F-31077 Toulouse Cedex 4, France
[6] Université de Toulouse, UPS, INSA, INP, ISAE, UT1, UTM, LAAS,
F-31077 Toulouse Cedex 4, France
owe@laas.fr

Abstract. The Internet is organized as a collection of networks called
Autonomous Systems (ASes). The Border Gateway Protocol (BGP) is
the glue that connects these administrative domains. Communication is
thus possible between users worldwide and each network is responsible
of sharing reachability information to peers through BGP. Protocol extensions are periodically added because the intended use and design of
BGP no longer fit the current demands. Scalability concerns make the
required internal BGP (iBGP) full mesh difficult to achieve in today's
large networks and therefore network operators resort to confederations
or Route Reflectors (RRs) to achieve full connectivity. These two options come with a set of flaws of their own such as persistent routing
oscillations, deflections, forwarding loops etc.

In this paper we present oBGP, a new architecture for the redistribution of external routes inside an AS. Instead of relying on the usual
statically configured set of iBGP sessions, we propose to use an overlay
of routing instances that are collectively responsible for (i) the exchange
of routes with other ASes, (ii) the storage of internal and external routes,
(iii) the storage of the entire routing policy configuration of the AS and
(iv) the computation and redistribution of the best routes towards Internet destinations to each router of the AS.

Keywords: routing, BGP, architecture, management.

J. Domingo-Pascual et al. (Eds.): NETWORKING 2011, Part I, LNCS 6640, pp. 420–431, 2011.

1   Introduction

The Border Gateway Protocol (BGP) is the actual standard that enables the computation of end-to-end paths in the Internet. BGP allows networks, called Autonomous Systems (ASes), to exchange their routing information and to implement independently customized routing policies. The Internet has reached a size of more than 35,000 ASes and roughly 350,000 blocks of IP addresses.

BGP and internal BGP in particular have been widely studied, and many extensions and improvements have been proposed to deal with matters like network convergence[2][3] or route diversity[7][8]. On the other hand, general design and architectural issues in iBGP have not been sufficiently confronted in our opinion.

In this paper, we propose a new solution for iBGP routing within an AS using a distributed overlay of routing software. The solution we elaborate is meant as a viable framework for replacing the current iBGP and we introduce a setting for scalable and flexible routing. By gathering the routing information in a platform, we aim to offer easier management of protocols and policies at the network level.

2   BGP Routing in a Nutshell

The Border Gateway Protocol[10] is in fact two protocols: internal BGP (iBGP) for handling messages inside an AS and external BGP (eBGP) for exchanging reachability information with other ASes in the Internet. This clear distinction makes it possible for an ISP to deploy a new iBGP in its network without any impact on neighboring ASes. We will further concentrate on describing some aspects of the iBGP mechanism.

Within a BGP router, the decision process takes into account the interactions with every neighbor. Roughly speaking, if n is the number of prefixes advertised in the Internet, an iBGP Routing Information Base (RIB) will contain about $n * m$ routes in the worst case, where m is the number of neighbors sending their full BGP table as seen in Fig. 1. The best path to a destination is selected, installed in the Forwarding Information Base (FIB) and actually used to perform packet forwarding. The router will also advertise its best path for a given prefix to the adjacent BGP peers.

BGP requires that entries be kept for each reachable network: this constraint leads to large routing tables. There are many routes that cannot be aggregated and even if the optimizations in vendor code reduce the size of memory needed,

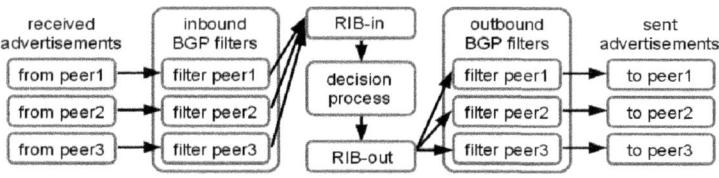

Fig. 1. The selection of the best route among the received routes

they still do not fix the problem. The natural growth of the Internet as well as its increasing connectivity and the tweaking of routing entries for traffic engineering purposes have inflated the size of BGP routing tables by a factor of more than 3 within the last decade [1]. The current trend of the routing table indicates continuous growth of the Internet and we expect future evolution to be similar, especially after the migration to the apparently inexhaustible IPv6 space.

Inside an AS, a single administrative entity manages all routers and distributes a consistent routing policy configuration. The goal of iBGP is to redistribute routing data inside the AS in accordance with the routing policy configured in each BGP router. iBGP routing originally required a full mesh between the routers within a single AS to guarantee that each router will be able to learn the best external route for forwarding IP packets.

The full mesh configuration can quickly turn out to be a scalability problem since the total number of sessions grows with the square number of involved peers. There are two alternatives for avoiding the processing overhead induced by full mesh: confederations and route reflectors (RRs) illustrated in Fig. 2.

Confederations are sub-ASes meant to divide a large network into more manageable areas. A route reflector is a router that takes the role of a central point where a subset of the other routers peer. These designs are both prone to unpredictable effects such as persistent routing oscillations and forwarding loops affecting network convergence [3][4]. They may face sub-optimal routing due to information masking or non-deterministic decisions influenced by the state of the network at the arrival time of the advertisements.

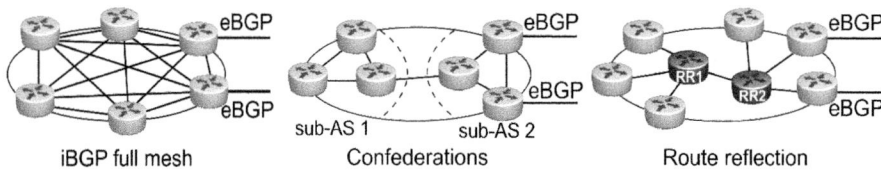

Fig. 2. ASes exemplifying a full mesh of iBGP sessions, confederations, route reflection

2.1   Plagues in Current iBGP

In specific architectures using route reflection, routing is victim of a series of aspects induced by the inherent design of iBGP.

Below is a brief display of the drawbacks in current iBGP:

- **scalability** is quantified by the number of protocol messages exchanged over time, the number of established sessions and especially by the size of the routing table. We estimate that the growth of the routing table can be handled by means of a partitioning model that we expose in this paper: achieve scalability through division of the control plane by placing subsets of prefixes in different locations and then performing the computation of the BGP decision process in a distributed manner.

- **network opacity** occurs in architectures where route reflection schemes are used for propagating routes. Incomplete knowledge of the set of routes advertised by neighbor ASes leads to inconsistencies and issues such as routing oscillations and deflections that can cause forwarding loops. Extensive studies [2][3][4][5][6] give conditions and methods for defining correct iBGP configurations that avoid anomalies and achieve full mesh optimality.
- **poor route diversity** is a direct consequence of network opacity. The fundamental design of BGP route redistribution demands that each peer advertise only its best route. Diffusing a single route impacts the available choices and there is a noticeable loss of route diversity when comparing border routers to internal routers [7].

  The graph in Fig. 3 presents the diversity of neighbor ASes and BGP next-hops for the received prefixes on 5 random routers. The data reveals the fact that there is a large diversity in the received routes but this diversity is severely reduced by the BGP selection mechanism of the best route.

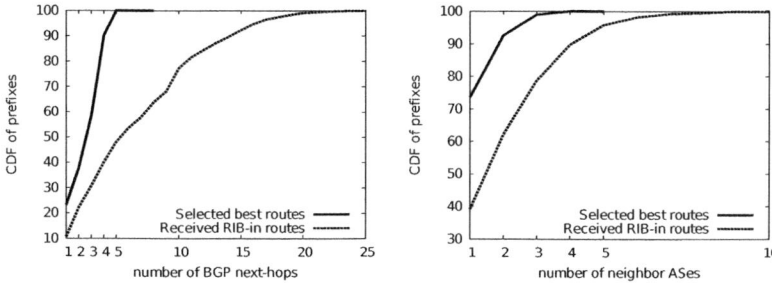

Fig. 3. Prefixes and routes on 5 random routers of a large ISP

  Redundancy in case of failure is highly desirable and a secondary path could also be used for extra features such as load balancing or multipath routing. Protocol extensions introduce new capabilities for adding paths to BGP in [8], but the impact on current networks is yet to be explored as done in [9].
- **management and troubleshooting** are often complex and challenging: inconsistency of the routing policies, path exploration meeting flap dampening [11] and difficulties in achieving network-wide traffic engineering are some of the issues encountered by network operators. A full view of the external routes and knowledge of the Interior Gateway Protocol (IGP) topology by one entity would make these processes easier. In oBGP, the interaction with the entire network is done through the overlay. Concentrating the BGP decision process in the overlay nodes increases control over the network behavior.

We should, however, separate theory from practice. Some of the presented issues are commonly avoided with engineering tricks and configuration tweaking. Network operators adapt to inadvertences by enforcing specific RR placement and building convenient topologies that behave correctly. Ideally, these aspects can be handled in an automated manner and this paper proposes an approach for better control over the network.

2.2   Previous Work

New routing paradigms like iBGPv2[12] or even PCE[13] propose different approaches for handling routing within an AS. LISP[14] tackles routing as a general problem and proposes a solution for the global Internet routing. In [15] Jing Fu exposes a centralized control scheme for IGPs with faster routing convergence than link-state routing protocols. His results show it is possible to conceive a routing platform reaching performances comparable to native routing.

The need for separating routing from the routers is emphasized also by N. Feamster et al. in [17]. The presented work is a design overview of a Routing Control Platform (RCP) that aims to offer separate selection of routes on behalf of the routers while maintaining backward compatibility.

M. Caesar et al. later offer an implementation to the RCP concept. The prototype described in [18] has three modules: the IGP Viewer to collect topology information, the BGP engine that learns the BGP routes, performs the decision algorithm and then communicates the best paths to the routers and finally the Route Control Server that processes messages received from the other two modules and makes it possible to store one single copy of each BGP route, keep track of the routers to which each route has been assigned and maintain an order of preference of the egress point for each router[19]. We extend this work by going a step further in reaching scalability: in our approach, the prefix table is split, making possible parallel computation of routes while in the RCP solution, all the BGP information is concentrated in one point, even if there are multiple replicas of it.

Our hybrid solution integrates the division of the routing table within a centralized routing platform. Other projects advocate the idea of downsizing the routing table: ViAggre (Virtual Aggregation) is a configuration-only method for shrinking the size of the routing table in the Internet default-free zone. It proposes a "dirty slate" technique for distributing routing within an ISP network so that routers maintain only a part of the global routing table. One of the negative impacts of ViAggre[20] is a stretch imposed on traffic, diverting it from the native shortest path. Another inconvenient is the difficulty of the configuration. This same approach is advanced in [21] and X. Zhang et al. elaborate similar work in [22], but CRIO seems to bring more benefit to VPN routing.

The work of S. Uhlig et al. [23][24] emphasizes the fact that network operators need a smarter way to do route reflection. In [25] a route server architecture is presented and in [26][27], C. Pelsser et al. aim to build distributed route servers. We go beyond these proposals by providing scalability through the distribution of the control plane in iBGP routing.

3   oBGP: A Scalable Overlay for iBGP Routing

In today's IP networks, routing is highly distributed: each router in the AS makes its own decisions. We propose to separate the selection of paths (routing plane) from the actual forwarding of traffic (data plane) on distinct equipments. Offloading the control plane from the routers can be seen as a remedy to the explosion of the routing table size.

When rethinking the current design, we place all the knowledge of routing data into a separate iBGP routing plane handled by an overlay of routing processes that do not forward user traffic. We propose to implement BGP routing engines called *oBGP*. The oBGP nodes act as the border routers of the domain and connect to the external peers through multi-hop eBGP sessions. This approach allows the overlay to receive all the routes from the neighboring ASes and gather the announced routes to achieve a unified complete view.

oBGP routing software is intended to be executed by additional servers running on commodity hardware. The logical overlay is composed of routing processes (or nodes) that are jointly responsible of:

- collecting, splitting and storing the complete set of routes received from eBGP and the internally originated routes,
- storing the routing policies and configurations of all the routers in the AS,
- redistributing the computed paths to the client routers.

One of the main concerns of an iBGP architecture is its ability to scale: support the growing routing table and handle protocol messages over time. To achieve scalability, we design an oBGP solution where the routing information is divided in several sub-planes. In this approach, distinct subsets of overlay nodes each handle only a fraction of the entire set of prefixes in the routing table.

3.1 Overview

The next paragraph explains the passage of a route advertisement in the oBGP overlay from the arrival in the AS to the installation of the best path in the RIB. Fig. 4 shows the chronological steps of a route announced to the oBGP overlay.

The oBGP acts as a border router and the neighboring ASes connect to an oBGP node through multi-hop eBGP sessions. When a route towards a destination (e.g. the prefix 1.2.3.0/24) is advertised in the Internet, it reaches the first oBGP node that determines the corresponding sub-plane in charge of the prefix. The oBGP node then forwards the information to the nodes handling the correct sub-plane. After running the BGP decision process and applying the according configuration and IGP topology constraints, the nodes output a best path per prefix for each router. The overlay distributes the best paths to the client routers connected through sessions and they can immediately install it in their RIBs. Upon reception of the best route, the native routing mechanism takes course and installs the path to the prefix in the FIB for actual packet forwarding.

The oBGP nodes need to be aware of the actual mapping of the reachable IP space within the overlay. To insure resiliency and avoid a single point of failure, a sub-plane is replicated on several oBGP nodes. Coordination between the copies of sub-planes is accomplished through an exchange of meta-data across the overlay. The following paragraphs depict the sub-plane concept.

3.2 Distributed Storage

A router learns routes toward a given prefix from its neighbors, and in the general case routers of the same AS do not learn the same exact set of routes or the

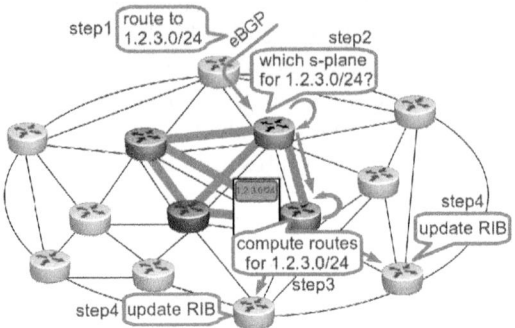

Fig. 4. The steps followed by an advertisement in the overlay network

same quantity. The full visibility of BGP routes received from external ASes can be assimilated to a sum of queries on all border routers of an AS. The total of routes received on the border routers is equivalent to the global view of the advertised Internet as seen by the domain.

oBGP manages to keep this external view intact by indexing it directly in the overlay according to a mapping mechanism. The oBGP nodes act as the collection of border routers of the AS and establish eBGP multi-hop sessions with neighbor ASes.

Storage of prefixes is distributed across the overlay and nodes divide between each other the computational load of the control plane. We define several chunks of the reachable address space that are allocated on distinct nodes. These large IP spaces are called routing sub-planes. The overlay is in charge of keeping a coherent state: no pair of sub-planes has overlapping prefixes and they are stored on different nodes. A structure similar to a distributed hash-table can be used for managing the sub-planes. The oBGP nodes guarantee the frontiers of the sub-plane, but another aspect to take into account is the replication of the information on the nodes covering the same sub-plane.

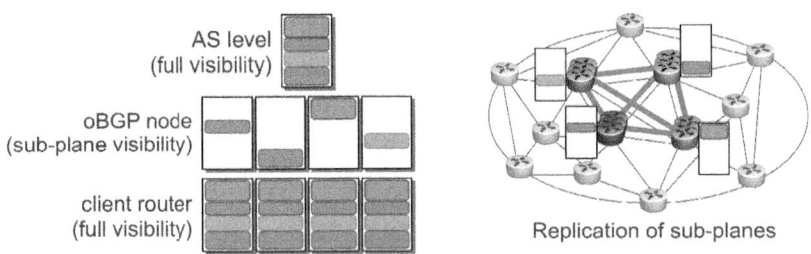

Fig. 5. The routing table is split between the $n = 4$ nodes of the overlay

Index of Virtual Prefixes: The mapping of the sub-planes on the oBGP nodes takes into account a split factor n (e.g. $n = 4$ as seen in Fig. 4) and allocates each chunk of $total/n$ prefixes to a corresponding sub-plane. This strategy turns

out to be very coarse grained and thus we introduce smaller containers for the
IP space called Virtual Prefixes as in [22].

Table 1 shows an example of a possible configuration of the sub-planes: the
reachable IP space is divided in $n = 4$ sub-planes and each sub-plane covers the
equivalent of a /2 prefix (consisting of roughly 2^{30} possible hosts). To better control the load incurred by the oBGP nodes handling the sub-planes, the network
operator may choose to define several virtual prefixes as is the case for sub-plane
1 that contains 2 virtual prefixes. The virtual prefixes may be swapped between
the oBGP nodes in order to achieve a balanced load on the sub-planes. Data[1] in
columns 3 and 4 shows that the density of prefixes advertised in the Internet can
be almost uniformly distributed across the previously defined sub-plane space. If
the distribution varies in time, we deem necessary to use a dynamic algorithm.

Table 1. Sub-planes containing virtual prefixes

sub-plane ID	virtual prefixes	# of prefixes	% of total
sub-plane 1	64.0.0.0/4	53250	17.85%
	32.0.0.0/3	21408	7.17%
sub-plane 2	160.0.0.0/3	38425	12.88%
	192.0.0.0/4	37667	12.62%
sub-plane 3	80.0.0.0/4	34552	11.58%
	96.0.0.0/3	35679	11.96%
sub-plane 4	208.0.0.0/4	40207	16.82%
	128.0.0.0/3	17719	5.93%
	0.0.0.0/3	9411	3.15%

We envision as future work to develop an on-line procedure that allocates
smaller virtual prefixes to the oBGP nodes to obtain a fine grain arrangement.
It is also possible to enforce a rule allowing for paths to popular prefixes to be
cached in the oBGP nodes based on a statistical computation of the frequency
of occurrence (i.e. cache the popular prefixes that are more stable as opposed to
swapping more often the less popular prefixes).

3.3   Selection and Propagation of BGP Routes

The main purpose for offloading the control plane into an overlay is to achieve
scalability of the routing table, but the separation of the decision process from the
actual forwarding of routes has several other benefits such as complete visibility
of the routes advertised to the AS.

The oBGP nodes gather information through eBGP and at the same time
they are part of the IGP topology which allows them to be aware of the IGP
metric from any router toward any BGP next-hop. This feature is important

---

[1] Private Tier-1 AS, dataset of November 2010, based on a total of 354682 prefixes.

because the customized computation of the best BGP route for a given prefix for a particular router will take into account the full view of the BGP routes and the interior cost for reaching the next-hop. The optimal routes are what the client router would choose if it had full view. Complete knowledge of both topologies allows the routing engines to make a correct selection and avoid situations like routing loops.

Having a federating entity makes policy management easier: a global policy can be configured on the oBGP nodes and then applied to all the routes entering or exiting the AS. The overlay ensures AS-wide coordination while still allowing for specific policies to be safely implemented on individual routers. The oBGP nodes run a neighbor-specific BGP implementation and provide the best output for every individual router connected to the overlay.

Propagation of routes from the overlay to the client routers relies on the classic iBGP sessions. Once the oBGP nodes compute the best routes for the alloted prefixes, the result is pushed to the connected routers very much in the same way that a client router would receive advertisements from RRs. Fig. 6 shows a router connected using an iBGP session to each oBGP node responsible of a specific sub-plane.

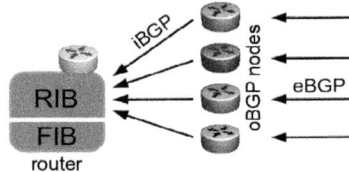

Fig. 6. The routers connect to the overlay through iBGP sessions

Once the best route to a given prefix reaches the router, the RIB is updated and the next-hop is programmed in the FIB. Client routers receive directly the announcements of the best routes that the oBGP has chosen, therefore the number of received routes is smaller and the size of the RIB is reduced.

4   Deployment of oBGP

An oBGP overlay network can be progressively deployed on top of an existing iBGP architecture, using a step by step approach. The overlay is a logical topology and can be optimized according to the underlying physical graph and the location of the oBGP nodes in the network.

As a first step, the network operator has to decide of an initial number of sub-planes and a mapping of each available oBGP node to one sub-plane while taking into account the expected redundancy of the overlay. An initial partitioning of the IP space specifies the granularity of the virtual prefixes and their correspondence to the defined sub-planes. The overlay nodes can be configured with specific policies to apply and finely tune the selection of routes for the individual client routers in the AS.

The second step is twofold: setting up the topology between the nodes of the same sub-plane and interconnecting the different sub-planes. Note that the oBGP approach does not specify a topology and several arrangements of the nodes can be studied in order to optimize the performances. The oBGP nodes participate to the IGP topology to retrieve knowledge of routing costs between any given pair of routers within the AS. The overlay implements redistribution of routes between the sub-planes and replication of the routes in a given sub-plane using reliable flooding mechanisms.

The final step is more delicate and consists of safely migrating the eBGP sessions to the overlay and removing iBGP sessions between routers. Integrity and coherence of routes must be guaranteed during the transition, until the oBGP manages to take over the route redistribution.

For each border router of the AS, establish one iBGP session with at least one oBGP node of each sub-plane, preferably the closest possible node. At this point, routers can receive the routes from the oBGP overlay. To enable the router to redistribute the eBGP-received routes to its iBGP neighbors, we turn all iBGP sessions of the router into a route reflector to client iBGP session. Once all the internal sessions are replaced by the overlay, the eBGP sessions migrate directly onto an oBGP node. As soon as a pair of border routers is migrated, the session between them can be removed.

When all border routers have been migrated, internal routers having only iBGP sessions are migrated the same way we migrate eBGP connected routers.

Concerning error scenarios, the construction of oBGP is redundant as to avoid single points of failure. Depending on the deployment topology (e.g. an oBGP platform per geographical region, per Point of Presence or per AS), the failure impact varies correspondingly. For a simple session loss between a client and one sub-plane node, there is a short interval (IGP re-convergence time) during which the disconnected router forwards packets according to a stale FIB, causing possible sub-optimal routing. Multiple failures (clients and oBGP nodes or several oBGP nodes) remain to be quantified in future work.

5 Gain through Design

This section summarizes some of the solution strengths that derive from the design and presents some feasible extensions.

The de-correlation of route selection from propagation allows routers to gain a broader visibility: the decision process takes into account both BGP and IGP information leading to an optimal selection at the AS-level. The overlay is aware of all external routes received through eBGP and of the IGP topology, which implies that the decision process is based on complete knowledge of the routes.

Another advantage is the increased diversity of routes that can be received by the client routers. Indeed, we first federate knowledge of the routes at the overlay level and then distribute the computational load according to sub-planes so that in the end, client routers connect to each of the sub-planes and receive the optimal routes. Controlling only a subset of the IP space means less memory needed to store a sub-plane instead of the entire RIB and less BGP reachability information to process on each oBGP node. The speedup in the selection of the best routes comes from the distribution of prefixes on several nodes, allowing for parallel computation.

The construction of the overlay leaves room for future additional features like flexible load sharing of routing data. An algorithm for dynamically partitioning and mapping the reachable space enables a re-organization of the various prefixes in the sub-planes.

6   Conclusions and Future Work

In this paper we present a new framework for scalable iBGP routing. The oBGP concept is illustrated: an overlay responsible for performing the BGP decision process on behalf of the client routers within the AS. We expose some of the major drawbacks in current iBGP and how the oBGP routing platform solves these issues. We provide the design principles and advantages of oBGP then reveal a possible scenario for deployment. Through the construction of this approach, oBGP provides ground for implementations of extra features and proposes a new direction in the study of the iBGP control plane scalability.

As previously mentioned, the advantage of splitting the routing table can be overshadowed by the computational overhead induced in the overlay. An important point of future work consists of determining the optimal threshold for which it is appealing to compute paths with oBGP. Research perspectives include refining the split algorithm and improving it to gracefully handle the dynamic re-organization of the virtual prefixes on the oBGP nodes. We will also evaluate the architecture, study the relevance of parameters and quantify scalability, convergence time, correctness and compliance to the routing policy.

References

1. Geoff Huston, http://www.potaroo.net/
2. Rawat, A., Shayman, M.: Preventing persistent oscillations and loops in iBGP configuration with route reflection. Comput. Netw. 50(18), 3642–3665 (2006)
3. Griffin, T., Wilfong, G.: On the correctness of iBGP configuration. In: Proc. of ACM SIGCOMM (2002)
4. Griffin, T., Wilfong, G.: Analysis of the MED Oscillation Problem in BGP. In: Proc. of IEEE International Conference on Network Protocols (2002)
5. Vutukuru, M., Valiant, P., Kopparty, S., Balakrishnan, H.: How to Construct a Correct and Scalable iBGP Configuration. In: IEEE INFOCOM (2006)
6. Buob, M.-O., Uhlig, S., Meulle, M.: Designing Optimal iBGP Route-Reflection Topologies. In: Proc. of IFIP Networking (2008)
7. Uhlig, S., Tandel, S.: Quantifying the BGP routes diversity inside a tier-1 network. In: Proc. of IFIP Networking (2006)
8. Walton, D., Retana, A., Chen, E., Scudder, J.: Advertisement of Multiple Paths in BGP. Internet draft, draft-ietf-idr-add-paths-04 (2010)
9. Van den Schrieck, V.: Improving internal BGP routing. PhD thesis (2010)

10. Rekhter, Y., Li, T., Hares, S.: A Border Gateway Protocol 4 (BGP-4). RFC 4271, IETF (2006)
11. Villamizar, C., Chandra, R., Govindan, R.: BGP Route Flap Damping. RFC 2439, IETF (1998)
12. Buob, M.-O.: Routage interdomaine et intradomaine dans les réseaux de coeur. PhD thesis (2008)
13. Farrel, A., Vasseur, J.P., Ash, J.: A Path Computation Element PCE-Based Architecture. RFC 4364, IETF (2006)
14. Hinden, R.: New Scheme for Internet Routing and Addressing (ENCAPS) for IPNG. RFC 1955, IETF (1996)
15. Fu, J., Sjödin, P., Karlsson, G.: Intra-domain routing convergence with centralized control. Comput. Netw. 53(18), 2985–2996 (2009)
16. IETF ForCES Working Group, http://tools.ietf.org/wg/forces/
17. Feamster, N., Balakrishnan, H., Rexford, J., Shaikh, A., van der Merwe, J.: The case for separating routing from routers. In: Proc. of ACM SIGCOMM Workshop on Future Directions in Network Architecture (2004)
18. Caesar, M., Caldwell, D., Feamster, N., Rexford, J., Shaikh, A., van der Merwe, J.: Design and implementation of a routing control platform. In: Proc. of NSDI (2005)
19. Zebra Route Server, http://www.zebra.org/zebra/Route-Server.html
20. Ballani, H., Francis, P., Cao, T., Wang, J.: ViAggre: Making Routers Last Longer! In: Proc. of Workshop on Hot Topics in Networks (2008)
21. Francis, P., Xu, X., Ballani, H., Jen, D., Raszuk, R., Zhang, L.: FIB Suppression with Virtual Aggregation. Internet draft, draft-ietf-grow-va-03 (2010)
22. Zhang, X., Francis, P., Wang, J., Yoshida, K.: Scaling IP Routing with the Core Router-Integrated Overlay. In: Proc. of IEEE International Conference on Network Protocols (2006)
23. Uhlig, S., Pelsser, C., Quoitin, B., Bonaventure, O.: Vers des réflecteurs de route plus intelligents. Colloque Francophone sur l'Ingénierie des protocoles (2005)
24. Bonaventure, O., Uhlig, S., Quoitin, B.: The case for more versatile BGP Route-Reflectors. Internet draft, draft-bonaventure-bgp-route-reflectors-00 (2004)
25. Bornhauser, U., Martini, P., Horneffer, M.: An Inherently Anomaly-free iBGP Architecture. In: Conference on Local Computer Networks (2009)
26. Pelsser, C., Masuda, A., Shiomoto, K.: Scalable Support of Interdomain Routes in a Single AS. In: Proc. of IEEE Globecom (2009)
27. Pelsser, C., Masuda, A., Shiomoto, K.: A novel internal BGP route distribution architecture. In: Proc. of the IEICE General Conference (2009)

Scalability of iBGP Path Diversity Concepts

Uli Bornhauser[1], Peter Martini[1], and Martin Horneffer[2]

[1] University of Bonn – Institute of Computer Science 4,
Roemerstr. 164 – D - 53117 Bonn
{ub,martini}@cs.uni-bonn.de

[2] Deutsche Telekom Netzproduktion GmbH - Fixed Mobile Engineering Deutschland,
Hammer Str. 216 – D - 48153 Muenster
Martin.Horneffer@telekom.de

Abstract. Improving the path diversity seems to be the next fundamental step in the iBGP evolution. Focusing the advantages an improvement of the path diversity implies, network protocol designers have disregarded the most critical drawback so far: The effect on the *scalability* of the iBGP routing, a fundamental requirement for production usage. This aspect is examined by the analyses discussed in our paper.

In this paper, we provide the theoretical groundwork for scalability analyses of four highly relevant path diversity schemes. Based on this groundwork, we exemplarily predict the information load the schemes induce in a system of a large ISP. Generalizing the system-specific results, we give an outlook on the load that can be expected in comparable ASs. We found that for two schemes currently in the standardization process, scalability problems in large ASs as they are operated by ISPs seem likely.

1   Introduction

From a global perspective, the Internet we know today is a network of interconnected Autonomous Systems (ASs). Global connectivity across these systems is realized by means of the Border Gateway Protocol (BGP) [1]. BGP distinguishes two operational modes, representing its basic tasks: *External BGP* (eBGP) terms the inter-AS mode of BGP, used to exchange routing information between ASs. To spread this information within ASs, *internal BGP* (iBGP), the intra-AS mode, is used. For the following discussions, being familiar with BGP [1] is most helpful.

1.1   Motivation

Today, iBGP is the de-facto standard to spread global routing information in ASs that route default-free. Since ASs may cover up to thousand routers, maybe even more, scalability is an important aspect. As a result of this fact, large ASs usually implement iBGP via Route Reflection [2] or AS Confederations [3]. However, the information reduction these architectures induce causes significant disadvantages in real life: Suboptimal or inconsistent routing decisions may be forced [4],

J. Domingo-Pascual et al. (Eds.): NETWORKING 2011, Part I, LNCS 6640, pp. 432–443, 2011.

routing processes may behave undesirable [5], and convergence may be slowed down [6]. A clear improvement with respect to these aspects can be achieved if the information exchange scheme improves *path diversity*: Instead of advertising only the best path for every address prefix, cf. [1], routers provide information on several known paths according to a certain scheme.

An improvement of path diversity seems to be the next important step in the iBGP evolution [7]. The path diversity is determined by the amount and kind of information speakers advertise to their peers. A classical tradeoff is given by the fact that providing more routing information principally improves the system behavior while it worsens scalability. However, even if the latter aspect is of high relevance in practice, proposals for new iBGP schemes only focus on the former aspect. Scalability issues are not adequately studied. This motivated our research.

1.2   Related Work

BGP as used today was specified as a consequence of scalability problems caused by class-based IP routing [8] in the mid 90ies. As scalability was a critical aspect at this time, the observed and expected behavior of BGP was thoroughly studied and documented [9,10]. In essence, it followed "that BGP should have no scaling problems in the area of link bandwidth and router CPU utilization". However, concerns rise with respect to iBGP, as its full-mesh design ties up many resources. This deficiency was remedied by Route Reflection and AS Confederations [2,3].

In the following years, several studies analyzed the expectable growth of the global routing table and average update rates per path [11,12]. But even if they led to highly important results, conclusions on iBGP were never drawn: As routers performed well in practice and an abrupt significant load growth did not have to be expected, this did not seem to be necessary. In 2002, the situation changed drastically with the publication of the first draft of *Add-path* [13]. Applying Add-path, routers may provide their internal peers with clearly more information than common iBGP allows. The ability to advertise several paths per prefix in parallel led to the development of several path diversity schemes [14,15,16,17]: schemes, that would cause a significant load growth. Analyses for the Route Server architecture showed that scalability can be evaluated efficiently [16]. However, even if several path diversity schemes are already in the standardization process [14,15], their effect on the scalability is still unknown. This is the starting point for our analyses.

1.3   Main Contribution and Paper Outline

Scalability is an essential linchpin for using an iBGP scheme productively. Nevertheless, the impact of increasing path diversity according to the proposed schemes is not well understood yet. Closing this gap is the main contribution of our paper. We focus on BGP Route Reflection, because this is the most common iBGP architecture in large ASs.

The rest of the paper is organized as follows: At first, in section 2, we give a basic outline on iBGP path diversity. After that, in section 3, we provide the theoretical groundwork to estimate the information load on routers in arbitrary ASs.

In section 4, we exemplary evaluate the information load path diversity schemes induce in a large AS. In section 5, we generalize the system-specific basic results. Finally, in section 6, we close with a short conclusion and sketch future work.

2   About Add-Path and iBGP Path Diversity

A core property of the today's iBGP routing information exchange is that routers only advertise their best path for a prefix. Since this limitation has proven to be a fundamental restriction in practice, Walton et al. started 2002 to specify a draft that relaxes this limitation, cf. [13]. The proposal is well-known as *Add-path*.

2.1   The Add-Path Concept

Even if technical aspects are beyond the scope of the paper, basically understanding the Add-path proposal seems helpful. Using classical iBGP, paths are internally keyed by the destination address-prefix they belong to and the peer router they are learned from. Consequently, as shown in figure 1.a), each peer can only provide one path per prefix that is stored and taken into account in the best path selection. As depicted in figure 1.b), the basic idea behind Add-path is to extend the key of a path by another attribute, the *path ID*. This ID is arbitrarily set by the router that announces the path. As known paths are only replaced if they have the same key, Add-path allows routers to advertise several valid paths in parallel.

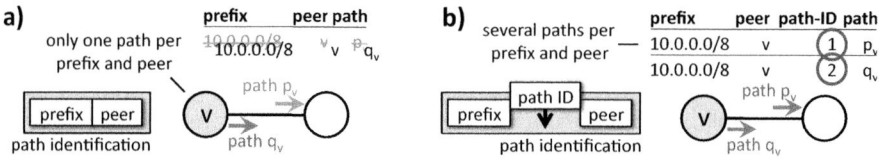

Fig. 1. Using classical BGP, routers can only store one path per prefix and peer speaker (a). The path ID removes this restriction (b).

The Add-path extension defines the semantics of the path ID and the required message exchange formats. However, while this defines the technical precondition for path diversity, Add-path does not specify which paths are to be announced. IBGP *routing information exchange schemes* are not defined. These schemes are specified in separate working documents [14,15] and iBGP architecture proposals, cf. [16] or [17], for example.

2.2   Routing Information Exchange Schemes

Motivated by different use-case scenarios, protocol designers have defined several different sets of paths routers shall be provided with. Due to the space limitation, we focus our studies on the most reasonable and from the scalability perspective most interesting proposals. This covers the following four concepts:

Advertisement of All Paths. The advertisement of all paths is one of the simplest schemes. It claims that routers advertise all known paths to all iBGP peers, provided that the export filters are passed. Usually, only paths are rejected which are already known at the receiver. Further details may be found in [15]. The basic idea is illustrated in figure 2.a).

Advertisement of n Global Neighbor-ASs Group Best Paths. Advertising global neighbor-AS group best paths is a more complex, but indeed also more scalable concept. It specifies to advertise the first n best paths with respect to every peer-AS group. Knowing these paths solves certain correctness problems [16]. The concept is illustrated in figure 2.b). See [15] for details.

Advertisement of Optimal and Local Neighbor-AS Group Best Paths. Another promising scheme is to provide every router with its optimal path and its AS-group best path for every AS it is neighbored to. Under adequate constraints, this scheme leads to consistent and optimal routing decisions. Even if this scheme is more complex than the second one, it further reduces the information routers are provided with. The idea is shown in figure 2.c). Details may be found in [16].

Advertisement of n Best Paths. The advertisement of n best paths generalizes the common iBGP information exchange scheme. Instead of advertising the best path for each prefix, routers advertise the best n paths, cf. figure 2.d), [15].

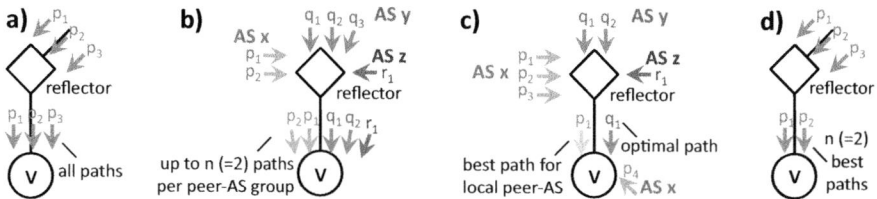

Fig. 2. The number of paths advertised to v may be unlimited (a), limited per prefix and peer-AS group (b), fitted to the receiver's view (c), or limited to n per prefix (d)

3   A Theoretical Framework for Scalability Analyses

As the schemes outlined in section 2.2 and proposed in [14,15,16,17] show, path diversity concepts may differ significantly. In this section, we derive an upper bound for the amount of ingoing routing information routers experience if the outlined concepts are applied.

3.1   Basic Model

Large ASs in the default-free zone where scalability is a critical aspect usually use Route Reflection to spread BGP data internally. As shown in figure 3.a), common routers in such architectures receive three different classes of paths: Firstly, there are those paths known due to external announcements. For router v, we refer to these paths as $\mathcal{P}_{ext}^v$. Secondly, paths may be known due to local configuration [1]. For v, we label these paths as $\mathcal{P}_{loc}^v$. The number of paths covered by both sets can be assumed as basically independent of the applied iBGP scheme. Thirdly, there is the group of internally learned paths advertised by the reflectors, labeled as set $\mathcal{P}_{int}^v$. The elements and size of this set depend on the AS-internally used iBGP scheme. IBGP architectures ensure that all three sets are disjoint [1]. Thus, the *information load* on a common router v using iBGP scheme i is given by

$$|\mathcal{P}^v|_i := |\mathcal{P}_{ext}^v| + |\mathcal{P}_{loc}^v| + |\mathcal{P}_{int}^v|_i =: |\mathcal{P}_{ext}^v| + |\mathcal{P}_{\neg ext}^v|_i. \qquad (1)$$

Problematic (CPU and memory) resource requirements can be expected, if the scheme increases the amount of ingoing information drastically in comparison to common iBGP. We derive the expectable load by predicting $|\mathcal{P}_{ext}^v|$ and $|\mathcal{P}_{\neg ext}^v|_i$.

The number of external paths a router maintains is determined by its external peering. For a fixed BGP session to a peer-AS, it receives one path for every prefix the peer-AS advertises. Generalizing this observation to all sessions a router $v \in V$ maintains, cf. figure 3.b), it holds that

$$|\mathcal{P}_{ext}^v| \leq \sum_{x \in peerAS(v)} |prefix(x, V)| \times |session(x, v)|, \qquad (2)$$

where V denotes the set of all routers in the AS, $peerAS(\mathcal{X})$ the neighbor ASs of a router or a set of routers $\mathcal{X}$, $prefix(x, V)$ the prefixes routers in AS x advertise into the own AS, and $session(x, \mathcal{X})$ the set of sessions kept between x and $\mathcal{X}$.

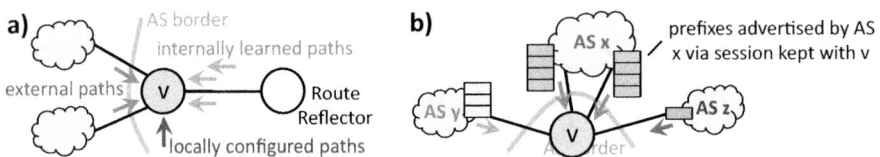

Fig. 3. Routers know paths due to local configuration, external, and internal announcements (a). The number of known external paths depends on the eBGP peer sessions (b).

The set of prefixes locally configured in the AS is called *internal routing table IT* in what follows. The vast majority of AS-internal prefixes is usually imported only once into the AS-internal BGP routing. Thus, for simplicity we assume that $\sum_{u \in V} |\mathcal{P}_{loc}^u| \approx |IT|$. Based on this assumption, we can now estimate $|\mathcal{P}_{\neg ext}^v|_i$.

3.2   Scheme-Specific Estimations for Non-external Paths $|\mathcal{P}_{\neg ext}^v|_i$

The number of non-external paths a router keeps depends on the iBGP scheme i used in the system. We now explain how to efficiently predict $|\mathcal{P}_{\neg ext}^v|_i$ for the path diversity proposals sketched in section 2.2. We assume that router v is connected to m redundant Route Reflectors.

Implementing the advertisement of all paths, a reflector passes on all available paths v does not already know. These are all paths external or local on any AS-internal router $u \in V \setminus \{v\}$. Thus, including the own local paths, v knows around

$$|\mathcal{P}^v_{\neg ext}|_1 \leq m \times \left(|IT| + \left(\sum_{u \in V \setminus \{v\}} \sum_{x \in peerAS(u)} |prefix(x, V)| \times |session(x, u)|\right)\right) \quad (3)$$

non-external paths. Note that multiplying $|IT|$ by $m > 1$ is not quite accurate, as routers learn their own local paths $\mathcal{P}^v_{loc}$ only once. However, since routers usually do not import more than a few hundreds of paths locally, this simplification can be neglected. We proceed analogously for the following estimations.

The main disadvantage of advertising all paths is that keeping several sessions to one neighbor AS generally bloats the amount of information reflectors provide. This problem is addressed if not more than n paths per prefix for every peer-AS group are advertised (scheme II). For each peer-AS x, Route Reflectors advertise n paths per prefix, provided that n or more eBGP sessions are kept between x and the border routers $u \in V \setminus \{v\}$. Also taking into account local paths, it holds that v knows around

$$|\mathcal{P}^v_{\neg ext}|_2 \leq m \times \left(|IT| + \sum_{x \in peerAS(V)} |prefix(x, V)| \times \min(n, |session(x, V \setminus \{v\})|)\right) \quad (4)$$

non-external paths. Recall that $peerAS(V)$ labels all neighbor ASs of the system.

Providing a router per prefix with its optimal path and n best paths for each of its local peer-ASs, the number of received paths becomes topology-dependent. Unknown optimal paths of v must all be provided internally. In the worst case, this covers one path for each globally routable and AS-internal prefix. The former prefixes are well-known and covered by the *global routing table GT* [18]. Latter ones are covered by IT. The unknown AS-group best paths for a router's peer-ASs have to be provided internally, too. For every peer-AS $x \in peerAS(v)$, in the worst case, up to n paths per prefix AS x advertises to v must be provided. Thus, applying this scheme, router v may learn up to

$$|\mathcal{P}^v_{\neg ext}|_3 \leq m \times \left(|GT| + |IT| + \sum_{x \in peerAS(v)} |prefix(x, V)| \times \min(n, |session(x, V \setminus \{v\})|)\right) \quad (5)$$

non-external paths via its reflectors and local configuration. Usually, $n = 1$ is the most reasonable configuration, cf. [16]. We assume this in what follows.

Generalizing the classical iBGP scheme, reflectors can simply advertise their n best paths per prefix. Applying this scheme, a reflector advertises the $|IT|$ AS-internal paths and up to n paths for every globally routable address-prefix. This results in up to

$$|\mathcal{P}^v_{\neg ext}|_4 \leq m \times \left(n \times |GT| + |IT|\right) \quad (6)$$

non-external paths v receives. If $n = 1$, we estimate the non-external information load the common iBGP information exchange scheme induces.

3.3   Required Performance Data

The methodology developed in the sections 3.1 and 3.2 reveals how system-specific estimations for the information load can be determined. To apply the estimations to a production system, information on the required basic AS-performance data must be available. Understanding how to gather this information properly finally completes the theoretical framework.

As already mentioned, the number of prefixes the global routing table covers, parameter $|GT|$, is well known [18]. The number of system-internal prefixes $|IT|$ should be known by the AS operator. This also holds for the number of reflectors in the system, parameter m, and the desired path diversity factor n many proposals support. Information on the peer-ASs $x \in peerAS(u)$ and the number of kept sessions $session(x, u)$ for all routers $u \in V$ can be derived from the routers' BGP configurations. Based on this information, also $peerAS(\mathcal{X})$ and $session(x, \mathcal{X})$ for all sets of routers $\mathcal{X} \subseteq V$ and peer-ASs x can be determined. The router configurations are usually available for internal use.

The most complex task is to determine the number of prefixes $|prefix(x, V)|$ each peer-AS x advertises paths for into the system. Gathering exact data for a peer-AS x requires analyzing the routing tables of all internal routers that peer with x. Consequently, to gather exact data for all peer-ASs, the routing tables of all internal eBGP speakers would have to be analyzed. This is hardly realizable in large ASs. However, ASs typically advertise paths for nearly the same prefixes to routers located in the same peer-AS. Consequently, a good approximation for $prefix(x, V)$ is usually already given by $prefix(x, \epsilon)$ for any $\epsilon \subseteq V \,|\, x \in peerAS(\epsilon)$. In practice, analyzing the routing tables of a small number of those routers that peer with the most ASs allows determining approximations for a large proportion of the peer-ASs. For the neighbor ASs not covered by this method, the number of advertised prefixes must be estimated. A simple estimation can be determined on the basis of the business relationship that is kept with the system: Provider ASs advertise paths for nearly all prefixes in the global routing table. For other ASs, the approximations derived for peer-ASs keeping the same business relationship can be averaged. A systematic underestimation is avoided at this point, if ϵ is chosen such that the large peer-ASs of each class are covered. They are typically known by the operator. This avoids that scalability problems remain undetected.

4   Performance Evaluation for a Reference System

The basic framework derived in section 3 allows operators to perform AS-specific scalability analyses for their systems. However, basis for such analyses should be a general understanding of the impact of the different schemes on the information load. For that purpose, we apply the framework to a reference system. Based on the achieved results, we identify the generalizable implications in section 5.

4.1   The Reference System AS3320

The reference system we study in this paper is the AS with the registered number 3320, considered at a snapshot in time of August 2009. The system is the Internet

Backbone of Deutsche Telekom AG, labeled as AS3320 in what follows. The AS is a comparatively large transit system. At the time the data have been taken, it maintained one customer relationship to a provider AS. All in all, the system peered with 588 neighbor ASs via 1198 sessions, kept by 238 AS-internal eBGP speakers. The relationships to these systems are well-known, too. Besides eBGP speakers, the system also covered 642 exclusive iBGP speakers. At the point in time of data acquisition, the system used around 63,000 internal prefixes. The global table covered around 300,000 prefixes. Most routers learned internal BGP routing information from two redundant Route Reflectors, meaning that $m = 2$.

4.2 Prefix Advertisement by Neighbor-ASs

To gather information on the number of prefixes peer-ASs advertise, we analyzed the routing tables of the ten routers keeping the most eBGP sessions. The basic results are shown in figure 4: The tables allowed us to approximate $|prefix(x, V)|$ for the only provider and over 70% of the peer ASs. The blue bar chart shows that the provider and, according to the number of advertised prefixes, large peer ASs are connected comparatively often. In addition to provider and peer ASs, we also covered around 20% of the customer ASs. They provide fewer paths and are less often connected. All in all, analyzing the BGP routing tables of ten of 238 eBGP speakers (around 4%) allowed us to cover over 28% of the peer-ASs.

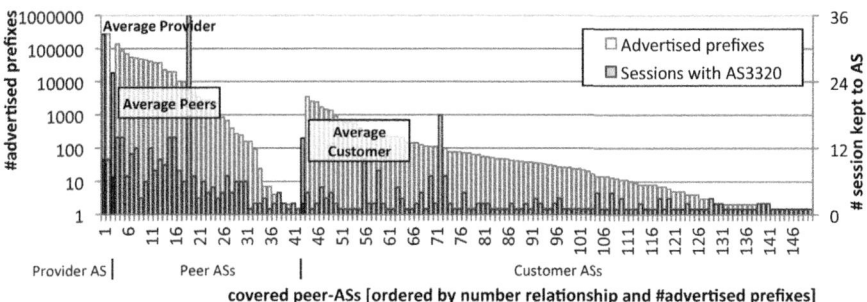

Fig. 4. Approximated number of prefixes neighbor-ASs of AS3320 advertise paths for

The provider AS advertises paths for around 281,000 prefixes and is connected via ten eBGP sessions. These are around 15 times more prefixes than a peer AS of AS3320 advertises on average (around 18,500). With around six eBGP sessions on average, peer ASs are also significantly less frequently connected than the provider AS. A customer AS in turn only provides paths for around 200 prefixes on average, which is only about 1.1% of an average peer AS. Customer ASs are also clearly less often connected via eBGP sessions: On average, the covered customer ASs keep around two sessions with AS3320. There also exist small customer and peer ASs which advertise paths only for a single prefix.

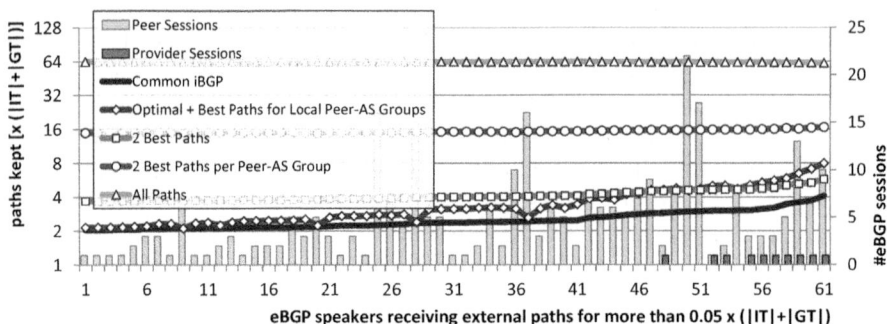

Fig. 5. Information load predictions for eBGP speakers receiving paths for more than 5% of the internal and global prefixes $|IT| + |GT|$

4.3   AS-specific Information Load Predictions

The expectable information load on common routers in AS3320 using the schemes outlined in section 2.2 is shown in figure 5. To keep the figure simple, we list only routers receiving more than $0.05 \times (|GT|+|IT|)$ paths externally. The predictions for unlisted routers are similar to those devices that receive little external information. The routers are ordered by the number of paths they receive externally. All values are normalized to multiplies of the routable prefixes $|GT|+|IT|$.

Applying classical iBGP, routers have to manage two to four times more paths than prefixes are routed (black curve). The exact value depends on the externally learned information. Internally, around $2 \times (|GT| + |IT|)$ paths are received, cf. equation 6 with $m=2$, $n=1$. A significantly higher load (up to the factor of two) arises on routers that learn many paths externally. Such routers keep a session to the provider AS or (several) large peer ASs, cf. the bars on the secondary y-axis.

Exchanging information on all available paths via iBGP, routers must manage 16 to 32 times more paths than today. As it can be expected, the information load would nearly be the same for all routers. Speakers which learn little information from external peers have to maintain approximately 64.1 times more paths than routable prefixes are exchanged. Routers that learn a lot of external information keep around 4% less information, since external paths are only learned once (and not twice via the reflectors). In practice, this load level would for sure increase the required resources significantly. Scalability problems seem likely.

Limiting the number of paths routers learn via an internal session to two per prefix and peer-AS group decreases the load significantly. In case of AS3320, the load could be reduced by a factor of 3.7 to 4.3 in comparison to the advertisement of all paths. The reason for this reduction becomes clear if we have a closer look at the number of sessions large peer-ASs keep, cf. figure 4: ASs advertising paths for many prefixes (i.e. the provider and the peer-ASs listed on the left hand) are mostly connected via clearly more than two eBGP sessions. This results in a lot of paths reflectors provide if the number of advertised paths per prefix and peer-AS group is not limited. Nevertheless, a load increase by a factor of four to eight compared to classical iBGP still defines a massive growth in practice.

Reflecting the two best paths, the information load in comparison to classical iBGP is only increased by a constant factor of $m \times |GT|$, cf. equation 6. As peer-ASs advertise many paths for the same prefixes, this scheme results in a further significant information reduction. Compared to the advertisement of two paths per peer-AS group and prefix, we observed a further reduction by a factor of two to four. Providing the optimal path and the best path for the local peer-ASs, the load is further reduced on the most speakers. For routers that receive only few external data, the expectable load is comparable to classical BGP. This is due to the fact that if router v keeps only sessions with few, mostly small ASs, equation 5 can be estimated by $m \times (|GT| + |IT|)$, which is equal to $|\mathcal{P}^v_{\neg ext}|_4$ for $n = 1$. For routers that keep eBGP sessions with the provider or (several) large peer-ASs, a load growth has to be expected. Here, besides the optimal path for every prefix, a wide range of AS-group best paths must be provided internally in the worst case. In case of AS3320, this increases the information load by a factor of up to two on some routers, cf. the routers on the right side of figure 5. Discussions with network engineers of Deutsche Telekom AG showed that for the latter two schemes, the expectable information load should be manageable in practice: Some additionally meshed routers in the reference AS manage the same amount of information today and scalability problems do not appear.

5   Generalization

Even if the absolute results obtained in section 4 are highly system-specific, they allow drawing general conclusions on the scalability of the studied path diversity schemes. We summarize the main aspects in what follows.

5.1   General Results

Large transit ASs in the default-free zone usually keep several sessions with large peer or provider ASs. In system-global terms, this from a forwarding perspective highly desirable property leads to a high number of paths available at the border of the AS. If all these paths are spread across a system, significant load increases in comparison to common iBGP must be expected in general. If the system peers with large-ASs clearly more often than n, exchanging only n paths per prefix and peer-AS group significantly lowers the expectable load. However, as long as paths for a comparatively high number of equal prefixes is received from different peer-ASs, a significant load growth in comparison to common iBGP must be expected. This effect is avoided if not more than a constant number of paths is advertised per prefix. The property that advertising n best paths per prefix leads to a constant load growth of $(n-1) \times (|IT| + |GT|)$ can be generalized. It is independent of AS- or other router-specific parameters, cf. equation 6. This makes the expectable load easily, reliably, and precisely predictable. The drawback of this performance behavior is that many routers may need hardware upgrades in the worst case.

Advertising the optimal path and best paths for the local peer-AS groups, the expectable load is heavily dependent on the router's peering. As long as routers peer only with few small ASs (as often the vast majority of routers in an AS do), the load is comparable to common iBGP. Routers that peer with the provider, large peer, or several mid-size peer ASs are provided with significantly more data. Even if the load increase on such routers may exceed the load induced by advertising n best paths, the limitation to few routers may be a significant advantage: Hardware updates on few devices are much easier to realize in practice. Finally, it shall be noted that equation 5 is a worst case estimation: Externally learned optimal and group best paths are not provided by the reflectors. For AS-border routers that peer with large ASs, this may relatively often be the case.

5.2   ASs without Route Reflection

A basic assumption for our load predictions is that the analyzed AS realizes the iBGP information exchange by means of Route Reflection. To finish our analyses, we sketch what should be observed if this precondition is relaxed.

Applying full-meshed iBGP (either natively or as part of AS Confederations), routers receive paths from more than m iBGP peers. Exchanging information on all paths, this has no direct influence on the information load, since routers learn the same paths. If internal peer routers limit the number of paths they announce, having more internal peers reduces the effectiveness of the information reduction. For example, if advertising n best paths is implemented, every internal peer may advertise n best paths for every prefix. Having more than m internal iBGP peers that peer with large neighbor ASs, a router may receive significantly more paths than specified by equation 6. Implementing the advertisement of the n best paths per peer-AS group in a full-meshed AS, a significant load reduction compared to exchanging all paths cannot be expected: Generally, eBGP speakers do not keep a high number of sessions to the same peer-AS. Similar reflections can be made for the other proposed path diversity schemes.

6   Conclusion and Future Work

In this paper, we provided the first in-depth analyses of scalability aspects important iBGP path diversity involves. For four schemes proposed by protocol designers and researchers, we developed the theoretical basis for load predictions. The results allow operators to estimate the expectable load on the basis of few basic parameters. Reference studies for a large carrier AS and their generalization led to an understanding for the pros and cons of the different schemes with respect to scalability. We found that a significant load growth has to be expected if routers spread all or the n best known paths per AS-group and prefix. Limiting the number of advertised paths on a per prefix basis or realizing a receiver-based information reduction improves the scalability usually drastically. However, it should be kept in mind that scalability is not the only relevant aspect: Every scheme comes along with specific (convergence and correctness) properties. Finally, we want to remark that we also studied the accuracy of the predictions in detail. Even if we could not present the data here, we can affirm that the results seem accurate.

Our analyses gave a first valuable insight into the scalability issues improving iBGP path diversity comes along with. However, due to the space limitation, several important and deeply interesting aspects could not be covered by our studies.

Analyzing and discussing these aspects are important aspects for future research. Examples are the composition of the load or details on the effect on the resulting CPU and memory requirements. Another area we could not discuss is iBGP schemes where the information reduction depends on topology independent path attributes as the Local Preferences or the AS-path length. Examples are *AS-wide* and *Best Local Pref.* [15], specified by Uttaro et al.. Understanding why certain path diversity schemes may cause scalability problems, concepts could be revised and improved. Complementary work covering these aspects is reasonable, too.

References

1. Rekhter, Y., Li, T., Hares, S.: A Border Gateway Protocol 4 (BGP-4), RFC 4271 (January 2006)
2. Bates, T., Chen, E., Chandra, R.: BGP Route Reflection - An Alternative to Full Mesh IBGP, RFC 4456 (April 2006)
3. Traina, P., McPherson, D., Scudder, J.: Autonomous System Confederations for BGP, RFC 5065 (August 2007)
4. Bornhauser, U., Martini, P., Horneffer, M.: The Scope of the iBGP Routing Anomaly Problem. In: Proceedings of the 17th Conference on Communication in Distributed Systems (KIVS 2011) (March 2011)
5. Griffin, T., Wilfong, G.: On the Correctness of IBGP Configuration. SIGCOMM Comput. Commun. Rev. 32(4), 17–29 (2002)
6. Walton, D., Retana, A., Chen, E., Scudder, J.: Advertisement of Multiple Paths in BGP, Internet Draft (August 2010)
7. Raszuk, R., Cassar, C., Aman, E., Decraene, B.: BGP Optimal Route Reflection (BGP-ORR), Internet Draft (October 2010)
8. Rekhter, Y., Gross, P.: Application of the Border Gateway Protocol in the Internet, RFC 1772 (March 1995)
9. Traina, P.: Experience with the BGP-4 Protocol, RFC 1773 (March 1995)
10. Traina, P.: BGP-4 Protocol Analysis, RFC 1774 (March 1995)
11. Bu, T., Gao, L., Towsley, D.: On Routing Table Growth. SIGCOMM Comput. Commun. Rev. 32, 77–87 (2002)
12. Elmokashfi, A., Kvalbein, A., Dovrolis, C.: On the scalability of bgp: the roles of topology growth and update rate-limiting (2008)
13. Walton, D., Retana, A., Chen, E., Scudder, J.: Advertisement of Multiple Paths in BGP, IETF IDR Internet Draft (May 2002)
14. Walton, D., Retana, A., Chen, E., Scudder, J.: BGP Persistent Route Oscillation Solutions, Internet Draft (May 2010)
15. Uttaro, J., Van den Schrieck, V., Francois, P., Fragassi, R., Simpson, A., Mohapatra, P.: Best Practices for Advertisement of Multiple Paths in BGP, Internet Draft (October 2010)
16. Bornhauser, U., Martini, P., Horneffer, M.: An Inherently Anomaly-free iBGP Architecture. In: Proceedings of the 34th IEEE Conference on Local Computer Networks (LCN 2009), pp. 145–152. IEEE Computer Society, Los Alamitos (2009)
17. Basu, A., Ong, C.H.L., Rasala, A., Shepherd, F.B., Wilfong, G.: Route Oscillations in I-BGP with Route Reflection. In: SIGCOMM 2002: Proceedings of the 2002 Conference on Applications, Technologies, Architectures, and Protocols for Computer Communications, pp. 235–247. ACM, New York (2002)
18. Huston, G.: BGP Routing Table Analysis Reports (Website) (November 2010), http://bgp.potaroo.net

MultiPath TCP: From Theory to Practice

Sébastien Barré, Christoph Paasch, and Olivier Bonaventure*

ICTEAM, Université catholique de Louvain,
B-1348 Louvain-la-Neuve, Belgium
firstname.lastname@uclouvain.be

Abstract. The IETF is developing a new transport layer solution, MultiPath TCP (MPTCP), which allows to efficiently exploit several Internet paths between a pair of hosts, while presenting a single TCP connection to the application layer. From an implementation viewpoint, multiplexing flows at the transport layer raises several challenges. We first explain how this major TCP extension affects the Linux TCP/IP stack when considering the establishment of TCP connections and the transmission and reception of data over multiple paths. Then, based on our implementation of MultiPath TCP in the Linux kernel, we explain how such an implementation can be optimized to achieve high performance and report measurements showing the performance of receive buffer tuning and coupled congestion control.

Keywords: TCP, multipath, implementation, measurements.

1 Introduction

The Internet is changing. When TCP/IP was designed, hosts had a single interface and only routers were equipped with several physical interfaces. Today, most hosts have more than one interface and the proliferation of smart-phones equipped with both 3G and WiFi will bring a growing number of multihomed hosts on the Internet. End-users often expect that using multihomed hosts will increase both redundancy and performance. Unfortunately, in practice this is not always the case as more than 95% of the total Internet traffic is still driven by TCP and TCP binds each connection to a single interface. This implies that TCP by itself is not capable of efficiently and transparently using the interfaces available on a multihomed host.

The multihoming problem has received a lot of attention in the research community and within the IETF during the last years. Network layer solutions such as shim6 [15] or the Host Identity Protocol (HIP) [14] have been proposed and implemented. However, they remain experimental and it is unlikely that they will be widely deployed. Transport layer solutions have also been developed, first as extensions to TCP [13,7,18,23]. However, to our knowledge these extensions have never been implemented nor deployed. The Stream Control Transmission Protocol (SCTP) [19] protocol was designed with multihoming in mind and supports

---

* This work is supported by the European Commission via the 7th Framework Programme Integrated Project TRILOGY.

J. Domingo-Pascual et al. (Eds.): NETWORKING 2011, Part I, LNCS 6640, pp. 444–457, 2011.

fail-over. Several SCTP extensions [8,12] enable hosts to use multiple paths at the same time. Although implemented in several operating systems [8], SCTP is still not widely used besides specific applications. The main drawbacks of SCTP on the global Internet are first that application developers need to change their application to use SCTP. Second, various types of middle-boxes such as NATs or firewalls do not understand SCTP and block all SCTP packets.

During the last two years, the MPTCP working group of the IETF has been developing multipath extensions to TCP [6] that enable hosts to use several paths possibly through multiple interfaces, to carry the packets that belong to a single connection. This is probably the most ambitious extension to TCP to be standardized within the IETF. As for all Internet protocols, its success will not only depend on the protocol specification but also on the availability of a reference implementation that can be used by real applications.

In this paper we explain the challenges in implementing MPTCP in the Linux kernel and evaluate its performance based on lab measurements. This is the first implementation of this major TCP extension in an operating system kernel.

The paper is organized as follows. In section 2, we briefly describe MPTCP and compare it to regular TCP. Then we describe the architecture of the implementation. In section 4, we describe a set of general problems that must be solved by any protocol wanting to simultaneously use several paths, together with the chosen solutions. Next, we measure the performance of the implementation in different scenarios to show how the implementation-choices taken do influence the results.

2   MultiPath TCP

MultiPath TCP [6] is different from existing TCP extensions like the large windows, timestamps or selective acknowledgement extensions. These older extensions defined new options that slightly change the reaction of hosts when they receive segments. MultiPath TCP allows a pair of hosts to use several paths to exchange the segments that carry the data from a single connection.

To understand the operation of MultiPath TCP, let us consider a very simple case where a client having two addresses, A.1 and A.2 establishes a connection with a single homed server. The client first sends a SYN segment from address A.1 that contains the MP_CAPABLE option [6]. This option indicates that the client supports MultiPath TCP and contains a token that identifies the Multi-Path TCP connection. The server replies with a SYN+ACK segment containing the same option and its own token. The client concludes the three-way handshake. Once the TCP connection has been established, the client can advertise its other addresses by sending TCP segments with the ADD_ADDR option. It can also open a second TCP connection, called a subflow, that will be linked to the first one by sending a SYN segment with the MP_JOIN option that contains the token sent by the server in the initial subflow. The server replies by sending a SYN+ACK segment with the MP_JOIN option containing the token chosen by the client in the initial subflow and the client terminates the three-way handshake. These two subflows are linked together inside a single MultiPath TCP connection and both can be used to send data. Subflows may fail and be added during the lifetime of a MultiPath TCP connection. Additional details about the establishment of MultiPath TCP connections and subflows may be found in [6].

The data produced by the client and the server can be sent over any of the subflows that compose a MultiPath TCP connection, and if a subflow fails, data may need to be retransmitted over another subflow. For this, MultiPath TCP relies on two principles. First, each subflow is equivalent to a normal TCP connection with its own 32-bits sequence numbering space. This is important to allow MultiPath TCP to traverse complex middle-boxes like transparent proxies or traffic normalizers. Second, MultiPath TCP maintains a 64-bits data sequence numbering space. Two TCP options use these data sequence numbers : DSN_MAP and DSN_ACK. When a host sends a TCP segment over one subflow, it indicates inside the segment, by using the DSN_MAP option, the mapping between the 64-bits data sequence number and the 32-bits sequence number used by the subflow. Thanks to this mapping, the receiving host can reorder the data received, possibly out-of-sequence over the different subflows. In MultiPath TCP, a received segment is acknowledged at two different levels. First, the TCP cumulative or selective acknowledgements are used to acknowledge the reception of the segments on each subflow. Second, the DSN_ACK option is returned by the receiving host to provide cumulative acknowledgements at the data sequence level. When a segment is lost, the receiver detects the gap in the received 32-bits sequence number and traditional TCP retransmission mechanisms are triggered to recover from the loss. When a subflow fails, MultiPath TCP detects the failure and retransmits the unacknowledged data over another subflow that is still active.

Another important difference between MultiPath TCP and regular TCP is the congestion control scheme. MultiPath TCP cannot use the standard TCP control scheme without being unfair to normal TCP flows. Consider two hosts sharing a single bottleneck link. If both hosts use regular TCP and open one TCP connection, they should achieve almost the same throughput. If one host opens several subflows for a single MultiPath TCP connection that all pass through the bottleneck link, it should not be able to use more than its fair share of the link. This is achieved by the coupled congestion control scheme that is discussed in details in [17,22]. The standard TCP congestion control [1] increases and decreases the congestion window and slow-start threshold upon reception of acknowledgments and detection of losses. The coupled congestion control scheme also relies on a congestion window, but it is updated according to the following principle:

- For each non-duplicate ack on subflow i, increase the congestion window of the subflow i by $min(\alpha/cwnd_{tot}, 1/cwnd_i)$ (where $cwnd_{tot}$ is the total congestion window of all the subflows and $\alpha = cwnd_{tot} \frac{\max_i(\frac{cwnd_i * mss_i^2}{RTT_i^2})}{(\sum_i \frac{cwnd_i * mss_i}{RTT_i})^2})$.

- Upon detection of a loss on subflow i, decrease the subflow congestion window by $cwnd_i/2$.

3   Linux MultiPath TCP Architecture

Our Multipath architecture for the Linux kernel is based on an important re-thinking of the building blocks that make the current Linux TCP stack (the current Linux patch has more than 12000 lines[1]). Of major importance is the separation between *user context* and *interrupt context*. A piece of kernel code runs in *user context* if it is performing a task for a particular application, and can be interrupted by the OS scheduler. All system calls (like `send, rcv`) are run in user context. Being under control of the scheduler, user context activities are fair to other applications. On the other hand, *software interrupts* are run with higher priority than any process, and follow an event (e.g. timer expiry, packet reception). They can be interrupted only by *hardware interrupts*, and must thus be as short as possible. Unfortunately the handling of incoming packets happens in a software interrupt, and the TCP reordering is normally done there as well, because it is needed to acknowledge as fast as possible. To allow spending less time in software interrupts, and hence gaining in overall system responsiveness, Van Jacobson proposed to force reordering in the user context when the application is waiting in a `receive` system call [10]. Instead of reordering itself, the interrupt then puts the segments in a so-called *prequeue*, and wakes up the process, that performs reordering and sends acknowledgements. More details on the Linux networking architecture can be found in [20].

As shown in figure 1, our proposed architecture for multipath is made of three elements. The first element is the **master subsocket**. If MPTCP is not supported by the peer, only that element is used and regular TCP is executed. The master subsocket is a standard socket structure that provides the interface between the application and the kernel for TCP communication. The second building block is the **Multi-path control bock (mpcb)**. This building block supervises the multiple flows used in a given connection, it runs the *decision algorithm* for starting or stopping subflows (based on local and remote information), the *scheduling algorithm* for feeding new application data to a particular subflow, and the *reordering algorithm* that allows providing an in-order stream of incoming bytes to the application. Note that reordering at the subflow-level is performed in the subflows themselves, this service being implemented already as part of the underlying TCP. The third building block is the **slave subsocket**. Slave subsockets are not visible to the application. They are initiated, managed and closed by the multipath control block. They otherwise share functionality with the master subsocket. The master subsocket and the slave subsockets together form a pool of subflows that the mpcb can use for scheduling outgoing data, and reordering incoming data. Should the master subsocket fail, it keeps its role of contact point between the application and the kernel, but simply it stops being used as an eligible subflow by the MPTCP scheduler, until it recovers.

Connection establishment: The initiator of a TCP connection uses the initial subflow to learn from the peer its set of addresses, as defined in [6]. It then combines those addresses to try establishing subflows on every known path to

---

[1] Available at `http://inl.info.ucl.ac.be/mptcp/`

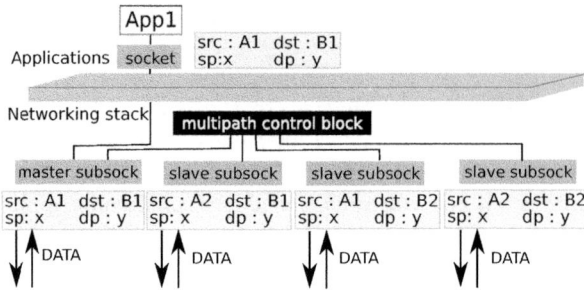

Fig. 1. Overview of the multipath architecture

the peer. The case of the receiver is more tricky: We want to be able to create new subsockets without relying on the usual *listen*. The reason is that the key to retrieve a listener is no longer the usual TCP 5-tuple, but instead a token attached in a TCP option. We solve this by attaching an accept queue to the multipath control block. The mpcb is found thanks to a token-based hash table lookup, and a new half-open socket is appended in the accept queue. Only when the new subsocket becomes established is it added to the list of active subsockets.

Scheduling and sending data: If more than one subflow is in *established* state, MPTCP must decide on which subflow to send data. The current policy implemented in the scheduler tries to fill all subflows, as described later in this paper. However, the scheduler is modular and other policies like preferring one interface over the other could be implemented in the future.

The scheduler must deal with the granularity of the allocations, that is, the number of contiguous bytes that are sent over the same subflow before deciding to select another subflow. This granularity influences the performance of a MultiPath TCP implementation with high bandwidth flows. The optimal use of all subflows would be an argument in favor of small allocation units. On the other hand, to spare CPU cycles and memory accesses, one would tend to allocate in large units (because that would result in fewer calls to the scheduler, and would facilitate segmentation offload). In this paper, we favor an optimal allocation of segments, deferring a full study of the performance trade-offs for another paper.

We have examined two major design options for the scheduler. In the first design, whenever an application performs a sendmsg() system call or equivalent, the scheduler is invoked and data is immediately pushed to a specific subflow. This implies per-subflow send buffers, and has the advantage of preformatting segments (with correct MSS and sequence numbers) so that they can be sent very quickly each time an acknowledgement opens up more space in the congestion window. However, there may be several hundreds of milliseconds between the time when a segment is enqueued on a subflow, and its actual transmission on the wire. The path properties may change during that time and what was a good choice when running the scheduler may reveal a very wrong choice when the data is put on the wire. Even worse is the case of a failing subflow, because a full buffer of segments needs to be moved to another subflow when this happens.

These major drawbacks lead us to a second design, that solves those problems. In the current implementation, the data pushed by the applications is not scheduled anymore, but instead stored in a connection-level send buffer. Subflows pull data from the shared send buffer whenever they receive an acknowledgement. This design is similar to the one proposed for pTCP in [7], but pTCP has only been simulated and not implemented. This effectively solves the problem of fluctuating path properties and allows to run the scheduler *when the segment is sent on the wire.*

Another problem that needs to be considered by an implementation is that different subflows may use different MSS. In this case, a byte-based connection-level send buffer would probably be needed. However, this would be inefficient in the Linux kernel that is optimized to handle queues of packets. To solve this problem, our implementation uses the minimum MSS over all subflows to send segments. This is slightly less efficient from an overhead viewpoint if one interface uses a much smaller MSS than the other, but in practice this is rarely the case.

To evaluate the impact of using the same MSS over all subflows, we performed a simple test by contacting the 10000 most popular web servers according to the Alexa top-10000 list. We sent a SYN segment advertising an MSS value of 1460, 4096 or 9148 bytes to each of these servers and analyzed the returned MSS. 97% of these servers returned an MSS of 1380 bytes without any influence of the MSS option that we included in our SYN segment. Most of the other web servers returned an MSS that was very close to 1380 bytes. Thus, using the same MSS for all subflows appears reasonable on today's Internet.

Receiving data: Data reception is performed in two steps. The first step is to receive data at the subflow level, and reorder it according to the 32-bits subflow sequence numbers. The second step is to reorder the data at the connection level by using the data sequence numbers, and finally deliver it to the application. As in regular TCP, each subflow maintains a COPIED_SEQ and a RCV.NXT [16] pointer, resp. to track the next byte to give to the upper layer (now the upper layer becomes the mpcb) and the next expected subflow sequence number. Likewise, the multipath control block maintains a connection level COPIED_SEQ and a RCV.NXT pointer, resp. to track the next byte to deliver to the application and the next expected data sequence number that is used when returning a DATA_ACK option.

To store incoming data until the application asks for it, we use a single connection-level receive queue. All subflow-receive queues are always empty, because as soon as a segment becomes in order at the subflow-level, it is enqueued in the connection-level receive queue, or out-of-order queue. The problem of charging the application for this additional processing time (to ensure fairness with other running processes) can be solved by using Van Jacobson's prequeues [10], just like regular TCP does. Enabled by default in current kernels, that option is even more important for MPTCP, because MPTCP involves more processing, especially with regards to reordering, as we will show in section 5.

4   Reaching Full Path Utilization

One obvious goal of MPTCP is to be able to consider all available paths as a shared resource, just behaving as the sum of the individual resources [21]. However, from an implementation viewpoint, reaching that goal requires to deal with several new constraints. We first study the implications of MPTCP on the receive buffers and then the coupled congestion control scheme.

4.1   Receive Buffer Constraints

To optimize the size of the receive buffer, Fisk et al. have proposed [4] to dynamically tune the TCP receive buffer to ensure that twice the Bandwidth-Delay Product (BDP) of the path can be stored in the buffers of the receiver. That value allows supporting reordering by the network (this requires one BDP), as well as fast retransmissions (one other BDP). This buffer tuning [4] is integrated in the Linux kernel.

In the context of a multipath transfer, additional constraints appear. The problem, described in details in [9] for SCTP multipath transfer, is that a fast path can be blocked due to the receive buffer becoming full. In single-path TCP the receive buffer can become full only if the application stops consuming data. In MultiPath TCP, it can become full as well if some data coming over a path with a high delay is needed for connection-level reordering, before to be able to deliver the bytes to the application. In the worst case, each path i is continuously transmitting at a BW_i rate, while the slowest path (that is, the path with highest RTT) is fast retransmitting. To accommodate for such a situation, the receive buffer must be dimensioned as [7,5]: $rbuf = 2 * \sum_{i \in subflows} BW_i * RTT_{max}$

Our implementation uses the technique proposed in [4] to estimate the RTT at the receiver by using the TCP timestamps, and computes the contribution of each path to the overall receive buffer. Whenever any of the contributions changes, the global receive buffer limit is updated in the mpcb. In practice, this dynamic tuning may reach the maximum allowed receive buffer configured on the system. This should be used as a hint to indicate that a subflow is underperforming and disable the slowest path.

4.2   Implementing the Coupled Congestion Control

When implementing in the Linux kernel the coupled congestion control described in section 2, several issues need to be solved. The main problem is that the Linux kernel does not support floating point numbers. This implies that increasing the congestion window by $1/cwnd_i$ is not possible, because $1/cwnd_i$ is not an integer. Our implementation counts the number of acknowledged packets and maintains this information inside a subflow-variable ($cwnd\_cnt_i$), as it is already done for other congestion control algorithms in the Linux kernel like New-Reno. Then, the congestion window is increased by one as soon as $cwnd\_cnt_i > tot_{cwnd}/\alpha$ [17].

The coupled congestion control involves the calculation of the α-factor described in section 2. Our implementation computes α as shown in equation 1.

As our implementation uses the same mss for all the subflows, the formula does not need the mss anymore. rtt_{max} and $cwnd_{max}$ are precalculated from the numerator, but rtt_{max} has been put into the denominator to reduce the number of divisions in the formula. As the Linux kernel only handles fixed-point calculations, we need to scale the divisions, to reduce the error due to integer-underflow ($scale_{num}$ and $scale_{den}$). Later, the resulting scaling factor of $scale_{num}/scale_{den}^2$ has to be taken into account.

$$\alpha = cwnd_{tot} \frac{cwnd_{max} * scale_{num}}{(\sum_i \frac{rtt_{max}*cwnd_i*scale_{den}}{rtt_i})^2} \qquad (1)$$

5 Evaluation

To evaluate the performance of our implementation, we performed lab measurements in the HEN testbed at University College London (http://hen.cs.ucl.ac.uk/). We used two different scenarios. The first scenario is used to evaluate the coupled congestion control scheme while the second analyses the factors that influence the performance of MultiPath TCP.

5.1 Measuring the Coupled Congestion Control

The first scenario is the *congestion testbed* shown in figure 2. In this scenario, we used four Linux workstations. The two workstations on the left use Intel® Xeon CPUs (2.66GHz, 8GB RAM) and Intel® 82571EB Gigabit Ethernet Controllers. They are both connected with one 1 Gbps link to a similar workstation that is configured as a router. The router is connected with a 100 Mbps link to a server. The upper workstation in figure 2 uses a standard TCP implementation while the bottom host uses our MultiPath TCP implementation.

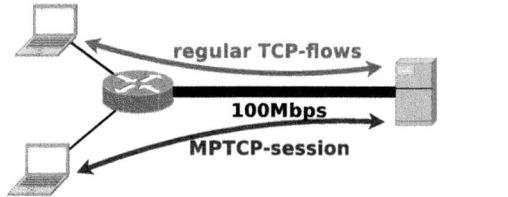

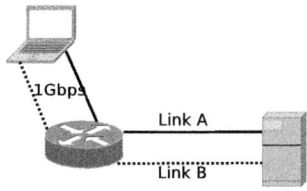

Fig. 2. Congestion testbed   **Fig. 3.** Performance testbed

Detailed simulations analyzed by Wischik et. al in [22] show that the coupled congestion control fulfills its goals. In this section we illustrate that our Linux kernel implementation of the coupled congestion control achieves the desired effects, even if it is facing additional complexity due to fixed-point calculations compared to the user-space implementation used in [22]. In the congestion testbed shown in figure 2, the coupled congestion control should be fair to TCP. This means that an MPTCP connection should allow other TCP sessions to take over the bandwidth on a shared bottleneck. Furthermore, an MPTCP connection that uses several subflows should not slow down regular TCP connections. To measure the fairness of MPTCP, we use the bandwidth measurement software iperf[2] to establish sessions between the hosts on the left and the server on the right part of the figure. We vary the number of regular TCP connections from the upper host and the number of MPTCP subflows used by the bottom host. We ran the iperf sessions for a duration of 10 minutes, to allow the TCP-fairness over the bottleneck link to converge [11]. Each measurement runs five times and we report the average throughput.

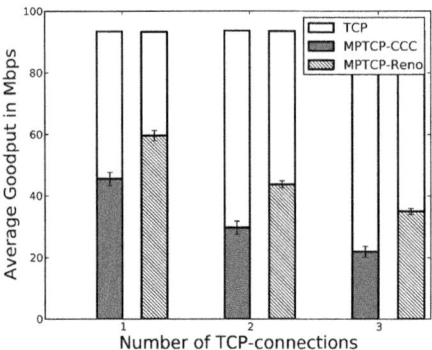

Fig. 4. MultiPath TCP with coupled congestion control behaves like a single TCP connection over a shared bottleneck with respect to regular TCP

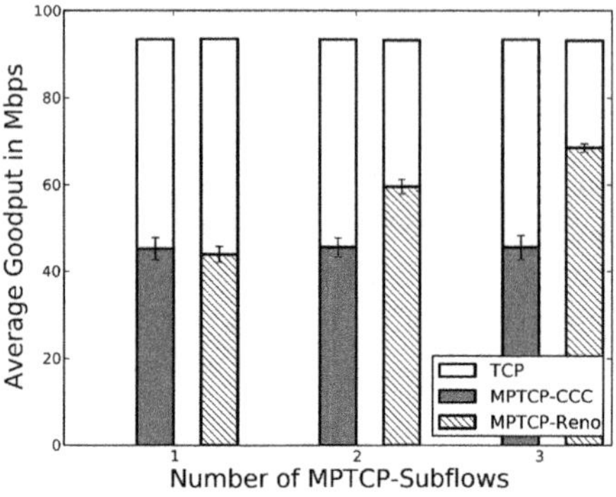

Fig. 5. With coupled congestion control on the subflows, MultiPath TCP is not unfair to TCP when increasing the number of subflows

Thanks to the flexibility of the Linux congestion control implementation, we perform tests by using the standard Reno congestion control scheme with regular TCP and either Reno or the coupled congestion control scheme with MPTCP.

The measurements show that MPTCP with the coupled congestion control is fair to regular TCP. When an MPTCP connection with two subflows is sharing a bottleneck link with a TCP connection, the coupled congestion control behaves as if the MPTCP session was just one single TCP connection. However, when Reno congestion control is used on the subflows, MPTCP gets more bandwidth because in that case two TCP subflows are really competing against one regular TCP flow (Fig. 4).

The second scenario that we evaluate with the coupled congestion control is the impact of the number of MPTCP subflows on the throughput achieved by a single TCP connection over the shared bottleneck. We perform measurements by using one regular TCP connection running the Reno congestion control scheme and one MPTCP connection using one, two or three subflows. The MPTCP connection uses either the Reno congestion control scheme or the coupled congestion control scheme. Figure 5 shows first that when there is a single MPTCP

---

[2] http://sourceforge.net/projects/iperf/

subflow, both Reno and the coupled congestion control scheme are fair as there is no difference between the regular TCP and the MPTCP connection. When there are two subflows in the MPTCP connection, figure 5 shows that Reno favors the MPTCP connection over the regular single-flow TCP connection. The unfairness of the Reno congestion control scheme is even more important when the MPTCP connection is composed of three subflows. In contrast, the measurements indicate that the coupled congestion control provides the same fairness when the MPTCP connection is composed of one, two or three subflows.

5.2   MPTCP Performance

Our second scenario, depicted in figure 3, allows us to evaluate the factors that influence the performance of our MultiPath TCP implementation. It is composed of three workstations. Two of them act as source and destination while the third one serves as a router. The source and destination are equipped with AMD Opteron™ Processor 248 single-core 2.2 GHz CPUs, 2GB RAM and two Intel® 82546GB Gigabit Ethernet controllers. The links and the router are configured to ensure that the router does not cross-route traffic, i.e. the packets that arrive from the solid-shaped link in figure 3 are always forwarded over the solid-shaped link to reach the destination. This implies that the network has two completely disjoint paths between the source and the destination. We configure the bandwidth on Link A and Link B by changing their Ethernet configuration.

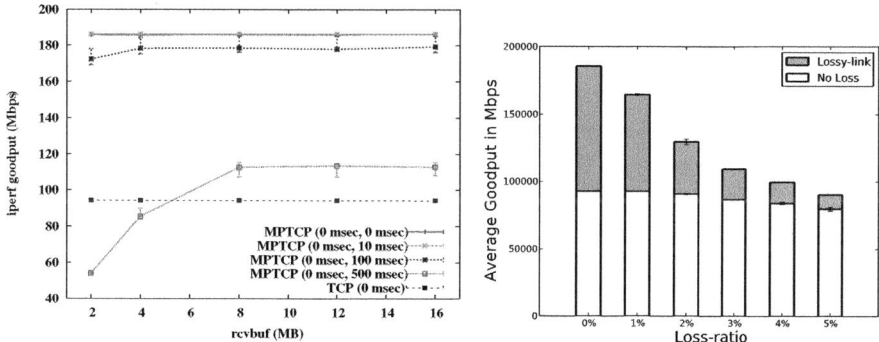

Fig. 6. Impact of the maximum receive buffer size   **Fig. 7.** Impact of the packet loss ratio

As explained in section 4.1, one performance issue that affects MultiPath TCP is that MultiPath TCP may require large receive buffers when subflows have different delays. To evaluate this impact, we configured Link A and Link B with a bandwidth of 100 Mbps. Figure 6 Shows the impact of the maximum receive buffer on the performance achieved by MPTCP with different delays. For these measurements, we use two MPTCP subflows, one is routed over Link A while the second is routed over Link B. The router is configured to insert an additional delay of 0, 10, 100 and 500 milliseconds on Link B. No delay is

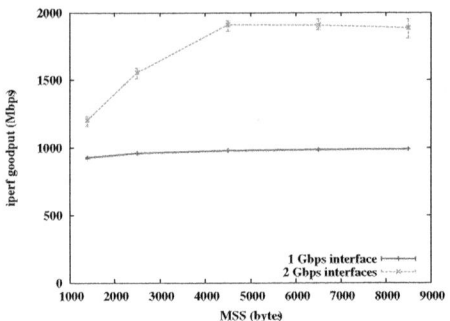

Fig. 8. Impact of the MSS on the performance

inserted on `Link A`. This allows us to consider an asymmetric scenario where the two subflows are routed over very different paths. Such asymmetric delays force the MPTCP receiver to store many packets in its receive buffer to be able to deliver all the received data in sequence. As a reference, we show in figure 6 the throughput achieved by `iperf` with a regular TCP connection over `Link A`. When the two subflows have the same delay, they are able to saturate the two 100 Mbps links with a receive buffer of 2 MBytes or more. When the delay difference is of only 10 millisecond, the goodput measured by `iperf` is not affected. With a difference of 100 milliseconds in delay between the two subflows, there is a small performance decrease. The performance is slightly better with 4 MBytes which is the default maximum receive buffer in the Linux kernel. When the delay difference reaches 500 milliseconds, an extreme case that would probably involve satellite links in the current Internet, the goodput achieved by MultiPath TCP is much more affected. This is mainly because the router drops packets and these losses cause timeout expirations and force MPTCP to slowly increases its congestion window due to the large round-trip-time.

A second factor that affects the performance is the loss ratio. To evaluate whether losses on one subflow can affect the performance of the other subflow due to head-of-line blocking in the receive buffer, we configured the router to drop a percentage of the packets on `Link B` but no packets on `Link A` and set the delays of these links to 0 milliseconds. Fig. 7 shows the impact of the packet loss ratio on the performance achieved by the MPTCP connection. The figure shows the two subflows that compose the connection. The subflow shown in white passes through `Link A` while the subflow shown in gray passes through `Link B`. When there are no losses, the MPTCP connection is able to saturate the two 100 Mbps links. As expected, the gray subflow that passes through `Link B` is affected by the packet losses and its goodput decreases with the packet loss ratio. However, the goodput of the other subflow remains stable with packet loss ratios of 1, 2 or 3 %. It is only when the packet loss ratio reaches 4% or 5% that the goodput of the white subflow decreases slightly.

The last factor that we analyze is the impact of the Maximum Segment Size (MSS) on the achieved goodput. TCP implementors know that a large MSS enables higher goodput [2] since it reduces the number of interrupts that needs to be processed. To evaluate the impact of the MSS, we do not introduce delays nor losses on the router and use either one or two Gigabit Ethernet interfaces on the source and the destination. Figure 8 shows the goodput in function of the MSS size. With a standard 1400 bytes Ethernet MSS, MPTCP can fill one 1Gbps link, and partly use a second one. It is able to saturate two Gigabit Ethernet links with an MSS of 4500 bytes. Note that we do not use TSO (TCP Segmentation Offload) [3] for these measurements. With TSO the segment size handled by the system could have grown virtually to 64KB and allowed it to reach the same goodput with a lower CPU utilization. TSO support will be added later in our MPTCP implementation. Finally, we have looked at the CPU consumption in the system. Given that the MPTCP scheduler runs for every transmitted segment, we were afraid that increasing the number of subflows (hence the cost of running the scheduler) could significantly impact the performance. *This is not true.* We have run MPTCP connections containing 1 to 8 subflows on a shared 1 Gbps bottleneck with various MSS. Increasing the number of concurrent subflows from 1 to 8 has no significant impact on the overall system charge. We also found that the server is globally more busy than the sender, which can be explained by the cost of reordering the received segments. MPTCP currently uses the same algorithm as TCP to reorder segments, but while TCP typically reorders only a few segments at a time, MPTCP may need to handle hundreds of them. A better reordering algorithm could probably improve the receiver performance. Regarding the repartition of the load between the software interrupts and the user context, we found that the majority of the processing was done in the software interrupt context in the receiver: around 50% of the CPU time is spent in soft interrupt, 8% in the user context with a 1400 bytes MSS and a single Gigabit Ethernet link. This can be explained by the fact that prequeues are not in use. The sender spends around 17% in soft interrupt and 25% in user context under the same conditions. Although more work is performed in user context, which is good, there is still some amount of work performed in soft interrupt because the majority of the segments are sent when an incoming acknowledgement opens more space in the congestion or sending window. This event happens in interrupt context.

6   Conclusion and Future Work

MultiPath TCP is a major extension to TCP that is being developed within the IETF. Its success will depend on the availability of a reference implementation. From an implementation viewpoint, MultiPath TCP raises several important challenges. We have analyzed several of them based on our experience in implementing the first MultiPath TCP implementation in the Linux kernel. In particular, we have shown how such an implementation can be structured and discussed how buffer management must be adapted due to the utilization of multiple subflows. We have analyzed the performance of our implementation in the HEN testbed and shown that the coupled congestion control scheme is more fair than the standard TCP congestion control scheme. We have also evaluated the impact of the delay on the receive buffers and the throughput and showed that losses on one subflow had limited impact on the performance of another subflow from the same MultiPath TCP connection. Finally, we have shown that our implementation was able to efficiently utilize Gigabit Ethernet links when using large packets.

Our implementation in the Linux kernel is a first step towards the adoption and the deployment of MultiPath TCP. There are however several research and implementation issues that remain open. Firstly, the MultiPath TCP protocol is not yet finalized and for example the security issues are still being developed. Secondly, MultiPath TCP currently uses the standard TCP retransmission mechanisms on each subflow while multipath-aware retransmission mechanisms could probably improve the performance in some cases. Thirdly, our current implementation uses all available subflows while better performance would probably be possible by adding and removing subflows based on their measured performance. Fourthly, the MultiPath TCP implementation should better interact with the network interfaces to benefit from the TCP segment offload mechanisms that some of them include. Finally, the performance of MultiPath TCP in the global Internet and its interactions with real middle-boxes should be evaluated. We expect that our publicly available implementation will allow other researchers to also contribute to this work.

Acknowledgements

We thank the members of the Trilogy project for their help and in particular Costin Raiciu for his help with the coupled congestion control, and Adam Greenhalgh for maintaining the HEN-Testbed and giving us the ability to execute the performance measurements.

References

1. Allman, M., Paxson, V., Blanton, E.: TCP Congestion Control. RFC 5681 (Draft Standard) (September 2009), http://www.ietf.org/rfc/rfc5681.txt
2. Borman, D.A.: Implementing TCP/IP on a cray computer. SIGCOMM Comput. Commun. Rev. 19, 11–15 (1989), http://doi.acm.org/10.1145/378444.378446
3. Currid, A.: TCP Offload to the Rescue. Queue 2, 58–65 (2004), http://doi.acm.org/10.1145/1005062.1005069
4. Fisk, M., chun Feng, W.: Dynamic right-sizing in TCP. In: Proceedings of the Los Alamos Computer Science Institute Symposium, pp. 1–5460 (2001)
5. Ford, A., Raiciu, C., Barré, S., Iyengar, J.: Architectural Guidelines for Multipath TCP Development, internet draft, draft-ietf-mptcp-architecture-03.txt, work in progress (December 2010)
6. Ford, A., Raiciu, C., Handley, M.: TCP Extensions for Multipath Operation with Multiple Addresses, internet draft, draft-ietf-mptcp-multiaddressed-02.txt, work in progress (October 2010)
7. Hsieh, H.Y., Sivakumar, R.: pTCP: An End-to-End Transport Layer Protocol for Striped Connections. In: ICNP, pp. 24–33. IEEE Computer Society, Los Alamitos (2002)

8. Iyengar, J., Amer, P.D., Stewart, R.R.: Concurrent multipath transfer using SCTP multihoming over independent end-to-end paths. IEEE/ACM Trans. Netw. 14(5), 951–964 (2006)
9. Iyengar, J., Amer, P., Stewart, R.: Receive buffer blocking in concurrent multipath transfer. In: IEEE Global Telecommunications Conference, GLOBECOM 2005, p. 6 (2005)
10. Jacobson, V.: Re: query about tcp header on tcp-ip (September 1993), ftp://ftp.ee.lbl.gov/email/vanj.93sep07.txt
11. Li, Y.T., Leith, D., Shorten, R.N.: Experimental evaluation of TCP Protocols for High-Speed Networks. IEEE/ACM Trans. Netw. 15, 1109–1122 (2007), http://dx.doi.org/10.1109/TNET.2007.896240
12. Liao, J., Wang, J., Zhu, X.: cmpSCTP: An extension of SCTP to support concurrent multi-path transfer. Communications (2008)
13. Magalhaes, L., Kravets, R.: Transport Level Mechanisms for Bandwidth Aggregation on Mobile Hosts. In: ICNP, pp. 165–171. IEEE Computer Society, Los Alamitos (2001)
14. Moskowitz, R., Nikander, P.: Host Identity Protocol (HIP) Architecture. RFC 4423 (May 2006), http://www.ietf.org/rfc/rfc4423.txt
15. Nordmark, E., Bagnulo, M.: Shim6: Level 3 Multihoming Shim Protocol for IPv6. RFC 5533 (June 2009), http://www.ietf.org/rfc/rfc5533.txt
16. Postel, J.: Transmission Control Protocol. RFC 793 (Standard), updated by RFCs 1122, 3168 (September 1981)
17. Raiciu, C., Handley, M., Wischik, D.: Coupled Multipath-Aware Congestion Control. Internet draft (work in progress), Internet Engineering Task Force (July 2010), http://tools.ietf.org/html/draft-ietf-mptcp-congestion-00
18. Rojviboonchai, K., Osuga, T., Aida, H.: R-M/TCP: Protocol for Reliable Multi-Path Transport over the Internet. In: AINA, pp. 801–806. IEEE Computer Society, Los Alamitos (2005)
19. Stewart, R.: Stream Control Transmission Protocol. RFC 4960 (September 2007)
20. Wehrle, K., Pahlke, F., Ritter, H., Muller, D., Bechler, M.: The Linux networking architecture: design and implementation of network protocols in the Linux kernel. Prentice Hall, Englewood Cliffs (2004)
21. Wischik, D., Handley, M., Braun, M.B.: The Resource Pooling Principle. SIGCOMM Comput. Commun. Rev. 38(5), 47–52 (2008)
22. Wischik, D., Raiciu, C., Greenhalgh, A., Handley, M.: Design, implementation and evaluation of congestion control for multipath TCP, USENIX NSDI (April 2011)
23. Zhang, M., Lai, J., Krishnamurthy, A.: A transport layer approach for improving end-to-end performance and robustness using redundant paths. In: USENIX 2004, pp. 99–112 (2004)

Stealthier Inter-packet Timing Covert Channels

Sebastian Zander, Grenville Armitage, and Philip Branch

Centre for Advanced Internet Architectures (CAIA),
Swinburne University of Technology,
Melbourne Australia
{szander,garmitage,pbranch}@swin.edu.au

Abstract. Covert channels aim to hide the existence of communication. Recently proposed packet-timing channels encode covert data in inter-packet times, based on models of inter-packet times of normal traffic. These channels are detectable if normal inter-packet times are not independent identically-distributed, which we demonstrate is the case for several network applications. We show that ~80% of channels are detected with a false positive rate of 0.5%. We then propose an improved channel that is much harder to detect. Only ~9% of our new channels are detected at a false positive rate of 0.5%. Our new channel uses packet content for synchronisation and works with UDP and TCP traffic. The channel capacity reaches over hundred bits per second depending on overt traffic and network jitter.

Keywords: Covert Channels, Steganography, Inter-packet Times.

1 Introduction

The very fact that two parties are communicating can be actionable information, even if an external observer cannot determine the contents of the messages being passed. Covert channels aim to hide the very existence of communication, enabling parties to convey information undetected [1]. Individuals and groups may have various reasons to utilise covert channels (e.g. for communication or bypassing access controls), often motivated by the existence of adversarial relationships. Examples include government agencies versus criminal organisations, hackers versus company IT departments, or dissenting citizens versus their governments.

The prisoner problem is the de-facto model for covert channel communication [1]. *Alice* and *Bob* are thrown into prison and intend to escape. To agree on an escape plan they need to communicate, but a *warden* monitors all their messages. If the warden finds any signs of suspicious messages she will place Alice and Bob into solitary confinement – making it impossible for them to escape. Alice and Bob must exchange innocuous messages containing hidden information that hopefully the warden will not notice.

Recently researchers proposed covert channels encoded in inter-packet times, also referred to as Inter Packets Gaps (IPGs). Gianvecchio *et al.* [2] and Sellke *et al.* [3] proposed channels that are hard to detect because they perfectly mimic the shape of IPG distributions of real applications. However, they are hard to detect only if IPGs are independent identically-distributed (iid). Paxson *et al.* showed that Telnet traffic exhibits this behaviour [4]. However, we found that IPGs of several applications, such

J. Domingo-Pascual et al. (Eds.): NETWORKING 2011, Part I, LNCS 6640, pp. 458–470, 2011.

as UDP-based game and VoIP traffic and TCP traffic, often are not iid because there is auto-correlation. The channel proposed in [2,3] is easy to detect with such applications.

The channel in [2,3] requires accessible sequence numbers in the overt traffic. Otherwise any lost packets desynchronise covert sender and receiver. TCP provides sequence numbers, but not all UDP-based traffic has sequence numbers, or they may not be accessible if the protocol is encrypted.

We propose an improved channel, based on techniques for information hiding in images (steganography), which is harder to detect when IPGs are not iid. Our new technique generates the random numbers needed for encoding from the packets themselves, which makes the channel robust enough for use with all UDP-based protocols.

First, we motivate our work by demonstrating that several applications have auto-correlated IPGs. We then present the improved covert channel. We show that for traffic with auto-correlated IPGs the existing timing channel is easy to detect with ~80% accuracy and a false positive rate of 0.5%[1]. Our new channel is much harder to detect. The detection accuracy reduces to only ~9% with a false positive rate of 0.5%. A drawback of our new channel is a reduced robustness against network jitter. However, based on a proof-of-concept implementation we show that the channel capacity is still high enough for practical use, even across uncongested Internet paths with more than 10 hops. The capacity ranges from a few bits per second to over hundred bits per second, depending on the overt traffic's packet rate and network jitter.

The paper is organised as follows. Section 2 outlines related work. In Section 3 we show that several applications often have auto-correlated IPGs. In Section 4 we propose our new channel. In Section 5 we analyse the detection accuracy for the channel in [2,3] and our improved channel using machine learning. In Section 6 we analyse the capacity of our improved channel. Section 7 concludes and outlines future work.

2   Related Work

The possibility of encoding covert channels in the timing of packets (or frames) was identified early by Padlipsky *et al.* [5]. However, all the timing channels proposed prior to Berk's work [6] were based on encoding in varying packet rates over time as opposed to encoding in IPG values directly. For space reasons we do not discuss them here and instead refer the reader to [1].

Berk *et al.* introduced packet-timing channels where the covert information is encoded in the IPGs of consecutive packets [6]. They compared channels with two IPGs and multiple IPGs, and developed a mechanism by which the sender can pick the optimal symbol distribution in multi-symbol channels. Sha *et al.* developed a "bug" that hooks into the connection between keyboard and computer and ex-filtrates all keystrokes by modulating the IPGs of network traffic send by the victim [7]. Gianvecchio *et al.* later showed that both of these channels are easy to detect [8].

Gianvecchio *et al.* [2] developed a stealthier IPG timing channel and evaluated its performance. They proposed to fit a model to the IPG distribution of real traffic and then use the model to generate a covert channel with identical distribution. If the IPGs

---

[1] In reality most flows are normal flows, so a detector's false positive rate must be low or covert channels are effectively masked (see Section 5).

of normal traffic are iid this channel cannot be detected. Sellke *et al.* proposed another scheme for encoding covert data in IPGs and evaluated its performance based on experiments across the Internet [3]. Similar to [2] they also showed that timing channels can be made indistinguishable from normal traffic by mimicking normal iid IPGs.

Luo *et al.* proposed to encode covert data into the length of TCP data bursts, where a data burst is a number of TCP segments sent before the next TCP ACK arrives [9]. The channel is robust against packet jitter, loss and reordering, but has very low capacity. Furthermore, its stealth is low, as flows with covert channel behave very different from normal flows [9]. Liu *et al.* introduced a channel that encodes covert data such that the normal distribution of IPGs is closely approximated and spreading techniques are used to provide robustness [10]. Their covert channel is hard to detect with simple shape and regularity tests, however these tests are known to be insufficient [8].

3   Inter-packet Times Analysis

We analyse the IPGs of several applications. Our results illustrate the existence of IPG auto-correlation for certain types of traffic.

3.1   Datasets and Methodology

We analysed two UDP-based applications, the First Person Shooter (FPS) game Quake III Arena (Q3) [11] and the VoIP application Skype. The Q3 client-to-server traffic was taken from trace files collected at our public game server. It contains traffic from 106 clients. The Q3 server-to-client traffic analysis is based on trace files captured at a client that connected to 224 different game servers. We analysed the IPGs of 44 traffic flows of Skype one-to-one calls collected by Branch *et al.* [12].

We also analysed a mix of UDP and TCP traffic from the Twente traces [13]. Since the trace does not contain payload data we do not know the applications. Based on the port numbers it seems most of the TCP flows were bulk-transfers: HTTP, FTP, or peer-to-peer file sharing applications, such as BitTorrent. A large part of the UDP flows were game traffic, for example Half-Life/Counterstrike and Quake. In total the dataset has 111 UDP and 214 TCP unidirectional traffic flows from different sources.

For each flow we computed the auto-correlation over the first 5 000 IPGs. Auto-correlation is the cross-correlation of a time series with itself. Let x_t be the IPG value at time t, let μ and σ be the mean and variance over all t, let τ be the time lag and $E(.)$ be the expectation value. The auto-correlation function (ACF) for one traffic flow is:

$$R(\tau) = \frac{E\left((x_t - \mu)(x_{t+\tau} - \mu)\right)}{\sigma^2} . \tag{1}$$

For a set of n traffic flows of one dataset we compute the ensemble average over all ACFs (referred to as *average ACF*). The average ACF at lag τ is defined as the mean of the absolute ACF values of each flow at τ:

$$\bar{R}(\tau) = \frac{1}{n} \sum_n |R_i(\tau)| . \tag{2}$$

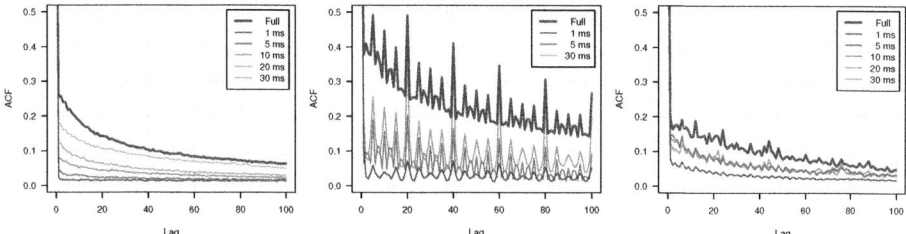

Fig. 1. Average auto-correlation for Q3 client-to-server (left), Skype (middle) and TCP (right) traffic (zoomed y-axis)

We analyse the auto-correlation of IPGs and least significant parts of IPGs. We define the least significant part as:

$$d_{\text{lsp}} = d \bmod l = d - \left\lfloor \frac{d}{l} \right\rfloor l \,, \tag{3}$$

where d is the IPG and l is the size of the least significant part. For example, if the IPG is 21.75 ms and $l = 1$ ms then $d_{\text{lsp}} = 0.75$ ms (sub-millisecond part).

3.2 Results

Figure 1 shows the average ACFs of Q3 client-to-server, Skype and TCP traffic for the full IPGs and decreasing least significant parts ($\tau \le 100$). The average ACF of the full IPGs decays more rapidly than individual ACFs (e.g. the one shown in Figure 2(left)), as individual ACFs have lows and highs at different places. Still it is significantly larger than the average ACF for small least significant parts.

For the UDP-based applications IPGs are often at least moderately correlated. However, smaller least significant parts of IPGs are largely uncorrelated. The size of the least significant part where correlation diminishes depends on the application. The TCP flows show less correlation, but many flows still have low to moderate correlation.

IPGs of applications where packet send times are human-driven and effectively random were shown to be iid (e.g. Telnet [4]). However, for other applications auto-correlation of IPGs tends to exist because of large-time-scale behaviours, such as the congestion window growth/collapse for TCP or the application's encoding for games or VoIP over UDP. Network jitter reduces existing correlations, but even after many hops often there is some correlation. Furthermore, if the warden is close to the covert sender, the IPGs observed are largely unaffected by network jitter.

The small time-scale behaviour (exposed by looking only at the least significant part of IPGs) is jittered by largely uncorrelated noise, for example queuing delays at each hop. Our new encoding technique exploits this effect.

4   Covert Channel

We first review the previous encoding scheme on which our improved scheme is based and then propose our improved technique.

4.1 Basic Encoding

In Gianvecchio's and Sellke's encoding scheme [2, 3] the covert sender (*Alice*) and receiver (*Bob*) share a model of the IPGs, which was previously generated based on an analysis of real traffic. The model can be a histogram of measured IPGs or a standard statistical distribution fitted to the observed IPGs.

Alice and Bob also need synchronised random number generators. They have to agree on a Cryptographically Secure Pseudo Random Number Generator (CSPRNG) and a seed value. The seed value can be derived from Alice's and Bob's shared secret.

Let $R = r_1, r_2, \ldots, r_n$ be the sequence of Uniform(0,1) random numbers generated and s be a symbol out of the set of possible symbols S. For example, a binary channel has two symbols: $S = \{s_1, s_2\} = \{0, 1\}$. Then Alice encodes covert bits as follows. Each discrete symbol is transformed into a continuous symbol:

$$F_{\text{cont}}(s, r) = \left(\frac{s}{|S|} + r\right) \bmod 1 = r_s . \tag{4}$$

The IPGs $d_1, d_2, \ldots, d_n$ are produced by the encoding function:

$$F_{\text{enc}} = F^{-1}_{\text{model}}(r_s) = d , \tag{5}$$

where F^{-1}_{model} is the inverse distribution function of the model. Bob decodes the bits from the observed IPGs $\tilde{d}_1, \tilde{d}_2, \ldots, \tilde{d}_n$, which is Alice's series modified by timing noise, such as timing inaccuracies at the sender, network jitter and measuring inaccuracies at the receiver. Bob decodes the continuous symbol by applying:

$$F_{\text{dec}} = F_{\text{model}}\left(\tilde{d}\right) = \tilde{r}_s , \tag{6}$$

where F_{model} is the distribution function of the model. The discrete symbol received is:

$$F_{\text{disc}}(\tilde{r}_s, r) = |S| \cdot ((\tilde{r}_s - r) \bmod 1) = \tilde{s} . \tag{7}$$

This encoding scheme generates an IPG distribution that is indistinguishable from the IPG distribution of normal traffic if normal IPGs are iid [2,3]. However, as we showed in the previous section this assumption is often not true. This makes the channel detectable. Figure 2 shows the ACF of the IPGs of normal Q3 client-to-server traffic on the left and the ACF of the IPGs of the covert channel in the middle. The IPG histograms of normal traffic and covert channel (not shown) look alike, but the difference of the ACFs is striking.

4.2 Improved Synchronisation

The basic encoding works only if the overt traffic has accessible sequence numbers. Otherwise any packet loss permanently desynchronises Alice and Bob. Furthermore, if Bob is unable to observe a transmission's start he can never synchronise with Alice.

Synchronisation can be improved by computing R from the packets themselves. Let H be a good hash function that maps the inputs as evenly as possible over the output

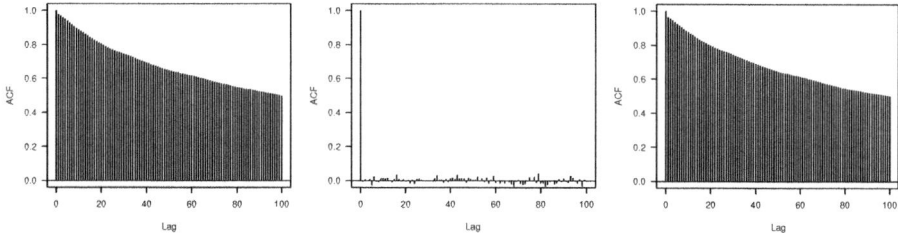

Fig. 2. Auto-correlation of IPGs of normal Q3 traffic (left), covert channel in [2, 3] (middle) and sub-band encoding with $l = 5$ ms (right)

range (uniform distribution). Let B denote some part of an overt packet that is immutable on the path from Alice to Bob, and let b_i be the value of B for the i-th packet. Random numbers are generated as follows:

$$r_i = H(b_i) \; . \tag{8}$$

We assume that the inputs b_i vary sufficiently so that the output of H is approximately uniformly distributed. Previous work on packet sampling and one-way delay measurement showed that generally this is the case if B is chosen properly, and suggested several suitable choices for H and B [14]. TCP retransmissions may cause the same inputs b_i repeatedly, but using TCP header information Alice and Bob can 'ignore' them.

Alice and Bob need to agree on H and B. It is possible for a warden to guess H and B and therefore to detect the covert channel, since the choices for H and B are limited. Alice and Bob can use keyed hash functions used for Message Authentication Codes (MACs) for higher security.

With our improved synchronisation technique lost packets can still cause lost bits, but never a permanent desynchronisation.

4.3 Sub-band Encoding

Now we present our new encoding scheme that is much stealthier if IPGs are not iid. The scheme encodes covert bits only into the least significant part of IPGs as defined in Equation 3.

Let l be the size of the least significant part of the IPGs. The parameter l determines the trade-off between stealth and robustness. Let D be the range of IPGs (maximum minus the minimum). Then an IPG distribution typically spans $m = \lceil D/l \rceil$ *sub-bands* of size l, although in the extreme case it may fit in only a single sub-band. The location of the sub-bands, the IPG value marking the start of each band, depends on the start value of the first band. The location must be selected carefully to minimise the error rate for a given IPG distribution.

For each of the sub-bands the basic encoding scheme is used. However, now we have one probability distribution of the least significant part of the IPGs for each sub-band j, resulting in $F_{enc}^{(j)}(.) = F_{model}^{-1(j)}(.)$ and $F_{dec}^{(j)}(.) = F_{model}^{(j)}(.)$. When building the model from the example traffic a distribution of the least significant part of the IPGs must

---

Algorithm 1. Sender and receiver algorithm for sub-band encoding

---

function encode(packet, prev_pkt_time, pkt_time)   **function** decode(packet, prev_pkt_time, pkt_time)

b = hash_input(packet) b = hash_input(packet)

orig_ipg = pkt_time − prev_pkt_time   ipg = pkt_time − prev_pkt_time

base_ipg = $\lfloor$ orig_ipg$/l\rfloor \cdot l$ base_ipg = $\lfloor$ ipg$/l\rfloor \cdot l$

band = BL $[d_{\text{start}}$ = base_ipg$].x$ lsp = ipg − base_ipg

bits = get_bits()   band = BL $[d_{\text{start}}$ = base_ipg$].x$

$r = H_R(b)$ $r = H_R(b)$

ipg = $F_{\text{enc}}^{(\text{band})}(F_{\text{cont}}(\text{bits}, r)) + $ base_ipg   bits = $F_{\text{disc}}\left(F_{\text{dec}}^{(\text{band})}(\text{lsp}), r\right)$

return ipg   **return** bits

---

be estimated for each sub-band. If the example traffic is only a small sample and the number of bands is large there may be only a few or even zero values for some bands. For such sub-bands as well as sub-bands outside the range covered by the example traffic our algorithm augments the data with uniformly distributed random values so that a minimum number of samples is reached.

Algorithm 1 shows the encoding algorithm. Let BL be a set of tuples $\left(d_{\text{start}}^{(j)}, x_j\right)$ that associate each sub-band index x_j with an absolute IPG value $d_{\text{start}}^{(j)}$ marking the start of the band (*base IPG*). First, the algorithm determines the actual sub-band based on the original unmodified IPG. Then the modified IPG is the sum of the least significant part as given by the sub-band model and the base IPG of the sub-band. Algorithm 1 also shows the decoding. The receiver selects the sub-band based on the observed IPG. Then it computes the least significant part of the IPG and decodes the bits as before.

Figure 2(right) illustrates the effectiveness of sub-band encoding. The covert channel only very slightly decreases the auto-correlations. The drawback of sub-band encoding is that it is less robust against network jitter, but network jitter is often small even on longer paths. Our results in Section 6 show that the channel capacity is still high enough for practical use, even across uncongested Internet paths with more than 10 hops. However, the scheme is also less robust against an active warden that can introduce artificial jitter.

For active channels, where the covert sender also generates the overt traffic, sub-band encoding requires models that model the shape of the distribution as well as the auto-correlations, for example autoregressive integrated moving average (ARIMA) models. The sender first uses the model to create a series of IPGs, and then uses the encode() function to encode the covert data. For passive channels, where the covert sender encodes the channel into existing traffic of unwitting hosts, the correlations are already present in the intercepted traffic.

5   Detection

We use classifiers constructed by a supervised Machine Learning (ML) algorithm. During training supervised ML techniques build classifiers based on data instances with class labels attached, so that the data instances are 'optimally' separated into the different classes based on characteristics (*features*) of the instances other than the class label. The classifiers are then used to classify data instances of unknown class.

5.1   Datasets and Features

As normal traffic we used the datasets from Section 3. The covert traffic was created based on these datasets using sub-band encoding. For each normal flow we generated one corresponding covert channel based on the same IPG distribution. We used the first 5 000 IPGs of each flow. The covert data was uniformly random distributed, as if Alice and Bob used encryption. Each dataset had the same number of covert and normal flows to avoid bias of the classifier towards a larger class.

Let $X = [X_1, \ldots, X_n]$ be a series of IPGs of a flow with values $x_1, \ldots, x_n$. For each X we computed the first order entropy (*Entropy*) and an estimate of the entropy rate (*EntropyRate*) using the corrected conditional entropy (CCE) [8]. The first order entropy is useful for comparing the shape of distributions of random variables, and the entropy rate is useful for comparing the regularity of time series. For computing the CCE we used equiprobable binning of the data as in [8]. To determine the number of bins Q we performed initial tests and found $Q = 5$ maximised the classification accuracy.

We also computed a feature based on the two-sample Kolmogorov-Smirnov (KS) test, which tests the hypothesis that two samples were drawn from the same distribution. A low KS test statistic means that the distributions are similar whereas a high KS test statistic means the distributions are different. The KS test is applicable to a variety of data with different distributions. Since we need a feature that reflects how different a distribution is from the set of distributions characterising normal traffic, we computed the set of KS test statistics between a data instance (covert or normal) and all instances of normal traffic (excluding tests of a normal instance with itself) and use the mean of all KS statistics (*MeanKS*).

5.2   Machine Learning

Previous research showed that for classification of network traffic the better ML techniques provide similar accuracy, but differ greatly regarding training time and classification speed [15]. We used the C4.5 decision tree classifier [16] – more precisely its implementation in the Waikato Environment for Knowledge Analysis (WEKA) [17], because it had performed well previously [15]. Using a decision tree algorithm also has the advantage that a human can interpret the resulting classifier, although with increasing size of the tree this becomes difficult.

C4.5 selects features in order of maximising an entropy-based gain ratio. The most useful features are always used at the top of the tree and irrelevant features are largely ignored. Hence, C4.5 is not adversely affected by irrelevant features like some other techniques and feature pre-selection is not necessary. C4.5 attempts to avoid over-fitting by removing some structure from a tree after it was built (tree pruning) [16]. In our experiments we used the default WEKA 3.4.4 settings for all parameters of C4.5.

We evaluate the classification accuracy based on the following metrics. A true positive (TP) is a class instance correctly classified. A false positive (FP) is a non-class member misclassified as class member and a false negative (FN) is a class member misclassified as non-class member. *Precision* is the number of class members classified correctly over the total number of instances classified as class members [17]:

$$\text{precision} = \frac{TP}{TP + FP} \ . \tag{9}$$

Recall (or *TP rate*) is the number of class members classified correctly over the total number of class members [17]:

$$\text{recall} = \frac{\text{TP}}{\text{TP} + \text{FN}} \cdot \tag{10}$$

We performed 10-fold cross-validation for each dataset. The data set was randomly divided into k subsets, and the classification was repeated k times. Each time, one of the k subsets was used as the test set and the other $k - 1$ subsets formed the training set. Then the average accuracy across all k trials was computed.

The classification accuracy varies depending on the random covert data, the random number series used for encoding R, and the random selection of instances during the cross-validation. Therefore, we performed the cross-validation 10 times and computed the average accuracy.

5.3   Results

Figure 3 shows the precision and recall of the covert traffic class averaged over all datasets for the basic covert channel proposed in [2, 3] and for sub-band encoding depending on the size of the least significant part l. For small l precision and recall drop to 60% or lower. For larger l they increase substantially, but even for $l = 25$ ms precision is still less than 80%, well below the 96% for the basic covert channel. We found the resulting classifiers rely mainly on EntropyRate, but MeanKS is also used occasionally at the bottom of the tree.

A recall of 70–80% for sub-band encoding still appears high. However, one needs to take into account the large FP rate as indicated by the precision of 60–70%. In reality, even when assuming the warden has already selected some traffic fraction of interest, most flows are normal flows. Thus even a FP rate of only a few percent very effectively masks the covert channel. To explore the detection accuracy at low FP rates we

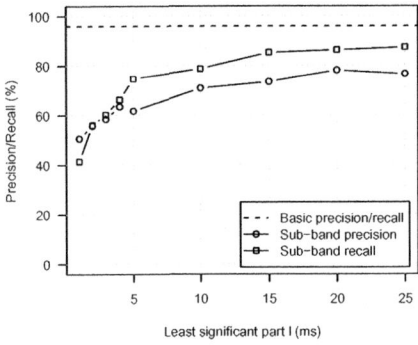

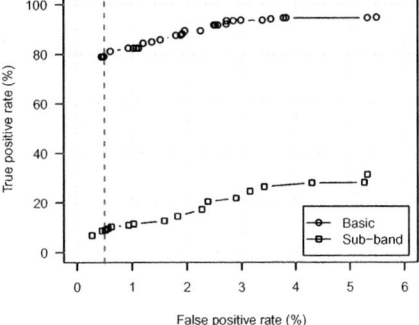

Fig. 3. Precision and recall of covert traffic class for basic encoding and sub-band encoding depending on least significant part l

Fig. 4. ROC curves of covert traffic class for the basic covert channel proposed in [2,3] and sub-band encoding with $l = 5$ ms

computed the Receiver Operating Characteristic (ROC) curves by using cost-sensitive classification and varying the cost for misclassified instances of either class.

Figure 4 shows the low FP-rate part of the ROC curves for the covert traffic class averaged over all datasets for the basic channel proposed in [2,3] and sub-band encoding with a sub-band size of 5 ms. It shows that sub-band encoding is much harder to detect. For a FP rate of 0.5% (indicated by the vertical dashed line) approximately 80% of the basic channels are detected, whereas for sub-band encoding the detection accuracy is greatly reduced to approximately 9%.

6   Channel Capacity

We propose a model to compute the channel capacity, describe our experimental setup, and present the measured channel capacities based on different network jitter.

6.1   Channel Model

We assume the channel's output only depends on the input and the errors but not on previous inputs (memoryless channel) and we focus on a binary channel (one bit encoded per IPG). Timing jitter causes bit substitution errors. Under timing jitter we subsume packet timing inaccuracies at the covert sender, timestamping inaccuracies at the covert receiver and network jitter.

In our testbed experiments the resulting error rates were approximately symmetric. Hence we model the channel as Binary Symmetric Channel (BSC) with a capacity [18]:

$$C = 1 - H(p) = 1 + p \cdot \log_2(p) + (1 - p) \cdot \log_2(1 - p) , \tag{11}$$

where $H(.)$ is the binary entropy and p is the probability of timing errors. The capacity C is in bits per IPG (bits per symbol). Given an average rate of IPGs f_S the average transmission rate in bits per second is:

$$R = C \cdot f_S . \tag{12}$$

6.2   Testbed and Methodology

We implemented a prototype of sub-band encoding, using the Covert Channels Evaluation Framework (CCHEF) [19], which was carefully designed to maximise Alice's packet timing accuracy. However, since it is a userspace program it competes with other userspace programs for CPU time. Other programs using a lot of CPU time decrease the timing accuracy. To avoid this we used real-time Linux 2.6.20 and ran Alice and Bob as real-time processes with high priority. We also set the kernel's tick frequency to 10 kHz to reduce the size of time slices.

Our testbed consisted of two computers connected via a Fast Ethernet switch. We used scp to perform file transfers capped at 2 Mbit/s, and SSH to perform remote interactive shell sessions. We generated Q3 with bots as players. Each experiment lasting 20 minutes was repeated three times and we report the average statistics. Network delay and jitter were emulated using Linux Netem [20]. The network delay was emulated using Pareto distributions with a mean of 25 ms and standard deviations (σ) of 0, 0.1, 0.2,

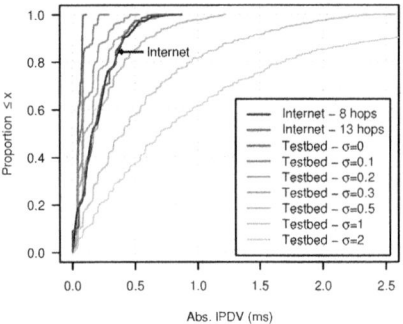

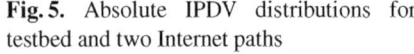

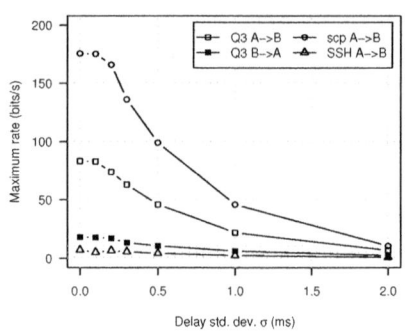

Fig. 5. Absolute IPDV distributions for testbed and two Internet paths

Fig. 6. Maximum transmission rates for sub-band encoding

0.3, 0.5, 1 and 2 ms in each direction, since previous research suggested that network jitter is heavy-tailed [21], and Netem only supports Uniform, Gaussian and Pareto jitter distributions. Setting the kernel tick timer frequency to 10 kHz ensured delay emulation was accurate to $\pm 100\,\mu s$.

Figure 5 shows CDFs of the absolute IP Delay Variation (IPDV [22]), both in the testbed with Pareto distributions with different σ and measured across two Internet paths. The 8-hop Internet path's RTT was approximately 32 ms, and the 13-hop path's RTT was approximately 46 ms. We estimated the one-way delay to half the measured RTT. Both Internet path's IPDV CDFs lie between the testbed CDFs for $\sigma = [0.3, 0.5]$.

The IPG models were built as follows. First, we measured the IPG distribution of each application at the source, unaffected by timing jitter. We then added a small amount of noise. Without the added noise the covert channel would not work well for applications with very narrow IPG distribution, such as Q3 client-to-server and scp traffic. The noise represents timing jitter caused by the network, or a high CPU or network interface load of the source host, which the warden would also encounter in reality.

Our models were histograms with small bin sizes of $100\,\mu s$, as our traffic sources cannot be modelled well with standard statistical distributions. For Q3 and SSH traffic the location of the sub-bands was chosen such that peaks in the distributions are approximately in the middle of bands.

6.3   Maximum Transmission Rates

We measured the error rates of the channel based on the network jitter and computed the maximum transmission rates using Equations 11 and 12. Figure 6 shows the maximum rates for sub-band encoding for a sub-band size of $l = 5$ ms. We selected this size because it provides acceptable noise levels for $\sigma \leq 0.5$ ms at a reasonably low detection accuracy (see Section 5). Note that for TCP traffic the channel can only be encoded in one direction due to the timing dependencies between packets in both directions. Multiple TCP flows could be used to achieve full-duplex communication.

Sub-band encoding has significantly higher error rates with Q3 client-to-server traffic or scp compared to Q3 server-to-client traffic or SSH (not shown). However, the transmission rates are still much higher for Q3 client-to-server traffic and scp because of the much higher packet rates. Sub-band encoding has low error rates and reasonably high capacities at low levels of jitter typical of uncongested paths. Transmission rates are up to over hundred bits per second. If the network jitter is high the capacity is severely reduced. However, larger sub-band sizes could be used to increase robustness at the cost of reducing the stealth.

Q3 and scp have narrow IPG distributions, and in our experiments even interactive SSH was limited because we played back a short recorded session repeatedly. Applications with wider IPG distributions can provide more robust channels. However, bit rates would still be relatively low since such applications also have lower packet rates.

7   Conclusions and Future Work

We showed that Gianvecchio's and Sellke's IPG timing channel [2, 3] is easy to detect with ~80% accuracy if the IPGs of normal traffic are not iid. Inspired by image steganography we developed an improved encoding scheme for IPG timing channels that is much harder to detect. The detection accuracy is reduced down to ~9%. The random numbers needed for encoding are generated from the packet content, which makes the new channel usable with all UDP-based traffic. Our new channel has reduced robustness, but it still provides a reasonably high channel capacity.

We performed experiments with a proof-of-concept implementation in a testbed with different emulated network jitter. The new channel has low error rates and reasonably high capacities for lower network jitter, comparable to measured jitter on short/medium uncongested Internet paths. Transmission rates are up to over one hundred bits per second, depending on the overt traffic. However, the capacity is severely reduced if the network jitter is high.

In future work our channel model could be extended to include the effects of packet loss. Our analysis could be extended to a wider range of applications generating the overt traffic, and to a larger range of network conditions including packet loss. The sender's timing accuracy could be potentially improved with a future kernel-based implementation. Another avenue for future studies are passive channels, where the covert sender encodes the channel into existing traffic of unwitting hosts.

References

1. Zander, S., Armitage, G., Branch, P.: A Survey of Covert Channels and Countermeasures in Computer Network Protocols. IEEE Communications Surveys and Tutorials 9(3), 44–57 (2007)
2. Gianvecchio, S., Wang, H., Wijesekera, D., Jajodia, S.: Model-based covert timing channels: Automated modeling and evasion. In: Lippmann, R., Kirda, E., Trachtenberg, A. (eds.) RAID 2008. LNCS, vol. 5230, pp. 211–230. Springer, Heidelberg (2008)
3. Sellke, S.H., Wang, C.-C., Bagchi, S., Shroff, N.B.: Covert TCP/IP Timing Channels: Theory to Implementation. In: Conference on Computer Communications (INFOCOM) (April 2009)

4. Paxson, V.: End-to-end Internet Packet Dynamics. IEEE/ACM Transactions on Networking 7(3), 277–292 (1999)
5. Padlipsky, M.A., Snow, D.W., Karger, P.A.: Limitations of End-to-End Encryption in Secure Computer Networks. Technical Report ESD-TR-78-158, Mitre Corporation (August 1978)
6. Berk, V., Giani, A., Cybenko, G.: Detection of Covert Channel Encoding in Network Packet Delays. Technical Report TR2005-536, Dartmouth College (November 2005)
7. Shah, G., Molina, A., Blaze, M.: Keyboards and Covert Channels. In: USENIX Security (August 2006)
8. Gianvecchio, S., Wang, H.: Detecting Covert Timing Channels: An Entropy-Based Approach. In: ACM Conference on Computer and Communication Security (CCS) (November 2007)
9. Luo, X., Chan, E.W.W., Chang, R.K.C.: TCP Covert Timing Channels: Design and Detection. In: IEEE/IFIP Conference on Dependable Systems and Networks (DSN) (June 2008)
10. Liu, Y., Ghosal, D., Armknecht, F., Sadeghi, A.-R., Schulz, S., Katzenbeisser, S.: Hide and Seek in Time — Robust Covert Timing Channels. In: Backes, M., Ning, P. (eds.) ESORICS 2009. LNCS, vol. 5789, pp. 120–135. Springer, Heidelberg (2009)
11. Quake, http://www.idsoftware.com
12. Branch, P., Heyde, A., Armitage, G.: Rapid Identification of Skype Traffic. In: ACM Network and Operating System Support for Digital Audio and Video (NOSSDAV) (June 2009)
13. M2C Measurement Data Repository (December 2003), http://traces.simpleweb.org/
14. Henke, C., Schmoll, C., Zseby, T.: Empirical Evaluation of Hash Functions for PacketID Generation in Sampled Multipoint Measurements. In: Moon, S.B., Teixeira, R., Uhlig, S. (eds.) PAM 2009. LNCS, vol. 5448, pp. 197–206. Springer, Heidelberg (2009)
15. Williams, N., Zander, S., Armitage, G.: A Preliminary Performance Comparison of Five Machine Learning Algorithms for Practical IP Traffic Flow Classification. SIGCOMM Computer Communication Review 36(5) (October 2006)
16. Kohavi, R., Quinlan, J.R.: Decision-tree Discovery, ch. 16.1.3, pp. 267–276. Oxford University Press, Oxford (2002)
17. Witten, I.H., Frank, E.: Data Mining: Practical Machine Learning Tools and Techniques, 2nd edn. Morgan Kaufmann, San Francisco (2005)
18. Cover, T.M., Thomas, J.A.: Elements of Information Theory. Wiley Series in Telecommunications. John Wiley & Sons, Chichester (1991)
19. Zander, S.: CCHEF - Covert Channels Evaluation Framework (2007), http://caia.swin.edu.au/cv/szander/cc/cchef/
20. Linux Foundation. Netem (2008), http://www.linuxfoundation.org/en/Net:Netem
21. Rizo, L., Torres, D., Dehesa, J., Muñoz, D.: Cauchy Distribution for Jitter in IP Networks. In: International Conference on Electronics, Communications and Computers, pp. 35–40 (2008)
22. Demichelis, C., Chimento, P.: IP Packet Delay Variation Metric for IP Performance Metrics (IPPM). RFC 3393, IETF (November 2002), http://www.ietf.org/rfc/rfc3393.txt

Author Index

GPSR Compliance

The European Union's (EU) General Product Safety Regulation (GPSR)
is a set of rules that requires consumer products to be safe and our
obligations to ensure this.

If you have any concerns about our products, you can contact us on
ProductSafety@springernature.com

In case Publisher is established outside the EU, the EU authorized
representative is:

Springer Nature Customer Service Center GmbH
Europaplatz 3
69115 Heidelberg, Germany

Batch number: 09478804

Printed by Printforce, the Netherlands